T0299217

MANIPULATING QUANTUM STRUCTURES
USING LASER PULSES

The use of laser pulses to alter the internal quantum structure of individual atoms and molecules has applications in quantum information processing, the coherent control of chemical reactions, and in quantum-state engineering. This book presents the underlying theory of such quantum-state manipulation for researchers and graduate students.

The book provides the equations, and approaches for their solution, that can be applied to complicated multilevel quantum systems. It also gives the background theory for application to isolated atoms or trapped ions, simple molecules, and atoms embedded in solids. Particular attention is given to the ways in which quantum changes can be displayed graphically to help readers understand how quantum changes can be controlled.

BRUCE W. SHORE worked as a theoretical physicist at the Lawrence Livermore National Laboratory for 30 years. His research dealt with numerical analysis and atomic physics, specializing in the theory of laser-induced atomic-vapor excitation.

The present volume is part of an informal series of books all of which originated as review articles published in *Acta Physica Slovaca*. The journal can be accessed for free at www.physics.sk/aps.

Vladimir Buzek, editor of the journal

MANIPULATING QUANTUM STRUCTURES USING LASER PULSES

BRUCE W. SHORE

CAMBRIDGE
UNIVERSITY PRESS

CAMBRIDGE
UNIVERSITY PRESS

University Printing House, Cambridge CB2 8BS, United Kingdom

One Liberty Plaza, 20th Floor, New York, NY 10006, USA

477 Williamstown Road, Port Melbourne, VIC 3207, Australia

314-321, 3rd Floor, Plot 3, Splendor Forum, Jasola District Centre, New Delhi - 110025, India

79 Anson Road, #06-04/06, Singapore 079906

Cambridge University Press is part of the University of Cambridge.

It furthers the University's mission by disseminating knowledge in the pursuit of education, learning and research at the highest international levels of excellence.

www.cambridge.org
Information on this title: www.cambridge.org/9780521763578

First published 2011

A catalogue record for this publication is available from the British Library

Library of Congress Cataloging in Publication data
Shore, Bruce W.
Manipulating quantum structures using laser pulses / Bruce W. Shore.
p. cm.
Includes bibliographical references and index.
ISBN 978-0-521-76357-8 (hardback)
1. Nuclear excitation. 2. Atomic structure. 3. Coherent states.
4. Perturbation (Quantum dynamics) 5. Laser interferometry. I. Title.
QC794.6.E9S56   2011
539.7′56–dc23   2011027979

ISBN 978-0-521-76357-8 Hardback

Contents

Preface

Manipulation of the internal structure of atoms and molecules – altering the quantum states of submicroscopic systems – makes an increasingly significant contribution to contemporary technology, as electronic circuits continue to shrink in size and new opportunities appear for applying abstract quantum theory to the creation of practical devices. The structural changes range from simple perturbative distortions of the electronic charge distribution to complete transformation into an excited energy state or the creation of superposition states whose properties cannot be fully described without quantum theory.

This monograph discusses ways of inducing such changes, primarily (but not only) with pulses of laser light, and ways of picturing the changes with the aid of suitable mathematical tools. Aiming at a level suitable for advanced undergraduates or researchers it explains the basic principles that underly the quantum engineering of devices used for such applications as coherent atomic excitation and quantum information processing.

Presupposing some familiarity with quantum mechanics, it first introduces notions of atoms (or other localized quantum systems) and quantum states, and of radiation (specifically laser pulses), defining thereby the essential observable quantities with which theory must deal. It presents the constructs – probabilities, probability amplitudes, wavefunctions, and statevectors – that serve as variables for the mathematics. It then discusses the differential equations that describe laser-induced changes to atomic structure. It contrasts the pre-laser incoherent absorption of energy, governed by rate equations, with the coherent regime of laser-induced changes, governed by the time-dependent Schrödinger equation that is the foundation for all descriptions of quantum-mechanical changes.

Subsequent sections present a variety of solutions to the differential equations governing coherent excitation: first for simple two-state systems, then for three-state and more elaborate multistate systems. Such characteristics as Rabi oscillations and adiabatic following, quantum-state superpositions, dark quantum states and population trapping receive attention, as do methods for measuring quantum-state changes and for evaluating alterations of radiation caused by the changes.

Illustrations of the time-varying excitation rely not only on plots of probabilities but also on various vectorial portraits of the dynamics, such as the Bloch vector (the quantum counterpart of the Stokes vector of optics), that offer valuable insights into coherent excitation processes. They answer the question: How can we think about laser-induced changes to atomic structure?

The present monograph is an enlarged and revised version of a 2008 review article:

B. W. Shore, "Coherent manipulations of atoms using laser light". Acta Physica Slovaka, **58**, 243–486 (2008). Available online at www.physics.sk/aps/pub.php?y=2008&pub=aps-08-03.

In turn, that was an updated condensation of a 1990 two-volume textbook, portions of which reappear in the present monograph:

B. W. Shore, *The Theory of Coherent Atomic Excitation. I. Simple Atoms and Fields* (Wiley, New York, 1990).
B. W. Shore, *The Theory of Coherent Atomic Excitation. II. Multilevel Atoms and Incoherence* (Wiley, New York, 1990).

As did those publications, the overall presentation here emphasizes theory, rather than experiment, stressing the mathematical formulation and tools for linking the equations both to observable properties and to graphical displays that assist understanding.

Acknowledgments

This monograph builds upon work begun decades ago in the Laser Program at the Lawrence Livermore National Laboratory and at Imperial College, London, when the study of laser-induced changes to atomic structure was comparatively new. The theory underlying the resulting coherent excitation of atoms was then relatively novel, and came together in the 1990 textbook. That work owed much to collaborations with Joe Eberly, Mike Johnson, Dick Cook, and Peter Knight. The subject has since expanded in many directions. I have continued to work on some of these and to give courses and review talks. I am indebted to many colleagues for opportunities to extend my own understanding, reflected in the present review, notably Klaas Bergmann, Leonid Yatsenko, Razmik Unanyan, Nikolay Vitanov, Michael Fleischhauer, Thomas Halfmann, and Axel Kuhn (connections formed at the Technical University of Kaiserslautern), to all of whom I am grateful for stimulating discussions and some support. The Alexander von Humboldt Foundation has supported several research stays in Germany. I am grateful to Vladimir Buzek for encouraging the 2008 review and the present work.

1

Introduction

The quantum world within an atom or molecule that once attracted explorations only by academic physicists now provides fertile sustenance for chemists seeking control of chemical reactions and for engineers developing ever smaller electronic devices or tools for processing information with greater security. Whereas the first pioneers could only discover the most elementary properties – the discrete energy levels that characterize the internal structures of atoms and the radiative transitions that link these structures to our external world – it is now possible to alter that structure at will, albeit briefly.

1.1 Objective

This monograph presents the physical principles that describe such deliberately crafted changes, namely how single atoms or molecules (or other simple quantum systems) are affected by coherent interactions, primarily laser light – a subject that has been regarded first as a part of quantum electronics and then quantum optics but is most generally described as coherent atomic excitation [Sho90].

This physics has relevance to such basic concerns as the detection and quantitative analysis of trace amounts of chemicals, the catalysis or control of chemical reactions, the alignment of molecules, and the processing of quantum information. The physics necessarily involves elementary quantum mechanics, but it has many associations to the classical dynamics that governs macroscopic objects – waves and particles. The mathematics that quantifies the changes is that of differential equations, specifically coupled ordinary differential equations (ODEs), whose parameters incorporate the controls of experimenters and whose solutions, appropriately interpreted, quantify the resulting changes.

The intended readership is working scientists, graduate students, or advanced undergraduates in physics, chemistry, or engineering. The physics relies on quantum theory and Maxwell's equations (summarized in App. C), and the mathematics uses linear algebra for solving coupled ODEs.

1.2 Background

The formalism presented in this monograph builds upon a foundation of classical dynamics and quantum theory to erect theory descriptive of deliberately produced quantum changes.

It begins, in Chap. 2, with notions (and definitions) of probability and expectation values and the theoretical constructs of wavefunctions and statevectors from which one calculates observables. From these come the coupled ODEs (primarily expressing the time-dependent Schrödinger equation of Chap. 7) with which one describes experimenter-induced changes. The many examples of solutions to these equations, for models that are appropriate to many realizable scenarios, form the remainder of the monograph. The discussion progresses from the simplest models to more complicated ones, often noting connections between the quantum description and more classical views, and pointing, when possible, to generalizations of the specific simple cases.

The mathematics used to describe coherent manipulation of quantum states (defined below) by radiation was first applied to studies of nuclear magnetic resonance (NMR) with radio-frequency (RF) fields. Although the theory used for NMR, and its applications to magnetic resonance imaging (MRI), differs in details from that needed for describing laser-induced excitation, the basic set of coupled ODEs, and their solutions, serves both communities. In particular, the Rabi oscillations of probabilities (Chap. 8), tailored pulse sequences (Chap. 11), and adiabatic passage (Sec. 8.4) discussed in the present monograph occur throughout atomic and molecular physics.

Another source of mathematical tools originates with the theoretical description of inelastic collisions between molecules, encounters in which portions of kinetic energy are converted into changes of internal energy of collision partners. From such work come the notions of adiabatic changes that have become an important part of coherent excitation (see Sec. 8.4, Chap. 14, and App. E).

1.3 Measurables, observables, and parameters

The theory used in the present exposition deals with several foundational concepts. The following paragraphs define these. Later sections and appendices elaborate upon the definitions.

Systems. The basic objects of interest in physics are physical *systems*: collections of particles or parts that move in accord with rules that can predict, from present conditions, the positions and motions of all system constituents at a prescribed future time. We are here interested primarily in *dynamical systems*, in which at least some of the constituents can change with time, as contrasted with static structures for which all measurable properties remain unchanged with passing time. The changes occur, at least in part, because of the controllable influences exerted by an experimenter.

The system may be a swarm of gas molecules or a complicated machine of linked parts, for which Newton's laws of motion apply, and for which the experimental influence is expressible as forces applied to the system. Or the system may be some collection of subatomic particles, for which the relevant rules are those of quantum theory and control occurs through adjustable interaction energies, as mediated by laser beams. Each of these two categories is describable by similar mathematical formalism.

Observables. At any specified time t a physical system has a state of motion that is uniquely and completely defined by a set of real numbers that are measurable. These are the basic *observables* associated with a set of variables that describe possible motions. For a classical system the observables are typically positions and velocities of parts. It is with these observables that theory must provide rules for changes induced by experimental controls. In the simple case of a point mass (i.e. an object of infinitesimal dimensions but finite mass) the observables are typically the three Cartesian coordinates x, y, z, collectively denoted by the vector $\mathbf{r}$, and the velocities $v_x = \dot{x}, v_y = \dot{y}, v_z = \dot{z}$, where the overdot denotes the time derivative.

State space. The values of the observables, a set of numbers, can be represented mathematically by a point in an abstract *state space* (mathematically, a *manifold*); the numbers are the *components* of the vector. The dimension of this space, N, i.e. the number of coordinates in the abstract space, is the number of observables needed to identify states uniquely.

With passing time the system changes, influenced by the experimenter. One refers to this change as the *time evolution* of the system. Mathematically, the change is a *mapping* of the points of state space at one time to those at a later time. The theory presented in this monograph provides examples of the rules governing that evolution, with examples of the state spaces (i.e. the various observables and their mathematical representation) and of the experimentally controlled time evolution – the picturing of geometrical properties of motion in the abstract state space.

Constrained motion. A solid nondeformable body requires three spatial coordinates and as many as three orientation angles to completely specify its position and orientation. The number of such coordinates or angles needed to unambiguously define the position and orientation is the number of *degrees of freedom* of the object.

Motions of classical particles, or collections of solid objects, are often constrained, for example by impenetrable surfaces or flexible joints. If the conditions of constraint (the forces exerted by the surfaces and joints) can be expressed as equations connecting the particle coordinates $\mathbf{r}_1, \mathbf{r}_2, \ldots$ (and time t), in the form of a functional relationship

$$F(\mathbf{r}_2, \mathbf{r}_2, \ldots, t) = 0, \tag{1.1}$$

then the constraints are termed *holonomic*; otherwise they are nonholonomic. The particles of a rigid body, or a particle constrained to move on a surface, move under holonomic constraint. Each equation of holonomic constraint reduces by one the number of degrees of freedom of the system [Gol50].

In describing motions of complicated arrangements of constrained particles it is useful to replace the three Cartesian coordinates x, y, z of each particle, and their time derivatives $\dot{x}, \dot{y}, \dot{z}$, with *generalized coordinates* that take into account the constraints. The generalized coordinates may be angles or energies, and so they generally cannot be arranged as orthogonal vectors [Gol50].

Quantum systems. An essential difference between a classical system and a quantum system is that the information about the latter, as embodied in the quantum observables, is subject to an essential constraint expressed by the Heisenberg *uncertainty principle*: it is not possible to obtain arbitrarily accurate simultaneous values for positions and velocities of a particle. This means that whereas a classical system of N unconstrained point particles requires $3N$ Cartesian coordinates and $3N$ corresponding velocities for complete characterization, an unconstrained quantum system of N particles requires only $3N$ variables.

Quantum states. The quantum systems treated in this monograph are collections of bound particles (electrons and nuclei, atoms, molecules) whose motions are quantized, e.g. only discrete energies are possible (see Sec. 2.2). Such a system can be fully described, within limitations imposed by quantum theory (e.g. the uncertainty principle), by a set of compatible measurements that produce a set of recordable numbers. The information contained in a set of such measurements, or *quantum observables*, defines a *quantum state*. The measured values are associated with the basic variables with which we describe the quantum dynamics – counterparts of the Cartesian coordinates used to describe classical motions.

The information that most completely describes a quantum system at an instant t (its quantum state) can be regarded as defining a point in an abstract vector space, where it is stored as a *statevector*, an element (a vector) of an N-dimensional abstract vector space, more specifically a Hilbert space, in which the statevector components along perpendicular (orthogonal) axes represent probability amplitudes and the statevector length is unity, as will be explained in Sec. 5.5.

Equations of motion. If the changes to the observables are small during a small time increment dt then the motion of the state-space point is slight, and the rules governing those incremental changes (the time evolution of the system) can be cast as (differential) *equations of motion*. For a classical system of slowly moving (i.e. nonrelativistic) parts these are Newton's equations of motion. For a quantum system the relevant equation is the time-dependent Schrödinger equation (see Chap. 7). These equations of motion predict, starting from any point in state space (a *system point*), the locus of all future points (the *trajectory* of the system point). For an isolated physical system (one free from irregular environmental influences) the rules for time evolution are *deterministic*: From any given state a unique future state will develop. They also provide a unique connection from past to future: From each state at time t there will evolve a single unique future state, and every state has a unique past.

Quantum observables; labels. Different quantum states are distinguishable by a set of compatible measurements that produce characterizing numbers, the *observables* of the quantum system, obtained from numbers displayed on dials or recorded as electrical signals. The measurements are physical operations upon the system, represented mathematically by operators acting on the abstract vectors that embody the information that identifies the state of motion.

Operator. An *operator* is a symbol for a set of rules that map elements of one set to another, e.g. the derivative of a function or a change of vector elements. A vector that under the action of an operator $\hat{M}$ only changes by a scalar multiple is said to be an *eigenvector* of the operator. The relationship corresponds to a physical measurement that leaves all observables unaffected. For the measurements to be compatible with one another the operators must commute – it must not matter which operation occurs first in a sequence. Such operators share eigenvectors. From a complete set of commuting operators one obtains the maximum possible information about a quantum system. The eigenvalues of these operators are real numbers that all together provide unique labels for the quantum states.

Parameter space. Changes of any dynamical system, whether classical or quantum, are to some extent under the control of an experimenter who, by adjusting various controls (e.g. valves or optical elements or the flow of electrical currents) alters the forces acting upon the unconstrained parts of the system. The settings of the controls at a moment t are describable by a set of numbers, the *parameters* that completely define (as far as possible) the momentary influence of the experimenter on the system. These parameters provide a second abstract space, a *parameter space* (or *control space*) in which the number of dimensions M equals the number of controls. The vector components are parameter values.

   The equations of motion, properly formulated, provide a link between points (or paths) in the controllable parameter space and trajectories in state space.

1.4 Notation and nomenclature

The literature on quantum theory exhibits a variety of conventions for the names and symbols of the mathematical entities that represent the physics. Following is a brief summary of some notation and nomenclature defined there and used in the present monograph. Most of these appear only in and after Chap. 5; definitions and explanations occur as the concepts are introduced.

   Complex numbers: The symbol i denotes the imaginary unit $i = \sqrt{-1}$. The notation Re Z and Im Z denote, respectively, the real and imaginary parts of the complex number $Z = X + iY$.

   Vectors are ordered lists of numbers or functions (the vector components), either in three-dimensional *Euclidean space* of everyday life or as column vectors in Hilbert space. They are denoted by boldface font, as $\mathbf{r}$ or $\mathbf{C}$, or with *Dirac notation* as $|C\rangle$. Their components are denoted by single subscripts, as r_j or C_n. The set of components of a vector are often displayed as a column, see eqns. (3.51) and (5.24). For any such column vector $\mathbf{A}$ or $|A\rangle$ there is a row vector, denoted $\mathbf{A}^\dagger$ or $\langle A|$, whose elements are complex conjugates of the elements of $\mathbf{A}$.

   Scalar products: The scalar product of vectors $\mathbf{A}$ and $\mathbf{B}$, denoted variously as $\mathbf{A} \cdot \mathbf{B}$ or $\langle A|B\rangle$, the projection of vector $\mathbf{B}$ onto vector $\mathbf{A}$, is the sum $\sum_n A_n^* B_n$, where the asterisk indicates complex conjugation. The length $|\mathbf{A}|$ of a vector $\mathbf{A}$ is the square root of the sum $\sum_n |A_n|^2$.

Unit vectors have unit length, meaning that the sum of the absolute squares of all their elements is unity. Several sorts of unit vectors occur: three-dimensional Euclidean unit vectors $\mathbf{e}_j$ and unit vectors in Hilbert space, $\boldsymbol{\psi}_n$. The subscript on such vectors identifies a particular vector from a set of possible similar vectors. The Dirac notation $|\psi_n\rangle$ or $|n\rangle$ is also used.

Statevectors: I will distinguish between a *statevector* – an N-component unit vector in Hilbert space (possibly time dependent), denoted $\boldsymbol{\Psi}$ or $|\Psi\rangle$ – and a *wavefunction*, a function of spatial coordinates $\mathbf{r}$ denoted $\Psi(\mathbf{r})$. Some authors use wavefunction as a synonym for statevector, i.e. a quantum state. In the narrow sense used in the present monograph the word wavefunction refers to spatial probability distributions, a function of three spatial coordinates for each particle of the system. Following the approach of Dirac, such a wavefunction $\Psi(x)$ is the projection $\langle x|\Psi\rangle$ of a statevector $|\Psi\rangle$ onto the Hilbert-space coordinate axis $|x\rangle$ representing particle positions x. In the sense of Dirac, a wavefunction can be considered as the *coordinate representation* of a statevector.

Matrices are ordered two-dimensional arrays of numbers or functions, denoted by sans-serif font (usually upper case), as W or U. Their elements are denoted by pairs of subscripts to an italic symbol, such as W_{nm} or U_{nm}. When it is desired to emphasize that the elements of a matrix vary with time, this is indicated as $\mathsf{W}(t)$. A matrix that has N rows and M columns is an $N \times M$ matrix. A matrix is a *square matrix* if $N = M$; it has an equal numbers of rows and columns. A $1 \times N$ matrix has just one row; it is a *row vector*. An $N \times 1$ matrix has just one column; it is a *column vector* of dimension N. The *Kronecker symbol* (or Kronecker delta) δ_{nm} is unity if $n = m$ and is zero otherwise.

Operators: Two classes of mathematical objects occur that act upon appropriate elements: Matrices act upon vectors or other matrices, and differential operators, such as the partial derivative $\partial/\partial z$, act upon functions. Operators are at times indicated by a circumflex, as in the Hamiltonian $\hat{H}$ or in the photon annihilation operator $\hat{a}$.

States and vectors: The terms (quantum) *state* and (Hilbert-space) *unit vector* are often regarded as synonyms, and I have accepted this common usage. Strictly speaking a unit vector in Hilbert space encapsulates the information that completely defines a possible state of a quantum system, together with a norm (unit length) that implies a bound on related probability; see Sec 5.5.2.

Units: Many authors blur the distinction between energy and angular frequency, embodied in the relationship $E = \hbar\omega$, by taking frequency and time units such that $\hbar$ does not appear (one says that $\hbar = 1$). Although this simplifies typography, I will not follow this convention. It is also common practice to measure electric charge in units of the electron charge $|e|$ and mass in units of the electron mass m_e, thereby setting $|e| = \mathsf{m}_e = 1$. The resulting *atomic units* [Sho90,§2.9] [Bay06a] measure lengths in units of the Bohr radius $a_0 \approx 5.3\,\text{Å} = 53$ pm and energies in units of twice the ionization energy of hydrogen, $E^{AU} \approx 27.2$ eV. The speed of light is $c \approx 137$ atomic units. I will not follow this convention, but will show $\hbar$, e, m_e and c explicitly.

The Hamiltonian is an operator (differential when acting on a wavefunction, matrix when acting on a statevector) that expresses the various energies (kinetic and potential) of the system. In transcribing the general formalism of quantum theory into a set of coupled ODEs the Hamiltonian operator $\hat{H}$ produces a matrix of coefficients, H. Its elements have dimensions of energy. Whereas this matrix refers to a fixed Hilbert-space coordinate system, the mathematics is best presented by using a rotating coordinate system and units of frequency (often done by setting $\hbar = 1$). I denote the resulting matrix by W. Its elements have dimensions of angular frequency. To be pedantically correct one should refer to $\hbar\mathsf{W}$ rather than W as a Hamiltonian matrix, but in adopting common practice I do not rigorously follow this guideline.

Terminology: states, vectors, and kets. Because the unit vectors of Hilbert space are associated with the information that defines a quantum state, the terms *vector, statevector,* and *quantum state* (or simply *state*) are commonly treated as synonyms. For *unit vector* or *basis vector* the synonym is *basis state* (or bare state). From the discipline of linear algebra comes the notion of an *eigenvector* of a matrix operator – a vector that remains unchanged in direction when acted on by the operator; the terminology *eigenstate* has become a synonym for Hilbert-space eigenvector of a Hamiltonian.

With Dirac notation a column vector in Hilbert space is denoted $|A\rangle$ or $|\Psi\rangle$, and termed a *ket vector* (or simply a *ket*). The Hermitian adjoint of this, a row vector, is denoted $\langle A|$ or $\langle\Psi|$ and is termed a *bra vector* (or simply a *bra*). The names come from the scalar product of two Hilbert space vectors, denoted with Dirac notation as a *bracket* $\langle A|B\rangle$ or $\langle\Psi_1|\Psi_2\rangle$. It is not uncommon to refer to the elements of a quantum-state Hilbert space as *kets* and *eigenkets*, rather than states and eigenstates.

1.5 Limitations of the theory

This monograph says little about the calculation of atomic or molecular structure – the province of quantum chemistry or atomic structure calculations. Instead it treats the energy values and transition strengths of these structures as empirical parameters.

Apart from Chap. 6, dealing with changes induced by *incoherent* interactions, this monograph deals almost entirely with aspects of *coherent* excitation. It assumes that the responsible interaction is laser radiation. However, it is possible, under appropriate conditions, to achieve coherent changes in other ways. A simple example occurs in the idealized model of two trapped atoms, such as might be held within an optical lattice (see Sec. 4.1.5). By altering the separation between the two atoms in a controlled way one can produce coherent changes of the sort governed by the time-dependent Schrödinger equation, but for quasistatic (DC) fields rather than pulsed laser fields. Basically one considers the controlled formation of a molecule from two atoms. Such situations are not considered explicitly here.

This monograph concentrates on changes induced by moderately intense coherent classical radiation, in contrast to effects produced by incoherent light sources. When the atom is in a small cavity, then each increment of energy transferred to and from the atom accounts for a significant portion of the field energy. It is then essential to incorporate the granularity of the field energy by quantizing the field, as discussed in App. D and illustrated with the

Jaynes–Cummings model of Sec. 22.2.3. At the opposite extreme, when the laser electric field becomes very strong (the averaged electric field of the radiation exceeds the Coulomb field that binds electrons to nuclei) then the formalism of coherent excitation presented in this monograph has little relevance. The laser radiation used for bulk-matter effects such as ablation, drilling, cutting, and welding serves as a very localized source of heat, and the quantum and coherent effects discussed here have not been of use.

1.6 Basic references

The literature relevant to laser-induced manipulation of quantum systems is spread over many areas of physics, chemistry, and mathematics. Numerous textbooks cover portions of the subject matter treated here, notably texts on quantum theory subtitled chemistry, computing, communication, electronics, information, logic, measurement, mechanics, optics, and physics. The present work includes a list of references, cited in the text by a bracketed keyword, e.g. [Sho90]. The list is alphabetized by the keyword. The primary sources of the present work are the references [Sho90; Ber98; Vit01a; Vit01b; Sho08], each of which includes a lengthy bibliography.

Nowadays, internet sources such as Wikipedia offer definitions of technical terms ranging from Abelian group to Zeeman shift. There one can find the mathematical properties of special functions (e.g. Bessel or hypergeometric), and so there is little need for detailed proofs or citations of older literature. Given a technical phrase, e.g. "coherent control" or "quantum information", a reader can readily find not only references but copies of the publications.

2

Atoms as structured particles

The philosophers who first hypothesized the existence of "atoms" had in mind the smallest particles of matter that could preserve identifiable chemical properties – building blocks that could be assembled into familiar substances. Little more than this definition – tiny masses that carry kinetic energy and undergo collisions – led to the fruitful quantitative explanation of vapor properties in the kinetic theory of gases, and to such devices as mass spectrometers and ion accelerators.

Atoms and molecules. As became clear during the early twentieth century, these "chemical atoms" from which materials are constructed have internal structure that endows them with their chemical attributes: one or more positively charged nuclei, each a few femtometers ($1\,\mathrm{fm} = 10^{-13}\,\mathrm{cm} = 10^{-15}\,\mathrm{m}$) in diameter, surrounded by one or more much lighter negatively charged electrons whose motion fills a volume of at least a few cubic angstroms ($1\mathrm{\AA} = 10^{-8}\,\mathrm{cm} = 10\,\mathrm{nm}$) in diameter. Nowadays we distinguish between particles having multiple nuclei (molecules) and those with a single nucleus (atoms); when the positive and negative charges are unbalanced these are ions (positive or negative). The simplest atom, hydrogen, has a single electron; the most complex atoms have more than a hundred electrons. Although I will often refer to "atoms", usually the discussion applies equally well to molecules or to any other structure whose constituents exhibit distinct quantum properties, as manifested by discrete energies.

Nuclei. The nuclei of atoms are, in turn, composed of protons and neutrons. The structures and motions of these nucleons are not altered by laser pulses (see App. H) and so their properties, expressed as static electric or magnetic moments (see App. B.2), are fixed parameters. The underlying structure of nucleons from quarks is even further removed from the observational realm under discussion here. The physics of such "fundamental particles" has no immediate connection with the world of atomic physics as discussed in the present monograph.

Isotopes. It is the electric charge of a nucleus that determines what chemical element (e.g. hydrogen or uranium) will be formed when a cloud of electrons, bound by Coulomb forces, neutralizes the nuclear charge. However, not all atoms of an element need be identical:

atomic nuclei of all but the very lightest element,[1] contain chargeless neutrons in various quantities (between two and three times as many neutrons as protons), and these are
responsible for different *isotopes* of an element. The isotopes of a neutral element, though
all having the same number of electrons and therefore the same chemical properties, differ
in neutron number and hence in their mass. The nuclei of different isotopes also differ in
their electric and magnetic multipole moments, attributed to an intrinsic angular momentum
(nuclear spin), see App. B.2. These affect details of their spectra that allow identification of
different isotopes, and allow selective laser manipulation of atomic structure of a chosen
isotope [Bok06].

2.1 Spectroscopy

Our understanding of the internal structure of atoms and molecules has historically come
through observation of light emitted from hot gases or transmitted through cool vapors or
condensates – samples in which many atoms contribute equally. Light from an incandescent
source, say sunlight as in the earliest observations by Newton, when passed through a slit
to produce a beam and then through a dispersing element (a prism or a grating) spreads
into a spectrum of colors. As noted in Chap. 3 the incremental elements of the spectrum
typically bear labels of either wavelength λ or frequency ν, expressed in hertz, Hz, a cycle
per second. In the present monograph a more convenient measure is the angular frequency
$\omega = 2\pi\nu$, expressed in radians per second. These radiation characteristics are related by
the formulas

$$\nu = \frac{c}{\lambda} = \frac{\omega}{2\pi},\tag{2.1}$$

where $c \approx 2.998 \times 10^8$ m/s is the speed of light in vacuum. The wavelength λ of an increment
of the spectrum is associated with the physiological sensation of color.

The distribution of radiation intensity within a typical spectrum is by no means uniform.
Studies of the features of the dispersed light developed into the science of spectroscopy.
In particular, studies by nineteenth-century spectroscopists showed that the radiation emitted from an optically thin source of hot atoms contained features centered around discrete
wavelengths, appearing as bright lines in spectra. By contrast, cooler vapor would absorb,
from the continuum of colors present in radiation from an incandescent solid (a blackbody
source), discrete wavelengths, appearing as dark spectral lines.[2] Work by Kirchhoff and
Bunsen established that spectral-line wavelengths were a unique signature of a particular
element: each wavelength, when accurately measured, is associated with one specific chemical species and the intensity of the line gives an indication of the relative concentration of
that species in the source. These properties of spectra underlie the vast contemporary use of
spectroscopy for quantitative chemical analysis, with application to such diverse disciplines

---

[1] A small portion of naturally occurring hydrogen nuclei contain a neutron (*deuterium*) or two neutrons (*tritium*).
[2] Although I have referred to spectral lines as having discrete frequencies, close examination reveals that the spectral features –
the frequency distribution of light added or removed – occur over a small range of frequencies: the spectral lines have a finite
width. These line widths convey useful information about the lifetime of the excited state and about the environment within
which the emitting or absorbing atom resides [Sob95].

as forensics and archeology. The dark lines in the solar spectrum observed by Fraunhofer have counterparts in the spectra of all stars; observation of their patterns has provided us with our knowledge of the constitution of stars and the size and motion of the universe.

The visible spectrum, with its familiar colors ranging from red through yellow to violet, is but a small portion of the full electromagnetic spectrum that ranges from gamma rays through x-rays, ultraviolet, infrared, microwave, and radio waves. Throughout this wider range of wavelengths there occur spectral lines, and branches of spectroscopy that use them.

All such applications of spectroscopy require no model of the internal structure of atoms. It is only necessary to have available a catalog of spectral-line wavelengths and some measure of their relative strengths. To some extent these are the data required to quantify the theory of coherent excitation discussed in the present monograph. The elementary constituents of matter that are responsible for the spectral lines might well be called "spectroscopic atoms" [Sho90, §1.1], defined by their spectroscopic properties.

2.1.1 Spectroscopic transitions

As spectroscopists built their catalogs of spectral-line wavelengths (or better, the *wavenumbers* – the inverses of vacuum-recorded wavelengths) they found that all spectral lines obeyed the Rydberg–Ritz *combination principle*: each wavenumber[3] $\tilde{\nu} \equiv 1/\lambda = \nu/c$ of a spectral line of frequency $\nu = c/\lambda$ could be expressed as the difference of two spectroscopic *term values* $\widetilde{E}$. That is, the wavenumbers were all expressible in the form

$$\tilde{\nu} = \widetilde{E}_n - \widetilde{E}_m, \tag{2.2}$$

where the integers n and m identify the term values in a suitably ordered list. Tables of spectral term values continue to be used in cataloging spectral lines.

The Bohr combination principle. The explanation of this empirical rule came with the early quantum theory of Bohr, who proposed that it expressed the conversion of discrete internal atomic energies E_n into radiation quanta of energy $h\nu$,[4]

$$h\nu \equiv \hbar\omega = E_n - E_m. \tag{2.3}$$

The scaling factor $h \approx 6.626 \times 10^{-34}$ J s is the *Planck constant*; the factor $\hbar = h/2\pi \approx 1.054 \times 10^{-34}$ J s is the rationalized Planck constant or *Dirac constant*. That is, the spectroscopic term values were scaled discrete internal energies of atoms, $\widetilde{E}_n = E_n/hc$. The discrete energy differences, scaled by $\hbar$, are *Bohr frequencies*,

$$\omega_{nm} = |E_n - E_m|/\hbar. \tag{2.4}$$

---

[3] Wavenumbers and term values have dimensions of inverse lengths, often expressed in cgs units of reciprocal centimeters, inverse centimeters, or kayser: $1\,\text{K} = 1\,\text{cm}^{-1}$ or kilokayser, kK, $1\,\text{kK} = 10^3\,\text{cm}^{-1}$. The symbol K is nowadays more frequently used for kelvin, a unit of temperature.

[4] The discrete electromagnetic energy $h\nu \equiv \hbar\omega$ needed to produce excitation of a single atom, or emitted during deexcitation, defines a *photon*; see App. D.

Atoms as structured particles

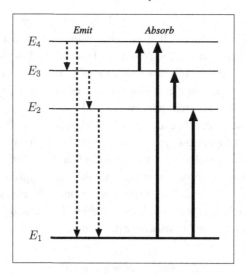

Figure 2.1 Schematic illustration of the Bohr atom: internal energies are discrete; transitions to lower energies (dashed left-side arrows) produce emission lines, while absorption of discrete frequencies (thick right-side arrows) produces excitation and absorption lines.

Figure 2.1 illustrates this quantum combination principle, showing transitions (dashed arrows) in which excitation energy decreases by radiation emission and other transitions (full lines) in which energy is absorbed from radiation.

The transitions that were first observed in optical spectra have counterparts throughout the electromagnetic spectrum: excited states of atomic nuclei emit gamma rays, electron bombardment of solids produces discrete frequencies of x-rays, atomic electron changes produce spectral lines in the visible, rotational changes of molecules produce microwave transitions, and radio-frequency lines are attributable to changes in orientation of atomic nuclei. In all cases the motions of bound particles are such that only discrete energies E_n are possible. The index n here identifies an entry in an ordered list of energies, in which the lowest value (typically $n = 1$ or $n = 0$) is termed the *ground-state* energy.

2.1.2 Transition selectivity

Transition strengths. Spectroscopists found that not all possible pairs of energies produced spectral lines; they devised various empirical *selection rules* to classify "allowed" and "forbidden" transitions. Seeking patterns, they found that the energy levels of free atoms could be classified as being one of two classes, termed *even parity* and *odd parity*. Transitions were generally not observed without a parity change. Appendix B.7 discusses the wavefunction origin of these selection rules. Drawing on the Lorentz model of electrons as charges subject to oscillator-like forces about equilibrium positions, spectroscopists parametrized the relative intensity of the spectral lines by *oscillator strengths*, see App. H. These remain a convenient parameter with which to quantify the strengths of radiative interactions.

Selectivity. The Bohr condition (2.3), when supplemented with symmetry-based constraints (e.g. angular momentum selection rules, see Sec. 12.2), makes possible the selective addition of energy (excitation) into not only a specified element but a chosen isotope of that element, without undesired changes to other species. Given radiation concentrated within a narrow frequency interval it is possible to affect only a very select class of atoms.

When used with carefully timed pulses, the Bohr condition underlies the procedures presented in this monograph for manipulating internal motions of atoms – for selecting a very specific transition and pair of energies. However, the theory to be presented here deals with radiation that differs in an essential way from the radiation used in traditional spectroscopy: it has coherence (see Sec. 3.5.5). This property makes possible a variety of improvements in the selectivity of traditional radiative processes [Per06] as well as the creation of quantum structures (superposition states) that are otherwise unobtainable; see Chap. 18.

2.2 Quantum states

Quantization; quantum theory. The Bohr interpretation of the spectroscopic combination principle requires that the internal energies of a "spectroscopic atom" be discrete: they are *quantized*. The theory needed to explain this quantization, now typically termed *quantum theory* or quantum mechanics, grew out of the wave mechanics of Schrödinger and the matrix mechanics of Heisenberg that provided the first comprehensive formalism for treating the motion of atomic electrons and later molecular structure. The remainder of this chapter introduces some aspects of quantum theory relevant for treating excitation.

Basic quantum theory requires that when some motion is constrained within a finite region of space, and the system is isolated from uncontrollable external influences, the energy of that motion can take only discrete values – it is quantized.[5] The separation between the discrete energy values, typically proportional to some simple function of an integer (a *quantum number*) is inversely proportional to both the mass of the bound particle and to the dimension of the confinement.

Quantization examples. Although atoms and molecules have dimensions of nanometers or less, many larger objects are affected by quantum-mechanical behavior and have discrete energies. Common examples of energy quantization include the following traditional systems:

- The motion of electrons bound to nuclei of atoms or molecules by Coulomb forces
- Vibrations of molecular parts
- Rotation of molecular frameworks
- The orientation energy of nuclear spins in an external magnetic field

---

[5] As discussed in textbooks on quantum theory, the discreteness is one consequence of the basic mathematical properties of the time-independent Schrödinger equation together with boundary conditions; see Sec. 5.4. Discreteness of orientation follows also from the noncommuting nature of the operators representing angular momentum.

- Energies of electrons and holes in quantum dots
- Center of mass motion of atoms or ions held in a trap
- Motions of nucleons (protons and neutrons) within an atomic nucleus.

Quantization is not limited to atomic-scale phenomena. Quantum dots are macroscopic objects, composed of millions of atoms, in which electrons are confined and hence have discrete energy states. Bulk superconductors are composed of many pairs of electrons that condense, as Cooper pairs, into a single quantum state; these allow the sort of discrete-state transitions, and quantum-state manipulation, discussed here. Josephson junctions between two different superconductors exhibit quantization of charge or phase (magnetic flux) degrees of freedom, controllable by gate voltages. The quantum properties are analogous to those of other systems in which energy is quantized.

In all these cases, and many others, the total energy of the constrained particles – the sum of kinetic and potential energies – is restricted to take only discrete values, between which radiative transitions can occur at discrete frequencies. It is these transitions, and the corresponding structure changes, that are the subject of the present monograph. Other quantum states do exist, corresponding to unbounded motions, for example a photoionized electron or photodissociated molecular fragments, but these are not considered here.

Quantum states. Classical states of motion particularize, for specified initial conditions, the dynamics of classical systems – those systems whose behavior follows the laws of motion first enunciated by Newton, e.g. trajectories of planets, missiles and atoms in beams. The discrete energies revealed by spectroscopy through the combination principle of eqn. (2.2) are associated with discrete states of motion of the electrons confined within atoms and molecules. Descriptions of such motions, distinguishable in part by energy, are the quantum states [Sho90, §1.3] with which we are here concerned. Specifically, a quantum state comprises *information* about the quantum system, as embodied in the wavefunctions and statevectors discussed in subsequent sections. By contrast, traditional classical descriptions of motion typically concern trajectories of particles: how positions vary with passing time under the influence of applied forces.

Labels. In the present monograph I typically number the discrete quantum states that are of particular concern with integers, $n = 1, 2, \ldots$. However, for some purposes it is useful to label the states with other integers, quantum numbers that have significance from quantum-mechanical models of the atomic or molecular structure. The strength of a spectral line, and the corresponding strength of the interaction responsible for the transition, depends on the quantum numbers associated with the two energy states. For most laser-induced transitions the interaction is that of an electric dipole moment in the presence of the electric field of the laser radiation, although other multipole transitions do occur, see App. B.2. Given a pair of quantum numbers, and their physical origin, selection rules specify what changes may occur, and hence what transitions may be induced from any specific energy level; see App. H.

Axiomatic quantum theory. The historical approach followed above deduces the existence of discrete quantum states from observational evidence – discrete spectral lines interpreted as transitions between discrete energies of atoms. It is an "operational" approach to physics that builds concepts based upon measurements. A common, more formal, approach deduces the spectral lines as an inevitable consequence of postulates that define quantum theory and that treat quantum states (or their representation in abstract vector spaces) as fundamental entities. That approach [von55; Dir58; Sty01], with its notion of "falsification" of postulates by measurement, has appeal akin to that of Euclid's axiomatic approach to geometry. It is deductive, based on analysis rather than synthesis.

2.3 Probabilities

Quantum theory, in common with classical statistical mechanics, deals with probabilities. Its predictions deal with either multiple observations of a single system, repeatedly reprepared, or with single observations of a large collection of identically prepared systems (an *ensemble*, see Chap. 16.1). In either case we perform a measurement of some observable attribute, say the energy.

For the ensemble we prepare a collection of $\mathcal{N}$ identical systems according to some procedure and then measure the number of atoms $\mathcal{N}_q$ that have attribute q. We convert this into the *probability*

$$P_q = \frac{\mathcal{N}_q}{\mathcal{N}} \tag{2.5}$$

of finding a single system with attributes q. Alternatively, for measurements on a single atom we repeatedly prepare the atom using an identical procedure and we measure the attribute. Again, we convert these multiple values into a probability by taking the ratio of successes to number of trials.

Probability and measure theory. Because the total number of systems $\mathcal{N}$ must include the set $\mathcal{N}_q$ the probability is bounded by zero and one:

$$0 \leq P_q \leq 1. \tag{2.6}$$

Mathematically, probability provides a mapping of a set (attributes) onto the closed unit interval of the number line. It is an example of a mathematical *measure*.

Alternative definition of probability. The word "probability" is used in two rather different senses. As presented here it refers to the frequency of events taken from a large sample. Alternatively it is regarded as a number that "quantifies the plausibility of a proposition given a state of knowledge" [Pre98]. That is, we devise a complete set of mutually exclusive alternatives and adopt a "principle of indifference", meaning that in the absence of any detailed knowledge that any particular outcome is more plausible we assume that the probabilities are equal. One enumerates the ways of obtaining a particular result to assign unequal probabilities. This approach allows evaluation of probabilities for drawing a straight flush or four of a kind in poker prior to engaging in play, and it leads to *Bayesian* interpretation of probabilities [Cou95; Jay03; Car09].

2.3.1 Time-dependent probabilities

To study time dependence, and temporal changes to the system, such as a change of energy, we note the time of preparation and we delay the measurement by a time t. We carry out a succession of such measurements (on either an ensemble or successively on a single atom), each with a different time t between preparation and observation. From those values we obtain probabilities $P_q(t)$ as a function of time t.

A fundamental principle of quantum theory is that any single measurement of the internal energy of an atom must be one of the values from the allowed list of quantized energies.[6] More generally, any single measurement of some attribute (e.g. an orientation) can only show a value associated with one of the allowed quantum states.

However, a succession of measurements will not always produce the same energy, or reveal the same quantum state. Typically the results will appear as a distribution of values, randomly occurring but with well-defined statistical characteristics such as mean and variance (see Sec. 16.2). This distribution of many results, when normalized by dividing by the number of cases, bears interpretation as a distribution of probabilities for finding the allowable quantum states. A key observable for describing the effect of laser radiation on an atom is the probability $P_n(t)$ of excitation into state n at a specified time t, defined by considering numerous ensembles, in each of which there can be found $\mathcal{N}_n(t)$ examples of state n. The probability that, in a series of measurements, each at time t following identical preparation, the observed system will be found in the quantum state bearing index n in some catalog is

$$P_n(t) = \frac{\mathcal{N}_n(t)}{\mathcal{N}_1(t) + \mathcal{N}_2(t) + \cdots}. \qquad (2.7)$$

By construction the probabilities are non-negative real numbers that sum to unity,

$$\sum_n P_n(t) = 1. \qquad (2.8)$$

I will refer to $P_n(t)$ either as an excitation probability at time t or as a time-dependent *population* in state n at time t, and to alterations of these probabilities as *population transfer*. Note that this operational definition of excitation probability rests on the implied assumption that every measurement occurs on a system that has undergone a specific common preparation procedure. Typically this preparatory step places the system in its lowest energy state, an assumption that holds for most of the present discussion. However, other preconditioning procedures are possible; one then must deal with conditional probabilities (transition probabilities) of finding the system in state n at time t, given that it was known to be in state m at time t_0.

---

[6] Unconstrained motion in free space has no limitation to discrete values of the energy; the possible values are associated with a *continuum* rather than the discrete sets discussed here.

Such probabilities form one of several key links between the formalism of quantum theory and the real world of experimental physics. Other links include the effect of laser-induced changes upon dipole moments, measurable by their effect on radiation propagating through a medium of altered atoms, see Sec. 21.1.

By their definition probabilities cannot be negative. However, situations occur when it is convenient to consider a subsystem of a larger system or a subset of all possible quantum states of a system. When probabilities are thus limited to an incomplete subset their sums may be less than unity. Section 7.2.2 discusses this situation.

2.3.2 Equations of motion for changes

Numerous questions come to mind when considering an atom exposed to radiation, most notably: What is the effect of a particular pulse of radiation? Alternatively, one might ask: Given some initial probability distribution of quantum states, how can one produce a desired final distribution? That is, what choice of radiation parameters – carrier frequency, pulse duration, peak intensity, and polarization (see Chap. 3) – will produce some desired change in the quantum state.[7]

As will be discussed in Chaps. 6 and 7, theory provides rules for calculating changes of probabilities, based on differential equations that prescribe the incremental change dP_n in an incremental time dt. There is one such equation defined for each probability, each of the form

$$\frac{d}{dt}P_n(t) = ? \tag{2.9}$$

in which the right-hand side depends upon controllable radiation parameters as well as a set of populations. Together the set of differentials lead to a set of coupled ODEs. Given such a set of differential equations we can, in principle, carry out the implied integration, starting from specified initial conditions, say $P_n(0)$, to obtain the probabilities $P_n(t)$ at any later time t. Chapters 6 and 16 provide examples of these equations. There are two regimes of excitation that have simple equations:

Incoherent: Historically, the first regime to be considered was that of thermal radiation or, more generally, *incoherent* light sources – those having a broad distribution of wavelengths, such as occur in the atmospheres of stars or laboratory plasmas. As discussed in Chap. 6, description is by a set of first-order ODEs (*rate equations*) for changes of probabilities [Sho90, §2.2].

Coherent: The second regime, that of *coherent excitation*, became of practical interest with the introduction of laser light sources. As discussed in Chap. 5, quantum theory expresses probabilities $P_n(t)$ as absolute squares of complex-valued probability amplitudes, $P_n(t) = |C_n(t)|^2$, and it is for these functions that the theory provides ODEs – the time-dependent Schrödinger equation [Sho90, Chap. 3].

---

[7] Traditionally the change was from an initially populated ground state, but more recent interest lies with other states, even superpositions of states.

Each of these sets of equations is an idealization, appropriate to different assumptions about the excitation. A more generally applicable approach requires a formalism that bridges these two extremes, a *density matrix*, see Chap. 16.

The present monograph centers attention on these differential equations and their connection with experimental observables. It discusses the elements of the equations (their relationship to parameters describing radiation), the initial conditions needed to complete the definition of the equations, and examples of solutions for various systems and excitation situations. Before presenting those equations it is useful to review the relevant properties of radiation, particularly the laser radiation used to manipulate quantum structures, done in the following chapter. Appendix C gives more details.

3

Radiation

From prehistoric times has come recognition that sunlight and firelight provide warmth, and that such illumination casts shadows. Expressed in more contemporary terms one would say that light travels in rays, and that this radiation has the potential to provide heat energy to absorbing material. From the time of Newton it has been known that light sources emit radiation comprising a distribution of colors. During the nineteenth century it was recognized that radiation had characteristics of transverse waves (with wavelength associated to color) but until the late twentieth century, when lasers became laboratory tools, it was hardly necessary to delve into the equations of electromagnetic theory to treat such experiments as were then possible; interest lay primarily with thermodynamic considerations of energy flow or with measurements of the dark or bright lines seen in the distribution of light that, after passing through a slit, was dispersed by a prism or grating into constituent colors. Although laser light sources are essential for the types of atomic excitation considered in this monograph, the legacy from thermal radiation still influences many interpretations of the interaction between radiation and matter, and it is therefore useful to summarize some of those concepts.

The mathematics needed for describing laser radiation, or polarized light in general, has much in common with the mathematics of quantum theory discussed starting in Sec. 3.5 and specialized to two-level atoms in Sec. 5.6. In both cases one deals with two complex-valued functions – independent electric field amplitudes or probability amplitudes – whose absolute squares are measurable. It is instructive to review, as will this chapter, the mathematical tools developed for this purpose in nineteenth century studies of optics.

3.1 Thermal radiation; quanta

The first recognized indications of quantum theory – the discreteness of radiation quanta – came with the successful mathematical description by Planck of the spectral distribution of blackbody radiation, idealized as the radiation visible through a pinhole in the walls of an enclosed cavity (in German a *Hohlraum* or hollow chamber). The radiation characteristics are independent of the chemical composition of the wall, and depend only on the temperature T.

To derive the Planck expression for the frequency distribution of radiation in the cavity one assumes that the radiation energy of angular frequency ω is carried by discrete quanta, each of energy $\hbar\omega$, and one assumes that these are in thermodynamic equilibrium at temperature T. One assumes that, as do particles, these quanta obey the Boltzmann distribution (see Sec. 16.2.2), so that the probability of finding energy $n\hbar\omega$ at temperature T is

$$p(n, T) = \exp(-n\hbar\omega/k_B T)/\mathcal{Z}(T), \tag{3.1}$$

where k_B is the *Boltzmann constant* and $\mathcal{Z}(T)$ is the *partition function*,

$$\mathcal{Z}(T) = \sum_{n=0}^{\infty} \exp(-n\hbar\omega/k_B T) = \frac{1}{\exp(\hbar\omega/k_B T) - 1}. \tag{3.2}$$

One next assumes an enclosure whose walls are perfect conductors, so that the electric field of the radiation vanishes on these surfaces; the allowed frequencies then form a discrete set such that an integral number of half-wavelengths fit within the cavity (see Chap. 22). The number of such discrete modes within a large volume V and wavelength increment $d\lambda$, of either of two orthogonal polarizations and traveling into all solid angles (4π steradians) is $d\mathcal{N} = 8\pi V d\lambda/\lambda^4$. It follows that the energy density (i.e. radiation energy flowing into all directions) within frequency increment $d\omega$ around ω,

$$u(\omega, T) = \sum_{n=0}^{\infty} n\hbar\omega \, p(n, T) \, d\mathcal{N}, \tag{3.3}$$

is the *Planck function*

$$u(\omega, T) = \left(\frac{\hbar\omega^3}{\pi^2 c^2}\right) \frac{d\omega}{\exp(\hbar\omega/k_B T) - 1}. \tag{3.4}$$

The success of the Planck function in fitting observations of the spectrum of blackbody radiation encouraged the interpretation of radiation as having granularity, quanta of energy increment $\hbar\omega$. It is such quanta (*photons*) that are emitted or absorbed in atomic transitions. However, the derivation of the Planck formula uses a model of radiation in thermodynamic equilibrium with its surroundings, flowing with equal probability in all directions. Such conditions do not fit the details of the radiation from thin samples of hot vapor, from which the spectrum shows discrete spectral lines on a background of thermal radiation, nor does it describe the near monochromatic directional radiation from a laser. Appendix D provides a brief discussion of the subsequent identification of quantum properties of radiation, evinced through photons as the field quanta.

3.2 Cavities

The hohlraum that inspired Planck to obtain the first indications of quantum-mechanical properties – the existence of discrete radiation quanta – have contemporary counterparts as cavities or other confined regions in which bounding surfaces enforce nodes of the electric field, thereby constraining the radiation to discrete frequencies (see Chap. 22).

One application of this enclosure-enforced spectral structure is to enhance spontaneous emission of atoms within the cavity when the frequency of the emitted radiation ω matches that of a cavity mode, ω_c. Under such resonance conditions the emitted radiation will persist within the cavity, and will be available to stimulate emission from other identical atoms. Macroscopic samples of excited atoms maintained within such a resonant cavity can thereby serve to amplify spontaneous emission, the mechanism embodied in the term *light amplification of stimulated-emission radiation* (laser). The light emerging from such a cavity provides the beams of coherent radiation discussed in the sections below.

A second application is to cold cavities devoid of matter. Unlike the hohlraum that Planck considered, the contemporary empty cavities equilibrate with walls that are at very low temperature, so that $k_B T \ll \hbar\omega$ for a frequency resonant with an atomic transition at visible wavelengths. The Boltzmann factor $\exp(-n\hbar\omega/k_B T)$ is then very small except for $n = 0$ or 1: The equilibrated cavity has at most one photon at any chosen frequency. These conditions allow investigations of single atoms and single photons: A single unexcited atom, moving into a cavity, can remove a single photon and an excited atom can deposit a photon [Mes85; Wal94]; see Chap. 22.

The spectral characteristics of cavity fields, and their interaction with single enclosed atoms, forms the subject of Chap. 22. The major portions of this monograph deal with radiation in free space, as formed by laser beams.

3.3 Incoherent radiation

The description of radiation during the nineteenth and early twentieth centuries dealt with a number of measurable attributes of radiation as flowing energy:

Intensity: The power per unit area passing through a surface is termed the *irradiance* (also called the flux density and, within the laser community, the *intensity*). The power per unit projected area passing into unit solid angle is the *radiance*.

Frequency: The distribution of power amongst the continuum of frequencies or wavelengths, made visible by a dispersing element such as a prism or grating, defines the *power spectrum*. Equation (2.1) defines the frequency ν, wavelength λ and angular frequency ω with which one identifies an element of this spectrum.

Polarization:[1] Passage of light through suitable optical devices (e.g. a polaroid) will produce polarized light, as will reflection from the surface of a pane of glass at an appropriate angle, the *Brewster angle*. Electromagnetic theory describes radiation as a traveling electric and magnetic field; the *polarization axis* is the axis of the electric field, transverse to the propagation direction.

The study of thermal radiation flow, and its interaction with bulk matter, has been the domain of astrophysics since the early twentieth century. Recognizing that radiation carries

---

[1] The nomenclature regarding polarization is confusing: one sees reference to the "plane of polarization" as the plane containing the **B** field and the propagation axis. This is perpendicular to the "axis of polarization" aligned with the **E** field.

energy, that this energy flows along straight-line paths in vacuum, and that at any location and time the energy can be regarded as distributed over wavelength (or frequency), one proposes as the basic physical variable the amount of energy per unit frequency interval per unit solid angle flowing across a unit area normal to the ray path. One then proposes that radiant energy is conserved in free space, but that the presence of cold matter will diminish the energy by absorption, and that hot matter will add to the energy by emission. As did Einstein in his first writings on radiation, one uses thermodynamic arguments to quantify the rates at which matter and radiation, in steady thermodynamic equilibrium, exchange energy. These concepts lead to the theory of radiative transfer, one of the foundations of astrophysics [Cha60; Ryb85].

Such pre-laser studies treated radiation as an incoherent superposition of many frequencies. By contrast, laser light offers possibilities for very monochromatic radiation; the traditional theory of radiative transfer, and of Einstein's view of absorption and emission of radiant energy, does not hold for such illumination. However, the frequencies of spectral lines, and their relative intensities, remain key parameters of the newer theory. The remainder of this section discusses the basic physics of radiation, with particular emphasis on laser radiation.

3.4 Laser radiation

Light from a laser has several properties that distinguish it from sunlight or light from a lamp. It is these attributes that make lasers such versatile tools for probing a variety of things, large and small [Sho03], and for inducing unique changes to quantum structures.

Directed: A laser can provide a highly directional beam of radiation that can be aimed with great precision. It is this property that makes the laser useful as a pointer during lectures, as an aid in leveling land or for probing selected portions of some distant target. It allows selection of microscopic spots of a surface or of lines (narrow beams) within a vacuum chamber, each localized within a dimension comparable to the wavelength of the light.

Monochromatic: Continuous wave (cw) lasers can emit radiation within an extremely narrow range of frequencies. This property makes them ideal references for standards of length and time. Such carrier waves, when modulated or pulsed, offer useful means for message communication; fibers carrying such waves have become an important component of our economies. This attribute allows selection of specific atomic or molecular species, or even specific isotopes, from a mixture of particles.

Intense: A laser can produce highly concentrated doses of energy. Applied to a surface, this deposited energy produces a mechanical distortion as the material temperature rises. When the dose is sufficiently large in a sufficiently small time, the surface is ablated. This attribute makes lasers useful for industrial machining such as welding or cutting and drilling, or precise cutting of tissues. In such commercial applications the laser serves as a localizable heat source.

Coherent: A laser beam has one property that even an intense focused beam of incoherent radiation lacks: the property of maintaining a definite pattern of phase modulation with time [Wol07]. This property, termed coherence (see Sec. 3.5.5 and App. C.3.2), leads to a very different transient nonlinear response of matter than would be predicted for incoherent light, such as that from a lamp or star. In particular, it permits one to create, to manipulate, and then to probe novel quantum states. It is this aspect of laser light that is needed for the quantum manipulations described in the present monograph.

Linear momentum: The radiation field of an intense laser can impart large amounts of momentum to a small surface. The field momentum, regarded as a pressure, can also be used to cool and trap atoms and molecules [Met99]. These radiative forces can be used to manipulate single molecules, as a kind of optical tweezers [Mof08].

Angular momentum: Traveling radiation, such as that in a laser beam, can carry not only linear momentum but also angular momentum and can impart a torque to matter. The angular momentum can come from a combination of spin (for circularly polarized light) and orbital angular momentum (for various structured beams) [All03; Auz10].

Laser light sources provide important tools for spectroscopic studies of matter and for measuring pre-existing properties, such as energy levels, but they also create opportunities to manipulate quantum structures – within individual atoms or molecules – in ways that would be impossible with incoherent light sources. It is with such deliberate changes of atomic and molecular structure that the present monograph deals. In turn, these laser-induced modifications permit one to produce, through emission of light from restructured atomic sources, tailored radiation fields, either propagating in free space (see Sec. 21.1) or in a cavity (see Chap. 22).

3.5 Laser fields

Visible light, in common with radio waves, x-rays and all other forms of electromagnetic radiation, is a manifestation of two electromagnetic fields: an electric field $\mathbf{E}(\mathbf{r}, t)$ and a companion magnetic field $\mathbf{B}(\mathbf{r}, t)$. Their mathematical description rests upon Maxwell's equations, expressed as wave equations, see App. C.2. In vacuum (i.e. free space) the wave equation for the electric field reads

$$\left[\nabla^2 - \frac{1}{c^2} \frac{\partial^2}{\partial t^2} \right] \mathbf{E}(\mathbf{r}, t) = 0, \tag{3.5}$$

from which the **B** field follows as a solution to the Maxwell's equation

$$\frac{\partial}{\partial t} \mathbf{B}(\mathbf{r}, t) = -\nabla \times \mathbf{E}(\mathbf{r}, t). \tag{3.6}$$

Each of these fields is *solenoidal*, meaning divergenceless,

$$\nabla \cdot \mathbf{E}(\mathbf{r}, t) = \nabla \cdot \mathbf{B}(\mathbf{r}, t) = 0. \tag{3.7}$$

From solutions to these equations, in various approximations, come the spatial and temporal variations of the fields that enter into the mathematical description of laser-induced quantum changes.

The two vector fields $\mathbf{E}$ and $\mathbf{B}$ are characterized by magnitude and direction at every location $\mathbf{r}$ and time t. In free space they are perpendicular to each other,

$$\mathbf{E}(\mathbf{r}, t) \cdot \mathbf{B}(\mathbf{r}, t) = 0, \tag{3.8}$$

and to the *Poynting vector* (see App. C.2.1),

$$\mathbf{S}(\mathbf{r}, t) = \frac{1}{\mu_0} \mathbf{E}(\mathbf{r}, t) \times \mathbf{B}(\mathbf{r}, t), \tag{3.9}$$

which measures the energy flow (the cycle-averaged Poynting vector is the intensity). The three vectors $\mathbf{E}$, $\mathbf{B}$, and $\mathbf{S}$ therefore define an orthogonal coordinate system that can provide local Cartesian axes x, y, z.

3.5.1 Monochromatic plane waves

The wave equations for the $\mathbf{E}$ and $\mathbf{B}$ fields (see eqn. (C.13) of App. C.2.1) have solutions in which the fields have constant values (magnitude and direction) over surfaces, either stationary (standing waves) or moving (traveling waves). The light emerging from a laser, and forming a beam, is an example of a traveling wave. Over the small size of an atom, laser fields can generally be regarded as monochromatic plane waves, meaning that at any fixed position the field is periodic in time at angular frequency ω, and all spatial variation occurs in a single direction, that of propagation.[2] In this (longitudinal) direction the electric and magnetic field vectors are periodic, with spatial period equal to the wavelength $\lambda = 2\pi c/\omega$. The planes of constant field move along a propagation direction (the optical axis) defined by the Cartesian components k_x, k_y, k_z of a propagation vector $\mathbf{k}$ of magnitude $k = |\mathbf{k}| = \omega/c$.

Traveling waves, plane polarization. The simplest mathematical description of a plane wave occurs when the directions of the $\mathbf{E}$ and $\mathbf{B}$ fields remain stationary, while their amplitudes move steadily in the $\mathbf{k}$ direction. Either of these vectors, taken with the Poynting vector $\mathbf{S}$, then defines a fixed plane in space; the radiation is *plane polarized* or *linearly polarized*. The axis of the electric vector $\mathbf{E}$, perpendicular (transverse) to the propagation direction, defines the *polarization axis* of the radiation [Bor99]. An example is the paired monochromatic fields

$$\mathbf{E}(\mathbf{r}, t) = \mathbf{e}\mathcal{E}\cos(\mathbf{k} \cdot \mathbf{r} - \omega t + \varphi), \quad \mathbf{B}(\mathbf{r}, t) = (\mathbf{k} \times \mathbf{e})\frac{\mathcal{E}}{ck}\cos(\mathbf{k} \cdot \mathbf{r} - \omega t + \varphi), \tag{3.10}$$

---

[2] As wavelengths become shorter, this approximation fails and transverse variation cannot be neglected. The physics of x-ray interaction differs in detail from the formalism presented here for optical excitation.

of frequency ω for constant amplitude $\mathcal{E}$. The unit vector $\mathbf{e}$ defines the axis of the $\mathbf{E}$ field (the polarization axis).[3] It is only constrained by the requirement that it have unit length and be orthogonal to the propagation direction $\mathbf{k}$. The Poynting vector for this field is

$$\mathbf{S}(\mathbf{r}, t) = \frac{\mathbf{k}}{\mu_0 ck} \mathcal{E}^2 \cos^2(\mathbf{k} \cdot \mathbf{r} - \omega t + \varphi). \quad (3.11)$$

When averaged over a cycle it gives the intensity

$$I(\mathbf{r}, t) = \tfrac{1}{2} c \epsilon_0 |\mathcal{E}|^2 \quad (3.12)$$

independent of position $\mathbf{r}$ and time t.

Standing waves. When the fields are those within a cavity, or between reflecting surfaces, they are subject to boundary conditions that select standing waves rather than running waves. Typically the boundaries impose nodes of the electric field at the cavity walls. Within such a bounded region, of length L in the z direction, plane-wave fields have the structure[4]

$$\mathbf{E}(z, t) = \mathbf{e}\sqrt{2}\mathcal{E} \sin(kz) \cos(\omega t - \varphi), \qquad \mathbf{B}(z, t) = \mathbf{e}_z \times \mathbf{e}\sqrt{2}\frac{\mathcal{E}}{c} \cos(kz) \sin(\omega t - \varphi),$$

$$(3.13)$$

where $\mathbf{e}_z$ is a unit vector in the propagation direction and k is restricted to discrete values $k = \pi m/L$ for integer m. Because $\omega = ck$ the allowed frequencies are also discrete. As with traveling waves, the vector $\mathbf{e}$ can be any unit vector in the transverse (x, y) plane. The Poynting vector for these fields is

$$\mathbf{S}(z, t) = \frac{\mathbf{e}_z}{2c\mu_0} \mathcal{E}^2 \sin(2kz) \sin(2\omega t). \quad (3.14)$$

The cycle average of this vector is zero: there is no net energy flow in a standing wave. However, the intensity defined in terms of the mean-square field by eqn. (3.12) is nonzero and equal to $\tfrac{1}{2} c \epsilon_0 |\mathcal{E}|^2$, as with the traveling wave of eqn. (3.13).

3.5.2 Pulses

The fields of interest for coherent excitation are pulsed, not of indefinite duration as is the idealization of eqn. (3.10). The pulsed nature is described by replacing the constant amplitude $\mathcal{E}$ with a *pulse envelope* $\mathcal{E}(t)$, localized in time, as in the construction

$$\mathbf{E}(z, t) = \mathbf{e}\mathcal{E}(t) \cos(kz - \omega t + \varphi). \quad (3.15)$$

---

[3] Treatments of classical optics typically take the propagation axis to be the Cartesian z axis. Evaluations of transition dipole moments typically orient the z axis along the $\mathbf{E}$ field.

[4] The factor $\sqrt{2}$ is introduced so that in all cases the mean value of $|\mathbf{E}|^2$, averaged over space and time, is $|\mathcal{E}|^2/2$.

Here ω serves as the *carrier frequency* of the pulse. The intensity of this pulse is

$$I(z,t) = \tfrac{1}{2} c \epsilon_0 |\mathcal{E}(t)|^2. \tag{3.16}$$

Although not explicitly shown, a general pulse may have varying polarization unit vector **e** as well as varying phase φ. Theory must provide a prescription for these several pulse characteristics that will accomplish a desired change to the quantum system.

3.5.3 Phase

In practice all laser radiation only maintains purely sinusoidal oscillations for a finite time. Inevitably fluctuations in the phase or frequency alter these appreciably after a finite *coherence time*. The inverse of this time, the laser bandwidth, characterizes the range of frequencies that contribute to the radiation. Various models and theoretical techniques exist for treating the random fluctuations that underlie this incoherence [Sho90, Chap. 23]. In this monograph I assume that the coherence time is much longer than any time interval of interest for radiation-induced changes (but generally also much shorter than the radiative lifetime).

The phase φ of eqn. (3.15) originates from several sources. When we idealize the radiation as a beam we typically express the traveling-wave nature of the field at a point with spatial coordinates x, y, z through a function such as $\cos[\xi(x, y, z, t)]$ or $\exp[i\xi(x, y, z, t)]$ where

$$\xi(x, y, z, t) = k_x(x - x_0) + k_y(y - y_0) + k_z(z - z_R) - \omega(t - t_0) + \chi \equiv \varphi - \omega t \tag{3.17}$$

describes the moving wavefront planes of the field, periodic in space and time.

The numerical value of this function, and hence the phase φ, depends, in part, on the choice of reference position x_0, y_0, z_0; away from this location there occurs a position-dependent phase. When one compares the field at the centers of mass for several atoms, there will occur a different phase for each atom. If the atoms are moving there will occur a time-dependent phase from the changing values of atom position. When the atoms are separated by distances comparable to a wavelength, then deviations from plane-wave behavior (e.g. the curved wavefronts of Gaussian beams, Sec. 3.7) introduce additional phase differences.

The constant χ of eqn. (3.17) together with the temporal phase ωt_0 defines the moment when the electric vector of a linearly polarized field passes through zero or, for a circularly polarized field, points in a specified direction. This moment is not controllable experimentally, although as long as the field remains that of a (modulated) sinusoid of the sort assumed here, this phase remains constant and can be taken as zero.

Such conditions, and the assumed null phase, can hold only for time intervals shorter than the coherence time of the laser. For longer times the phase φ must be regarded as random. Such is the case for the phases of pulses that are well separated in time. It is also the case for fields of different carrier frequency; it is not possible to assign both fields the same moment of zero crossing.

Although an experimenter is unable to fix the absolute value of the constant phase φ, it is possible to impose phase modulation upon the laser field, e.g. sinusoidal modulation. Such manipulation alters the frequencies present in the laser spectrum.

Phase measurement. It should be recognized that the overall phase of the field, here $\xi(x, y, z, t)$, is not directly measurable. One can only determine phase *differences* between a specified field and a reference. At a fixed position one can count the number of field cycles that occur during a set time interval, thereby measuring a *temporal* phase change. One can compare fields at different positions, as is done in interferometry (see Sec. 20.4), thereby measuring a *spatial* phase change. But in all such measurements the reference phase remains unmeasurable and can be taken as zero.

3.5.4 Complex-valued fields; the positive-frequency part

The electric fields associated with radiation are observable, and therefore real-valued functions of space and time. However, the mathematical analysis often proceeds more simply by introducing complex-valued functions. The observable fields are the real parts of these complex-valued functions (often called *analytic signals*). For example, in place of eqn. (3.15) we write

$$\mathbf{E}(t) = \frac{1}{2}\left\{\mathbf{e}\mathcal{E}(t)\,\mathrm{e}^{-\mathrm{i}\varphi}\,\mathrm{e}^{\mathrm{i}\omega t} + \mathbf{e}\mathcal{E}(t)\,\mathrm{e}^{\mathrm{i}\varphi}\,\mathrm{e}^{-\mathrm{i}\omega t}\right\}. \tag{3.18}$$

We separate the field in this way into a *positive-frequency part* $\mathbf{E}^{(+)}(t)$ and a negative-frequency part $\mathbf{E}^{(-)}(t)$, each of which is complex-valued:

$$\mathbf{E}(t) = \mathbf{E}^{(+)}(t) + \mathbf{E}^{(-)}(t). \tag{3.19}$$

By tradition the positive-frequency part is associated with the carrier $\exp(-\mathrm{i}\omega t)$, and so, for the field above, the complex-valued fields are

$$\mathbf{E}^{(+)}(t) = \frac{1}{2}\mathbf{e}\mathcal{E}(t)\,\mathrm{e}^{\mathrm{i}\varphi}\,\mathrm{e}^{-\mathrm{i}\omega t}, \qquad \mathbf{E}^{(-)}(t) = \frac{1}{2}\mathbf{e}\mathcal{E}(t)\,\mathrm{e}^{-\mathrm{i}\varphi}\,\mathrm{e}^{+\mathrm{i}\omega t}. \tag{3.20}$$

The combination $\mathcal{E}(t)\,\mathrm{e}^{\mathrm{i}\varphi}$ provide a complex-valued pulse envelope (the analytic signal)

$$\check{\mathcal{E}}(t) \equiv \mathcal{E}(t)\,\mathrm{e}^{\mathrm{i}\varphi}, \qquad \check{\mathcal{E}}(t)^* \equiv \mathcal{E}(t)\,\mathrm{e}^{-\mathrm{i}\varphi}, \tag{3.21}$$

characterized by a real-valued magnitude $\mathcal{E}(t)$ and a phase φ. Typically we assume that φ is chosen to make $\mathcal{E}(t)$ positive at $t = 0$. If we hold φ constant then changes in sign of the envelope may occur with passing time. Alternatively, the envelope may be chosen to remain always positive; the changes then require a change of phase.

3.5.5 *Fourier components; coherence and bandwidth*

The frequency content of any complex-valued time-varying signal $F(t)$ is defined by means of Fourier analysis, expressible as the complex integral (the Fourier transform)

$$F(t) = \int d\omega \exp(-i\omega t)\widetilde{F}(\omega), \tag{3.22}$$

where the Fourier components $\widetilde{F}(\omega)$ are obtained as the Fourier transform (FT) of the signal:

$$\widetilde{F}(\omega) = \frac{1}{2\pi} \int dt \exp(+i\omega t)F(t). \tag{3.23}$$

Such transform pairs, applied to the positive-frequency part of the electric field of a radiation pulse, completely characterize its temporal properties.

An important case is the Gaussian pulse of unit peak value, for which the transform pairs are the real-valued functions

$$F(t) = \exp[-(t/\tau)^2/2], \qquad \widetilde{F}(\omega) = \frac{\tau}{\sqrt{2\pi}} \exp[-(\omega\tau)^2/2]. \tag{3.24}$$

Because the Gaussian function $F(t)$ extends indefinitely in time (it is a function of *infinite support*) it can only represent an idealization of pulses that have finite duration. Another idealized example is the rectangular pulse of unit value within the time interval $|t| \leq \tau/2$,

$$F(t) = \begin{cases} 1 & \text{if } |t| \leq \tau/2 \\ 0 & \text{if } |t| > \tau/2 \end{cases}, \qquad \widetilde{F}(\omega) = \frac{\sin(\omega\tau/2)}{\pi\omega}. \tag{3.25}$$

A third example is the field at the Bohr frequency ω_B that undergoes exponential decay starting from $t = 0$, a model of spontaneous emission,

$$F(t) = \begin{cases} 0 & \text{if } t < 0 \\ F_0 \exp[-i\omega_B t - \Gamma t/2] & \text{if } t \geq 0 \end{cases}, \qquad \widetilde{F}(\omega) = \frac{i F_0}{2\pi(\omega - \omega_B - i\Gamma/2)}. \tag{3.26}$$

The power spectrum of the radiation has a *Lorentz profile* of full-width at half-maximum (FWHM) Γ,

$$I(\omega) = \frac{I_{max}}{(\omega - \omega_B)^2 + (\Gamma/2)^2}. \tag{3.27}$$

Bandwidth. Various measures of the pulse duration and its frequency content have use. Typically the *bandwidth* of laser radiation is taken as some measure of the range of frequencies where the Fourier transform $\widetilde{\mathcal{E}}(\omega)$ has power above some specified value, for example the range bounded by a fall of $1/2$ or of $1/e$ or by 3 dB from the peak value.

The FWHM of the intensity is denoted $\Delta t_{1/2}$. The corresponding FWHM of the square of the transform $|\widetilde{F}(\omega)^2|$ is $\Delta\omega_{1/2}$. For the Gaussian pulses these are

$$\Delta t_{1/2} = 2\sqrt{\log 2} \times \tau, \qquad \Delta\omega_{1/2} = 2\sqrt{\log 2}/\tau \approx 1.665/\tau, \tag{3.28}$$

while for the rectangular pulse they are

$$\Delta t_{1/2} = \tau, \qquad \Delta \omega_{1/2} \approx 5.6/\tau. \qquad (3.29)$$

Other measures come from various moments of the power distribution, when they exist. One such measure is the bandwidth $\Delta \omega$ defined by

$$(\Delta \omega)^2 = \int d\omega \, \omega^2 |\tilde{\mathcal{E}}(\omega)|^2 \Big/ \int d\omega \, |\tilde{\mathcal{E}}(\omega)|^2, \qquad (3.30)$$

paired with a similar measure of pulse duration,

$$(\Delta t)^2 = \int dt \, t^2 |\mathcal{E}(t)|^2 \Big/ \int dt \, |\mathcal{E}(t)|^2. \qquad (3.31)$$

For the Gaussian pulses of eqn. (3.24) the two parameters are

$$\Delta t = \frac{\tau}{\sqrt{2}}, \qquad \Delta \omega = \frac{1}{\tau \sqrt{2}} \approx 0.707/\tau. \qquad (3.32)$$

For any pulse the time–bandwidth product of these parameters satisfies the inequality

$$\Delta \omega \, \Delta t \geq \frac{1}{2}. \qquad (3.33)$$

This product has its smallest value, $1/2$, for Gaussian pulses. In all cases as the pulse duration τ increases the distribution of frequencies becomes proportionally more narrow; the width $\Delta \omega$ of the distribution in frequencies (the bandwidth) is inversely proportional to the duration of the signal Δt.

Coherence time. The electric fields discussed above are monochromatic, comprising a single frequency and hence a signal of infinite duration. Deviations from this idealized approximation occur for several reasons, in addition to having finite pulse duration.

The light originated with spontaneous emission from individual atoms, possibly supplemented with additional stimulated emission, and so there is an inevitable temporal variation in phase. The direction of the polarization axis will vary randomly. For each of these changes there is a characteristic *coherence time* τ_{coh} during which changes are negligible and the field remains as parametrized in eqn. (3.18), but after which there is increasingly less memory of past values. The inverse of the bandwidth provides one estimate of the coherence time,

$$\tau_{coh} = 1/\Delta \omega. \qquad (3.34)$$

For times well after the coherence time the polarization state can only be parametrized by statistical measures, such as mean values for two independent components (see Sec. 3.6.1); the radiation is *partially polarized* or, when randomization is complete, *unpolarized*. Even from well-stabilized lasers there is some uncontrollable drift of the phase, although suitable filters can restrict the polarization. It is with such fields, of unchanging phase and well-defined polarization, that the theory of coherent excitation deals.

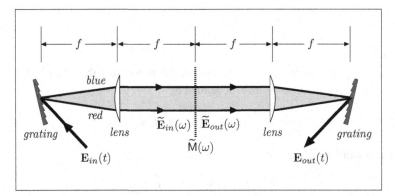

Figure 3.1 Pulse structuring technique for femtosecond pulses: a dispersive element combined with a lens spreads the Fourier components $\widetilde{\mathbf{E}}_{in}^{(+)}(\omega)$ of an incoming pulse for manipulation by a mask $\widetilde{\mathsf{M}}(\omega)$. The optical transformation of this modified spectrum $\widetilde{\mathbf{E}}_{out}^{(+)}(\omega)$ produces the output pulse $\mathbf{E}_{out}^{(+)}(t)$.

A *bandwidth-limited pulse* (also known as a transform-limited pulse) is one that has the minimum possible bandwidth for a given duration (or maximum duration for a given spectral bandwidth). Such a pulse can have no fluctuations in phase or amplitude: the coherence time cannot be shorter than the pulse duration.

3.5.6 Pulse structuring

The techniques available for shaping pulses – for crafting amplitude and phase – depend on the time scale of interest. For pulses whose duration is a microsecond the experimenter can work with tools that alter the time-varying amplitude or add controllable frequency shifts to the carrier. These techniques, of direct manipulation of time-dependent signals, are not available for very short pulses. The structuring of femtosecond pulses is typically accomplished by manipulating the Fourier components [Pra03; Wol07; Wol10]. Figure 3.1 presents a schematic diagram of a typical scheme.

Here the basic pulsed field has the complex-valued positive-frequency part

$$\mathbf{E}_{in}^{(+)}(t) = \tfrac{1}{2}\mathbf{e}_{in}\, \check{\mathcal{E}}_{in}(t)\, e^{-i\omega t}. \tag{3.35}$$

With the aid of a grating or prism the spectral components of this pulse are distributed in angle. The grating lies at the focus of a lens of focal length f that produces, at subsequent distance f, the Fourier components of the incident field, $\widetilde{\mathbf{E}}_{in}^{(+)}(\omega)$. These are spread transversely, so that each spectral component can be manipulated, in phase, amplitude, and polarization, by a suitable frequency and polarization-dependent masking operator $\widetilde{\mathsf{M}}(\omega)$. The result of this mask is the transform field function

$$\widetilde{\mathbf{E}}_{out}^{(+)}(\omega) = \widetilde{\mathsf{M}}(\omega)\widetilde{\mathbf{E}}_{in}^{(+)}(\omega). \tag{3.36}$$

A second lens and a second dispersing element then produce a Fourier transform of this frequency function thereby constructing the output temporal field. This can be expressed as a unit vector, an amplitude, and a phase, see eqn. (3.17),

$$\mathbf{E}_{out}^{(+)}(t) = \tfrac{1}{2}\mathbf{e}(t)|\mathcal{E}_{out}(t)|\exp[i\,\xi(t)]. \tag{3.37}$$

The carrier frequency of this envelope, often time dependent, is defined as the time derivative of the phase,

$$\omega(t) = -\frac{d}{dt}\xi(t). \tag{3.38}$$

By means of such linear manipulation one can alter the Fourier components present in an incident pulse. For example, the addition of a phase that varies quadratically in frequency will appear as a linear chirp of the output carrier frequency.

Spectral components can be eliminated, but manipulations of the sort described here cannot produce new components. Thus the bandwidth of the pulse can be reduced, but not increased: the pulse can be lengthened, but not shortened. This constraint places some limitations on the pulses that can be formed. In particular, the sweep of frequency cannot be larger than the range of frequencies present in the initial pulse.

In general the temporal pulse shapes produced by spectral shaping are very complicated, sometimes very asymmetric. It is then not useful to try to characterize the pulse duration by a FWHM. Instead one relies on various moments [Wol07].

3.6 Field vectors

Specification of the direction of any three-dimensional vector field requires three independent unit vectors as coordinate axes. Usually these are taken to be orthogonal, as with Cartesian unit vectors. The following sections discuss some conventional choices. The discussion is relevant not only to the characterization of laser light used for producing changes in quantum systems but also for presenting a two-dimensional classical system whose mathematical description has much in common with the two-state quantum system – the two-level atom – of Sec. 5.6.

Because the **E** and **B** fields of a traveling wave in free space are each orthogonal to the propagation direction **k** (the direction of the Poynting vector **S**) as well as to each other it is convenient to take the propagation axis as one unit vector $\mathbf{e_k}$. The remaining two vectors must lie in a plane transverse to this axis but they are otherwise arbitrary. A common choice of coordinates augments the unit vector $\mathbf{e_k}$ with a second vector $\mathbf{e}_H$ that is perpendicular to $\mathbf{e_k}$ and lies in the horizontal plane. The third unit vector $\mathbf{e}_V$ is perpendicular to each of these (so that $\mathbf{e_k}$ and $\mathbf{e}_V$ therefore define the vertical plane). The three vectors form an orthogonal triad (the labels H, V may equally well read x, y or 1, 2)

$$\mathbf{e}_H \cdot \mathbf{e}_H = \mathbf{e}_V \cdot \mathbf{e}_V = 1, \qquad \mathbf{e}_H \cdot \mathbf{e}_V = 0, \qquad \mathbf{e}_H \times \mathbf{e}_V = \mathbf{e_k}. \tag{3.39}$$

Generalizing eqn. (3.15) we express an arbitrary field that travels along the z axis as

$$\mathbf{E}(z,t) = \mathbf{e}_H \mathcal{E}_H \cos(kz - \omega t + \varphi_H) + \mathbf{e}_V \mathcal{E}_V \cos(kz - \omega t + \varphi_V). \tag{3.40}$$

The vector properties of the field in the two-dimensional H, V plane are established by the relative phase

$$\varphi = \varphi_V - \varphi_H \tag{3.41}$$

between the two orthogonal components. To simplify the presentation below I set $\varphi_H = 0$.

To describe circularly polarized light most simply it is convenient to introduce complex-valued field envelopes $\check{\mathcal{E}}$ and complex unit vectors

$$\mathbf{e}_R = [\mathbf{e}_H - i\mathbf{e}_V]/\sqrt{2}, \qquad \mathbf{e}_L = [\mathbf{e}_H + i\mathbf{e}_V]/\sqrt{2}, \tag{3.42}$$

with properties

$$\mathbf{e}_R^* \cdot \mathbf{e}_R = \mathbf{e}_L^* \cdot \mathbf{e}_L = 1, \qquad \mathbf{e}_R^* \cdot \mathbf{e}_L = 0, \qquad \mathbf{e}_R \times \mathbf{e}_L = \mathbf{e}_{\mathbf{k}}, \tag{3.43}$$

and to write the electric field as

$$\mathbf{E}(z,t) = \mathrm{Re}\left\{ \left[\mathbf{e}_R \check{\mathcal{E}}_R + \mathbf{e}_L \check{\mathcal{E}}_L \right] \exp[i\,(kz - \omega t)] \right\}. \tag{3.44}$$

Right-circularly polarized light then has the property $\check{\mathcal{E}}_L = 0$ while linearly polarized light has the property $|\check{\mathcal{E}}_L| = |\check{\mathcal{E}}_R|$. A linearly polarized beam can be treated as a coherent superposition of right- and left-circular polarization having equal amplitudes, for example

$$\mathbf{e}_H \check{\mathcal{E}}_H = [\mathbf{e}_R \check{\mathcal{E}}_H + \mathbf{e}_L \check{\mathcal{E}}_H]/\sqrt{2}. \tag{3.45}$$

Helicity vectors. All three-dimensional vector fields have an intrinsic spin of unity (see App. A). The cylindrical symmetry of a typical laser beam brings with it a conservation of angular momentum around the optical axis. Treatments of angular momentum proceed best by introducing complex unit vectors (helicity vectors, see App. D.2.1)

$$\mathbf{e}_{\pm 1} = \mp[\mathbf{e}_H \pm i\mathbf{e}_V]/\sqrt{2} \tag{3.46}$$

that are eigenstates of spin $\mathbf{S}$ projected onto the propagation direction:

$$\mathbf{k} \cdot \mathbf{S}\,\mathbf{e}_{\pm 1} = \pm k\,\mathbf{e}_{\pm 1}. \tag{3.47}$$

The field associated with *positive helicity*, $\mathbf{e}_{+1}$, undergoes rotation of a positive screw as it propagates. When the field travels toward the observer this appears as a counterclockwise rotation, i.e. left-circular polarization. The negative helicity field, associated with unit vector $\mathbf{e}_{-1}$, corresponds to right-circularly polarized light. A linearly polarized beam can be treated as a coherent superposition of equal amplitude helicity states, for example

$$\mathbf{e}_H \check{\mathcal{E}}_H = [\mathbf{e}_{+1} \check{\mathcal{E}}_H + \mathbf{e}_{-1} \check{\mathcal{E}}_H]/\sqrt{2}. \tag{3.48}$$

3.6.1 Parametrizing the electric field

Given an optical axis we can completely describe the field by specifying two complex-valued electric-field amplitudes (or two real amplitudes along with their two phases), either $\check{\mathcal{E}}_H$, $\check{\mathcal{E}}_V$ or else the pair $\check{\mathcal{E}}_R$, $\check{\mathcal{E}}_L$. Any two independent complex amplitudes (or two independent magnitudes and their two phases) suffice to define the field, for example

$$\mathbf{e}\check{\mathcal{E}} = \mathbf{e}_R\check{\mathcal{E}}_R + \mathbf{e}_L\check{\mathcal{E}}_L = \mathbf{e}_H\check{\mathcal{E}}_H + \mathbf{e}_V\check{\mathcal{E}}_V = \sum_\lambda \mathbf{e}_\lambda\check{\mathcal{E}}_\lambda. \tag{3.49}$$

The labeling index λ can denote any pair of independent axes. The effect upon the field components after passing through an optical instrument is an alteration to which each original component may contribute linearly, as in the expressions

$$\check{\mathcal{E}}'_H = M_{HH}\check{\mathcal{E}}_H + M_{HV}\check{\mathcal{E}}_V, \tag{3.50a}$$

$$\check{\mathcal{E}}'_V = M_{VH}\check{\mathcal{E}}_H + M_{VV}\check{\mathcal{E}}_V, \tag{3.50b}$$

in which the array of four numbers $M_{\lambda,\lambda'}$ may be complex valued; they can be regarded as elements of a matrix M.

The Jones calculus. The two complex-valued amplitudes $\check{\mathcal{E}}_\lambda$ may be regarded as coordinates in an abstract vector space, a two-dimensional *Hilbert space*. The two components $\check{\mathcal{E}}_H$ and $\check{\mathcal{E}}_V$, when presented as a column, are known as a *Jones vector* [Jon41; Jon47; Hec87]. By specifying a point in this Hilbert space – a Jones vector – we completely characterize the monochromatic plane wave. The mathematics of this representation of the polarization state reappears when we treat a two-state quantum system, and so the optical system offers a way to visualize that quantum system.

The Jones vector offers a useful way to describe the effects of various optical elements acting on a beam of light. The formalism is known as Jones calculus, recognizing the series of papers by Jones [Jon41; Jon56] in which he developed the mathematics. The basic elements of the Jones calculus are two-dimensional column vectors $\mathbf{V}$ whose components are complex-valued electric field envelopes

$$\mathbf{e}\check{\mathcal{E}} = \begin{bmatrix} \check{\mathcal{E}}_H \\ \check{\mathcal{E}}_V \end{bmatrix} \equiv \mathbf{V}. \tag{3.51}$$

The effect upon the field, after passing through an optical instrument, is represented as multiplication by a 2×2 matrix M,

$$\begin{bmatrix} \check{\mathcal{E}}'_H \\ \check{\mathcal{E}}'_V \end{bmatrix} = \begin{bmatrix} M_{HH} & M_{HV} \\ M_{VH} & M_{VV} \end{bmatrix} \begin{bmatrix} \check{\mathcal{E}}_H \\ \check{\mathcal{E}}_V \end{bmatrix} \quad \text{or} \quad \mathbf{V}' = \mathsf{M}\mathbf{V}. \tag{3.52}$$

The action of a succession of N instruments is represented by the product of their matrices,

$$\mathsf{M} = \mathsf{M}(N)\cdots\mathsf{M}(2)\mathsf{M}(1). \tag{3.53}$$

Missing from the Jones calculus is the reflected wave at a surface; it does not treat cavities or multilayer coatings. Treatment of such systems requires both forward- and backward-traveling waves [Mac10].

As an example of the Jones calculus, an instrument that introduces a relative phase shift δ of one component has the matrix representation

$$M = \begin{bmatrix} e^{i\delta} & 0 \\ 0 & 1 \end{bmatrix}.$$ (3.54)

An idealized filter that passes only one component of the field (a projection operator) has a matrix representation such as

$$M(H) = \begin{bmatrix} 1 & 0 \\ 0 & 0 \end{bmatrix}, \qquad M(V) = \begin{bmatrix} 0 & 0 \\ 0 & 1 \end{bmatrix}.$$ (3.55)

In keeping with all projection operators these two matrices are *idempotent*, meaning

$$M(H)M(H) = M(H), \qquad M(V)M(V) = M(V),$$ (3.56)

and they have the exclusion property (they commute),

$$M(H)M(V) = M(V)M(H) = 0.$$ (3.57)

The experimentally induced changes to a polarization state represented by these mathematical operations have a counterpart in the quantum mechanics of two-state systems altered by pulses of laser light, as discussed in Sec. 5.5.2.

Propagation. Rather than treat a succession of discrete optical elements, we can consider propagation of a beam through anisotropic material whose optical properties vary slowly with distance, so that backward waves can be neglected. The change to the Jones vector in propagating an incremental distance dz is then represented by the differential equation

$$\frac{d}{dz}\mathbf{V}(z) = N(z)\mathbf{V}(z),$$ (3.58)

in which the matrix elements of the 2×2 matrix $N(z)$ are expressible in terms of a spatially varying complex-valued anisotropic dielectric tensor [Jon56], see App. C.4.3. This calculus of polarized beams is equivalent to the excitation dynamics of two-state quantum systems with the replacement of propagation distance z by time t. The matrix $N(z)$ then becomes the Hamiltonian $H(t)$, see Chaps. 8–11, and the various population changes have counterparts in changes of polarization properties.

The Poincaré sphere. An alternative parametrization of the radiation descriptors uses two angles to define a complex unit vector

$$\mathbf{e}(\vartheta, \varphi) = \cos(\vartheta/2)\,\mathbf{e}_R + e^{i\varphi} \sin(\vartheta/2)\,\mathbf{e}_L.$$ (3.59)

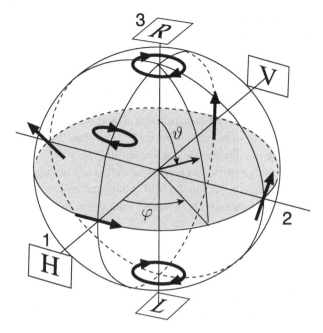

Figure 3.2 The Poincaré sphere, with labels of particular polarizations: right circular (R), left circular (L), linear horizontal (H) and linear vertical (V). The axes are 1, 2, 3. Arrows show the direction of the electric-vector motion as viewed looking into a beam moving out of the coordinate origin.

The description of the field then requires the magnitude of the field, or the intensity, and two angles.[5] This information suffices to specify a point on a sphere, the *Poincaré sphere* [Bor99; Auz10] shown in Fig. 3.2. The north pole ($\vartheta = 0, 2\pi, 4\pi, \ldots$) of this sphere represents right-circular polarization, the south pole ($\vartheta = \pi, 3\pi, \ldots$) represents left-circular polarization. Points around the equator, for which $\vartheta = \pi/2, 3\pi/2, \ldots$, represent linear polarization at differing angles. In all cases two points at opposite sides of the sphere represent independent, orthogonal unit vectors.

The coherency matrix. For monochromatic radiation the motion of the electric-field vector, along a line or on a more general ellipse, occurs at the carrier frequency of the light, ω. However, when an experimenter introduces modulation of the field this exactly periodic motion need not occur. For example, the modulation may cause the electric vector, while predominantly linearly polarized, to undergo a slow, controlled, rotation of the plane of polarization. This appears as a change of Poincaré angles ϑ and φ.

When the field is not strictly monochromatic the phases φ_λ associated with each component $\breve{\mathcal{E}}_\lambda$ may vary irregularly and uncontrollably, thereby negating, in part, any fixed phase relationship between the two polarization components. To characterize such *partially*

---

[5] A full description of the field also requires a phase; see Sec. 3.6.4.

polarized light it is useful to introduce a *coherency matrix* [Bor99] formed by time averages over products of complex-valued amplitudes,

$$\mathcal{J}_{\lambda,\lambda'} = \langle \breve{\mathcal{E}}_{\lambda}^* \breve{\mathcal{E}}_{\lambda'} \rangle = \mathcal{J}_{\lambda',\lambda}^*. \tag{3.60}$$

The complex-valued elements of this 2×2 Hermitian matrix provide the needed parameters to characterize the transverse field of a propagating field. The two diagonal elements are each proportional to the intensity of radiation that would pass through an idealized filter that selects $\mathbf{E}$ fields aligned with the unit vector $\mathbf{e}_\lambda$, and thus they are measurable. The off-diagonal elements are not directly measurable by filtering.

The trace of the coherency matrix, independent of the chosen orthogonal basis vectors, is proportional to the total intensity. When the field is unpolarized the off-diagonal elements vanish and the two diagonal elements are equal to each other. Such fields are not suitable for producing coherent excitation.

The Stokes vector. The two Poincaré angles ϑ and φ provide a portion of the information required to define the field. We can supplement these with a measure of the intensity to complete the description (apart from an overall phase). An alternative useful parametrization of the beam descriptors is the set of four real numbers (*Stokes parameters*)[6] [Bic85; Bor99; Auz10]

$$\begin{aligned} s_0 &= \mathcal{J}_{HH} + \mathcal{J}_{VV} = \mathcal{J}_{RR} + \mathcal{J}_{LL}, \\ s_1 &= \mathcal{J}_{HH} - \mathcal{J}_{VV} = 2\mathrm{Re}\,\mathcal{J}_{RL} & = s_0 \sin\vartheta\cos\varphi, \\ s_2 &= 2\mathrm{Re}\,\mathcal{J}_{HV} = 2\mathrm{Im}\,\mathcal{J}_{RL} & = s_0 \sin\vartheta\sin\varphi, \\ s_3 &= 2\mathrm{Im}\,\mathcal{J}_{HV} = \mathcal{J}_{RR} - \mathcal{J}_{LL} & = s_0 \cos\vartheta. \end{aligned} \tag{3.61}$$

These completely characterize the transverse field of monochromatic radiation, apart from an overall phase. For a general partially polarized field they satisfy the inequality

$$(s_0)^2 \geq (s_1)^2 + (s_2)^2 + (s_3)^2. \tag{3.62}$$

The equality holds for monochromatic radiation. Unpolarized light has $s_1 = s_2 = s_3 = 0$.

The three numbers s_1, s_2 and s_3 may be regarded as Cartesian components of a three-dimensional vector $\mathbf{S}$ whose tip touches the Poincaré sphere. The vertical component (axis 3) quantifies the excess of R over L polarization while the excess of H over V polarization is along the 1 axis. Table 3.1 summarizes the connection between polarization state and Stokes parameters.

State space. The Stokes parameters and their display on the Poincaré sphere are an example of an abstract vector space in which the coordinates are real-valued numbers that completely characterize a system, a state space or *system space*.[7] A point on this sphere defined by

---

[6] These are variously denoted S_0, S_1, S_2, S_3 or I, Q, U, V.

[7] As noted below and in Chap. 20, this state space is *incomplete*: it provides no information about the overall phase of the field.

Table 3.1. *Polarizations and Stokes parameters*

ϑ, φ	λ	Polarization	$(s_1, s_2, s_3)/s_0$
$\frac{\pi}{2}, 0$	H	Linear H	$+1, 0, 0$
$\frac{\pi}{2}, \pi$	V	Linear V	$-1, 0, 0$
$\frac{\pi}{2}, \frac{\pi}{2}$	$\nearrow$	Linear $+45°$	$0, +1, 0$
$\frac{\pi}{2}, -\frac{\pi}{2}$	$\nwarrow$	Linear $-45°$	$0, -1, 0$
$0, \varphi$	R	Circular R	$0, 0, +1$
π, φ	L	Circular L	$0, 0, -1$

particular Stokes parameters is a system point. With inclusion of the intensity or s_0 it is four dimensional; those four elements are components of the *Stokes vector*. For fixed intensity it has three dimensions. The effect of any optical device can be described by its effect on a vector in this system space. The effect of an optical element upon the Stokes vector is represented by a 4×4 matrix, the *Mueller matrix* [Mue48; Bic85]. Alterations of polarization by a sequence of optical elements are represented by products of Mueller matrices, chained to produce a matrix that represents the overall change.

Propagation. Just as with the Jones calculus, changes to Stokes vectors need not occur in discrete steps, as produced by optical elements, but can be regarded as slow changes occurring as the beam propagates through inhomogeneous material. As noted by Kubo and Nagata [Kub80], the differential equation describing the incremental change to the three-dimensional Stokes vector S, as it propagates distance dz through nonabsorbing weakly-inhomogeneous anisotropic optically-active material, has the form of a torque equation,

$$\frac{d}{dz}\begin{bmatrix} s_1 \\ s_2 \\ s_3 \end{bmatrix} = \begin{bmatrix} 0 & -\Upsilon_3 & \Upsilon_2 \\ \Upsilon_3 & 0 & \Upsilon_1 \\ -\Upsilon_2 & \Upsilon_1 & 0 \end{bmatrix}\begin{bmatrix} s_1 \\ s_2 \\ s_3 \end{bmatrix} \quad \text{or} \quad \frac{d}{dz}S = \Upsilon \times S, \tag{3.63}$$

where the elements of the torque vector Υ obtain from the dielectric tensor. Here too there is an analogy with coherent excitation, either by a two-state system, see Chap. 10, or by a three-state system, see Chaps. 13 and 14.

Projective space. Intensity-preserving changes of a directed laser beam are representable by unitary transformations of two complex-valued electric field amplitudes associated with two selected orthogonal polarization directions, i.e. representations of the two-dimensional unitary group $U(2)$. The association of the electric field vector for a plane wave as a point on the Poincaré sphere (or by the Stokes parameters) does not include information about the overall phase of the field: The position on the sphere and the corresponding values of the Stokes parameters is unaffected if we multiply the complex field by $e^{i\phi}$, an example of a *gauge transformation*.

The reduction of information, from the description requiring two complex-valued electric field amplitudes for orthogonal directions to a description offering only three real-valued parameters, is an example of using a *projective space* [Ana97; Bha97]. A trajectory in the projective space (a path on the Poincaré sphere) is linked to a trajectory in the larger space, where the path may accumulate an additional phase; see Chap. 20. Whereas intensity-preserving optical transformations of the electric field amplitudes are representations of $U(2)$, the linked changes of a trajectory on the Poincaré sphere are representations of the special unitary group $SU(2)$, as are changes to two-state quantum system.

3.6.2 *Measuring the field parameters*

Propagating beams can undergo a variety of changes as they pass through transmissive optical devices that are aligned with the propagation direction (the optical axis). Amongst those devices are polarization filters that selectively transmit only a particular polarization, say that for which the axis of the electric field is along the unit vector $\mathbf{e}$, and wave plates that alter the two phases φ_λ differently for two orthogonal directions $\mathbf{e}_\lambda$. By combining these elements in succession it is possible to produce a beam of arbitrary polarization.

It is also possible to determine the polarization state of an unknown beam by measuring only intensities, i.e. absolute squares of field components $\check{\mathcal{E}}_\lambda$. The key to such characterization is to filter the light prior to detecting the intensity, so that only one component $\check{\mathcal{E}}_\lambda$ enters the detector. Preceding this filter one places a wave plate that adds a phase increment $\Delta\varphi$ to one of the two orthogonal components. By suitable choice of phase and filter one can obtain a set of signals from which to deduce parameters that fully characterize the magnitude and direction of the field.

Traditionally the measurements are designed to provide the elements of the Stokes vector. These are the "observables" of the field. One set of four measurements that will provide the Stokes parameters is the following [Bic85; Bor99; Auz10]:

s_0 = Total intensity.
s_1 = Excess of H polarization over V polarization.
s_2 = Excess of $45°$ polarization over $135°$ polarization.
s_3 = Excess of R polarization over L polarization.

To fully characterize the beam by means of Stokes parameters we must make four measurements, each one comparing the intensity through a pair of spatially orthogonal filters, e.g. H and V or R and L. The full set includes directions based on unit vectors that are not orthogonal, e.g. H and R. Such quantities have counterparts in quantum information theory as *positive operator-valued measures* (POVM) [Per95; Bra99; Bar09].

Phase. It should be noted that although the Stokes parameters and the Poincaré sphere provide a complete description of polarization by means of real-valued parameters, they carry no information about the field phase. This missing information, needed to complete the description of the field, cannot be supplied by measuring filtered intensities of a beam, as is done when evaluating the Stokes parameters. Instead it is necessary to compare the beam with some reference, as is done in interferometry; see Sec. 20.4.

3.6.3 Polarization states and quantum states

For a given direction of propagation an arbitrary monochromatic field can be specified by two complex-valued field envelopes or by the magnitude of the envelope, a phase, and two angles on the Poincaré sphere. These characteristics change as the beam propagates through optical elements. A polarizing filter can eliminate the component of the electric vector along a chosen transverse axis, or wave plates can augment the phase of a field component along a selected transverse axis. With various optical elements an experimenter can alter the state of polarization, and with it the location of the descriptive point on the Poincaré sphere.

The two states of polarization with which we describe a laser beam have much in common with the pairs of quantum states of atoms that are discussed in Chap. 8. Indeed, the information content is the same for the field polarization states and the atom states. Each physical system is characterized by two complex numbers and by changes to these numbers that are representations of the group, $U(2)$. This analogy forms the basis for treatments of quantum information contained in atom states and photon-polarization states.

3.6.4 Poincaré-sphere loops and geometric phase

The Poincaré sphere can represent not only the state of field polarization at a particular location along a beam, or at a particular moment at a fixed position, it can also provide a picture of polarization changes that occur along a beam as it passes through optical elements, or that occur at a fixed position as various optical elements are placed into the beam. For lossless optical elements, and hence constant intensity, the changes in polarization state effected by an experimenter appear as a curve on the surface of the Poincaré sphere.

It is possible to construct a succession of optical elements that will return the Stokes vector to its initial position on the Poincaré sphere. In so doing the locus of points on the surface of the sphere forms a closed loop (a cycle). However, although the measurable Stokes parameters have returned to their original values, these parameters and the system point on the Poincaré sphere do not provide information about the overall phase of the field, $\xi(x, y, z, t)$. During the manipulation of the optical beam this phase will increase as time increases (a contribution ωt) and as the beam propagates through the optical chain (a contribution $\mathbf{k} \cdot \mathbf{r}$). But after the system point returns to its original location the phase may have acquired, in addition to this *dynamical phase* increase, an increment that depends only on the geometry of the loop.

Pancharatnam [Pan 56] discovered that if the polarization state is taken around a closed loop on the Poincaré sphere (the state space of polarization) comprising a sequence of discrete geodesic arcs symbolized by the sequence of Stokes vectors

$$|A\rangle \to |B\rangle \to |C\rangle \to |A'\rangle \tag{3.64}$$

then, as discussed in Chap. 20, the electric field acquires, in addition to the dynamical phase, phase increment equal to half the solid angle subtended at the center of the sphere by the surface loop. This *geometric phase* or *Pancharatnam phase* [Sha89; Ana97] is equal to minus one-half the solid angle subtended on the Poincaré sphere by the curve generated as the system point moves through its closed cycle. It can be measured by means of interferometry, using a beam splitter to create a reference beam with which the altered beam is subsequently combined. The recombination adds field amplitudes, and therefore is sensitive to phase.

The geometric phase produced by moving the system point in a closed loop on the Poincaré sphere is an example of a *holonomy*: a change in one system characteristic when all others return to their initial value. In particular it is an example of *parallel transport* of a reference frame constrained to the surface of a sphere; see Sec. 20.2. It has a counterpart in changes to the overall phase of a statevector undergoing adiabatic change; there it is known as the *Berry phase*; see Chap. 20.

3.7 Laser beams

The idealization of laser radiation as a plane wave is adequate only over transverse distances comparable to atomic or molecular dimensions. Viewed over larger distances the radiation will be seen as a beam, typically circular in cross section and diminishing in magnitude away from the beam axis. Let this propagation direction be the Cartesian z axis and write the field as

$$\mathbf{E}(\mathbf{r},t) = \mathrm{Re}\left\{ \mathbf{e}\check{\mathcal{E}}(\mathbf{r}) \exp(\mathrm{i}kz - \mathrm{i}\omega t) \right\}, \tag{3.65}$$

where $k = \omega/c$ and the amplitude $\check{\mathcal{E}}(\mathbf{r})$ is now allowed to be complex valued. The relevant equation is the *paraxial wave equation* (see App. C.2.4), in which transverse changes, in the x, y plane, occur over a much larger distance than a wavelength. Expressed in Cartesian coordinates the equation for the scalar $\check{\mathcal{E}}(\mathbf{r})$ reads [Mil88]

$$\left[\frac{\partial^2}{\partial x^2} + \frac{\partial^2}{\partial y^2} + 2\mathrm{i}k\frac{\partial}{\partial z} \right]\check{\mathcal{E}}(\mathbf{r}) = 0. \tag{3.66}$$

The most common form of the beam is one in which the field magnitude has a Gaussian dependence upon x, y, a *Gaussian beam*. For such a field to satisfy the paraxial wave equation the transverse dimensions must change with z, passing through a minimum at the focal position $z = 0$, as expressed by the construction

$$\check{\mathcal{E}}(\mathbf{r}) = \check{\mathcal{E}}_0 e^{-\mathrm{i}\varphi(z)} \frac{w_0}{w(z)} \exp[-(x^2 + y^2)/w(z)^2] \exp[\mathrm{i}k(x^2 + y^2)/2\mathcal{R}(z)]. \tag{3.67}$$

Here the beam waist at position z is, for wavelength $\lambda = 2\pi/k$,

$$w(z) = w_0\sqrt{1 + (z/z_R)^2}, \qquad z_R \equiv \pi w_0^2/\lambda. \tag{3.68}$$

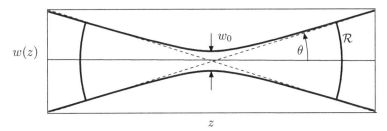

Figure 3.3 Gaussian beam profile, showing variation of waist $w(z)$ with distance and a wavefront of curvature $\mathcal{R}$. The angle θ is $\theta = \lambda/\pi w_0$.

The wavefront curvature at position z is

$$\mathcal{R}(z) = z[1 + (z_R/z)^2], \tag{3.69}$$

and the *Gouy phase* is

$$\varphi(z) = \arctan(z/z_R). \tag{3.70}$$

The parameter z_R, the *Rayleigh range*, establishes the longitudinal scale of the focusing: The beam expands from its minimum radius w_0 to $\sqrt{2}w_0$ in one Rayleigh range away from the focus at $z = 0$. The intensity of this Gaussian beam is

$$I(\mathbf{r}) = \frac{c\epsilon_0}{2} |\check{\mathcal{E}}_0|^2 \left(\frac{w_0}{w(z)} \right)^2 \exp[-2(x^2 + y^2)/w(z)^2]. \tag{3.71}$$

As with the description of plane waves, the complete electric field must include some unit electric field vector $\mathbf{e}$ and the time dependence $\exp(-i\omega t)$. Figure 3.3 shows how the beam waist varies with distance.

Cylindrically symmetric nodeless Gaussian beams are routinely produced; they are the simplest (zero-order) solutions to the paraxial wave equation. Combinations of suitable lenses and arrays of phase and amplitude modifying elements over a transverse plane can also create beams in which the profile has sharp edges, e.g. a "top hat" having uniform intensity over an area of cylindrical or rectangular symmetry.

It is also possible to produce nodal patterns in the transverse plane [Sie86; Mil88]. These are higher-order modes of Gaussian beams. *Gaussian–Hermite modes* have nodes along transverse Cartesian axes; they have rectangular symmetry in the transverse plane. *Gaussian–Laguerre modes* have circular or radial nodes; they represent higher-order modes in circular coordinates.

3.8 Photons

The discrete incremental changes of internal atom energy are directly observable, by probing the presence of excitation. To raise the internal energy of an atom from E_1 to E_2 requires a discrete increment of energy. When this energy $E_2 - E_1 = \hbar\omega$ is supplied to a single atom by an electromagnetic field, it is both convenient and customary to refer to these changes as events in which a single photon is absorbed or emitted. That photon is a discrete energy packet of radiation having frequency ω.

An atom, having absorbed radiation energy to place it in an excited state, will eventually give this energy back to electromagnetic radiation: it will spontaneously emit an energy increment $\hbar\omega$ into a field of frequency ω. This *spontaneous emission* radiation can be regarded as a wavepacket – a photon – traveling outward from the atom.

Such descriptions of radiation interactions suggest a granular, particulate nature of radiation (photons as particles). The discreteness of photons has important implications, particularly for atoms in cavities [Hin90; Har06]; see Chap. 22. However, ordinary laser pulses contain such large numbers of photons that the granularity of their quanta can be neglected. For example, a red laser beam (wavelength $\lambda = 650$ nm) of irradiance 1 GW/cm^2 comprises some 3×10^{28} photons per cm^2 per second. This means that every nanosecond there will pass through a disk of radius 10^{-9} m some 10^5 photons. In essence, the laser beams of concern here contain such large numbers of photons that their discreteness is immaterial. It is rather like the flow of water from a tap: The fluid is known to be composed of individual molecules, but the equations of fluid dynamics needed to describe such flow make no use of atoms. Thus it is also possible, and often more useful, to regard the radiation, as it acts on or is generated by an atom, as a wave train – a classical entity whose time dependence can be specified without considerations of the underlying quantum nature. This "wave picture" is in keeping with classical views of light as a traveling electromagnetic field. It is the viewpoint that will underlie most of the presentation in this monograph. Only when we discuss discrete excitation or deexcitation events does the concept of photons become useful, although it is not essential.

Quantum theory of radiation. The preceding reasoning argues for the existence of photons from the discreteness of energy states of atoms. An alternative line of reasoning regards electromagnetic fields as dynamical objects, subject to the laws of quantum mechanics [Lou73; Bia75]. Within the resulting quantum field theory, fields can be seen to have a granularity: Their changes occur through indivisible quanta of fixed energy increment. For the electromagnetic field the quantum is the photon. For traveling waves of radiation, moving at the speed of light c, the linear momentum has similar granularity; the photons of energy $\hbar\omega$ carry linear momentum $\hbar\omega/c$ in the direction of propagation. Photons, as quanta of vector fields, have unit intrinsic spin, associated with an irreducible angular momentum of $\hbar$. Appendix D discusses more completely the notion of photons as radiation quanta.

Photon localization. Although it is often useful to attribute particulate attributes to photons, a photon is localized in space only upon being absorbed (destroyed) at the location of a detector. In emission, or in free space, it is delocalized, either as a wavepacket [Bia96; Coo82] or an infinite train. Appendix D.5 discusses these alternative operational definitions of a photon.

3.9 Field restrictions

Coherence. To obtain coherent excitation, as is the central concern of the present mono-graph, it is necessary that field phases not undergo appreciable random changes during the excitation process. As a minimum requirement this means that the inverse of the band-width from random fluctuations should be much longer than the duration of the pulse. But bandwidth alone does not fully characterize the coherence properties of a field. The full description of coherence requires a succession of correlation functions (of various orders) relating the field to itself at various times, see App. C.3.2. Bandwidth measures the first-order correlation time. Higher-order correlation times must also not be shorter than the pulse duration. A monochromatic field is coherent to all orders and will produce coherent excitation. By contrast a single-mode thermal field will not produce coherent excitation: It has large fluctuations in photon number and hence in amplitude.

Wavelength. The limitation to discrete quantum states implies a limit on the radiation wave-length of the laser excitation: Bound particles become unbound as their excitation energy increases beyond some limit, the *binding energy*. For an electron bound within an atom the result is *photoionization*, see Sec. 4.2. For an atom vibrating within a molecule the result is *photodissociation*. Energies beyond the ionization or dissociation limit are not constrained to be discrete – they can take any value within a continuum. In turn, the particle is no longer localized; it is not constrained to be found within a finite volume.

To avoid such situations, and deal only with discrete quantum states, the wavelength of the radiation must be sufficiently long that the photon energy $\hbar\omega$ does not exceed the binding energy of the excited state (or states). The limitation to long wavelengths (i.e. laser radiation in the visible or infrared regions of the spectrum), and correspondingly low-energy photons, allows us to employ a nonrelativistic description of the system: particle velocities are all much slower than the speed of light. This approximation is implicit in all of the present discussion.

Intensity. We must also require that the electric field of the radiation be weaker than the Coulomb field that holds the electron within the atom, so that the behavior of the bound electron will be dominated by the forces that provide the structure of the free atom. In the opposite extreme, of a very intense pulse, the behavior is predominantly that of independent charges moving under the influence of the laser field and perturbed by Coulomb forces. A relativistic description is then needed.

4

The laser–atom interaction

As light passes an atom, it exerts forces on the charges, electrons, and nuclei that alter the atomic structure. These may be slight distortions (perturbations) of the electron cloud or they may be more severe, as described in subsequent sections of this monograph. We wish to determine those changes, given the radiation field, or to devise a radiation field that will produce specified changes.

The changes to the atomic structure also affect the radiation that subsequently passes the atom. To describe those effects we must consider wave equations for radiation in the presence of altered atomic structure. More generally we must find self-consistent equations for the atoms and the field together, as discussed in Chaps. 21 and 22. Here we consider mathematical descriptions of the influence of coherent radiation on individual atoms, molecules, or other single quantum systems.

4.1 Individual atoms

Traditional sources of emission and absorption spectra, though revealing energy states of the constituent atoms and molecules, are macroscopic samples. One observes averaged characteristics of many individual particles, see Chap. 16. Quantum theory offers the basic formalism for dealing with individual atoms exposed to controlled radiation fields. Several experimental techniques provide acceptable approximations to this ideal. The following paragraphs note some of these examples.

4.1.1 Vapors

Neutral atoms or molecules in a vapor move freely along straight-line paths, interrupted by brief collisions that redirect the two collision partners. When the kinetic energies of the two partners are small, there can be no transfer of kinetic energy into internal energy of either particle – the collision is *elastic*. Between the collisions the individual atoms behave as undisturbed free particles. This idealization lasts, on average, the mean free time between collisions, τ.

Large volumes of vapors are present in equilibrium with heated surfaces, or they may be created as small plumes of material ablated by laser pulses from a surface. In all cases there

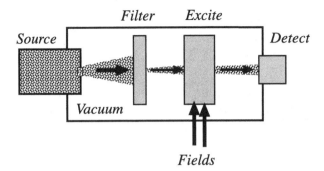

Figure 4.1 Schematic presentation of atomic beam excitation: Atoms emerge from a vapor source into vacuum, through apertures (a spatial filter), forming a collimated beam. This passes through a laser beam (the excitation region), which produces changes in the atom internal structure. A beam detector monitors the result. [Redrawn from Fig. 7 of B. W. Shore, Acta Physica Slovaka, **58**, 243–486 (2008) with permission.]

are present within the vapor individual atoms, observable collectively, and whose internal quantum structure can be coherently manipulated.

4.1.2 Particle beams

A simple experimental realization of single atoms or molecules as a part of an ensemble occurs in particle beams. These are produced by allowing a suitable vapor source – say a noble gas or vaporized solid – to expand through a nozzle into a vacuum chamber. The atoms pass through a series of apertures, from which emerges a collimated beam, as sketched in Fig. 4.1. Typically the density of the beam is sufficiently low that collisions are infrequent, and so the result is an ensemble of essentially free particles, localized in position and momentum, moving on average with velocity v.

This atomic beam then passes across one or more laser beams, usually at right angles. In moving with velocity v across the spatially varying laser-beam profile (typically a Gaussian) of electric field $\mathcal{F}(x)$, an atom experiences a time-varying electric field $\mathcal{E}(t) = \mathcal{F}(x_0 + vt)$.

Because there is a distribution of velocities within the atomic beam, not all atoms experience the same field duration. Furthermore, because of velocity-imparted Doppler shifts, proportional to the velocity $v_\perp$ transverse to the beam, the atoms experience different laser frequencies. In consequence, one observes an average over various atomic conditions. Section 16.2 discusses these and other aspects of beam interactions that require ensemble averages.

4.1.3 Cavities

Various schemes make possible the placement of an atom into a cavity, there to interact with the standing-wave field enclosed. Because the field is constrained by boundary surfaces,

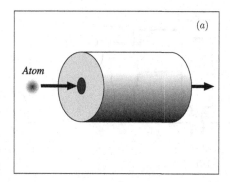 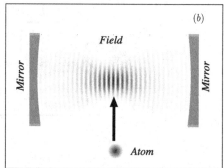

Figure 4.2 (*a*) Microwave cavity bounded by conducting surfaces through which an atom moves. (*b*) An optical cavity bounded by mirrors into which an atom passes (or is brought).

the frequencies of the cavity field are restricted to discrete values.[1] Figure 4.2 illustrates two examples of cavities.

Frame (*a*) shows a microwave cavity through which a stream of atoms pass. Typically the flow is such that only a single atom is within the cavity at one time, and the transitions are between highly excited and therefore closely spaced energy states (Rydberg states [Fen75; Ste83; Fil85]). Frame (*b*) shows a cavity appropriate to optical transitions. The cavity is bounded only along one optical axis, and the atoms are brought, by suitable trapping and transport, into the center of the optical cavity. In each of these cases the field within the cavity is that of a standing wave (see App. D.2.5). Typically the experiments deal with cavity fields of only a few photons, and it is essential to consider the quantized nature of these fields. Chapter 22 discusses the dynamics of such systems.

4.1.4 Trapped particles

Numerous techniques exist to obtain single isolated ions or atoms, localized (trapped) by suitable forces [Deh90; Wal94; Xie98; Lei03; Mes06; For07]. Beams of charged particles can, by use of suitably controlled electromagnetic fields, be slowed and brought to rest, held by electric and magnetic forces within a confined region. Alternatively, neutral beams can be photoionized within regions where trapping fields are present, and held there by suitable fields. Conceptually simpler are the various magneto-optical traps that can, when present within a cold sample of gas, trap an individual atom. Section 4.1.5 below mentions arrays of traps for neutral particles.

Once a particle is trapped, it can be placed within the relatively narrow waist of a focused laser beam, and there be subject to controlled radiation. It is possible, with suitable pulse-shaping techniques, to irradiate a single trapped atom with a pulse whose temporal and frequency characteristics are crafted to produce a desired result. Figure 4.3 shows the

---

[1] Like spectral lines, the discrete values have finite spectral width.

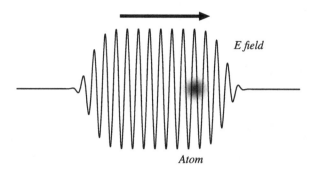

E field

Atom

Figure 4.3 Schematic presentation of trapped atom excitation: A traveling pulse of near-monochromatic radiation passes a stationary (trapped) atom. The traveling field appears, to the fixed atom, as a time-dependent electric field transverse to the propagation direction.

essential elements of the interaction: a traveling electromagnetic wave and a stationary atom.

By means of such techniques it is possible to make repeated observations upon a single atom. Quantum theory provides predictions of time averages of such observations – of sets of histories that provide the counterpart of ensemble averages.

4.1.5 Optical lattices

An electric dipole $\mathbf{d}$ in an inhomogeneous electric field $\mathbf{E}(\mathbf{r})$ possesses a spatially varying potential energy $-\mathbf{d} \cdot \mathbf{E}(\mathbf{r})$. Thus it experiences a force (the negative gradient of the energy) in the direction of the field gradient. When the dipole is an induced moment, proportional to the electric field, then the force is proportional to the gradient of the square of the electric field, i.e. to the gradient of the intensity. Suitably constructed, such fields can trap single atoms [Jes96; Ash07]. They can then move the individual particles from one interaction region to another [Kuh01; Sch01; Nus05].

The required potential energy of the induced dipole at position $\mathbf{r}$ is obtained from the cycle-averaged square of the electric field as

$$V^{pol} = -\frac{1}{2} X(\omega) \{|\mathbf{E}(\mathbf{r})|^2\}_{av}, \tag{4.1}$$

where $X(\omega)$ is the frequency-dependent (dynamic) polarizability, see Sec. 13.4.5. The force on a point particle at position $\mathbf{r}$ resulting from this energy, the *optical-gradient force*, is

$$\mathbf{F} = -\nabla V^{pol} = \frac{1}{2} X(\omega) \nabla \{|\mathbf{E}|^2\}_{av}. \tag{4.2}$$

Like the ponderomotive force of eqn (5.6) it is proportional to the local intensity. Macroscopic dielectric particles are typically attracted into regions of high field. It is this behavior that makes possible the use of tightly focused laser beams as "optical tweezers". For an atom or molecule the frequency dependence of the polarizability changes sign as the frequency varies from red to blue about a resonance, and so the force can either repel the particle from regions of intense field or attract the particle toward a high-field region. A free electron experiences a similar intensity-dependent force, from the gradient of the ponderomotive energy, that forces the electron out of the field.

For light near resonance at a Bohr frequency ω_B, the induced dipole will be in phase with the electric field if the carrier frequency ω is smaller than the resonant frequency, i.e. it is "red-detuned". The gradient force will then be in the direction of increasing field, and the particle will be drawn toward regions of high intensity. A "blue-detuned" laser field induces a force that repels the particle from the high-field region. A standing-wave optical field will, through this mechanism, provide a periodic array of potential troughs in which cold atoms will be trapped [Chu86]. A combination of two orthogonal standing waves will create an optical lattice [Jes96; Mur06] of wells in which cold atoms, or a Bose–Einstein condensate, can be trapped.

4.1.6 Atoms in solids

Stationary quantum systems described by a few discrete quantum states can be obtained in various other ways. For example, suitable chemical preparations can create impurity atoms held within a solid matrix. The electrons of such atoms are held by both Coulomb attraction to a nucleus and repulsion from nearby atomic electrons. Although experiments may not distinguish between individual atoms, each one can have well-characterized discrete energies, and can be regarded as a simple atom of the sort discussed here [Wyb07]. To describe the aggregate collection of atoms one must include an average over the various atom environments, as discussed in Sec. 16.2.

Other techniques are used to fabricate small solid regions in which electron-hole pairs are localized – quantum dots. Because the excitations involve confined motion, the energies are discrete. The theory presented here provides a description of the laser-induced changes to such excitations.

4.1.7 Surface layers

Vapor in contact with a cold surface forms a condensate, first of individual atoms and then of a monolayer. Although these atoms no longer move freely in all directions, they retain many of their free-atom properties [Aro06]. Their internal structure responds to the presence of the static field produced by the surface atoms, but they retain discrete quantum states of electronic excitation, and these are subject to manipulation by coherent radiation [Saa06; Mat06].

4.1.8 Collisions

The energy acquired by an atom to produce excitation can come from any external source, not only radiation. A passing charged particle (a projectile ion) creates a transient electric field at the atom (the target). The direction of this electric field, the counterpart of the polarization axis of a laser pulse, is along the changing vector between the coordinate origin at the atom center of mass and the position of the charged particle as it moves past the atom. As the projectile passes the target this vector sweeps out an area in a plane; its magnitude increases as the projectile draws near, reaches a maximum when the projectile reaches the distance of closest approach, and then diminishes to zero as the projectile recedes. This pulsed electric field has much in common with the laser pulses considered in the present monograph, and one can use the time-dependent Schrödinger equation of Chap. 7 to describe the effect of a single projectile on a single atom (the field is treated as having carrier frequency zero).

However, many of the projectile characteristics are not controllable. Even with a monoenergetic projectile beam there will occur a statistical distribution of distances of closest approach. The ensemble average of projectile characteristics will blur evidence of coherent excitation.

When the projectiles are neutral, the interaction is treated, when the particles are well separated, as that of two electric (or magnetic) multipole moments, separated by changing distance R. The potential varies as some power of R; that power exponent depends on the orders of the target and projectile multipoles, as shown in the following table:

Interaction	monopole–monopole	monopole–dipole	dipole–dipole	...
Potential	R^{-1}	R^{-2}	R^{-3}	...

The interaction also varies with the relative orientation of the two multipoles. As with the interaction between an atom and an ion, this multipole form of the interaction remains valid only when the projectile is outside the electron cloud of the atom, and its effect must be averaged over an ensemble of projectile characteristics.

The energy acquired by the atom in a collision comes from the kinetic energy of the projectile, together with any internal energy of the projectile. Together these two sources must exceed the excitation energy acquired by the target atom. By adjusting the kinetic energy of the projectile an experimenter can choose to allow or prevent target excitation. However, this controlled energy variation does not have the selectivity of a laser field: inevitably there are ranges of projectile energies.

When a projectile penetrates the electron cloud of the target, the multipole form of the interaction is no longer applicable. Then one must consider all of the system constituents – the electrons and nuclei of the target and the projectile – as forming one molecule, whose structure must be determined from the time-independent Schrödinger equation by taking into account the Coulomb interaction between each charge.

In a vapor collisions are inevitable, although they can be reduced to infrequent random events by reducing the density. Such random collisions typically do not impart excitation energy to the target, but they do produce brief pulses of electric field. These produce small shifts of energies that, in turn, appear as phase shifts of wavefunctions. Because these are random, their averaged effect is to prevent continuation of coherence of the atomic state. Idealization of undisturbed coherent interactions, as described by the time-dependent

Schrödinger equation (see Chap. 7) applies only for times shorter than the mean time between collisions.

4.1.9 The isolated-atom idealization

The theory presented in this monograph, that of *coherent excitation*, is based on several assumptions about quantum systems interacting with optical radiation. The requirements are that, during the interval when the excitation occurs to produce coherent excitation, there are:

- No radiative decays
- No random collisions
- No random interruptions of strict laser periodicity
- No random changes of the environment.

Given these assumptions, we deal with coherent excitation of a single atom or other structured particle.[2] The relevant equation of motion is the time-dependent Schrödinger equation for the statevector, discussed in Chap. 7.

4.2 Detecting excitation

Many experimental techniques exist for detecting single atoms or molecules that have specified characteristics. The simplest examples, conceptually, are those in which the particles form a beam traveling through a vacuum chamber, see Sec. 4.1.2. As these impinge upon a suitably designed detector the impacts produce a steady electric current. When the flow of particles is sufficiently low, individual impacts (i.e. single atoms) can be detected. It is possible to devise detectors that register only the arrival of particles whose internal excitation energy, supplementing the particle kinetic energy, exceeds some chosen amount. In this way it is possible to detect the arrival of atoms in excited states.

One can pass the beam through a static electric field that counters the Coulomb field that binds the electrons. Exposed to this field the atoms become ionized; either the ions or the free electrons can provide a signal [Gol06].

Other techniques make use of radiation as a probe. One method uses ultraviolet light to photoionize a sample of atoms, producing a reaction symbolized as

$$A + \hbar\omega \rightarrow A^+ + e. \tag{4.3}$$

By detecting the photoelectrons and determining their kinetic energy (i.e. electron spectroscopy) it is possible to determine the difference between the internal energy of the initial atom A and the ion A^+. By selecting only those photoelectrons whose kinetic energy indicates ejection from an excited state it is possible to determine the relative population in that state. When the atoms of this reaction are part of a beam, it is possible to detect the ions by

---

[2] It proves convenient to refer to the system as an "atom", though the descriptions apply equally well to other systems that have discrete bound energies.

separating them from the uncharged flow with the aid of a static magnetic field, as is done in a mass spectrometer.

Alternatively, one can observe the fluorescent radiation that spontaneously carries away excitation energy, as symbolized by the reaction

$$A^* \rightarrow A + \hbar\omega. \tag{4.4}$$

The spectrum of such radiation reveals the presence of excitation energy; the amount of radiation in the relevant spectral line is proportional to the excited population.

When the quantum system is a molecule of constituent fragments A and B the absorption of a photon may produce photodissociation, symbolized as

$$(AB) + \hbar\omega \rightarrow A + B. \tag{4.5}$$

The fragments may be detected directly or indirectly by means of further photon-induced changes.

With appropriately chosen multiple frequencies of radiation it is possible to produce a succession of excitations, first to one excited state and then from this to a more energetic state in a multiple-photon chain of excitations.[3] Each stage is selective. The final product may be an ion, detectable either directly or by photoelectrons.

In each of these scenarios – detection of atoms, ions, or photons – there occurs a signal proportional to the population of an excited state. It is the challenge of an experimenter to relate the strength of such electrical signals to the quantities of interest – the excitation probabilities – taking into account such factors as the solid angle subtended by the detector and the efficiency with which the detector responds to individual events.

Other types of detection are related to applications. A laser beam, passing through material in which the atoms have undergone quantum changes, will be affected in measurable ways; see Chap. 21. An excited atom, brought into a cavity, may leave its excitation energy there as a photon; see Chap. 22.

Aggregates of atoms and molecules also reveal the consequences of coherent excitation. Typically we treat the transmission of light through matter – a gas or a condensed phase – as an encounter with a continuous distribution of quantum structures, each of which responds to the field. Their collective effects are incorporated into the electric polarization field $\mathbf{P}(\mathbf{r}, t)$ and magnetization field $\mathbf{M}(\mathbf{r}, t)$ that appear, through Maxwell's equations, as sources (and modifiers) of electromagnetic radiation, see App. C.1. By altering the structure of the numerous atoms that constitute bulk matter we alter the propagation characteristics of the radiation. Pulses can become reshaped and energy can be coherently stored, all as a consequence of coherent excitation. Thus these effects serve as indicators of quantum changes. Chapter 21 discusses some of these effects.

---

[3] One should distinguish between *multiple-photon transitions*, in which a succession of incoherent excitation events occur, and *multiphoton transitions*, produced by the coherent action of several photons, distinguishable by frequency or polarization.

4.3 The interaction energy; multipole moments

The bulk of this monograph concerns the effect on atoms of a given field, and the equations that describe those effects, taking the field as known. The relevant equations generalize those that describe the classical motion of charged particles subject to forces (the Lorentz force) exerted by specified electric and magnetic fields.

The effects of radiation upon *unbound* particles are dominated by momentum considerations; one can regard each photon of a monochromatic directed beam, of wavelength λ and angular frequency $\omega = 2\pi c/\lambda$, as carrying energy $\hbar\omega$ and linear momentum, of magnitude $\hbar\omega/c$, in the direction of propagation. The absorption of a photon therefore induces a momentum change $\Delta p = \hbar\omega/c$ that alters the velocity of an absorbing particle and which thereby acts on bulk matter as radiation pressure. The momentum change can also cool and trap individual particles [Met99]. Section 17.2 mentions aspects of such interactions that have relevance to coherent excitation, particularly the correlation of internal excitation with center of mass motion.

Momentum transfer plays a less important part in the changes to bound particles (e.g. those responsible for internal structure and excitation of atoms and molecules). Instead, energy considerations dominate the discussion. We convert the Lorentz force into an energy of bound particles by evaluating the work done by the force in moving the individual charges into positions defined with respect to the center of mass (cm). The result, discussed in App. B, is a multipole interaction [Cra84; Muk99][Sho90, Chap. 9], in which the energy appears as series of expressions for interactions of electric or magnetic multipole moments with a field (or a field derivative) evaluated at the center of mass. The atomic multipole moment of order k pairs with the $k - 1$ derivative of a field, starting with $k = 1$ (dipole), $k = 2$ (quadrupole), $k = 3$ (octupole), etc.

For bound electrons the dominant interaction, for transitions induced by radiation of optical frequencies, is that of an electric dipole moment in an electric field: it gives the interaction energy $-\mathbf{d} \cdot \mathbf{E}$ proportional to the component of the electric dipole moment $\mathbf{d}$ along the direction of the electric field, evaluated at the atomic center of mass. This electric-field interaction is responsible for the Stark shift [Dav76c; Cow81; Sob92; Gol06; Auz10] of energies in the presence of static electric fields; the dipole moment for such interactions is typically an *induced* dipole moment, proportional to the electric field strength.

The corresponding magnetic interaction, $-\mathbf{m} \cdot \mathbf{B}$, is that of a magnetic moment $\mathbf{m}$ and a magnetic field. This interaction produces the Zeeman shift [Dav76c; Cow81; Sob92; Gol06; Auz10] of energies in the presence of static magnetic fields. Transitions involving this interaction typically occur in the radio-frequency regime and are observed as nuclear magnetic resonance (NMR). The magnetic moment has two sources. One is the motion of the charge: a moving charge generates a magnetic field around its trajectory, and these combine into magnetic moments (an infinitesimal bar magnet oriented perpendicular to the plane of motion). Collectively this *orbital* angular momentum of the electrons sums to the vector $\hbar\mathbf{L}$. The second source is the *intrinsic magnetic moment* that electrons, protons, and neutrons carry, a property associated with an *intrinsic angular momentum* (*spin*). Collectively the electron spin angular momentum sums to the vector $\hbar\mathbf{S}$.

In each of these cases the interaction energy is expressed as the product of some function of atomic variables (a multipole moment), whose value will be fixed by the polarization axis and the two states involved in a transition, and a numerical factor, periodic at frequency ω, whose magnitude depends on the square root of the intensity, i.e. on $\mathcal{E}$.

Each of these moments has application both to coherent excitation and to diverse disciplines of astronomy, chemistry, and technology. The magnetic dipole is responsible for the 21 cm line observed by radio astronomers from interstellar hydrogen. It is used terrestrially for NMR. Most of the line emission of flames and fluorescent lights originates with electric dipole moments. They are the most commonly used moments for laser-induced structure changes. The electric quadrupole has been used for excitation of trapped calcium ions [Lei03].

General properties. Appendix B expresses the atom–field interaction energy as a set of multipole interactions

$$V^{E1}(t), \ V^{E2}(t), \ V^{M1}(t), \ V^{pol}(t),$$

illustrated by formulas appropriate to the fields

$$\mathbf{E}(\mathbf{r}, t) = \mathbf{e}_z \mathcal{E} \cos(kx - \omega t), \qquad \mathbf{B}(\mathbf{r}, t) = -\mathbf{e}_y \frac{\mathcal{E}}{c} \cos(kx - \omega t). \tag{4.6}$$

Several properties of these formulas are worth noting:

- The electric and magnetic dipole interactions V^{E1} and V^{M1} involve components of vectors; the required components (z and y) differ in the two cases.
- Typically matrix elements of V^{M1} will be smaller than those of V^{E1} by the value of half the fine-structure constant $\alpha/2 \simeq 1/274$. (The ratio is larger for ions and smaller for a Rydberg state.)
- The electric quadrupole and polarizability interactions V^{E2} and V^{pol} involve second-rank tensors; the required components (xz and zz) differ in the two cases.
- Typically matrix elements of V^{E2} will be smaller than those of V^{E1} by the value $a_0/2\pi\lambda_0$, and those of V^{pol} will be smaller than those of V^{E1} by the value $a_0^2 \mathcal{E}/c$.

In each of these examples the polarization axis of the field dictates the particular components of the vector or tensor that we require. (The particular combinations of x, y, and z components result from our choice of coordinate axes.) Each of the multipole interactions involves harmonic variation at the laser frequency ω (although the $E2$ interaction differs in phase), whereas the induced interaction involves both a static portion and a portion that varies at twice the laser frequency. The multipole interactions are proportional to the field amplitude $\mathcal{E}$ whereas the induced interaction is proportional to $\mathcal{E}^2$.

4.4 Moving atoms

The atoms and molecules undergoing coherent excitation by laser radiation typically fall into two categories. Either they are stationary, or nearly stationary, and are acted on by a controlled laser pulse, or else they are part of an atomic or molecular beam that cuts across a stationary laser beam. In either situation it is important to evaluate the field at the moving center of mass. For this purpose we consider a center of mass at position $\mathbf{R}$, acted on by the electric field of a traveling plane wave, of wavevector $\mathbf{k}$ and frequency $\omega = ck$,

$$\mathbf{E}(\mathbf{R}, t) = \mathrm{Re}\left\{ \mathbf{e}\check{\mathcal{E}} \exp(\mathrm{i}\mathbf{k} \cdot \mathbf{R} - \mathrm{i}\omega t) \right\}. \tag{4.7}$$

Two cases have particular interest: oscillatory motion and steady straight-line motion.

Harmonic motion. A particle bound in a potential well undergoes vibrations. The simplest to treat are those in which the motion is harmonic, at a frequency ω_h, so that the center of mass motion is expressible as

$$\mathbf{R}(t) = \mathbf{R}_0 + \mathbf{s}\sin(\omega_h t). \tag{4.8}$$

The exponentiation of the sinusoid is expressible as a series of harmonics of ω_h, whose amplitudes are Bessel functions $J_n(z)$. From the *Jacobi–Anger expansion* formula

$$\exp(\mathrm{i}z\sin x) = \sum_{n=-\infty}^{\infty} \exp(\mathrm{i}nx) J_n(z) \tag{4.9}$$

we obtain the expression

$$\mathbf{E}(\mathbf{R}(t), t) = \mathrm{Re}\left\{ \mathbf{e}\check{\mathcal{E}} \exp[\mathrm{i}\mathbf{k} \cdot \mathbf{R}_0 + \mathrm{i}\mathbf{k} \cdot \mathbf{s}\sin(\omega_h t) - \mathrm{i}\omega t] \right\} \tag{4.10}$$

$$= \mathrm{Re}\left\{ \mathbf{e}\check{\mathcal{E}} \exp(\mathrm{i}\mathbf{k} \cdot \mathbf{R}_0) \sum_{n=-\infty}^{\infty} \exp[-\mathrm{i}(\omega - n\omega_h)t] J_n(\mathbf{k} \cdot \mathbf{s}) \right\}.$$

The original sinusoidal variation now appears as a series of equidistant *sidebands*, separated by the harmonic frequency ω_h. Their magnitudes depend on the projection of harmonic displacement $\mathbf{s}$ on the propagation direction (they are absent if $\mathbf{k} \cdot \mathbf{s} = 0$) and diminishes with increasing order n.

Doppler shift. Let the atom be steadily moving, with constant nonrelativistic velocity $\mathbf{v}$, so that the center of mass position is

$$\mathbf{R}(t) = \mathbf{R}_0 + \mathbf{v}t, \tag{4.11}$$

where $\mathbf{R}_0$ is the position at $t = 0$. Then the interaction involves the field

$$\mathbf{E}(\mathbf{R}_0 + \mathbf{v}t, t) = \mathrm{Re}\left\{ \mathbf{e}\check{\mathcal{E}} \exp[\mathrm{i}\mathbf{k} \cdot \mathbf{R}_0 - \mathrm{i}(\omega - \mathbf{k} \cdot \mathbf{v})t] \right\}. \tag{4.12}$$

This field is that of a plane wave that is Doppler shifted to the frequency

$$\omega' = \omega - \mathbf{k} \cdot \mathbf{v}. \tag{4.13}$$

A more accurate expression, taking into account the theory of special relativity, is

$$\omega' = \omega - \frac{\mathbf{k} \cdot \mathbf{v}}{\sqrt{1 - (v/c)^2}}. \tag{4.14}$$

The shift is largest, for given speed, when the atom moves along the direction of propagation; it is least (only a second-order relativistic correction) when the atom moves perpendicular to the propagation axis. Thus variation of the angle between the atom beam and the laser beam will alter the effective frequency.

Motion across a laser beam. When a beam of atoms passes across a laser beam, the field at the moving center of mass of each atom undergoes a temporal change as the atom moves into, and then out of, the region in which the laser field is concentrated; in the reference frame of the moving atom the field appears as a temporal pulse. As an example, consider the field of a Gaussian beam propagating along the z axis, as presented in eqn (3.67) of Sec. 3.7,

$$\mathbf{E}(x, y, z, t) = \mathrm{Re} \left\{ \mathbf{e} \check{\mathcal{E}}_0 \frac{w_0}{w(z)} \exp\left[-\frac{x^2 + y^2}{w(z)^2} \right] \exp\left[\mathrm{i}k \frac{(x^2 + y^2)}{2\mathcal{R}(z)} - \mathrm{i}\omega t - \mathrm{i}\varphi(z) \right] \right\}, \tag{4.15}$$

where $w(z)$ is the waist and $\mathcal{R}(z)$ is the curvature. Let the center of mass move at constant velocity $\mathbf{v}$ directly across the laser beam, in the x direction, maintaining fixed offset y from the laser axis,

$$\mathbf{R}(t) = \mathbf{R}_0 + \mathbf{e}_x v t. \tag{4.16}$$

Then the field at the moving center of mass is

$$\mathbf{E}(vt, y, z, t) = \mathrm{Re} \left\{ \mathbf{e} \check{\mathcal{E}}_0 \frac{w_0}{w(z)} \exp\left[-\left(\frac{vt}{w(z)} \right)^2 \right] \exp\left[-\left(\frac{y}{w(z)} \right)^2 \right] \right.$$
$$\left. \times \exp\left[\mathrm{i}k \frac{y^2}{2\mathcal{R}(z)} - \mathrm{i}\varphi(z) \right] \exp\left[-\mathrm{i}k \frac{v^2 t^2}{2\mathcal{R}(z)} - \mathrm{i}\omega t \right] \right\}. \tag{4.17}$$

This is a temporal pulse whose amplitude varies as a Gaussian, $\exp[-(vt/w(z))^2]$, to a peak value $\exp[-(y/w(z))^2]$ that is set by the offset y of the trajectory above or below the axis of the laser beam. Unless the trajectory is close to the focus of the beam, where the wavefront curvature $\mathcal{R}(z)$ is very large, there will occur a quadratic variation of the phase with time. The time derivative of this phase is the effective frequency,

$$\omega' = \omega + \omega[v^2/c\mathcal{R}(z)]t. \tag{4.18}$$

That is, the frequency at the center of mass is *chirped* – it varies linearly with time, at a rate proportional to the kinetic energy of the atom and inversely proportional to the wavefront curvature.

Transverse beam structure. The Gaussian profile presented here is the simplest of many spatial structures that can be prepared by passing a beam through suitable optical elements. Many such fields have analytic expressions. There are, for example, various orders of Gaussian–Hermite modes, in which nodes (in the x, y plane) have rectangular symmetry, or Gaussian–Laguerre modes, where the nodes have circular symmetry [Sie86]. Alternatively, one can create beams that, over short propagation distances, have sharp edges so that atoms passing across the beams encounter sudden changes of field strength. Such fields offer opportunities to subject beam atoms to a variety of pulse shapes.

5

Picturing quantum structure and changes

Recognizing that atoms and molecules are constructed from electrons and nuclei, it is natural to ask: What can we know about the motions of these constituents? This chapter offers some suggestions for depicting electronic structure that fit within the framework of quantum theory. It also presents several ways of picturing the effect of optical radiation, starting with a rather classical portrait relating this to the wavefunction portraits that form the basis of quantum chemistry, and concluding with a rather abstract description based on the use of statevectors in Hilbert space. The discussion serves as a precursor to the examples of laser-induced change, and the equations that govern these changes, starting with Chap. 7.

5.1 Free electrons: Ponderomotive energy

Our understanding of systems of unconstrained particles, such as pitched baseballs or planets encircling a star, builds upon equations of motion for localizable particles (mass points). These allow one to define and plot trajectories of particles – loci of positions as a function of time.

The Lorentz force. All charged particles, bound or free, are affected by electric and magnetic fields. The force exerted on a point charge e at position $\mathbf{r}$, moving with velocity $\mathbf{v}$, is the *Lorentz force*

$$\mathbf{F}(\mathbf{r}) = e\mathbf{E}(\mathbf{r}) + (e/c)\mathbf{v} \times \mathbf{B}(\mathbf{r}). \tag{5.1}$$

Except for particles whose speeds are an appreciable fraction of the speed of light (relativistic motion) the dominant force on a charge is that of the electric field.

The free electron. The nonrelativistic motion of a point-charge electron, of mass m_e and charge $-e$, exerted by a uniform but time-varying electric field, obeys the Newtonian equation

$$m_e \frac{d}{dt}\mathbf{r} = -e\mathbf{E}(\mathbf{r}). \tag{5.2}$$

For application to a laser field, consider a uniform field of intensity I and carrier frequency ω:

$$\mathbf{E} = \mathbf{e}\mathcal{E}\cos(\omega t), \qquad I = \tfrac{1}{2} c \epsilon_0 |\mathcal{E}|^2. \tag{5.3}$$

The solution to the equation of motion for an electron initially at rest is

$$\mathbf{r}(t) = \mathbf{e}\frac{e\mathcal{E}}{m_e \omega^2}\cos(\omega t), \qquad \mathbf{v}(t) \equiv \dot{\mathbf{r}}(t) = -\mathbf{e}\frac{e\mathcal{E}}{m_e \omega}\sin(\omega t). \tag{5.4}$$

The motion of the electron follows the oscillatory electric field (a wiggling or quivering motion), tracing a trajectory along the field direction $\mathbf{e}$. The cycle averaged kinetic energy, termed the *ponderomotive energy* [Ebe91], is[1]

$$E^{pond} = \left\{ \tfrac{1}{2}m_e \mathbf{v}^2 \right\}_{av} = \frac{e^2\mathcal{E}^2}{4m_e\omega^2} = \frac{\alpha\hbar}{m_e c^2}\frac{\lambda^2}{2\pi}I. \tag{5.5}$$

Ponderomotive force. The approximation of uniform intensity cannot hold over large distances. Changes of intensity cause a *ponderomotive force*,

$$\mathbf{F} = -\mathbf{\nabla}E^{pond}, \tag{5.6}$$

that produces a drift of the electron away from the most intense region of the laser field. The jittering motion, at the carrier frequency ω, wavelength $\lambda = 2\pi c/\omega$, supplements any slower motion of the electron as it drifts into regions of lower ponderomotive energy, and must be included when evaluating the energy of a photoelectron in an intense laser field, as is typically required for multiphoton ionization.

5.2 Picturing bound electrons

Uncertainty principle. The dominant forces acting on systems of charges that are bound together as atoms and molecules come from electrostatic Coulomb forces between them; for moderate intensities the external laser field distorts but does not dominate the dynamics.

When the dimensions of a system and the masses of the particles are sufficiently small, as they are for electrons within an atom, then the Heisenberg uncertainty principle imposes constraints on the possibility of determining, simultaneously, the position and velocity of a bound particle as it follows a trajectory. For one-dimensional motion of a particle of mass m with the uncertainty Δx in position x and the uncertainty Δp in momentum $p = mv$ for velocity v, the uncertainty principle imposes the inequality (see Sec. 16.5)

$$\Delta x\, \Delta p \geq \frac{\hbar}{2}. \tag{5.7}$$

Because the uncertainty principle limits the simultaneous measurement of position and momentum (hence velocity), it is not possible to follow any details of particle trajectories within the small confines of an atom; only statistical properties of motion have possibility

---

[1] This ponderomotive energy is typically denoted U_p.

of observation. The atomic and molecular structures between which transitions occur are inherently quantum mechanical, and their description requires some version of quantum theory. Nevertheless, it is often possible to find classical descriptions of the motion that help visualize the changes.

Constants of motion. Although the uncertainty principle prevents us from following details of electron trajectories within an atom, it does not prevent accurate measurement of various constants of the motion. Chief of these is the total energy, the Hamiltonian function of position and momentum that sums kinetic and potential energies. For quantum systems we seek general observables that can be simultaneously measured, unrestricted by the uncertainty principle.[2] These inevitably take only discrete values, parametrized by integer quantum numbers.

Models. The regularities of spectral lines and the causative patterns of energy levels often bear interpretation as originating with classical motion. For example, the Bohr model of hydrogen drew upon the experiments of Rutherford that showed all atoms were composed of light electrons surrounding a much more massive nucleus, held by Coulomb attraction. The similarity between Coulomb and gravitational attraction allows one to picture the motion of the hydrogen electron as akin to the model of planetary motion first codified by Kepler.

A somewhat different model of multielectron atoms came earlier from Lorentz, who proposed that the electromagnetic fields seen as spectral lines were emitted and absorbed from the oscillating motion of light charged particles displaced from equilibrium within the atom. These Lorentz electrons offered a classical picture of the internal changes to quantum states.

In many instances it is possible to recognize the patterns of energy levels as being associated with the vibrations of a harmonic oscillator or the spinning of a symmetric top. These are collective motions of the electrons and nuclei, fully consistent with quantum theory but allowing classical portrayal.

Planetary example. The uncertainty principle forces us to consider statistical properties of atomic constituents. A simple example will illustrate the probabilistic concepts and their application to picturing atomic structure. Consider a small but classical planet bound to a massive central star by gravitational attraction. In accord with Kepler's laws it will move in a plane about the star, along an elliptical orbit, whose semi-major axis is set by the total planetary energy E, kinetic plus potential, and whose shape (ellipticity) is set by the magnitude of the angular momentum. The direction of the orbit plane is perpendicular to the direction of the angular momentum vector, and the direction of the ellipse axis in the plane (the *line of apsides*) is the direction of the *Runge–Lenz vector*. The total energy, the angular momentum vector, and the Runge–Lenz vector remain constant during the motion.

---

[2] Quantum theory represents these by commuting operators.

Now consider an ensemble of such solar systems, each with specified constants of the motion. In the preparation of that collection there is no distinction of the time when the planet reaches closest approach, and so the ensemble will contain planetary systems with all possible planet locations along an ellipse. We then deal with a probability distribution that is an elliptical curve. If there is some uncertainty in the energy or orbital angular momentum then the curve gains finite thickness – the probability distribution becomes a tube. If, in preparing the ensemble, we do not specify the Runge–Lenz vector then we deal with a planar area as the probability density. If, further, we do not specify the direction of the angular momentum vector the probability distribution becomes a smeared-out volume distribution of all possible ellipses of given shape.

It is such classical probability distributions that we must keep in mind when attempting to visualize motion of electrons within atoms. We can only consider a *probability distribution*: The elliptical curve of planetary motion becomes a fuzzy tube of probability distribution, spread further within a volume by unspecified directions of the angular momentum and Runge–Lenz vector. We are led, in this way, to visualize the electrons within an atom as a probability distribution or a cloud of negative charge. As with classical statistical mechanics, the probability distribution of electrons (the charge cloud) has meaning only when we consider observations upon many atoms, specifically an ensemble of atomic systems – multiple identically prepared systems, see Sec.16.1.

Expectation values. The quantum theory with which we describe the properties of atomic-scale systems is a probabilistic theory: Like statistical mechanics its predictions apply to repeated observations of systems (say an atom) drawn from an ensemble of systems prepared according to some prescription. From repeated observations of many atoms we are able to deduce such properties as the average number of atoms in the ground state, or the mean internal energy of the atoms at time t after they are first exposed to radiation. These observable ensemble averages, or *expectation values*, are amongst the predictions of quantum theory, see Sec. 16.4. Equally well, the theory can describe results of multiple measurements upon a single trapped atom or ion.

Atom size and shape. The internal energy of an atom or molecule, observed through spectroscopy, derives from motions of the constituent electrons and nuclei: as this energy increases the motions become more vigorous and occupy larger regions of space. The increase in atomic size is not necessarily an increase in random motions, as would occur if the source of energy were heat: Laser-induced excitation produces very organized motions. Although it is not possible to follow trajectories of individual electrons as they move within atoms, aspects of the differing motions are observable. Primarily the observations of atoms are of various moments of the charge and current distributions within the atom, see App. B.2. Observations of molecules may include, in addition to these, the geometrical arrangement of the constituent atoms and kinetic energies of vibrations and rotations.

Of the several moments of charge distribution the electric dipole moment has particular importance. This moment represents a (small) separation of the mean values of positive and negative electric charge – in an atom a shift of the electron cloud relative to the nucleus. This is a vector quantity, having both a magnitude and a direction. Higher multipole moments may exist, characterizing nonspherical shapes or nonplanar currents. All of these moments are controllable and may be modified by crafting suitable laser pulses. Their transient presence produces identifiable patterns of spontaneously emitted fluorescence radiation. Their expectation values are observable by their effect upon radiation that passes the atoms, see Chap. 21.

5.3 The Lorentz force

Lorentz force from laser beam. The force exerted by a linearly polarized plane-wave field moves a charge back and forth along a line, in a plane perpendicular to the laser propagation axis. The alternating changes of transverse linear momentum average to zero over one optical cycle. By contrast, the electric vector associated with circular polarization moves in a spiral rather than in a plane as does the electric vector of linearly polarized light. This field produces a force that twists a charge distribution, at the steady rate of the carrier frequency ω; the field exerts a torque. When the field has left-circular polarization (positive helicity) the twisting force is in a clockwise direction. The twisting motion tends to impart angular momentum to the charge, as it moves in a circular path under the influence of such a field. Unlike the to-and-fro linear motion forced by linear polarization, the torque and the ensuing circular motion is always in the same sense – clockwise or counterclockwise.

Action of the Lorentz force. A simple semi-classical description[3] of laser radiation acting upon an atom provides a useful picture. With this approach we regard the laser radiation as providing a periodic electric field, of angular frequency ω. This periodic field in turn exerts a periodic force upon charges. Because the atom nucleus is much more massive than any electron, it remains essentially stationary; the primary effect is upon the more easily moved electrons.[4]

The inevitable wavelike properties of a bound electron, imposed by quantum theory,[5] force us to picture a distribution of negative electric charges – an electron cloud. In the absence of laser radiation the electrons maintain steady motion around the nucleus; the electron cloud is held by Coulomb forces in an equilibrium position. The electric field of the laser disturbs this equilibrium, exerting a periodic force.

The simplest illustration is the force produced by linearly polarized light, say in the x direction, as in eqn. (3.10) with unit vector $\mathbf{e}_x$. This field exerts a periodic force, alternately in the $+x$ direction and then in the $-x$ direction, see Fig. 5.1. Responding to this force,

---

[3] The term "semi-classical" appears often; it refers to physics in which the atom or other quantum system obeys laws of quantum mechanics, while the radiation field obeys only the classical Maxwell field equations (see App. C) and is not quantized.

[4] As discussed in Sec. 4.3, the laser interaction affects multipole moments, and these include nuclear moments.

[5] To be more correct, quantum theory associates wavelike properties with the *information* we have about an electron, embodied in the electron wavefunction.

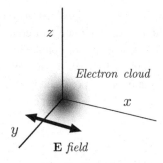

Figure 5.1 The oscillating electric field of the laser forces oscillations of the electron cloud. [Redrawn from Fig. 9 of B. W. Shore, Acta Physica Slovaka, **58**, 243–486 (2008) with permission.]

the electron cloud must move, periodically toward $+x$ and then toward $-x$. This motion appears as a "shaking" of the charge cloud[6] at the radiation frequency ω.

The effect of a classical periodic force upon harmonic motion is well known, familiar to every child who has mastered the art of a playground swing: When the force is in phase with the motion, then there occurs an increase in the amplitude of the motion – an increase of swing energy. When the force is timed to oppose the velocity, then the motion diminishes – the swing energy decreases.

Similar effects are to be expected, and are observed, in the action of a periodic electric field upon the motion of a charge cloud. Specifically the atom, having gained energy from the field, then proceeds to lose it and return to the initial unexcited state. This gain and subsequent loss of excitation energy occurs periodically; the frequency of this population oscillation (a Rabi oscillation) is known as the *Rabi frequency*[7] [Kni80] [Sho90, §3.4]. Similar oscillations occur, between degenerate states, when the excitation is constant (DC) rather than oscillatory; the corresponding frequency then is the *Majorana frequency* [Maj32] [Sho90, §3.4].

However, unlike classical systems, the energy of bound electrons is quantized, and the distortion of the electron cloud must exhibit associated quantization. To describe the quantum nature of the excitation – transitions between discrete states of internal energy of the atom – it is necessary to consider the spatial properties of the bound states of a single electron. Here I summarize the key points.

5.4 The wavefunction; orbitals

The charge-cloud picture of an atomic transition is most directly presented by means of a single-electron *wavefunction* $\Psi(\mathbf{r}, t)$ such that the probability $P(\mathbf{r}, t)d\mathbf{r}$ of an electron being in a volume increment $d\mathbf{r}$ around position $\mathbf{r}$ at time t, i.e. the *probability density*, is

---

[6] Circularly polarized light forces a "spinning" motion of the charge cloud, about an axis through the center of mass. This motion is not evident for a charge distribution that has cylindrical symmetry about the rotation axis, as does the $1s$ state of hydrogen, but it becomes evident for excited states whose charge distributions have angular nodes.

[7] Some authors refer to half the frequency of population oscillation as "the Rabi frequency" .

proportional to the absolute square,

$$P(\mathbf{r}, t)d\mathbf{r} = |\Psi(\mathbf{r}, t)|^2 d\mathbf{r}. \tag{5.8}$$

The electron must be somewhere, and so the spatial integral of the probability density, at any time, must be unity,

$$\int d\mathbf{r} |\Psi(\mathbf{r}, t)|^2 = 1. \tag{5.9}$$

The electron carries an electric charge, and so the probability density $|\Psi(\mathbf{r}, t)|^2$ times the electron charge e bears interpretation as the charge density.

In the absence of external fields (or spontaneous emission) the exact distribution of charge probability remains constant. The relevant wavefunction, often termed an electron *orbital* [Sho90, §19.1], $\psi(\mathbf{r})$ is a solution to the time-*independent* Schrödinger equation, a second-order partial differential equation, supplemented by boundary conditions, whose multiple solutions I here label by a simple integer n,[8]

$$\left[-\frac{\hbar^2}{2m_e} \nabla^2 + V(\mathbf{r}) - (E_n - E_\infty) \right] \psi_n(\mathbf{r}) = 0. \tag{5.10}$$

The second-order spatial derivative, the differential operator ∇^2, expresses kinetic energy; the potential energy – responsible for the binding force – is $V(\mathbf{r})$. Here E_n is the energy of state n above the ground state ($n = 1$) and E_∞ is the minimum energy required to move the electron to infinite separation from the nucleus (photoionization). To complete the mathematical definition of the quantum system we require boundary conditions. For bound states (those for which $E_n < E_\infty$) these are that the wavefunction should be confined in space,

$$\psi_n(\mathbf{r}) \to 0 \text{ as } \mathbf{r} \to \infty. \tag{5.11}$$

With this constraint the solutions for bound-state energies exist only for particular discrete values (the quantized *eigenenergies*).

5.4.1 Two-state example

From the Bohr picture of transitions we expect that when the frequency of radiation matches a Bohr transition frequency from the ground state to an excited state we can expect energy to flow into the atom, producing a transition between the two discrete quantum states. In many situations such as this only two quantum states need be considered. Conditions are such that the wavefunction $\Psi(\mathbf{r}, t)$ can be regarded as constructed from only two orbitals.

For a particle that can exist in two, and only two, quantum states, we describe the possible probability distributions with two single-particle wavefunctions (orbitals), $\psi_1(\mathbf{r})$ and $\psi_2(\mathbf{r})$,

---

[8] The presence of the constant E_∞ here is in keeping with the convention that bound energies E_n should be positive, measured from the smallest value E_1. The combination $E_n - E_\infty$ is therefore negative for bound states; positive values represent scattering states, available for any energy.

such that, if the particle is known to be in state n and is free from external disturbances, its spatial distribution is (independent of time),

$$P_n(\mathbf{r}, t) = |\psi_n(\mathbf{r})|^2. \tag{5.12}$$

Because, by assumption, the particle can only be found in one of these two quantum states, then even when external forces are present the wavefunction must always be some superposition,

$$\Psi(\mathbf{r}, t) = c_1(t)\psi_1(\mathbf{r}) + c_2(t)\psi_2(\mathbf{r}), \tag{5.13}$$

with time-varying complex-valued coefficients $c_n(t)$. The probability of the electron being in state n at time t is

$$P_n(t) = |c_n(t)|^2, \tag{5.14}$$

and hence $c_n(t)$ is a *probability amplitude*. The probability of the electron being at position $\mathbf{r}$ at time t is

$$P(\mathbf{r}, t) = P_1(t)|\psi_1(\mathbf{r})|^2 + P_2(t)|\psi_2(\mathbf{r})|^2 + 2\mathrm{Re}\left\{c_1(t)c_2(t)^*\,\psi_1(\mathbf{r})\psi_2(\mathbf{r})^*\right\}. \tag{5.15}$$

The first two terms are recognizable as expressing the individual charge distributions – these one expects from classical probabilities. The final term expresses quantum-mechanical interference; it makes possible nodes of zero probability from individual orbitals whose probabilities are everywhere nonzero.

Single-electron orbitals are typically classified according to their spatial symmetries, such as are labeled with orbital angular momentum and parity.[9] These labels permit statements of selection rules governing transitions; see Sec 12.2.

5.4.2 Semi-classical distortion and excitation

Over time a periodic force adds to, or subtracts from, energy of the bound electrons – a transition occurs between discrete states of internal energy of the atom. Subsequently a reversing transition occurs as energy leaves the bound electrons. Frames showing the distortion and change induced by a linearly polarized field resonant with the $1s-2p$ transition of hydrogen appear in the text by Auzinsh *et al.* [Auz10]. A very nice graphical illustration of this process has been created by Falsted, available from his web site, http://www.falstad.com/qmatomrad/. There one can see, for example, the effect of a periodic electric field, directed along the x axis and resonant with the $1s-2p$ transition of hydrogen, on the two relevant hydrogenic orbitals: the ground-state $1s$ orbital (null orbital angular momentum, even parity) and the excited-state $2p_x$ orbital (unit orbital angular momentum, odd parity), selected by electric field polarization from the three possible degenerate $2p$ orbitals p_x, p_y, and p_z appropriate to linear polarization; the orbitals p_{-1}, p_0, and p_{+1} are appropriate to angular momentum.[10]

---

[9] By definition, odd parity wavefunctions undergo a sign change when the coordinates are reversed; even functions do not.
[10] The wavelength of the $1s-2p$ transition in hydrogen, responsible for the "Lyman α" spectral line, lies in the far ultraviolet, at 121.6 nm; it has not been possible to demonstrate Rabi oscillations with this short wavelength. Amongst other problems,

One sees, in the Java applet, at first a small periodic wiggling of the spherically symmetric charge cloud of the $1s$ state. These are small perturbations of the otherwise stationary charge distribution at the radiation frequency ω. During this initial time interval the electron behaves like a driven harmonic oscillator, moving about an equilibrium position under the influence of a periodic force. It is the counterpart of the motion responsible for the ponderomotive energy of a free electron. Over many cycles at the driving frequency ω one observes a distortion of the cloud: A nodal surface appears, and the cloud takes on an increasingly bimodal appearance, though still subject to a continual small oscillation along the x axis. The charge distribution is distinctly asymmetric. The internal energy increases steadily and in time the charge distribution becomes that of the $2p_x$ orbital of hydrogen: two lobes of equal probability but opposite sign. At this moment a transition has been completed, $1s \rightarrow 2p$, and the internal energy is that of a $2p$ state.

But, with continued presence of the periodic driving field, the wavefunction continues to respond. It continues to undergo small (wiggling) oscillations together with a gradual change of form, until it appears once more as the spherical $1s$ distribution, and the energy has returned to that of the ground state.

These effects continue as long as the field is present: a rapidly varying perturbation at twice the driving frequency ω (i.e. a shaking of the charge cloud) together with a slower periodic variation of excitation probability (a distortion of the charge cloud), at the Rabi frequency. As will be noted in Sec. 8.1.4, the latter is a measure of the cycle-averaged interaction energy between atom and field, and can be altered by changing the intensity of the light. Subsequent sections of the present monograph provide the theoretical basis for understanding the quantitative properties of Rabi oscillations and other manifestations of coherent excitation.

The picture presented here, of periodic linear displacements of the electron charge-cloud and slower distortions, holds for an electric field that is linearly polarized. When the field is circularly polarized there is no linear displacement. Instead the atom undergoes rotation under the influence of the field-induced torque. Again slow distortions occur, at the Rabi frequency, but these changes occur in a framework that rotates steadily at the carrier frequency.

A presentation of single-electron probability distribution requires only a three-dimensional space, but when one considers more complicated systems, such as multi-electron atoms or molecules, then such wavefunction displays are impractical. Instead, we plot the spatially integrated probability $P_n(t)$ that at time t the system is in state n. Figure 8.1 shows such a plot, appropriate to the two-state excitation by linearly polarized light, as discussed above.

Periodic distortion of an electron cloud, as discussed here, relies on the two-state approximation. When the electric field becomes sufficiently strong (i.e. the interaction energy

---

the radiation will simultaneously photoionize the excited state. Rydberg atoms, those with large principal quantum number, offer better choices for demonstrations but not for simplest picturing of wavefunctions because they have more nodes [Fen75; Ste83; Fil85] .

exceeds the binding energy), this simple picture fails. Instead, one must include an infinite number of additional quantum states, including an ionization continuum, in order to depict more extensive distortion of the wavefunction (see Sec. 13.4.5). These additional states allow a portrait in which, for example, a single cycle of the field forces the electron cloud to leave the vicinity of the nucleus and move away – photoionization occurs [Hay03; Kul06].

5.5 The statevector; Hilbert spaces

The wavefunction of a single particle offers a simple interpretation of the particle as a probability cloud – a distribution of electric charge when the particle is an electron. When one considers atoms heavier than hydrogen an N-electron wavefunction catalogs information about $3N$ independent spatial coordinates (and N spin variables). It is not possible to display such a multidimensional wavefunction, other than by selected two-dimensional slices, and so it is desirable to find alternative approaches to cataloging the information.[11] An alternative approach has therefore found favor, one based on the use of an abstract vector space to represent probability amplitudes, The abstract space differs from a state space by allowing complex numbers as components of vectors. This statevector approach forms the basis of the present monograph.

5.5.1 Quantum states and Hilbert spaces

The needed quantum mechanics is based upon the recognition that, whatever the electron trajectories may be within the confines of an atom, whatever may be the rotational and vibrational motions of the molecular framework, the internal energies of atoms and molecules form a discrete set that can be ordered, in part, by increasing value, E_1, E_2, ..., together with a continuum for energies above some limit. For any atom or molecule there are many discrete internal energy states – in idealized models, an infinite number. Each can often be assigned various identifying quantum-number labels, such as those associated with angular momentum. The single index 1 or 2 would then be understood as standing for the more complete list of identifying labels.

Because the quantum states of interest here are discrete, they can be associated with an abstract vector space, a mathematical construct in which each individual coordinate is associated with one of the possible quantum states. There are as many dimensions as there are quantum states; in many situations there is an infinite number of these.

The multidimensional vector spaces of interest in quantum mechanics have several distinguishing properties that mark them as *Hilbert spaces* [Sze04; Deb05]. In such a space lengths and angles are well defined, and the coordinates, projections of a vector along the coordinate axes, are complex-valued numbers, i.e. with real and imaginary parts. At the

---

[11] An important exception occurs with descriptions of wavepackets formed from molecular vibrational states. Pictures of these provide valuable insight into the dynamics of laser-induced unimolecular reactions, such as photodissociation; see Sec. 18.2.1.

heart of the mathematics of quantum mechanics there is therefore the connection[12]

$$\text{quantum states} \leftrightarrow \text{Hilbert-space unit vectors.} \qquad (5.16)$$

As it turns out, many examples of coherent excitation by lasers, and the consequent state manipulation, involve relatively few quantum states – primarily only those for which the laser frequencies satisfy (at least approximately) the Bohr resonance condition of eqn. (2.3). Thus the Hilbert space of interest has finite dimension, say N.

A basic postulate of quantum theory is that, because the possible internal structures of the atom – the quantum system – must be expressible as one of the allowed quantum states, any particular quantum state can be represented as a single point in this abstract space. This *system point* can be regarded as a vector from the coordinate origin: the *statevector* $\mathbf{\Psi}$.

The connection between the abstract mathematics of a Hilbert space and the world of experimental physics occurs through the coordinates of the statevector: the projection of vector $\mathbf{\Psi}$ onto the unit vector $\boldsymbol{\psi}_n$ of the nth Hilbert-space axis, denoted $\langle \psi_n | \Psi \rangle$ or $\langle n | \Psi \rangle$, when squared absolutely, is the probability P_n of finding the system in state n,

$$P_n = |\langle \psi_n | \Psi \rangle|^2. \qquad (5.17)$$

This equation lies at the heart of the association of quantum theory with the mathematics of Hilbert space – the mapping of physically observable sets of atomic properties that distinguish each quantum state, onto the mathematics of an abstract vector space in which the coordinates are complex numbers. It is notable that this postulated connection leaves the phase of the probability amplitude unspecified; only its magnitude has direct connection with measurable probability.

5.5.2 Hilbert-space coordinates and probability amplitudes

As is the case with any vector, in any mathematical space, the statevector must be expressible by its coordinates along independent axes. With the present notation this construction reads

$$\mathbf{\Psi} = c_1 \, \boldsymbol{\psi}_1 + c_2 \, \boldsymbol{\psi}_2 + \cdots, \qquad (5.18)$$

where c_n is a complex-valued number, the *probability amplitude*, whose absolute square is the probability P_n,

$$P_n = |c_n|^2. \qquad (5.19)$$

Completeness. It is customary to assume, as I do, that the N quantum states associated with the N-dimensional Hilbert space are complete, meaning that certainly, with unit probability,

---

[12] For greatest clarity one should distinguish the unit vectors in Hilbert space with the quantum states with which they are associated, but common usage applies the term "state" to both concepts, quantum state and Hilbert-space unit vector, speaking of the latter as *basis states*.

the system will be found in one of these states. This requirement means that the individual-state probabilities sum to unity,[13]

$$\sum_{n=1,N} P_n = \sum_{n=1,N} |c_n|^2 = 1. \tag{5.20}$$

It follows that the statevector has unit length:

$$|\Psi|^2 \equiv |\langle\Psi|\Psi\rangle|^2 = \left| \sum_{n=1,N} \langle\Psi|\psi_n\rangle\langle\psi_n|\Psi\rangle \right|^2 = \sum_{n=1,N} P_n = 1. \tag{5.21}$$

The unit sum property is often expressed as a *completeness relationship*, in which N orthonormal unit vectors provide an expansion (or decomposition) of unity,

$$\sum_{n=1,N} |\psi_n\rangle\langle\psi_n| \equiv \sum_{n=1,N} |n\rangle\langle n| = 1. \tag{5.22}$$

Though the statevector has unit magnitude (unit norm), $|\Psi| = 1$, its phase is arbitrary; it is a *ray* in Hilbert space, see Sec. 20.1.

Phase. It may happen that the system is known to be in a single quantum state, say state 1. Then the statevector has the form

$$\Psi = e^{i\phi_1}\psi_1, \tag{5.23}$$

where ϕ_1 is a real-valued phase.[14] In all other cases the statevector is a superposition of basis states. Because the complex-valued probability amplitudes c_n require not only magnitudes $|c_n|$ but phases ϕ_n for complete definition, the statevector generally incorporates phase relationships between its constituent coordinates. These have no counterpart in classical probability theory: the statevector generally is a *coherent superposition* of quantum states, as expressed by eqn. (5.18).

When one deals with degenerate quantum states, i.e. states sharing a common energy, the choice of unit vectors is somewhat arbitrary.[15] This flexibility in choosing a Hilbert-space coordinate system means, for example, that what appears to be a superposition of two states may, by suitably redefining basis states, appear as a single quantum state. Such coordinate transformations are an example of a traditional goal of physics: to find a simple way of expressing complicated behavior. I will discuss several examples.

---

[13] Section 7.2.2 discusses situations that violate this requirement.

[14] The overall phase of the statevector, ϕ_1 in this case, is not usually observed (but see Chap. 20). Thus it is usually convenient to omit it. As will be noted, this overall phase choice is related to the arbitrariness in setting the zero of energy.

[15] When external fields act on the atom the degeneracy is less (it is at least partially "lifted"). It is common to use the symmetry properties of such a field, expressed using the mathematics of group theory, to label the resulting quantum states. In the absence of any external field we are at liberty to use any convenient group representation to classify the degenerate states; each such classification leads to a different Hilbert-space basis of unit vectors.

Column vectors. It is often useful to present the vectors of Hilbert space as column vectors, writing

$$
\psi_1 = \begin{bmatrix} 1 \\ 0 \\ \vdots \end{bmatrix}, \quad \psi_2 = \begin{bmatrix} 0 \\ 1 \\ \vdots \end{bmatrix}, \quad \ldots, \quad \Psi = \begin{bmatrix} c_1 \\ c_2 \\ \vdots \end{bmatrix}. \tag{5.24}
$$

The scalar product $\langle A|B \rangle$ of two vectors, **A** and **B** (the projection of **A** onto **B**) is then the product of a row vector and a column vector. For vectors having complex numbers as components this construction reads

$$
\langle A|B \rangle = [A_1^*, A_2^*, \ldots] \begin{bmatrix} B_1 \\ B_2 \\ \vdots \end{bmatrix} \equiv A_1^* B_1 + A_2^* B_2 + \cdots = \langle B|A \rangle^*. \tag{5.25}
$$

The effect on the N probability amplitudes of a pulse of radiation can be expressed in matrix form as the action of an $N \times N$ matrix U,

$$
\begin{bmatrix} c_1' \\ c_2' \\ \vdots \end{bmatrix} = \begin{bmatrix} U_{11}c_1 + U_{12}c_2 + \cdots \\ U_{11}c_1 + U_{12}c_2 + \cdots \\ \vdots \end{bmatrix} = \begin{bmatrix} U_{11} & U_{12} & \cdots \\ U_{21} & U_{22} & \cdots \\ \vdots & \vdots & \ddots \end{bmatrix} \begin{bmatrix} c_1 \\ c_2 \\ \vdots \end{bmatrix} = \mathsf{U} \begin{bmatrix} c_1 \\ c_2 \\ \vdots \end{bmatrix}. \tag{5.26}
$$

The matrix U must preserve the overall probability $|c_1|^2 + |c_2| + \cdots$ and so it must be unitary. A succession of N pulses can be represented as the product of N such unitary matrices,

$$
\mathsf{U} = \mathsf{U}(N) \cdots \mathsf{U}(2)\mathsf{U}(1). \tag{5.27}
$$

5.6 Two-state Hilbert spaces

The simplest example of a Hilbert-space description is that of a two-state system (or two-level atom) [All87], as in the hydrogenic excitation example discussed in Sec. 5.4.1 above. We have two fixed unit vectors ψ_1 and ψ_2, and at any time the statevector must be some superposition of these, say

$$
\Psi = c_1 \psi_1 + c_2 \psi_2. \tag{5.28}
$$

The two corresponding probabilities P_1 and P_2 are the squared projections of the statevector upon the Hilbert-space axes,

$$
P_1 = |c_1|^2, \qquad P_2 = |c_2|^2. \tag{5.29}
$$

When dealing with quantum information such a system is known as a *qubit*; the states (unit vectors) are conventionally denoted $|0\rangle$ and $|1\rangle$. When treating spin-half particles the states are typically denoted $|\uparrow\rangle$ and $|\downarrow\rangle$, as is appropriate to spin pointing up or down. Figure 5.2 illustrates this abstract space. However, a two-dimensional picture cannot fully

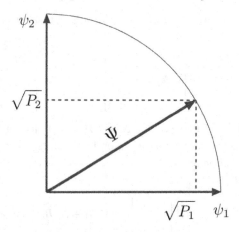

Figure 5.2 The statevector for two dimensions, showing projections onto the two unit vectors $\boldsymbol{\psi}_1$ and $\boldsymbol{\psi}_2$ that provide probability amplitudes $\sqrt{P_1}$ and $\sqrt{P_2}$. Although two real numbers – a single angle and fixed unit length – often suffice to describe a statevector, more generally a third variable is needed, to describe the relative phase between two projections, as discussed in the following section.

describe the full generality of the two complex-valued coordinates – it does not show the phase relationship between them. A three-dimensional picture, as in Sec. 5.6.1 below, can incorporate information about the relative phases of the two coordinates, though not an overall phase.

When we limit the number of quantum states to two the resulting description of pulsed excitation,

$$\begin{bmatrix} c_1' \\ c_2' \end{bmatrix} = \mathsf{U} \begin{bmatrix} c_1 \\ c_2 \end{bmatrix}, \tag{5.30}$$

is reminiscent of the Jones calculus of Sec. 3.6.1 used to describe the electrical field of an optical beam. The two complex-valued electric fields $\hat{\mathcal{E}}_R$ and $\hat{\mathcal{E}}_L$ used there are the counterparts of the complex-valued probability amplitudes c_1 and c_2.

5.6.1 The Bloch sphere

The Hilbert-space coordinates c_n are generally complex valued; we therefore require four real numbers to specify these projections of the statevector. However, the condition of unit norm, eqn. (5.20), provides one real number, so that only three real-valued parameters are required.

The challenge of finding a useful set is the same as that of representing a complex-valued electric field vector in the two-dimensional plane transverse to the propagation direction (see Sec. 3.6). As with eqn. (3.59) for mapping the complex unit vector $\mathbf{e}(\vartheta, \varphi)$ onto the Poincaré sphere we write the statevector of eqn. (5.28) as

$$\boldsymbol{\Psi} = \mathrm{e}^{-\mathrm{i}\zeta} \Big[\cos(\theta/2)\,\boldsymbol{\psi}_1 + \mathrm{e}^{-\mathrm{i}\varphi} \sin(\theta/2)\,\boldsymbol{\psi}_2 \Big], \tag{5.31}$$

by defining two angles through the expressions

$$c_1 = e^{-i\zeta}\cos(\theta/2), \quad c_2 = e^{-i\zeta-i\varphi}\sin(\theta/2). \tag{5.32}$$

The probabilities associated with this expression,

$$P_1 = \cos^2(\theta/2) = \frac{1}{2}[1+\cos\theta], \qquad P_2 = \sin^2(\theta/2) = \frac{1}{2}[1-\cos\theta], \tag{5.33}$$

are independent of the relative phase φ (and of the overall phase ζ), and sum to unity. The connection between the angle parametrization and the probability amplitudes can be seen from the following table.

θ	c_1	c_2
0	1	0
$\pi/2$	$1/\sqrt{2}$	$e^{i\varphi}/\sqrt{2}$
π	0	$e^{i\phi}$
$3\pi/2$	$-1/\sqrt{2}$	$e^{i\varphi}/\sqrt{2}$
2π	-1	0
4π	1	0

The two angles θ, φ represent a statevector as a point on a unit sphere, known in this context as the *Bloch sphere*[16] [Blo45; Ebe06; Sho90, §8.1].The south pole of this sphere represents population entirely in state 1, while the north pole represents population in state 2; the vertical axis is the excess of population in the excited state – the *population inversion*. The equator represents a phased 50:50 coherent superposition of the two states. Section 10.1 discusses this vector and its depiction; see Fig. 10.1.

The connection between the Bloch angles and the probability amplitudes c_n is often expressed through bilinear products of the two probability amplitudes (see also Chap. 16):

$$\begin{aligned} 2c_1c_1^* &= 2\cos^2(\theta/2) & = 1+\cos\theta = 2P_1, \\ 2c_2c_2^* &= 2\sin^2(\theta/2) & = 1-\cos\theta = 2P_2, \\ 2c_1c_2^* &= 2e^{i\varphi}\cos(\theta/2)\sin(\theta/2) = e^{i\varphi}\sin\theta. \end{aligned} \tag{5.34}$$

The Cartesian components of the system point on the Bloch sphere (the *Bloch vector*), defined by the angles used in constructing the statevector of eqn. (5.31), are

$$\begin{aligned} r_1 &= \sin\theta\cos\varphi = 2\mathrm{Re}\,c_1c_2^*, \\ r_2 &= \sin\theta\sin\varphi = 2\mathrm{Im}\,c_1c_2^*, \\ r_3 &= \cos\theta & = |c_2|^2 - |c_1|^2. \end{aligned} \tag{5.35}$$

---

[16] This abstract vector space has real-valued coordinates; it is an example of a state space rather than a Hilbert space. Mathematicians refer to the Bloch sphere, or the equivalent Poincaré sphere, as *Riemann spheres* that provide a "compactification of the complex-number line".

Chapter 10 illustrates the usefulness of the Bloch sphere for presenting a simple visualization of various pulse effects.

Although the picturing of the statevector and the elliptical polarization unit vector uses the same angular parameters, they produce dynamic changes differently. Polarization changes are produced by passing a beam through a set of discrete optical elements, each of which induces a finite incremental change in state space. Changes to the statevector, by contrast, take place continuously, under the dictates of a Hamiltonian operator, as discussed in Chap. 7. However, one can design a discrete set of pulses, each with a different Hamiltonian, that mimic the discrete optical elements.

5.6.2 Multiple states and the Bloch sphere

The general construction of a statevector involves N quantum states and a corresponding N-dimensional Hilbert space,

$$\Psi = c_1\psi_1 + c_2\psi_2 + \cdots + c_N\psi_N. \tag{5.36}$$

This specification involves N complex-valued amplitudes c_N, meaning $2N$ real numbers. The overall normalization requirement of eqn. (5.20) provides one number (unity), and the overall phase, unobservable, provides a second number. Thus we require $N-1$ pairs of real numbers. For this purpose it is possible to generalize the two Bloch-sphere angles (θ, φ) to a set of $N-1$ angle pairs,

$$(\theta_1, \varphi_1), \ldots, (\theta_{N-1}, \varphi_{N-1}).$$

In this way we expect that a statevector in N dimensions can be represented by $N-1$ points on the Bloch sphere. One way to associate the states of certain classes of N-state systems with sets of two-state systems follows from the use of a Morris–Shore (MS) transformation, as described in App. F. In the MS basis each point on the Bloch sphere moves independently, in accord with well-defined two-state equations. But their connection with the original N-state description is through a transformation that makes evolution indirectly associated with the original Hamiltonian.

More generally one can use a mathematical device used by Majorana to express the connection between a set of $2S$ particles, each a two-state system (and hence representable as spin one-half, see App. A.1.3), and a statevector having $2S+1$ Hilbert-space coordinates (and hence a representation of total spin S) [Mec58; Sch77; Mak10]. Applied to the present case the Majorana approach considers the $N = 2S+1$ roots x_n of the polynomial equation[17]

$$\sum_{n=1}^{N} c_n h_{n-1} x^{n-1} = 0. \tag{5.37}$$

---

[17] The statevector coefficients c_n have indices $n = 1, \ldots, N-1$. These are associated in the polynomial with powers $x^{N-1}, \ldots, x^0$.

Majorana chose the parameters h_n to be

$$h_n = \frac{(-1)^{n+1}}{\sqrt{(N-n)!(n+1)!}},$$ (5.38)

but other choices can be used (e.g. $h_n = 1$). Each complex-valued root Z_n can be expressed as a pair of angles,

$$Z_n = \tan(\theta_n/2) \, e^{i\varphi_n}.$$ (5.39)

Each of these defines a point on a unit sphere – the Bloch sphere. The complete description of the statevector consists of $N-1$ points on the Bloch sphere. This association, between the N probability amplitudes c_n and the set of $N-1$ Bloch angles, holds for any statevector.

The Majorana decomposition, eqn. (5.37) taken with the parameters of eqn. (5.38), is particularly useful when the N-state Hamiltonian is expressible using matrix representations of angular momentum, i.e. spin matrices, as discussed in Sec. 19.2 and App. A.1.3. With such a Hamiltonian the motion of the statevector in N dimensions has portrayal as a rotation, defined by three Euler angles.[18] The motions of the points on the Bloch sphere therefore undergo rotations defined by these same Euler angles. Thus for this class of N-state Hamiltonians, having SU(2) symmetry, there is a simple portrait of time evolution of $N-1$ points on the Bloch sphere [Hio87].

5.7 Time-dependent statevectors

When excitation occurs, the state of motion changes. There must be a corresponding time dependence of the statevector, indicated by including a time argument and writing

$$\mathbf{\Psi}(t) = c_1(t)\,\boldsymbol{\psi}_1 + c_2(t)\,\boldsymbol{\psi}_2 + \cdots + c_N(t)\boldsymbol{\psi}_N.$$ (5.40)

The changing statevector will have changing projections $c_n(t) = \langle \psi_n | \Psi(t) \rangle$ on the fixed Hilbert-space axes, thereby providing the time-dependent probabilities

$$P_n(t) = |\langle \psi_n | \Psi(t) \rangle|^2 = |c_n(t)|^2.$$ (5.41)

The individual-state probabilities must sum to unity,[19]

$$\sum_{n=1,N} P_n(t) = \sum_{n=1,N} |c_n(t)|^2 = 1.$$ (5.42)

It follows that the statevector has unit length at all times,

$$|\mathbf{\Psi}(t)|^2 \equiv |\langle \Psi(t) | \Psi(t) \rangle|^2 = \left| \sum_{n=1,N} \langle \Psi(t) | \psi_n \rangle \langle \psi_n | \Psi(t) \rangle \right|^2 = \sum_{n=1,N} P_n(t) = 1.$$ (5.43)

---

[18] In this context the Euler angles have no connection with rotations of objects in ordinary physical (Euclidean) space; they are simply three parameters.

[19] Section 7.2.2 discusses situations in which the sum is less than unity, but it can never exceed unity.

Thus any change to the statevector must be characterizable as a generalized rotation in Hilbert space.[20]

5.7.1 Moving coordinates for Bloch vectors

It is usually desirable to introduce a Hilbert space of moving (rotating) unit vectors $\psi'_n(t)$ in order to exhibit dynamical behavior most simply. Section 7.2.1 discusses this procedure. Expressed in such a rotating framework the statevector of eqn. (5.31) reads

$$\Psi(t) = C_1(t)\,\psi'_1(t) + C_2(t)\,\psi'_2(t)$$
$$= e^{-i\zeta}\left[\cos(\theta/2)\,\psi'_1(t) + e^{-i\varphi}\sin(\theta/2)\,\psi'_2(t)\right]. \qquad (5.44)$$

In this frame the Bloch vector has components

$$r_1(t) = 2\mathrm{Re}\,C_1(t)C_2(t)^*, \quad r_2(t) = 2\mathrm{Im}\,C_1(t)C_2(t)^*, \quad r_3(t) = |C_2(t)|^2 - |C_1(t)|^2,$$
$$(5.45)$$

and the formulas of the preceding paragraphs apply with the replacement $c_n \to C_n(t)$. It is this replacement that subsequent discussions of the Bloch vector, and its motion, will assume.

5.7.2 Geometric phase and the Bloch sphere

As with the Poincaré portrait of the polarization state of a laser beam, the Bloch sphere provides a picture of a two-state quantum system. This can be a snapshot at a particular instant, when the system is depicted as a point on the Bloch sphere. We can also follow changes to the system, as controlled by the experimenter manipulating a laser field. The changes to the system appear as a curve on the surface of the sphere. Under some conditions the system point can be brought back to its initial location on the Bloch sphere, having traced a closed curve. Although all of the Bloch variables have returned to their initial values, these give no information about the overall phase of the statevector. Indeed, during the process this acquires a *geometric phase* proportional to the solid angle subtended on the Bloch sphere by the tip of the Bloch vector. This geometric phase (known as the *Berry phase* when the evolution is adiabatic) is observable by interferometry, in which the two-state system is extracted from a larger system, modified, and then returned to the larger system. Under such manipulations the overall phases of probability amplitudes have measurable effects. Chapter 20 discusses the statevector phase and its measurement.

---

[20] The motion may appear as a rotation in the full N-dimensional Hilbert space, but the overall effect of pulsed change may appear as a reflection within a subspace.

5.7.3 Typical goals

Traditional goals for pulsed excitation include the production of complete population transfer, from the initially populated state 1 to the excited state 2,

$$\Psi = \psi_1 \rightarrow e^{i\phi} \psi_2. \tag{5.46}$$

When the phase ϕ is zero this corresponds to a rotation of the statevector by 90 degrees (a rotation of the Bloch angle θ by π). Alternatively, we might wish to maximize the induced dipole moment (for use with nonlinear optics, see App. C.2), by producing the coherent superposition

$$\Psi = \psi_1 \rightarrow \beta\psi_1 + \alpha\psi_2, \qquad |\alpha|^2 + |\beta|^2 = 1. \tag{5.47}$$

For null phase this corresponds to a rotation of the statevector by $45°$ (a rotation of the Bloch angle by $90°$).

Each of these objectives can be obtained by producing a predetermined portion of a Rabi oscillation. However, other procedures are possible, often providing advantages. In particular, alternative schemes based upon adiabatic changes are much less sensitive to details of the pulse; they are basically independent of the time integral of the Rabi frequency, for example, and are therefore termed *robust*.[21] Later sections will discuss some of these alternatives.

Although population changes have traditionally drawn attention, the more detailed manipulation of quantum states needed for quantum information processing requires the ability to alter phases. One might then wish to produce the change

$$\Psi = \psi_1 \rightarrow e^{i\phi} \psi_1. \tag{5.48}$$

Such a phase change only becomes observable when there are several states; it might be implemented in a two-state system as the alteration

$$\Psi = c_1\psi_1 + c_2\psi_2 \rightarrow e^{i\phi}c_1\psi_1 + c_2\psi_2. \tag{5.49}$$

This is a special case of the general two-state manipulation indicated in eqn. (5.30),

$$\Psi = c_1\psi_1 + c_2\psi_2 \rightarrow c_1'\psi_1 + c_2'\psi_2. \tag{5.50}$$

Fidelity. To quantify the success of any pulse in producing a particular result we require a measure of how close the actual statevector Ψ_{act} is to the desired statevector Ψ_{des}. A common measure is the *fidelity*

$$\mathcal{F} = |\langle\Psi_{act}|\Psi_{des}\rangle|^2. \tag{5.51}$$

---

[21] Barabasi [Bar03] notes that the word robustness is rooted in the Latin word *robus*, meaning "oak", the symbol of strength and longevity in the ancient world. In the present context robustness implies relative stability despite disruptive events, a resilient design that is able to survive in a wide range of conditions despite high error rate.

The *infidelity* is $\mathcal{I} = 1 - \mathcal{F}$. Although traditional applications of coherent excitation to chemistry often need only infidelities of 10^{-1}, more recent applications to quantum-information processing require infidelities of 10^{-4} or less.

5.7.4 Equations for changes

Changes to a quantum system are described by differential equations, either for probabilities or for probability amplitudes. Several forms of these equations have merit; the choice depends, in part, upon which of several regimes of laser intensity holds interest. The coherent-excitation regimes, and the associated idealized physical descriptions, fit generally into three classes:

Weak: In this regime an individual pulse has very little effect upon the initial probability distributions; it represents a *perturbation* of the initial state, and can be treated by time-dependent perturbation theory [Sho90, §4,5] [Won76; Lan77; Pal06b], as mentioned in Chap. 9.

Strong: In this regime there occurs a significant change in probability distributions; Rabi oscillations, discussed in the following sections, are an example. Perturbation theory fails, and one must find alternative approaches. This is the regime treated in the present monograph, based on equations of motion for a few selected quantum-state probabilities.

Ultrastrong: When the pulse is sufficiently intense, the electric field of the laser overwhelms the Coulomb field that binds the electrons to the nucleus, and photoionization can occur within a few optical cycles.[22] The use of a few essential bound quantum states, and their equations of motion, no longer provides a satisfactory description and new models, alternative approaches, are needed [Kul06], often based on numerical evaluation of wavefunctions. I will not discuss this regime.

5.8 Picturing quantum transitions

Probabilities governed by differential equations change continuously with passing time. Observable transitions can be said to take place over an interval of time. If the interval is short, the probability of observing a transition is negligible, while if the interval is sufficiently long it is certain that a transition will have occurred. At intermediate times there will be some intermediate probability for observing a transition. The scale of times – the meaning of "short" or "long" time – is set by the coefficients of the differential equations. When change occurs by spontaneous emission (see Chap. 6) this is the Einstein A coefficient. For coherent radiation the time scale for change is set by the Rabi frequency (see Chap. 7). In order to observe coherent excitation it is necessary that the Rabi frequency be much larger than any incoherent rates.

---

[22] Under suitable conditions stabilized bound states may exist even in such very strong fields [Ebe93; Pro97].

5.8.1 Quantum jumps

The quantum theory presented in this monograph allows the prediction of population histories $P_n(t)$, i.e. the probability that, in an ensemble of atoms viewed at time t after their initial preparation, an atom will be found to have been excited into the nth discrete excited state. This is a continuous function of time, exhibiting such characteristics as damped oscillations. However, any measurement of single-atom energy can reveal only one of the discrete values E_n, and so the changes in energy of any atom can only occur as discrete increments. One can therefore regard the changes, as did earlier proponents of quantum theory, as discrete *quantum jumps*, in which an atom changes from one discrete quantum state to another [Zol87; Bla88; Coo90; Ple98]. In examining this viewpoint we must keep in mind that what quantum theory deals with is information, in the form of probabilities. The statevector embodies this information. Our information about an atom changes suddenly when we make a measurement – this altered information is the quantum jump. By stressing the probability $P_n(t)$ rather than the measurement process the present monograph has no call to discuss quantum jumps.

5.8.2 Wavefunction collapse

There are various ways to interpret quantum-mechanical change, whether coherent or incoherent. One can, as I have above, regard the process as being described by differential equations whose solutions predict observables at a given time. These vary continuously, and describe, for example, the number of atoms that have undergone a discrete change of energy during time t.

Alternatively, one can regard the quantum system as being in the initial state until a measurement is made, at which time it undergoes an instantaneous transition. This viewpoint underlies treatments involving "the collapse of the wavefunction": a wavefunction embodies information about a quantum system and as that information changes by the arrival of measurement data so too does the wavefunction.

6

Incoherence: Rate equations

Prior to the widespread use of laser light for manipulation of atomic structure the notion of local thermodynamic equilibrium (LTE) – a topic within astrophysics – dominated theoretical treatments of atoms illuminated by radiation. In strict thermodynamic equilibrium all macroscopic properties of radiation remain unchanging and deducible from knowledge of temperature. Of more practical interest are systems in which parts of the system may have LTE properties (say radiation) while other parts (say atoms) change with time as their characteristics move towards those of thermodynamic equilibrium.

The simplest of the atomic properties to characterize are the excitation probabilities. The present section, drawing on [Sho90, Chap. 2], discusses both the equilibrium values and the equations that describe changes – *rate equations*. Although these equations do not apply when the excitation is coherent, as occurs for short pulses of laser light, they provide an important context for radiative excitation in general and for various parameters with which the strength of the radiative interaction with matter has been characterized.

6.1 Thermalized atoms; the Boltzmann equation

When a quantum system is in thermal equilibrium with surroundings at temperature T then the probability of finding it in a state with energy E_n is constant and given by the *Boltzmann equation*,

$$P_n = \varpi_n \exp(-E_n/k_B T)/\mathcal{Z}(T), \tag{6.1}$$

where k_B is the Boltzmann constant, ϖ_n is the *statistical weight* (the degeneracy, or number of quantum states that share the same energy), and $\mathcal{Z}(T)$ is the partition function,

$$\mathcal{Z}(T) = \sum_n \varpi_n \exp(-E_n/k_B T). \tag{6.2}$$

For atoms that are in thermodynamic equilibrium two energy states are equally populated only if they have equal energies or if the temperature is very high. An excited state can never have more population than a lower-energy state of equal statistical weight (a population inversion): this would require a negative temperature. Laser excitation, though possibly producing an unchanging population, is not thermodynamic equilibrium. As will be shown with numerous examples, it can produce population inversion.

6.2 The radiative rate equations

The early discussions of radiative transitions followed the approach of Einstein, who postulated that one could describe the effects of resonant radiation by means of simple rate equations that express rates of change of number densities as being the difference between gains and losses [Sho90, §2.2]. The resulting rate equations provide a set of coupled ODEs for the probabilities $P_n(t)$, having the form

$$\frac{d}{dt}P_n(t) = \sum_m P_m(t)\mathcal{R}_{m\to n}(t), \qquad (6.3)$$

where the *rate coefficient* $\mathcal{R}_{m\to n}(t)$, is the rate of change in state n produced by population arriving from state m, i.e. the transition probability per unit time for $m \to n$. The rate coefficients, possibly time dependent, incorporate the three radiative mechanisms postulated by Einstein and itemized in Sec. 6.3 below. These coupled linear ODEs take simplest form when we place the probabilities into a column vector

$$\mathbf{P}(t) = \begin{bmatrix} P_1(t) \\ P_2(t) \\ \vdots \end{bmatrix}. \qquad (6.4)$$

The equations then appear as a matrix equation

$$\frac{d}{dt}\mathbf{P}(t) = \mathsf{R}(t)\mathbf{P}(t), \qquad (6.5)$$

where $\mathsf{R}(t)$ is the square matrix of time-dependent rate coefficients.[1]

These equations pertain to reactions in which only a single species is affected. The right-hand side involves only sums of probabilities; the equations are linear. Chemical reactions involve processes in which changes occur proportional to products of two or more populations; the resulting rate equations are nonlinear.

6.3 The Einstein rates

Einstein postulated that three radiative mechanisms contribute to population changes:

I. *Absorption* of radiation, proportional to the radiation energy density;

II. *Stimulated emission*, proportional to the radiation energy density;

III. *Spontaneous emission* of radiation, independent of existing radiation and parametrized by the Einstein A coefficient A_{21}.

Rather than deal with radiation beams, and irradiance or intensity, Einstein considered steady broadband spectral radiation energy density in a cavity, $u(\nu)$ (with dimension of

---

[1] The matrix element R_{nm} is the rate $\mathcal{R}_{m\to n}$.

energy per unit volume per Hz), and wrote the rate of excitation, from state 1 to state 2, in the form

$$\mathcal{R}_{1 \to 2} = B_{12} u(v). \tag{6.6}$$

The radiative deexcitation transitions, $2 \to 1$, combining stimulated and spontaneous emission, were postulated to occur at the rate

$$\mathcal{R}_{2 \to 1} = B_{21} u(v) + A_{21}. \tag{6.7}$$

The connection between the Einstein A and B coefficients, deduced originally from thermodynamic arguments, is

$$B_{21} = \frac{(\lambda_{12})^3}{4\hbar} A_{21}, \tag{6.8}$$

where $\lambda_{12} = 2\pi c / \omega_{12}$ is the resonance wavelength of the transition.

Beam equations. Although rate equations, and A and B coefficients, remain widely used, the original equations require modification for use with narrow-bandwidth and directed radiation. This proceeds by considering the absorption of a directed beam, such as eqn. (3.10). The relevant rate equation for the incremental reduction of steady intensity along the propagation axis z, through cold matter, is

$$\frac{d}{dz} I(z) = -\kappa I(z), \tag{6.9}$$

where κ is the absorption (or attenuation) coefficient. This linear attenuation rate gives intensity that diminishes exponentially with distance,

$$I(z) = \exp(-\kappa z) I(0), \tag{6.10}$$

a behavior known variously as *Beer's law*, the *Beer–Lambert law*, or the *Beer–Lambert–Bouguer law*. The product of absorption coefficient and distance, κz, is the *optical depth*. The inverse of the absorption coefficient of eqn. (6.9), $1/\kappa$, is sometimes termed the *Beer's length*.

Absorption cross section. Resonant excitation from state 1 to state 2 occurs when a pulse, of carrier frequency ω, passes through matter. Let the number density of absorbers (atoms in state 1 per unit volume) be $\mathcal{N}_1$. In the absence of any previous excitation (i.e. $\mathcal{N}_2 = 0$), the linear absorption coefficient $\kappa(\omega)$, with dimensions of inverse length, is expressible in terms of an *absorption cross section* $\sigma(\omega)$, with dimensions area per atom,

$$\kappa(\omega) = \mathcal{N}_1 \frac{\sigma(\omega)}{\hbar\omega}. \tag{6.11}$$

In turn, the explicit frequency dependence can be placed into a single function, either $s(\omega)$ or $s(v) = s(\omega)/2\pi$, by writing

$$\sigma(\omega) = \sigma^{tot} s(v), \qquad \text{with} \quad \int_0^\infty dv \, s(v) = 1. \tag{6.12}$$

The frequency-integrated cross section can be written in terms of Einstein A and B coefficients as

$$\sigma^{tot} = \frac{(\lambda_{12})^2}{8\pi} A_{21} = \frac{\hbar\omega_{12}}{c} B_{21}. \tag{6.13}$$

Appendix H defines additional parameters, including oscillator strength, with which to express the ability of an atom to absorb or emit radiation.

6.4 The two-state rate equations

The radiative rate equations can be expressed in several ways. Expressed in terms of laser intensity $I(t)$ the relevant equations for a two-state atom subjected to narrow-bandwidth near-resonant radiation read

$$\frac{d}{dt} P_1(t) = -B_{12}s(v)\frac{I(t)}{c} P_1(t) + \left[A_{21} + B_{21}s(v)\frac{I(t)}{c} \right] P_2(t), \tag{6.14a}$$

$$\frac{d}{dt} P_2(t) = -\frac{d}{dt} P_1(t). \tag{6.14b}$$

The radiative rates appearing here can be expressed in several ways, e.g.

$$\mathcal{R}_{1\rightarrow 2}(t) = B_{12}s(v)\frac{I(t)}{c} = A_{21}\bar{n}(t) = \sigma(\omega)\frac{I(t)}{\hbar\omega}, \tag{6.15a}$$

$$\mathcal{R}_{2\rightarrow 1}(t) = A_{21} + B_{21}s(v)\frac{I(t)}{c} = A_{21}[1+\bar{n}(t)] = A_{21}\left[1 + \frac{I(t)}{I^{sat}}\right]. \tag{6.15b}$$

Here $\bar{n}(t)$ is the mean number of interacting photons, i.e. the photon flux $I(t)/\hbar\omega$ within the cross section $\sigma(\omega)$,

$$\bar{n}(t) = \frac{B_{21}s(v)I(t)}{cA_{21}} = \frac{s(v)I(t)\lambda^2}{4\hbar\omega} = \sigma(\omega)\frac{I(t)}{\hbar\omega}, \tag{6.16}$$

and I^{sat} is the *saturation intensity* appropriate to the carrier frequency ω,

$$I^{sat} = \frac{\hbar\omega A_{21}}{2\sigma(\omega)} = \frac{cA_{21}}{2B_{21}s(v)} = \frac{2\hbar c}{\lambda^3 s(v)}. \tag{6.17}$$

6.5 Solutions to the rate equations

The effects of radiation take simpler form when expressed in terms of the population inversion[2]

$$w(t) \equiv P_2(t) - P_1(t). \tag{6.18}$$

---

[2] The notation $w(t)$ for population inversions comes from the Feynman–Vernon–Hellwarth model of two-state dynamics, involving variables $u(t)$, $v(t)$, and $w(t)$, as discussed in Chap. 10.

With the assumption $B_{12} = B_{21}$ appropriate for nondegenerate transitions ($\varpi_n = 1$) the rate equation for population inversion reads

$$\frac{d}{dt} w(t) = -\mathcal{R}(t)\, w(t) - A_{21}, \tag{6.19}$$

where

$$\mathcal{R}(t) = \mathcal{R}_{1 \to 2}(t) + \mathcal{R}_{2 \to 1}(t) = A_{21}[1 + 2\bar{n}(t)]. \tag{6.20}$$

The solution to this equation is

$$w(t) = [w(0) - w(\infty)] \exp\left[-\int_0^t dt'\, \mathcal{R}(t')\right] + w(\infty). \tag{6.21}$$

For steady illumination this solution monotonically approaches the asymptotic value

$$w(\infty) = \frac{-A_{21}}{A_{21} + 2B_{21}s(v)I(t)} = \frac{-1}{1 + I(t)/I^{sat}} = \frac{-1}{1 + 2\bar{n}(t)}. \tag{6.22}$$

That is, the populations *saturate* at constant values. The inversion approaches the final equilibrium value $w(\infty)$ at a rate that increases with intensity $I(t)$ (or mean photon number $\bar{n}$) but which is never less than the spontaneous emission rate A_{21}. The instantaneous excitation, at time t, depends on the time-integrated intensity (known as pulse *fluence*),

$$\int_0^t dt'\, \mathcal{R}(t') = A_{21}t + \frac{\sigma(\omega)}{\hbar\omega} \int_{-\infty}^t dt'\, I(t'). \tag{6.23}$$

This behavior contrasts with that of coherent excitation, for which the relevant parameter is the time integral of the electric field amplitude; see Sec. 8.3.1.

Saturated inversion. The final saturated inversion $w(\infty)$ depends upon the ratios of the several rates, but if $B_{12} = B_{21}$, as is the case for nondegenerate transitions, then the inversion can never exceed $w(\infty) = 0$, meaning in turn that the excitation probability P_2 can never exceed $P_2 = 0.5$. This value is only approached if the stimulated rate is much larger than the spontaneous rate, $B_{21}s(v)I(t) \gg A_{21}$ or, equivalently, $\bar{n} \gg 1$, as will occur for sufficiently high intensity $I(t)$. Figure 6.1 illustrates the behavior.

The final equilibrium ratio of the two populations is

$$\frac{P_2(\infty)}{P_1(\infty)} = \frac{1 + w(\infty)}{1 - w(\infty)} = \frac{\bar{n}}{\bar{n} + 1}. \tag{6.24}$$

This can never exceed unity: there can never be more population in the excited state than in the ground state when rate equations apply.

We can attribute the equilibrium populations to a Boltzmann distribution by writing, for nondegenerate energy levels,

$$\frac{P_2(\infty)}{P_1(\infty)} = \exp[-(E_2 - E_1)/k_B T]. \tag{6.25}$$

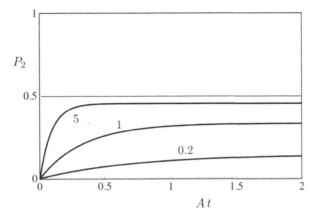

Figure 6.1 Excitation probability P_2 vs. decay lifetimes At following sudden exposure of an unexcited atom to steady illumination for various intensities of the radiation, as measured by the mean photon number, $\bar{n} = 0.2, 1, 5$. The populations saturate at a value dependent on $\bar{n}$. Once the radiation ceases, all populations return to the ground state, at the spontaneous emission rate A. [Redrawn from Fig. 2.2-2 of B. W. Shore, *The Theory of Coherent Atomic Excitation*. Copyright Wiley-VCH Verlag GmbH & Co. 1990. Reproduced with permission.]

Here too we see that, because $E_2 > E_1$, the equilibrium population cannot be inverted. When the excitation is by laser pulses inversion can occur; the relevant equations are not rate equations but the time-dependent Schrödinger equation of Chap. 7.

6.6 Comments

Radiative rate equations, supplemented with additional rates for collision-induced population changes, have been adequate for the description of numerous environments in which the radiation is incoherent. Examples include hot dense gases or the atmospheres of stars [Mih78; Car07]. But they are not applicable to excitation of isolated atoms by laser radiation. Such radiation has a long coherence time; during shorter intervals its effects cannot be described adequately by rate equations that assume random environmental fluctuations. Instead we require the apparatus of quantum mechanics, not only to explain the existence of discrete energy levels but also to describe the changes induced by coherent radiation; see Chap. 7.

Although the Einstein B coefficients play no direct role in the description of excitation by coherent light, the A coefficient descriptive of spontaneous emission remains of importance. For isolated atoms in free space the possibility of radiative excitation to a selected quantum state implies an ever-present possibility of spontaneous emission, with a lifetime equal to the inverse of the sum of all possible spontaneous emission rates from that state. Only laser-induced procedures that are significantly shorter in duration than the spontaneous-emission lifetime will be considered in most of the present monograph (an exception is the discussion of optical pumping, Sec. 12.4). Because the spontaneously radiated field is that of a single photon – a single increment of an electromagnetic field mode – these rates can be altered by modifying the field modes, say by placing the atom in a suitable cavity [Pur46; Goy83], or into a solid whose structure has been designed to allow only selected radiation modes to propagate (a *photonic crystal*) [Joa95; Ang01]. Such techniques allow extension of coherent excitation to various discrete states of solids.

7

Coherence: The Schrödinger equation

Information about a quantum system that remains free of randomizing disturbances[1] resides not with probabilities but within a statevector Ψ, as discussed in Sec. 5.5. Changes to the system produced by a pulse of radiation correspond to changes of the statevector in its Hilbert-space setting. Because the statevector must maintain unit length,[2] the changes can be expressed as the linear relationship

$$\Psi \rightarrow \Psi' = \mathsf{U}\Psi, \tag{7.1}$$

in which U is a unitary matrix (e.g. rotations or reflections). When the change occurs with passing time U is the matrix representation of the *time-evolution operator* or *propagator*.

This chapter presents the basic formalism with which we describe these coherent changes, starting from the time-dependent Schrödinger equation. The discussion is quite general, applicable to any coherent interaction. Subsequent chapters examine particular systems, particular multipole interactions, and particular classes of pulses.

The differential equation. To obtain the elements of U it is best to follow incremental changes and to consider a time-dependent statevector $\Psi(t)$. The effects of coherent changes are then described by an equation of motion for $\Psi(t)$. The operator responsible for these incremental changes to the statevector with passing time is the Hamiltonian energy-operator $\hat{H}(t)$. The relevant equation of motion, governing the changes induced in the system by laser pulses, as represented by motion of the statevector in Hilbert space, is the time-dependent Schrödinger equation (TDSE),[3]

$$\frac{d}{dt}\Psi(t) = -\frac{i}{\hbar}\mathsf{H}(t)\Psi(t), \tag{7.2}$$

---

[1] The succession of measurements on any quantum system produces random results, unless the system is in an eigenstate of the measurement operator. The randomness mentioned here is additional beyond that intrinsic randomness of quantum measurement, and refers to the Hamiltonian itself having random fluctuations.

[2] Section 8.3.2 discusses situations that violate this requirement.

[3] Traditionally the time derivative appears in the TDSE as a partial derivative, $\partial/\partial t$, to allow the inclusion of spatial coordinates; the equation then applies to a wavefunction $\Psi(\mathbf{r}, t)$.

involving, on the right-hand side, the imaginary unit i and the time-dependent Hamiltonian, the matrix $H(t)$. The Dirac constant $\hbar$ serves merely to convert energy units into angular frequency units.[4]

We wish to integrate this differential equation, i.e. to obtain the time-dependent statevector $\Psi(t)$, an N-dimensional column vector in the Hilbert space of physical states. To do so we must solve the differential equation subject to initial conditions that specify the statevector at some time t_0, typically taken either as $t_0 = 0$ or $t_0 \to -\infty$, before the laser field appears. Usually the required condition is that the statevector should initially align with one of the unit vectors (i.e. it should represent a single unperturbed state), typically ψ_1 associated with the quantum state having energy E_1. More generally the initial condition may require specifying the statevector as a coherent superposition of basis states.

The Hamiltonian. When used to describe laser-induced changes, the Hamiltonian has two parts:

$$H(t) = H^{at} + V(t). \tag{7.3}$$

The first, constant, contribution, H^{at}, incorporates the energies of the (*bare*) atom in the absence of the laser radiation. The Hilbert-space unit vectors are eigenvectors of this operator; they correspond to energy eigenstates of the (bare) isolated-atom Hamiltonian,

$$H^{at}\psi_n = E_n\psi_n. \tag{7.4}$$

Its eigenvalues are the observable energies, $E_1, E_2, \ldots$, of the free, undisturbed atom. Its eigenvectors ψ_n correspond with physically distinguishable quantum states, i.e. *physical states*. Thus, by construction, the matrix H^{at} is *diagonal* (in the basis of physical states); the elements of H^{at} are the unperturbed energies of the system, E_n.

The matrix elements of the interaction $V(t)$ can, in principle, be obtained from the wave-functions of the system. Suppose, for example, that we deal with a single electron, and that the interaction energy of this electron with the field at position $\mathbf{r}$ and time t is $V(\mathbf{r}, t)$. Using the discrete wavefunctions $\psi_n(\mathbf{r}) \equiv \langle \mathbf{r}|\psi_n\rangle$ one can evaluate the matrix elements of the interaction Hamiltonian as the integral

$$V_{nm}(t) = \int d\mathbf{r}\,\psi_n(\mathbf{r})^* V(\mathbf{r}, t)\psi_m(\mathbf{r}). \tag{7.5}$$

More generally the integral will involve three spatial coordinates for every particle, as well as summation over spin degrees of freedom. As noted in the following section, the interaction typically allows parametrization in terms of experimentally measurable quantities (e.g oscillator strengths and intensities) and so it is not necessary to have any wavefunctions.

The diagonal elements of the interaction matrix $V(t)$ embody shifts of the observable energies. Those that derive from electric fields are Stark shifts; when they vary with time they are dynamic Stark shifts. The energy shifts produced by magnetic fields are Zeeman

---

[4] Many authors choose time scales such that they set $\hbar = 1$.

shifts. The off-diagonal elements of $V(t)$, for which $n \neq m$, are responsible for transitions between quantum states.

It is important to recognize that, unlike rate equations, the Schrödinger equation treats single quantum states, not degenerate states. To treat degeneracy when describing coherent excitation it is necessary to treat each quantum state separately; see Sec. 16.2.

The Bohr transition frequency. To produce significant excitation the energy of a photon must match the energy difference of two quantum states. That is, to induce a transition from state n to state m the carrier frequency of the field, ω, must match a Bohr transition frequency ω_{nm}, as defined by the equation

$$\hbar\omega_{nm}(t) = |E_n + V_{nn}(t) - E_m - V_{mm}(t)|. \tag{7.6}$$

This definition incorporates into the transition frequency the effect of any energy shifts $V_{nn}(t)$ (Stark or Zeeman) produced by electric or magnetic fields. When the interaction V is time dependent, then the Bohr frequency shares that time dependence. The equality of the two frequencies is termed the resonance condition,

$$\omega(t) = \omega_{nm}(t). \tag{7.7}$$

7.1 Essential states; effective Hamiltonians

Only if the carrier frequency nearly matches a Bohr frequency will any appreciable population ever appear in a quantum state other than those initially populated (see Sec. 10.2.4). Thus although there may be an infinite number of quantum states, very few of those participate in the excitation dynamics. We can therefore restrict attention to a finite subspace of the infinite Hilbert space, say one of N dimensions – the N *essential states*. In this subspace the statevector is the summed superposition of eqn. (5.40),

$$\Psi(t) = c_1(t)\psi_1 + c_2(t)\psi_2 + \cdots + c_N(t)\psi_N. \tag{7.8}$$

The probability $P_n(t)$ of finding the system in state n at time t is the absolute square of the projection $\langle \psi_n | \Psi(t) \rangle$ of the statevector onto the Hilbert-space coordinate axis ψ_n,

$$P_n(t) = |\langle \psi_n | \Psi(t) \rangle|^2 = |c_n(t)|^2. \tag{7.9}$$

The expansion coefficient $c_n(t)$ is therefore the probability amplitude.

The expansion displayed in eqn. (7.8) holds in a fixed Hilbert space, spanned by N unit vectors ψ_n whose components are the time-varying complex numbers $c_n(t)$. Although we speak of N dimensions, each of the coordinate vectors can be considered as comprising a two-dimensional space in the complex plane; we require $2N$ real numbers to specify the location of the statevector.

Nonessential states. Although the Bohr resonance condition of eqn. (2.3) may restrict appreciable population to only two "essential" states, this does not mean that the remaining "nonessential" states have no effect on the dynamics. Even with minuscule excitation probability these nonessential states can have two important effects, each based on regarding them as nonresonant intermediate states in a chain of transitions that is overall resonant. Both effects can be treated by means of an "effective" Hamiltonian H^{eff} [Sho90, §14.8]. Section 13.4.4 discusses how such an effective two-state Hamiltonian arises from a three-state system when the field carrier frequency differs appreciably from the Bohr frequency (i.e. the detuning is large). Section 13.4.5 discusses more general examples. In all cases the elements of H^{eff} have the following interpretation.

Off-diagonal elements. The off-diagonal elements of H^{eff} describe *multiphoton transitions,* the simplest of which are two-photon transitions, such as occur with stimulated Raman scattering (SRS) when the excited state is nonresonant. Much effort dealt with calculations of these multiphoton transition rates during the early decades of laser usage [Fai87; Mai91; Del00].

Diagonal elements. The diagonal element H_{nn}^{eff} of H^{eff} can be regarded as the result of transitions from an initial state n into an excited state, say m, and back again. Let the difference of the Bohr frequency ω_{nm} for this transition and the laser frequency ω be $\Delta = \omega_{nm} = \omega$. From a version of the energy–time uncertainty principle,

$$\Delta \times \tau > \hbar/2, \tag{7.10}$$

one expects that, as the detuning Δ grows large (in units of Rabi frequency), the duration of excitation τ must become small (in units of Rabi cycles). These processes, though producing negligible excitation, produce an induced dipole moment. In turn, this induced moment, parametrized as a *polarizability* of the atom (see Sec. 13.4.5), interacts with the laser electric field to produce an energy shift – a dynamic Stark effect.

7.2 The coupled differential equations

The probability amplitudes $c_n(t)$ of eqn. (7.8) must be chosen such that the resulting state-vector satisfies the TDSE, eqn. (7.2). To assure this behavior we substitute the construction (7.8) into eqn. (7.2) and require that the resulting equation be fulfilled along each of the N independent coordinates. We thereby obtain a set of N coupled ODEs for the probability amplitudes,

$$\frac{d}{dt}c_n(t) = -\frac{i}{\hbar}\sum_m H_{nm}(t)c_m(t), \qquad n, m = 1, \ldots N. \tag{7.11}$$

Notably the equations are *linear* in the unknowns $c_n(t)$; nonlinear equations would involve higher powers of the unknowns on the right-hand side. The solutions are to be normalized such that the sum of the individual probabilities sums to unity,

$$\sum_n |c_n(t)|^2 = 1. \tag{7.12}$$

To complete our definition of the dynamics described by the Schrödinger equation we must specify initial conditions for the dynamical variables $c_n(t)$. We will usually suppose that at an initial time $t = t_i$, prior to the presence of the interaction $V(t)$, the atom is in state 1, as specified by the initial probability amplitudes

$$c_1(t_i) = 1, \qquad c_2(t_i) = \cdots = c_N(t_i) = 0. \tag{7.13}$$

More generally there may occur several nonzero probability amplitudes – the quantum state is then a superposition state. It is these equations, (7.11)–(7.13), and their solutions, that will occupy our attention throughout the remainder of this monograph.

Like eqn. (6.5), the coupled equations of the TDSE comprise a set of linear ODEs. These differ from the rate equations by the presence of the imaginary unit i. This difference appears in the solution forms: rate equations have exponentials where the Schrödinger equation has sinusoids.

7.2.1 Rotating coordinate system

In the absence of any interaction $V(t)$ the equations for the probability amplitudes are uncoupled. Each has the form

$$\frac{d}{dt} c_n(t) = -\frac{i}{\hbar} E_n c_n(t). \tag{7.14}$$

These have the simple solutions

$$c_n(t) = \exp(-i E_n t/\hbar) \, c_n(0), \tag{7.15}$$

and thus in this situation the N-dimensional statevector at time t is expressible as the superposition

$$\Psi(t) = \sum_{n=1}^{N} \exp(-i E_n t/\hbar) \, c_n(0) \, \psi_n. \tag{7.16}$$

This construction can be regarded as expressing a fixed superposition,

$$\Psi(t) = \sum_{n=1}^{N} c_n(0) \, \psi'_n(t), \tag{7.17}$$

of N time-dependent basis vectors $\psi'_n(t)$,

$$\psi'_n(t) \equiv \exp(-i E_n t/\hbar) \, \psi_n. \tag{7.18}$$

Each of these unit vectors rotates, with constant angular velocity $E_n/\hbar$, in a complex-number plane. To visualize the resulting Hilbert-space motion we can express the unit vectors as two-dimensional axes in the complex plane $\psi_n = \psi_n^R + i\psi_n^I$ and, with the abbreviation $\omega_n = E_n/\hbar$, write

$$\psi'_n(t) = \left[\cos(\omega_n t)\psi_n^R - \sin(\omega_n t)\psi_n^I \right] - i \left[\sin(\omega_n t)\psi_n^R - \cos(\omega_n t)\psi_n^I \right], \tag{7.19}$$

thereby displaying explicitly the steady turning of the rotating coordinate axes. In such a rotating coordinate system the expansion coefficients $c_n(0)$ remain fixed, in the absence of any interaction $V_{nm}(t)$.

More generally, for a Hamiltonian that includes time-varying interaction it is often advantageous to use a rotating Hilbert-space reference frame to present the description of the statevector. We introduce these by writing the N-dimensional statevector as the superposition[5]

$$\Psi(t) = \sum_{n=1}^{N} \exp[-i\zeta_n(t)] \, C_n(t) \, \boldsymbol{\psi}_n, \tag{7.20}$$

where the $\zeta_n(t)$ are phases, generally time dependent, to be fixed. The probabilities are independent of the phases,

$$P_n(t) = |\langle \psi_n | \Psi(t) \rangle|^2 = |C_n(t)|^2 = |c_n(t)|^2, \tag{7.21}$$

but other observables, such as dipole moments, exhibit them, as will be noted below.

The statevector construction of eqn. (7.20) can be expressed as a superposition of rotating coordinates,

$$\Psi(t) = \sum_n C_n(t) \, \boldsymbol{\psi}'_n(t), \qquad \boldsymbol{\psi}'_n(t) \equiv \exp[-i\zeta_n(t)] \boldsymbol{\psi}_n, \tag{7.22}$$

that, at any time t, retain the orthonormality properties of the fixed basis,

$$\langle \psi'_n(t) | \psi'_m(t) \rangle = \langle \psi_n | \psi_m \rangle = \delta_{nm}. \tag{7.23}$$

Expressed as column vectors the construction uses the identifications

$$\boldsymbol{\psi}'_1(t) = \begin{bmatrix} 1 \\ 0 \\ \vdots \end{bmatrix}, \quad \boldsymbol{\psi}'_2(t) = \begin{bmatrix} 0 \\ 1 \\ \vdots \end{bmatrix}, \quad \dots, \quad \Psi(t) = \begin{bmatrix} C_1(t) \\ C_2(t) \\ \vdots \end{bmatrix}. \tag{7.24}$$

It usually is convenient to refer all such coordinate rotations, in the absence of interaction, to that of the initial state, axis $\boldsymbol{\psi}'_1(t)$. This we do most simply by choosing $\zeta_n(t) = E_1/\hbar$ and taking the undisturbed initial-state energy E_1 to be the zero point of energies,

$$E_1 = 0. \tag{7.25}$$

The coupled equations. With this expansion the TDSE becomes the set of (linear) ODEs

$$\frac{d}{dt}C_n(t) = -i \sum_m W_{nm}(t) \, C_m(t), \qquad n, m = 1, \dots N, \tag{7.26}$$

or, in vector and matrix notation,

$$\frac{d}{dt}\mathbf{C}(t) = -i\mathbf{W}(t)\mathbf{C}(t), \tag{7.27}$$

---

[5] The symbol $C_n(t)$ will refer to a rotating coordinate, while $c_n(t)$ refers to a fixed coordinate system.

where $\mathbf{C}(t)$ is a column vector comprising amplitudes $C_n(t)$. The diagonal elements of the Hamiltonian matrix $\hbar\mathbf{W}(t)$ are

$$\hbar W_{nn}(t) = E_n + V_{nn}(t) - \hbar\dot{\zeta}(t), \qquad (7.28)$$

while the off-diagonal elements are

$$\hbar W_{nm}(t) = V_{nm}(t)\exp[\mathrm{i}\,\zeta_n(t) - \mathrm{i}\,\zeta_m(t)]. \qquad (7.29)$$

The interaction matrix elements $V_{nm}(t)$ for laser-induced changes are obtained from multipole moments, as given in Sec. 4.3.

Phases and pictures. The phases $\zeta_n(t)$ of eqn. (7.20) are arbitrary functions of time that we introduce in order to simplify the subsequent differential equations. For example, we can always choose them to make the diagonal elements of eqn. (7.28) vanish, a choice known as the *Dirac picture* or *interaction picture*. In the Dirac picture the off-diagonal elements have exponential phases that vary with time. Alternatively, when the Hamiltonian involves interaction with a field of frequency ω we can choose the phases to make the off-diagonal elements of eqn. (7.29) slowly varying, a choice known as the *rotating-wave picture*.[6] With this choice the Hamiltonian matrix has, as diagonal elements, differences between Bohr frequencies and carrier frequencies, i.e. *detunings*. The choice $\zeta_n(t) = 0$ is the *Schrödinger picture*.

Expectation values. The choice of phases $\zeta_n(t)$ has no effect on the probabilities. However, expressions for other observables, expectation values, are affected by the phase choice (see Sec. 16.4). For example, the expectation value of the single-atom dipole moment is

$$\langle \mathbf{d}(t)\rangle = \langle \Psi(t)|\mathbf{d}|\Psi(t)\rangle = 2\mathrm{Re}\left\{ \sum_{n<m}\langle n|\mathbf{d}|m\rangle\, C_n(t)^* C_n(t)\exp[\mathrm{i}\,\zeta_n(t) - \mathrm{i}\,\zeta_n(t)]\right\}. \qquad (7.30)$$

Although our phase choice may lead to slowly varying probability amplitudes $C_n(t)$, the expectation value restores whatever rapid variation is induced by the Hamiltonian. A specific example occurs with treatment of quasiperiodic Hamiltonians, where the phases incorporate carrier frequencies that then appear in expectation values; see Chap. 21.

7.2.2 Probability loss

All excited states – states whose energy exceeds that of the ground state – will eventually impart this energy to surroundings. When collision partners are present, these can carry off the internal energy as increased kinetic energy of the projectile, or can themselves lose kinetic energy to produce excitation. Whenever radiative excitation is possible between two states, then spontaneous emission can transfer excitation energy from the atom to radiation. Often the result of such emission, observable as fluorescence, is one of several possible

---

[6] The rotating-wave *picture* should be distinguished from the rotating-wave *approximation* (RWA) discussed below.

final states. These need not be amongst the set of essential states – they may be unaffected by the particular active laser fields. Because they are not included amongst our limited set of basis states, any transition to one of them represents a loss of probability for finding the atom in an essential state. Treatments rely on the following observations.

Spontaneous emission. When the time scale of interest – that of pulsed excitation – is much shorter than the lifetime of any of the essential states, then we can neglect spontaneous emission. But when we consider short-lived species, or long pulse durations, then the effects of spontaneous emission must be included in the dynamics. This can be done rigorously with the use of a density matrix (see Chap. 16). However, a simpler description, using only the statevector, is possible under some circumstances: If nearly all spontaneous-emission transitions take the system out of the limited Hilbert subspace spanned by the essential states, then these events can be regarded as simple probability losses.

Photoionization. Although we here treat an idealized system comprising a single isolated atom, having only a few discrete states of interest, in reality this atom has additional states that form a continuum, e.g. ionization or dissociation states. To treat these we imagine the atom (an ion and its associated bound electron) to be contained within a very large but finite box, so that its wavefunction is constrained by boundary conditions. It then has discrete but closely spaced energies rather than a continuum (the spectrum is a *quasicontinuum*, see Sec. 15.4.5). By superposing such states we can construct a state whose associated wavefunction is localized at a given time; the electron, though possessing energy above the ionization limit, has not yet departed from the ion. With passing time this wavepacket will move into other portions of the box, spreading as it does so. When we view only a small volume around the original location of the wavepacket we have the impression that the particle has vanished; its probability, measured by the square of a wavefunction, is spread throughout other regions of the box.

It may very well happen that, after a *recurrence time*, the particle is once again seen to be localized.[7] But until then the electron behaves as a free particle. This simple picture preserves overall probability – the electron that is removed from the atom is still within the large box. But the probability of finding it in the immediate vicinity of the ion core has diminished. By directing attention to only a small region of the large confining box we regard this as "lost" probability. More precisely, the electron is not in one of the small set of discrete states (e.g. two) that we have taken as our Hilbert space.

Complex energy. There exist straightforward mathematical techniques for treating situations in which a continuum of energies occur and in which, as a consequence, there appears to be probability loss from a set of discrete states. In essence, we deal with a weak interaction linking a discrete state with a near-continuum of excited states. We regard the time

---

[7] This behavior occurs when intense laser radiation removes a bound electron and places it, as a wavepacket, onto an elliptical trajectory that returns regularly to the ion [Hay03; Kul06]. Modelings of intense field response often rely on simulations within a finite box; recurrences then are avoided by making the boundary have absorbing walls.

evolution as unidirectional, e.g. we include photoionization but not its inverse, radiative recombination. Under these conditions the continuum structure retains no memory; changes are irreversible. The portion of the system that is described by a few discrete states then appears to undergo exponential decay. That is, they cause probability decays described by the equation

$$P_n(t) = \exp(-\Gamma_n t) P_n(0), \tag{7.31}$$

where Γ_n is the decay rate for state n. Such processes, irreversible probability loss from the system of interest, make this an example of an *open system* [Dav76a; Bre02]. Although proper treatment requires a density matrix (see Chap. 16), they can be treated by replacing the real-valued energy E_n by the complex-valued number $E_n - i\Gamma_n/2$,

$$E_n \rightarrow E_n - i\Gamma_n/2. \tag{7.32}$$

7.3 Classes of interaction

Many examples of pulsed coherent excitation permit idealization as one of several classes, distinguished by temporal properties of the interaction:

Impulsive initiation: There occurs a sudden change of the Hamiltonian, expressing the sudden imposition of an interaction.

Quasistatic interactions: The interaction Hamiltonian $V(t)$ varies only slowly, if at all, once the interaction is present.

Quasiperiodic interactions: The interaction Hamiltonian involves terms that are proportional to sinusoidally varying functions whose amplitudes vary slowly with time. A pulsed laser field, monochromatic apart from a pulse envelope, produces such an interaction.

In each of these cases there are particularly useful theoretical approaches to evaluating the changes.

7.4 Classes of solutions

Solutions to the TDSE can be obtained in various forms, each with considerable literature. The following sections describe some of these that are relevant for the description of coherent excitation.

7.4.1 Analytic solutions

Traditionally solutions to ordinary differential equations were sought as expressions involving the standard mathematical functions, such as sines and cosines, Bessel functions, Legendre polynomials, etc. Analytic solutions are known for two-state equations in a variety of cases, and for N-state equations in some special cases. Some procedures make direct use of matrix forms for the N coupled equations, while others use the fact that N coupled

first-order differential equations can be recast as a single Nth order equation. This is particularly useful for $N = 2$ when, for a variety of closed-form pulse expressions, it leads to
well-studied second-order equations [Hio85a] [Sho90, Chap. 5].

Closed-form expressions typically require that pulse shapes be idealized with simple
analytic forms. However, for such shapes they offer opportunities to examine various limiting cases, and to evaluate the dependence of the *pulse aftermath* $P_n(\infty)$ upon various
parameters. They serve as valuable checks upon numerical solutions.

Solutions for constant pulses (i.e. time-independent interactions) and any N are readily
treated by means of Laplace transforms. These convert the ODE, with initial conditions,
into a set of algebraic equations, and hence they lead to problems of linear algebra.

7.4.2 Perturbative solutions

Prior to the widespread availability of personal computers, the traditional approach to theoretical treatments of time dependence in quantum mechanics relied on time-dependent
perturbation theory. This approach fixed the initial amplitude at unity, e.g. $C_1(0) = 1$. The
resulting equations read

$$C_1(t) = 1, \qquad \left[\frac{d}{dt} + i\, W_{nn}(t)\right] C_n(t) = -i\, W_{n1}(t), \qquad n = 2, \ldots N. \qquad (7.33)$$

For simple explicit functional forms of the time dependence these are readily integrated to
obtain closed-form solutions,

$$C_n(t) = -i\, e^{-i\chi_n(t)} \int_0^t dt'\, e^{+i\chi_n(t')}\, W_{n1}(t'), \qquad (7.34)$$

where the exponential phase is

$$\chi_n(t) = \int_0^t dt'\, W_{nn}(t'). \qquad (7.35)$$

Such an approach provides, through Fermi's famous Golden Rule (see Sec. 9.6), conventional expressions for the coefficients appearing in rate equations, for example
photoionization rates, but they are not generally satisfactory for treating coherent excitation.

7.4.3 Numerical solutions

The widespread availability of software packages for solving coupled ODEs makes numerical solutions the primary procedure nowadays. The typical numerical techniques, those
of predictor–corrector or Runge–Kutta, use polynomial approximations to integrate, step
by step, along the time variation. Possible difficulties appear when the solution has both a
slow variation and very rapid changes – the equations are *stiff* [Sha79; Hai10]. Elaborate
algorithms, involving adjustment of step size and polynomial order, are incorporated into
the more robust software packages, but even so the resulting solutions may not be reliable.

Thus even with numerical solutions it is still useful to utilize approximations such as those based upon adiabatic time evolution (see Sec. 8.4).

7.5 The time-evolution matrix; transition probabilities

When developing either analytic or numerical techniques it is desirable to work, not with solutions to a specific initial condition, such as $C_1(t_0) = 1$, but with arbitrary initial conditions at $t = t_0$. Then one expresses the subsequent function by means of a time-evolution matrix, or propagator, $U(t, t_0)$ such that

$$\mathbf{C}(t) = \mathsf{U}(t, t_0)\mathbf{C}(t_0), \quad \text{or} \quad C_n(t) = \sum_m U_{nm}(t, t_0)C_m(t_0). \tag{7.36}$$

By definition the time-evolution matrix $\mathsf{U}(t_0, t_0)$ at time $t = t_0$ is the unit matrix, while for later times it satisfies the Schrödinger equation,

$$\frac{d}{dt}\mathsf{U}(t, t_0) = -i\mathsf{W}(t)\mathsf{U}(t, t_0), \qquad \mathsf{U}(t_0, t_0) = \mathbf{1}. \tag{7.37}$$

The propagator has the property

$$\mathsf{U}(t_2, t_1)\mathsf{U}(t_1, t_0) = \mathsf{U}(t_2, t_0). \tag{7.38}$$

That is, the evolution process can be expressed as a succession of incremental evolutionary steps. Chapter 11 exploits this property.

The absolute square of the matrix element $U_{nm}(t, t_0)$ is the *transition probability*: the probability that the system, known to be in state m at the initial time t_0, will be found in state n after the time interval $\tau = t - t_0$,

$$w_{nm}(\tau) = |U_{nm}(t_0 + \tau, t_0)|^2. \tag{7.39}$$

The unsquared matrix element $U_{nm}(t, t_0)$ is the *transition amplitude*. As discussed in Sec. 9.5 the time derivative of the transition probability, suitably averaged, gives the transition rate $\mathcal{R}_{m\to n}$ used with rate equations.

Often we express times from the initial value $t = 0$. The propagator matrix then requires only a single argument, $\mathsf{U}(t, 0) = \mathsf{U}(t)$. For example, if the initial state, at time $t = 0$, has population entirely in state m, so that $|C_m(0)| = 1$, then the population in state n at time t is

$$P_n(t) = |U_{nm}(t)|^2. \tag{7.40}$$

More specifically, the absolute squares of the first column of $\mathsf{U}(t)$ are the probabilities that will occur starting from state 1.

Integral form. The solution to the defining equations for $\mathsf{U}(t, t_0)$, eqn. (7.37), can be written in integral form as

$$\mathsf{U}(t, t_0) = \mathbf{1} - i \int_{t_0}^{t} dt' \mathsf{W}(t')\mathsf{U}(t', t_0). \tag{7.41}$$

For short time intervals the matrix $U(t', t_0)$ under the integral can be approximated by its initial value, $U(t', t_0) \approx \mathbf{1}$, to give the first-order perturbative approximation

$$U(t, t_0) \approx \mathbf{1} - i \int_{t_0}^{t} dt' W(t'). \tag{7.42}$$

Further iteration of the solution, placing each successive approximation under the integral of eqn. (7.41) leads to successively more elaborate multiple definite integrals. Only for a few analytic pulse shapes is it possible to proceed further than with the first-order approximation of eqn. (7.42).

Exponential form. When the Hamiltonian is constant the Laplace transform technique provides an algorithm for relating the elements of the evolution matrix to those of the Hamiltonian. Alternatively, we can write the solution to eqn. (7.37) as an exponential,

$$U(t + \tau, t) = \exp[-iW\tau], \tag{7.43}$$

where the exponential of a matrix M is defined by the power series that defines the exponential of a scalar,

$$\exp[M] = \mathbf{1} + \frac{M}{1!} + \frac{M^2}{2!} + \cdots. \tag{7.44}$$

The exponential form can also be used when the Hamiltonian commutes with itself at different times,

$$[W(t), W(t')] \equiv W(t)W(t') - W(t')W(t) = 0. \tag{7.45}$$

Then we can express the evolution matrix as the exponential of an integral,

$$U(t, t_0) = \exp\left[-i \int_{t_0}^{t} dt' W(t') \right]. \tag{7.46}$$

By constructing the evolution matrix one has at hand the propagator for any arbitrary initial superposition state. It can be specialized to the traditional interest in excitation of state 1, but it offers useful generality, often with little added effort.

8

Two-state coherent excitation

The simplest examples of coherent excitation are those in which only two quantum states undergo appreciable change, an idealization that has been termed a two-level atom [Ber74; All87]. These two states may, in principle, be any pair of discrete states n, m between which the interaction Hamiltonian has nonzero matrix elements $V_{nm}(t)$. I will here label them 1 and 2,[1] with state 1 having the lower energy, $E_1 < E_2$. I will refer to state 1 as the ground state; state 2 is the excited state. Later sections treat more general pairs of states.

Interest in the dynamics of two-state systems dates from the rate equations of Einstein. Decades of quantum theory exposition dealt with relatively weak and incoherent fields, for which perturbation theory sufficed. Here I begin with the full TDSE for coherent excitation and various "exact" solutions, both analytic and numerical. Following this introduction Chap. 9 discusses the traditional time-dependent perturbation theory, upon which rests the derivation of rate equations appropriate to incoherent excitation. Chapter 10 presents a number of specific examples of two-state excitation, accompanied by Bloch-sphere pictures of the statevector motion.

8.1 The basic equations

The Hilbert-space representation of a quantum state by means of a statevector leads from the TDSE to a set of coupled ODEs, eqn. (7.26). For a two-state system the statevector construction is

$$\Psi(t) = e^{-i\zeta_1(t)}C_1(t)\psi_1 + e^{-i\zeta_2(t)}C_2(t)\psi_2 \tag{8.1}$$

and the resulting coupled differential equations read

$$\frac{d}{dt}C_1(t) = -i\, W_{11}(t)C_1(t) - i\, W_{12}(t)C_2(t), \tag{8.2a}$$

$$\frac{d}{dt}C_2(t) = -i\, W_{21}(t)C_1(t) - i\, W_{22}(t)C_2(t). \tag{8.2b}$$

---

[1] Other common notations in place of $\{1, 2\}$ for two-state systems include $\{0, 1\}$, $\{g, e\}$, $\{-, +\}$, $\{\uparrow, \downarrow\}$, $\{a, b\}$ and $\{b, a\}$.

In matrix form these appear as

$$\frac{d}{dt}\left[\begin{array}{c} C_1(t) \\ C_2(t) \end{array}\right] = -iW(t)\left[\begin{array}{c} C_1(t) \\ C_2(t) \end{array}\right],\tag{8.3}$$

where $W(t)$ is a 2×2 matrix expression of the Hamiltonian in frequency units. Its elements $W_{nm}(t)$ depend upon details of the interaction $V(t)$ and our (arbitrary) choice of the phases $\zeta_n(t)$. When treating excitation from state 1, as has been customary, it is convenient to choose the phase of that state, $\zeta_1(t)$, such that $W_{11}(t)$ vanishes,

$$\hbar W_{11}(t) = E_1 + V_{11}(t) - \hbar\dot{\zeta}_1(t) = 0.\tag{8.4}$$

The requirement is that, apart from an arbitrary constant, this phase is

$$\hbar\zeta_1(t) = E_1 t + \int_{t_0}^{t} dt'\, V_{11}(t').\tag{8.5}$$

The second diagonal element is

$$\hbar W_{22}(t) = E_2 + V_{22}(t) - \hbar\dot{\zeta}_2(t) \equiv \hbar\Delta(t).\tag{8.6}$$

It is possible to choose the second phase $\zeta_2(t)$ to make this zero (as is done in the Dirac picture), but it is often more convenient to choose it to simplify the off-diagonal element,

$$\hbar W_{12}(t) = V_{12}(t)\exp[i\zeta_1(t) - i\zeta_2(t)] \equiv \frac{\hbar}{2}\Omega(t)e^{-i\varphi}.\tag{8.7}$$

The general form for the matrix $W(t)$ is then

$$W(t) = \left[\begin{array}{cc} 0 & \frac{1}{2}\Omega(t)e^{-i\varphi} \\ \frac{1}{2}\Omega(t)e^{i\varphi} & \Delta(t) \end{array}\right].\tag{8.8}$$

In the examples discussed following Sec 8.2.3 the phase $\zeta_2(t)$ will be chosen so that, although the Hamiltonian is oscillatory, the matrix $W(t)$ is slowly varying: the functions $\Omega(t)$ (the Rabi frequency) and $\Delta(t)$ (the detuning) are slowly varying real-valued functions of time (dimensionally frequencies) that parametrize the interaction.

8.1.1 Initial and final conditions

The two probabilities $P_1(t)$ and $P_2(t)$ must, at all times, sum to unity – the system must always be in one of the two states. This conservation of probability translates into the requirement that the solutions to eqn. (8.3) should have the normalization

$$|C_1(t)|^2 + |C_2(t)|^2 = 1.\tag{8.9}$$

We will usually suppose that at an initial time $t = t_0$, prior to the presence of the interaction $V(t)$, the atom is in state 1, as specified by the initial probability amplitudes

$$C_1(t_0) = 1, \qquad C_2(t_0) = 0.\tag{8.10}$$

Of particular interest is dynamics that leads to complete population transfer after cessation of the interaction at a final time t_f,

$$C_1(t) = 0, \qquad |C_2(t)| = 1, \qquad t > t_f. \tag{8.11}$$

More generally, we may wish to produce a coherent superposition, such as described by the magnitudes

$$|C_1(t)| = \cos(\theta/2), \qquad |C_2(t)| = \sin(\theta/2), \qquad t > t_f. \tag{8.12}$$

This construction may then serve as the initial state for subsequent manipulation, as discussed in Chap. 11.

The time-evolution matrix. The allowance for arbitrary initial conditions is most simply treated by means of the propagator, or time-evolution matrix, of Sec. 7.5. For the two-state system and the phase choice of eqn. (8.5) this obeys the equation

$$\frac{d}{dt}\mathsf{U}(t) = -i \begin{bmatrix} 0 & \frac{1}{2}\Omega(t)e^{-i\varphi} \\ \frac{1}{2}\Omega(t)e^{i\varphi} & \Delta(t) \end{bmatrix} \mathsf{U}(t), \qquad \mathsf{U}(t_0) = \begin{bmatrix} 1 & 0 \\ 0 & 1 \end{bmatrix}. \tag{8.13}$$

As noted above, the functional forms of $\Omega(t)$ and $\Delta(t)$ depend upon the choice of the phases $\zeta_n(t)$ as well as on the Hamiltonian. Often it is useful to parametrize the propagator for a completed pulse by Cayley–Klein parameters a, b as

$$\begin{bmatrix} C_1(+\infty) \\ C_2(+\infty) \end{bmatrix} = \begin{bmatrix} a & b \\ -b^* & a^* \end{bmatrix} \begin{bmatrix} C_1(t_0) \\ C_2(t_0) \end{bmatrix}. \tag{8.14}$$

These can be evaluated analytically for a number of pulse shapes [Kyo06].

8.1.2 Integral form of the equations

Formally, one can integrate the differential equations to obtain integral expressions that incorporate the initial conditions. We readily verify, by differentiation, that the integral expressions

$$C_2(t) = C_2(t_0) - \frac{i}{2}e^{-i\chi(t)} \int_{t_0}^{t} dt' \, e^{i\chi(t')} \left[\Omega(t')e^{i\varphi} C_1(t') + \Delta(t')C_2(t_0) \right], \tag{8.15a}$$

$$C_1(t) = C_1(t_0) - \frac{i}{2} \int_{t_0}^{t} dt' \, \Omega(t')e^{-i\varphi} C_2(t'), \tag{8.15b}$$

reproduce the differential equations of eqn. (8.8). Here the excited-state phase $\chi(t)$ is the integrated detuning,

$$\chi(t) = \int_{t_0}^{t} dt' \Delta(t'). \tag{8.16}$$

Although these expressions, when differentiated, yield the TDSE, they are not of immediate use because the integrands involve the functions that we wish to determine: the unknowns

$C_n(t)$ appear under the integrals. However, like the differential form of the TDSE, they serve as the starting point for various approximations, such as those of Chap. 9.

8.1.3 General two-state analytic solutions

For various choices of the elements $W_{nm}(t)$ the TDSE has solutions that are expressible in terms of well-documented functions [Car86; Kyo06]. From the two coupled first-order equations for two-state probability amplitudes we obtain two uncoupled second-order equations. The equation for $C_2(t)$ is

$$\frac{d^2}{dt^2}C_2(t) + \left[i[W_{11}(t) + W_{22}(t)] - \frac{d}{dt}\ln W_{21}(t) \right] \frac{d}{dt}C_2(t) \tag{8.17}$$

$$+ \left[W_{12}(t)W_{21}(t) - W_{11}(t)W_{22}(t) + i\,\dot{W}_{22}(t) - i\,W_{22}(t)\frac{d}{dt}\ln W_{21}(t) \right] C_2(t) = 0.$$

A similar equation applies to $C_1(t)$, but with subscripts 1 and 2 interchanged. For the parametrization of eqn. (8.8) the resulting second-order ODE is

$$\frac{d^2}{dt^2}C_n(t) - \left[(\dot{\Omega}/\Omega) - i\Delta \right] \frac{d}{dt}C_2(t) + \left[(\Omega/2)^2 + i\dot{\Delta} - i(\dot{\Omega}/\Omega)\Delta \right] C_2(t). \tag{8.18}$$

Many choices of the functional forms of the time dependences $W_{nm}(t)$ lead to differential equations studied by mathematicians during the nineteenth century. Their solutions define what are commonly called the *special functions* of mathematical physics [Luk69; Leb72; Olv10]. A perusal of textbooks and handbooks will provide numerous applicable examples involving such special functions. Amongst the pulses that have been studied for application to coherent excitation are those in Table 8.1. For many of these it is possible to evaluate the Cayley–Klein parameters analytically [Kyo06].

Initial conditions. Because the differential equation is of second order there are two standard solutions (e.g. two cases of a special function); the amplitude $C_n(t)$ is a superposition of these two functions such that the combination satisfies the initial condition, say $C_1(t_0) = 1$. The analytic solutions are most useful when it is possible to associate each $C_n(t_0)$ with a single one of the special functions.

However, often that is only possible for idealized limits, such as $t \to \pm\infty$, and so the existence of exact analytic solutions may then be of limited value. Other analytic approaches, such as that of adiabatic approximations (see App. E), then become useful.

8.1.4 The electric-dipole interaction

Appendix B gives examples of multipole interactions, specifically those of an electric dipole, V^{ME1}, magnetic dipole, V^{M1} and electric quadrupole V^{E2}, as well as the induced moment V^{pol}. Each of these can be parametrized by a Rabi frequency $\Omega(t)$ that has as one factor

Table 8.1. *Examples of soluble pulsed interactions*

Pulse shape	$\Omega(t)$	$\Delta(t)$	function
Rabi[a]	Ω_0	Δ_0	trig
Piecewise constant	$\Omega_1, \Omega_2, \ldots, \Omega_{N-1}$	$t_0, t_1, \ldots t_N$	piecewise trig
Demkov[b] $(t > 0)$	$\Omega_0 \, e^{-t/\tau}$	Δ_0	Bessel
Landau–Zener–Stückelberg[c]	Ω_0	$r \, t$	parabolic cylinder
Rosen–Zener[d]	$\Omega_0 \operatorname{sech}(t/\tau)$	Δ_0	hypergeometric
Allen–Eberly[e]	$\Omega_0 \operatorname{sech}(t/\tau)$	$B \tanh(t/\tau)$	hyperbolic
Demkov–Kunike[f]	$\Omega_0 \operatorname{sech}(t/\tau)$	$\Delta_0 + B \tanh(t/\tau)$	hypergeometric
Bambini–Berman[g]	$\dfrac{\Omega_0}{\tau} \dfrac{\sqrt{x(1-x)}}{1 + \lambda x}$ $t = \tau \ln[x/(1-x)^{1+\lambda}]$	$\dfrac{\Delta_0}{\tau}$	hypergeometric
Gaussian[h]	$\Omega_0 e^{-(t/\tau)^2}$	$\Delta_0 e^{-(t/\tau)^2} \tan(z)$ $z = (\pi/2)\operatorname{erf}(t/\tau)$	hypergeometric
Carroll–Hioe[i]	$\dfrac{\Omega_0}{1 + (t/\tau)^2}$	$\Delta_0 \dfrac{(t/\tau)}{1 + (t/\tau)^2}$	hypergeometric

[a] [Rab37] [b] [Dem64; Bur82; Rad82a; Vit92] [Sho90, §5.4] [c] [Lan32a; Lan32b; Zen32; Stu32] [Sho90, §5.6] [d] [Ros32; Vit94] [Sho90, §5.5] [e] [All87; Hio84b] [f] [Dem69; Hio85a; Zak85; Yat02a; Vit03] [g] [Bam81; Hio85a; Zak85; Vit95d; Vit95c; Kyo06] [h] [Hio85a; Vas04] [i] [Hio85a]

a measure of the atomic response (the multipole moment), and as another factor the field amplitude $\mathcal{E}(t)$.

The interaction of bound particles with laser light most often originates with the electric-dipole interaction. This means that the interaction energy operator is the projection of the electric dipole moment $\mathbf{d}$ onto the electric field:

$$V(t) = -\mathbf{d} \cdot \mathbf{E}(t). \qquad (8.19)$$

The electric field vector $\mathbf{E}(t)$ appearing here is the time-varying electric field at the center of mass of the atom. For a single laser beam we write the field $\mathbf{E}(t)$ as a unit vector $\mathbf{e}$, a complex-valued envelope $\check{\mathcal{E}}(t)$, and a carrier frequency ω as

$$\mathbf{E}(t) = \operatorname{Re}\left\{ \mathbf{e}\check{\mathcal{E}}(t) \exp(-\mathrm{i}\,\omega t) \right\}. \qquad (8.20)$$

The dipole transition moment between states m and n, projected onto the field unit vector $\mathbf{e}$, is

$$d_{nm} = \langle \psi_n | \mathbf{d} \cdot \mathbf{e} | \psi_m \rangle. \qquad (8.21)$$

For a single-electron orbital this is obtainable from the spatial integral

$$d_{nm} = -e \int d\mathbf{r}\, \psi_n(\mathbf{r})^*\, \mathbf{r} \cdot \mathbf{e}\, \psi_m(\mathbf{r}), \qquad (8.22)$$

where $|e|$ is the elementary charge. Alternatively, the magnitude of the dipole moment (but not the phase) can be extracted from experimental spectroscopic data such as oscillator strengths or Einstein A coefficients; see App. H.1.

Because orientations are quantized, there occur discrete values for this interaction, see Sec. 12.2. Typically there exist selection rules (see Sec. 12.2) such that, for a given pair of states n, m, only one polarization direction $\mathbf{e}$ gives a nonzero transition moment [Few93]. I refer to the possible nonzero array of values for the interaction Hamiltonian, fixed by nonzero dipole transition moments, as a *linkage pattern*.[2] The possible dipole transition moments – the possible linkages – are fixed for any given atom or molecule; statevector manipulation takes place through control of the magnitude and direction of the electric field $\mathbf{E}(t)$.

The directional properties of the electric field, relative to a coordinate system fixed with the atom, appear here embodied in the scalar product of $\mathbf{d}$ with $\mathbf{E}$. For linear polarization we can take the unit vector $\mathbf{e}$ to be real and write the two-state interaction as

$$V_{21}(t) = -d_{21}\mathcal{E}(t)\cos(\omega t - \varphi) \equiv \hbar\Omega(t)\cos(\omega t - \varphi), \qquad (8.23)$$

where, by suitable choice of phase φ, the field amplitude $\mathcal{E}(t)$ can be taken as real.[3] The function $\Omega(t)$, presumed slowly varying, is the *Rabi frequency*[4]

$$\Omega(t) = -d_{21}\mathcal{E}(t)/\hbar. \qquad (8.24)$$

When the field is that of a standing wave, e.g. eqn. (3.13), rather than a traveling wave, the field structure leads to spatial dependence of the Rabi frequency, which becomes

$$\Omega(x, t) = -d_{21}\mathcal{E}(t)\sqrt{2}\sin(kx)/\hbar. \qquad (8.25)$$

A stationary atom positioned at a field node will experience no excitation, whereas an atom at an antinode will be subject to maximum interaction. Excitation by circularly polarized light exerts a steady torque that produces the Rabi oscillations but not the rapid variation at the carrier frequency.

For circular polarization and a field of helicity q the interaction matrix element is a complex number, and it is useful to treat the Rabi frequency as a complex-valued function of time, $\check{\Omega}(t)$. The needed expression is the sum of two terms,

$$V_{21}(t) = \hbar\check{\Omega}_q(t)\exp(-i\omega t) + \hbar\check{\Omega}_{-q}(t)\exp(+i\omega t). \qquad (8.26)$$

---

[2] This pattern has the properties of a mathematical *graph*.

[3] The absolute phase of the field, as parametrized with the time of electric field zero crossing, is not controllable. During time shorter than the coherence time of the laser field it is permissible, and useful, to regard the phase as zero. However, it is then essential to maintain this convention during subsequent time evolution, particularly if controlled phase changes occur, as in Sec. 11.4.

[4] Some authors refer to $\Omega(t)/2$ as the "Rabi frequency" and denote this (perhaps with a minus sign) with the symbol $\Omega(t)$. I follow the original terminology in which the Rabi frequency referred to the frequency of population oscillations of a resonant two-state system. The oscillations of probability amplitudes occur with half this frequency.

Typically selection rules pick just one of these [Mar06b]; see Secs. 12.2 and 12.3. More generally, for elliptical polarization, the interaction is the sum of two helicity components.

Quasistatic fields. This monograph emphasizes excitation by laser radiation, and hence the appropriate form of the interaction is one in which there is a carrier at the laser frequency ω. All of the mathematical machinery presented below, with rotating coordinate frames and the rotating-wave approximation, holds for quasistatic (DC) fields ($\omega = 0$) if we take the Bohr frequency to be zero, $E_2 - E_1 = 0$, i.e. if we consider degenerate states. Then the interaction strength $d_{12}\mathcal{E}(t)/\hbar$ is termed the *Majorana frequency* and the population oscillations are termed *Majorana oscillations* [Sho90, §3.4].

8.1.5 Measures of interaction strength

The model of excitation by monochromatic light is an idealization that neglects any fluctuations in the phase or amplitude of the field, behavior that affects the frequency content of the field. The traditional measures of temporal variations rely on Fourier transforms. These quantify the distribution of frequencies present in the illumination.

For treatments of coherent excitation it is the electric field, not the intensity, that must be analyzed. To examine the connection between electric field frequencies and excitation we introduce the Fourier transform (FT) of the interaction Hamiltonian expressed via the FT of the pulsed Rabi frequency. As frequency variable we take the detuning Δ:

$$\widetilde{\Omega}(\Delta) = \frac{1}{2\pi} \int_{-\infty}^{\infty} dt \, e^{i\Delta t} \, \Omega(t). \tag{8.27}$$

The inverse Fourier transform gives the time-varying Rabi frequency,

$$\Omega(t) = \int_{-\infty}^{\infty} d\Delta \, e^{-i\Delta t} \, \widetilde{\Omega}(\Delta). \tag{8.28}$$

The peak value of the Rabi frequency $\Omega_0 \equiv \Omega(t=0)$ provides one measure of the strength of the interaction; a second measure is the *temporal pulse area*, expressible as the resonant ($\Delta = 0$) contribution to this Fourier transform,[5]

$$\mathcal{A}_{\infty} \equiv \int_{-\infty}^{\infty} dt \, \Omega(t) = 2\pi \widetilde{\Omega}(0). \tag{8.29}$$

The *pulse duration* τ can be regarded as the ratio of temporal pulse area to peak Rabi frequency,

$$\tau \equiv |\mathcal{A}_{\infty}/\Omega_0| = 2\pi |\widetilde{\Omega}(0)/\Omega_0|. \tag{8.30}$$

---

[5] Note that this quantity can be positive, negative, or zero.

8.2 Abrupt start

The following subsections consider two examples of suddenly applied interactions, one in which the interaction $V(t)$, once initiated, maintains a constant phase, as occurs when there are quasistatic fields (Sec. 8.2.1), the other when $V(t)$ is the quasiperiodic interaction responsible for laser-induced excitation, in the so-called *rotating-wave approximation* (RWA) [Rab54], discussed in Sec. 8.2.3. In each case we will parametrize the Hamiltonian with two real-valued functions of time, $\Delta(t)$ and $\Omega(t)$. Their physical significance, and hence their control, depends upon the particular implementation of the two-state system. The two cases discussed below, in Secs. 8.2.1 and 8.2.3, have particular importance and occur frequently.

8.2.1 Constant phase; quasistatic interactions

The simplest example of two-state excitation occurs when the phase of the interaction $V_{12}(t) = V_{21}(t)^*$ remains constant. Typically this occurs when all elements of the interaction Hamiltonian $V(t)$ vary only slowly, if at all, subsequent to the onset of the interaction, but that is not a requirement. We choose the two phases $\zeta_n(t)$ to be equal, so that the statevector expansion reads

$$\Psi(t) = e^{-i\zeta_1(t)}\Big[C_1(t)\psi_1 + C_2(t)\psi_2 \Big]. \qquad (8.31)$$

We choose the phases by the requirement

$$\hbar\zeta_1(t) = E_1 t + \int_{t_0}^{t} dt'\, V_{11}(t'), \qquad \zeta_2(t) = \zeta_1(t), \qquad (8.32)$$

so that $\hbar W(t)$ has the elements

$$\hbar W_{11} = 0, \qquad \hbar W_{22}(t) = E_2 + V_{22}(t) - E_1 - V_{11}(t),$$
$$\hbar W_{12}(t) = \hbar W_{21}(t)^* = V_{12}(t) \equiv \frac{\hbar}{2}\Omega(t)e^{-i\varphi}, \qquad (8.33)$$

with real-valued $\Omega(t)$. The resulting coefficient matrix $W(t)$ is that of eqn. (8.8) of Sec. 8.1. The sole nonzero diagonal element of $W(t)$ quantifies the difference between the two quasistatically shifted energies (the *diabatic energies*) $E_n + V_{nn}(t)$,

$$\hbar\Delta(t) = E_2 + V_{22}(t) - E_1 - V_{11}(t) \equiv \hbar\omega_{12}(t). \qquad (8.34)$$

That is, here $\Delta(t)$ represents a time-dependent Bohr frequency, $\omega_{12}(t)$. The two equal off-diagonal elements are parametrized by the real-valued Majorana frequency,

$$\Omega(t) = 2e^{i\varphi}V_{12}(t)/\hbar, \qquad (8.35)$$

associated with the quasistatic interaction energy.

8.2.2 Degeneracy

When the two energies $E_1 + V_{11}(t)$ and $E_2 + V_2(t)$ are equal at all times then the phases can be chosen so that the diagonal matrix elements $W_{nn}(t)$ both vanish and the off-diagonal elements are real. The coupled equations of the resulting TDSE

$$\frac{d}{dt}C_1(t) = -\frac{i}{2}\Omega(t)e^{-i\varphi}C_2(t), \qquad \frac{d}{dt}C_2(t) = -\frac{i}{2}\Omega(t)e^{+i\varphi}C_1(t), \qquad (8.36)$$

have solutions that are expressible as definite integrals. Specifically the time-evolution matrix (or propagator) is

$$\mathsf{U}(t) = \begin{bmatrix} \cos[\mathcal{A}(t)/2] & -i\,e^{-i\varphi}\sin[\mathcal{A}(t)/2] \\ -i\,e^{i\varphi}\sin[\mathcal{A}(t)/2] & \cos[\mathcal{A}(t)/2] \end{bmatrix}, \qquad (8.37)$$

where the argument of the trigonometric functions is the integral

$$\mathcal{A}(t) = \int_0^t dt\,\Omega(t'). \qquad (8.38)$$

From these formulas one obtains an exact expression for the probability amplitudes, for any time dependence $\Omega(t)$, either analytically when the integral is available from suitable tables [Gra07], or by numerical quadrature. When the population is initially in state 1 the later probabilities are

$$P_1(t) = \cos^2[\mathcal{A}(t)/2], \qquad P_2(t) = \sin^2[\mathcal{A}(t)/2], \qquad (8.39)$$

and complete population inversion ($P_2 = 1$) will occur whenever $\mathcal{A}(t)$ is an odd integer multiple of π (a *pi pulse*).

Example. Section 8.2.4 discusses an example of degenerate solutions obtained with constant $\Omega(t)$. Tables of integrals offer alternative choices for analytic functions for use as $\Omega(t)$. One such example, with $t_0 = 0$, is

$$\Omega(t) = \Omega_0\cos(\omega t), \qquad \mathcal{A}(t) = \frac{\Omega_0}{\omega}\sin(\omega t), \qquad (8.40)$$

where Ω_0 is a positive real number. This example, with a periodic Hamiltonian, is best treated by Floquet theory, see App. G. It permits exact solution of the TDSE albeit the expressions involve infinite sums. From the Jacobi–Anger expansion of eqn. (4.9).

$$\exp[-iz\sin(\omega t)] = \sum_{n=-\infty}^{+\infty}\exp(-in\omega t)J_n(z), \qquad (8.41)$$

where $J_n(z)$ is the Bessel function of order n, we obtain, for use with eqn. (8.27), the formulas

$$
\cos[\mathcal{A}(t)/2] = \sum_{n=-\infty}^{+\infty} \cos(n\omega t) J_n(\Omega_0/2\omega),
$$
$$
\sin[\mathcal{A}(t)/2] = \sum_{n=-\infty}^{+\infty} \sin(n\omega t) J_n(\Omega_0/2\omega). \tag{8.42}
$$

These show that the system response occurs at harmonics of the driving field. When the field is weak, such that $\Omega_0 \ll \omega$, the series is dominated by the lowest-order Bessel functions,

$$
J_n(z) \approx \frac{z^n}{2^n n!}, \qquad z \ll 1. \tag{8.43}
$$

8.2.3 Quasiperiodic interactions

For an idealized monochromatic interaction the function $V(t)$ changes periodically at the carrier frequency ω. As discussed in Sec. 5.4.2, when the polarization is circular the effect is a steady torque – a steady rotation at the carrier frequency ω – that tends to alter angular momentum, first adding and then subtracting angular momentum during a succession of transitions. When the polarization is linear the relatively rapid field variation, at the carrier frequency ω, produces a shaking of the charge distribution; much more slowly there occurs a distortion of the charge distribution attributable to a transition. It is this latter aspect of the motion that concerns us because it is associated with permanent structural change.

Consider the interaction of atomic electrons with a pulsed linearly polarized laser field whose electric vector, evaluated at the atomic center of mass, is

$$
\mathbf{E}(t) = \mathbf{e}\,\mathcal{E}(t)\cos(\omega t - \varphi), \tag{8.44}
$$

where the real-valued function $\mathcal{E}(t)$ is the magnitude of the electric field, ω is the carrier frequency, and the unit vector $\mathbf{e}$ defines the laser polarization axis. The relevant quasiperiodic interaction energy is that of a dipole transition moment $\mathbf{d}$ in this electric field, $V(t) = -\mathbf{d} \cdot \mathbf{E}(t)$. We choose the phase $\zeta_1(t)$ as in eqn. (8.31) but we alter the definition of the second phase to include the carrier frequency ω,

$$
\hbar\zeta_1(t) = E_1 t + \int_{t_0}^{t} dt'\, V_{11}(t'), \qquad \zeta_2(t) = \zeta_1(t) + \omega t. \tag{8.45}
$$

This linearly varying phase increment we incorporate into a rotating Hilbert-space coordinate system by introducing the unit vector

$$
\boldsymbol{\psi}_2'(t) = \mathrm{e}^{-\mathrm{i}\omega t}\boldsymbol{\psi}_2, \tag{8.46}
$$

which rotates at the carrier frequency. In this rotating reference frame the state vector appears as

$$
\boldsymbol{\Psi}(t) = \mathrm{e}^{-\mathrm{i}\zeta_1(t)}\left[C_1(t)\boldsymbol{\psi}_1 + C_2(t)\boldsymbol{\psi}_2'(t)\right], \tag{8.47}
$$

where $\zeta(t)$ is given by eqn. (8.32). The off-diagonal elements of $\mathsf{W}(t)$ obtained with this statevector expansion are

$$W_{12} = W_{21}^* = -\mathbf{d_{12}} \cdot \mathbf{e}\,\mathcal{E}(t)\cos(\omega t - \varphi)\,e^{-i\omega t} = \tfrac{1}{2}\Omega(t)\left[e^{-i\varphi} + e^{-2i\omega t + i\varphi}\right], \quad (8.48)$$

where the strength of the interaction is parametrized by the Rabi frequency,

$$\Omega(t) = -d_{12}\mathcal{E}(t)/\hbar. \quad (8.49)$$

Here d_{12} is the projection onto the field direction $\mathbf{e}$ of the dipole transition moment. The sole nonzero diagonal element of $\mathsf{W}(t)$ is the *detuning*, the difference between the Bohr transition frequency ω_{12} and the laser carrier frequency ω, each of which may be time dependent,

$$\Delta(t) = \omega_{12}(t) - \omega(t). \quad (8.50)$$

The time dependence of $\Delta(t)$ may come either from alteration of the carrier frequency (say with a frequency chirp) or from a controlled alteration of the Bohr frequency (say by Zeeman or Stark shifts), such as is produced by a laser-induced dynamic Stark shift [Yat99a; Ric03; Ran05]. Thus the pair of coupled ODEs has, as the coefficient matrix,

$$\mathsf{W}(t) = \begin{bmatrix} 0 & \tfrac{1}{2}\Omega(t)e^{-i\varphi}[1 + e^{-i2\omega t + i2\varphi}] \\ \tfrac{1}{2}\Omega(t)e^{i\varphi}[1 + e^{+i2\omega t - i2\varphi}] & \Delta(t) \end{bmatrix}. \quad (8.51)$$

These equations are exact (in a coordinate frame rotating with angular velocity ω), within the idealization of an isolated two-state atom that undergoes no spontaneous emission.

8.2.4 Example: Resonant excitation

Solutions to the two-state equations characterized by the coefficient matrix $\mathsf{W}(t)$ of eqn. (8.51) underly the calculations used to display wavefunction changes discussed in Sec. 5.4.2. To illustrate that example we consider resonant excitation, $\Delta = 0$, of a two-state atom that is initially, at $t = 0$, in state 1,

$$C_1(0) = 1, \qquad C_2(0) = 0, \quad (8.52)$$

and we set the field phase to zero, $\varphi = 0$, so the coefficient matrix becomes

$$\mathsf{W}(t) = \begin{bmatrix} 0 & \tfrac{1}{2}\Omega(t)[1 + e^{-i2\omega t}] \\ \tfrac{1}{2}\Omega(t)[1 + e^{+i2\omega t}] & 0 \end{bmatrix}. \quad (8.53)$$

Figure 8.1 shows one example of the resulting probabilities, appropriate to excitation of a two-state system by linearly polarized monochromatic radiation. In this example one sees very clearly the rapid linear oscillations associated with the carrier frequency ω, together with the slower Rabi oscillations, at the constant Rabi frequency Ω.

Excitation by circularly polarized light exerts a steady torque that produces the Rabi oscillations but not the rapid variation at the carrier frequency. The relevant interaction

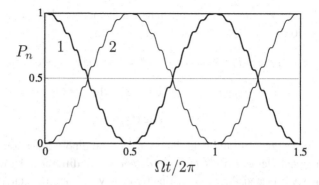

Figure 8.1 Plot of probabilities $P_n(t)$ vs. Rabi cycles $\Omega\, t/\pi$ for linearly polarized monochromatic excitation with $\omega = 20\Omega$. Initially the system is in state 1. Note the relatively rapid small oscillations (these are at the carrier frequency) that slowly lead to a transition into state 2. The population oscillates periodically between state 1 and state 2 at the Rabi frequency Ω.

terms do not have the exponential $e^{-2i\omega t}$,

$$W_{12} = W_{21}^* = \tfrac{1}{2}\Omega(t). \tag{8.54}$$

8.3 The rotating-wave approximation (RWA)

The exponential $\exp(-2i\,\omega t)$ appearing in eqn. (8.51) produces sinusoidal variation at twice the laser carrier frequency, 2ω. For optical radiation of commonly used laser pulses, the Rabi frequency Ω is typically four or five orders of magnitude smaller than the carrier frequency ω. That is, the photon energy $\hbar\omega$ is much larger than the interaction energy $\hbar\Omega$. Typical optical frequencies (the carrier) are

$$\omega = 2\pi c/\lambda \approx 5\times 10^{15}\ \text{s}^{-1}, \tag{8.55}$$

while typical Rabi frequencies (the interaction energy divided by $\hbar$) are much less,

$$|\Omega_0| = |d\mathcal{E}|/\hbar = 2.2\times 10^8\ \text{s}^{-1}\sqrt{I[\text{W cm}^{-2}]}\times \frac{|d|}{(ea_0)}$$

$$\approx 2\times 10^{11}\ \text{s}^{-1}\ \text{for}\ I = 1\ \text{GW cm}^{-2}\ \text{and}\ |d| = ea_0\,. \tag{8.56}$$

Thus for optical excitation the inequality $\omega \gg \Omega_0$ applies. Therefore the small carrier-frequency oscillations hold no interest; we are instead concerned with activity that takes place only over very many optical cycles.

Eliminating counter-rotating terms. We therefore consider probability amplitudes that are averaged over many optical cycles. In the Schrödinger equation we neglect the terms that vary as $2\omega t$ (*counter-rotating* terms) when added to constant terms

$$\cos(\omega t - \varphi)\, e^{-i\omega t} = \tfrac{1}{2}\left[e^{-i\varphi} + e^{-2i\omega t + i\varphi} \right] \rightarrow \tfrac{1}{2}e^{-i\varphi}. \tag{8.57}$$

This approximation, the (two-state) *rotating-wave approximation* (RWA)[6] [Sho90, §3.9], produces the two-state *RWA Schrödinger equation* in which the coefficient matrix $W(t)$ is the RWA Hamiltonian, shown in eqn. (8.8). The Rabi frequency $\Omega(t)$ appearing here is the slowly varying real-valued function of time given in eqn. (8.49),

$$\Omega(t) = -d_{12}\mathcal{E}(t)/\hbar. \tag{8.58}$$

There occurs a phase φ associated with the interaction, shown here explicitly.[7] At times it proves convenient to incorporate the phase into the Rabi frequency, and regard that as a complex-valued quantity $\hat{\Omega}(t) \equiv e^{i\varphi}\Omega(t)$. Alternatively, one can place the phase into the Hilbert-space unit vector, by defining $\hat{\psi}_2 \equiv e^{-i\varphi}\psi_2$. The present choice of exhibiting a phase explicitly emphasizes the possibility of altering it, as is discussed in Chap. 11.

The basic controls appearing here are the time-dependent Rabi frequency $\Omega(t)$ of eqn. (8.58) and the detuning $\Delta(t)$, defined through the equation

$$\hbar\Delta(t) \equiv E_2 + V_{22}(t) - E_1 - V_{11}(t) - \hbar\omega. \tag{8.59}$$

It is through manipulation of the functions $\Omega(t)$ and $\Delta(t)$ that we control the time evolution of the statevector. The remainder of this section offers insights into the connection between statevector motion in Hilbert space, as characterized by coordinates $C_n(t)$, and deliberate manipulation of these control functions.

8.3.1 Resonant population oscillations

When the excitation is resonant (meaning $\Delta(t) = 0$), the two coupled RWA equations read

$$\frac{d}{dt}C_1(t) = -\frac{i}{2}\Omega(t)\,e^{-i\varphi}\,C_2(t),$$

$$\frac{d}{dt}C_2(t) = -\frac{i}{2}\Omega(t)\,e^{i\varphi}\,C_1(t). \tag{8.60}$$

Analytic solutions are readily found, for any temporal variation of the pulsed Rabi frequency $\omega(t)$, by introducing a new time scale

$$d\tau = \Omega(t)dt. \tag{8.61}$$

The result, for a system known to be in state 1 at time $t = 0$, as specified by the initial condition $C_1(0) = 1$, is

$$C_1(t) = \cos[\mathcal{A}(t)/2], \qquad C_2(t) = -i\,e^{i\varphi}\,\sin[\mathcal{A}(t)/2]. \tag{8.62}$$

The amplitude of state 2 acquires the phase φ of the field and an amplitude set by the field amplitude: $\mathcal{A}(t)$ is the *Rabi angle*, the integral to time t of the (real-valued) Rabi frequency,

---

[6] The use of Hilbert-space coordinates that rotate at a carrier frequency dates back to early work with microwave resonances [Rab54].

[7] The absolute value of the phase is not controllable, and can be taken as zero at any convenient time. Only when one compares two pulses, both within a coherence time, or pulses affecting two locations, does one need to keep track of phases.

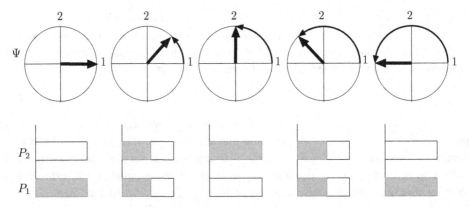

Figure 8.2 Rabi oscillations: top frames show the two-dimensional statevector Ψ at a succession of times; bottom frames show histograms of the populations P_1 and P_2 at these times. The populations return to their initial values when the statevector has completed one-half a revolution. [Redrawn from Fig. 15 of B. W. Shore, Acta Physica Slovaka, **58**, 243–486 (2008) with permission.]

$$\mathcal{A}(t) = \int_{-\infty}^{t} dt\, \Omega(t). \tag{8.63}$$

The Rabi angle for a pulse of finite duration, $\mathcal{A}_\infty \equiv \mathcal{A}(\infty)$, is the (temporal) *pulse area*.[8] Note that for resonant excitation, as shown here, the phased probability amplitude $C_2(t)' \equiv$ i $e^{i\varphi}C_2(t)$ is real, and the two probability amplitudes can therefore be displayed as motion of a unit vector, having components $C_1(t)$ and $C_2(t)'$, in a two-dimensional space. These expressions do not depend on details of the pulse shape, only on its integral: many different pulses can produce identical results.

The corresponding populations read, independent of the field phase φ,

$$P_1(t) = \tfrac{1}{2}[1 + \cos\mathcal{A}(t)], \quad P_2 = \tfrac{1}{2}[1 - \cos\mathcal{A}(t)]. \tag{8.64}$$

When the Rabi frequency remains constant, the Rabi angle is $\mathcal{A}(t) = \Omega t$, and the populations undergo periodic Rabi oscillations at the Rabi frequency Ω. Figure 8.2 illustrates this behavior.

The trigonometric functions are expressible as power series in their arguments, and so the probability amplitudes have the Taylor series expressions

$$C_1(t) = 1 - \frac{\mathcal{A}(t)^2}{8} + \cdots, \qquad C_2(t) = -i\, e^{i\varphi}\left[\frac{\mathcal{A}(t)}{2} - \frac{\mathcal{A}(t)^3}{24} + \cdots\right]. \tag{8.65}$$

---

[8] Often no distinction is drawn between the integral over a finite time, producing $\mathcal{A}(t)$, and the integral over a complete pulse, $\mathcal{A}_\infty = \mathcal{A}(\infty)$, and both are called the temporal pulse area. I refer to the incomplete integral $\mathcal{A}(t)$ as the Rabi angle and the completed integral $\mathcal{A}(\infty)$ as the pulse area.

For weak or short pulses we need only retain the first terms. The full series is an example of a perturbation expansion in which $\mathcal{A}(t)$ is regarded as small. Although such series are useful for short times, they do not make apparent the periodicity of the probability amplitudes.

Amplitude periodicity. In general, for any t, the angle $\mathcal{A}(t)$ can be positive, negative, or zero, as can the temporal pulse area $\mathcal{A}_\infty$. When the Rabi angle (or pulse area) is π (a pi pulse), or any odd-integer multiple of π, there occurs complete population transfer; there occurs complete population return (CPR)[9] when $\mathcal{A}(t) = 0$ or when $\mathcal{A}(t)$ is an even-integer multiple of π.

$$\text{If } P_2(0) = 0 \text{ then } \begin{cases} P_2(t) = 1 & \text{when } \mathcal{A}(t) = \pi, 3\pi, 5\pi, \ldots \\ P_2(t) = 0 & \text{when } \mathcal{A}(t) = 0, 2\pi, 4\pi, \ldots \end{cases} \tag{8.66}$$

These conclusions are valid for any pulse shape, but they hold only for resonant excitation.

It is noteworthy that the period of the probability amplitudes is half the period of the populations: During one Rabi oscillation of the population the probability amplitudes change sign; restoration of the initial state of the system requires two Rabi cycles. The sign change of the amplitudes becomes evident only when the two-state system is linked to other states.

The area rule. The factorization of a common time-dependent $f(t)$ out of the matrix $\mathsf{W}(t)$ applies to systems of equations with any number of components (i.e. any number of basis states). It requires that each element of $\mathsf{W}(t)$ – each Rabi frequency and each detuning in a general N-state atom – have exactly the same time dependence, $f(t)$. Because the Rabi frequency depends upon pulse intensity whereas the detuning depends upon pulse frequency, the two types of matrix elements generally have different time dependence. For resonant excitation (zero detuning) in the RWA we can be sure that this requirement will hold. Thus we have the rule

Resonant excitation by a single laser depends only upon the temporal pulse area.

This conclusion holds more generally for resonant multistate excitation when all fields have the same pulse shape: the behavior of the system is governed by the integral of the common pulse envelope.

8.3.2 Resonant excitation with loss

The extension of the resonant two-state TDSE for the excited-state probability amplitude provides the equations

$$\frac{d}{dt} C_1(t) = -\frac{i}{2} \Omega e^{-i\varphi} C_2(t),$$

$$\frac{d}{dt} C_2(t) = -\frac{i}{2} \Omega e^{i\varphi} C_1(t) - \frac{\Gamma}{2} C_2(t). \tag{8.67}$$

---

[9] Also known as *coherent population return* or restoration because it relies on coherent excitation.

These two coupled first-order equations can be replaced by a single second-order differential equation; each amplitude $C_n(t)$ obeys the same equation, recognizable as that of a damped harmonic oscillator,

$$\frac{d^2}{dt^2}C_n(t) + \frac{\Gamma}{2}\frac{d}{dt}C_n(t) + \frac{1}{4}\Omega^2 C_n(t) = 0. \tag{8.68}$$

The system is completely described by this ODE together with the initial values $C_n(0)$ of the two probability amplitudes (fixing $C_2(0)$ fixes also the derivative $\dot{C}_1(0)$ and *vice versa*).

To find solutions we test the trial function $C_n(t) = A_n \exp(-\mathrm{i}\,Zt)$ and find that Z must satisfy the quadratic equation

$$Z^2 + \mathrm{i}\frac{\Gamma}{2}Z - \frac{1}{4}\Omega^2 = 0. \tag{8.69}$$

Two regimes exist [Sho90, §3.10]: a regime of *underdamped* oscillation when the loss rate is small, $\Omega > \Gamma/2$, and a regime of *overdamping* when the loss rate is large, $\Omega < \Gamma/2$. The two regimes are separated by the condition of *critical damping*, when $\Omega = \Gamma/2$. Figure 8.3 shows examples of the three regimes. I will treat primarily situations in which the damping is slight.

8.3.3 Fixed Rabi frequency and detuning

When steady nonzero detuning occurs, along with steady intensity, then the solutions are again oscillatory. For the constant two-state RWA Hamiltonian matrix of eqn. (8.8) with real-valued Δ and Ω the propagator is expressible as trigonometric and exponential functions,[10]

$$U_{11}(t) = \mathrm{e}^{+\mathrm{i}\,\Delta t/2}\Big[\cos(\Upsilon t/2) + \mathrm{i}\,(\Delta/\Upsilon)\sin(\Upsilon t/2)\Big],$$

$$U_{12}(t) = -\mathrm{i}\,\mathrm{e}^{+\mathrm{i}\,\Delta t/2}(\Omega\,\mathrm{e}^{-\mathrm{i}\varphi}/\Upsilon)\sin(\Upsilon t/2), \tag{8.70}$$

$$U_{21}(t) = -\mathrm{i}\,\mathrm{e}^{-\mathrm{i}\,\Delta t/2}(\Omega\,\mathrm{e}^{\mathrm{i}\varphi}/\Upsilon)\sin(\Upsilon t/2),$$

$$U_{22}(t) = \mathrm{e}^{-\mathrm{i}\,\Delta t/2}\Big[\cos(\Upsilon t/2) - \mathrm{i}\,(\Delta/\Upsilon)\sin(\Upsilon t/2)\Big],$$

where Υ is the (generalized) nonresonant Rabi frequency (the root-mean-square (rms) value of the Rabi frequency Ω and the detuning Δ),

$$\Upsilon = \sqrt{\Omega^2 + \Delta^2}. \tag{8.71}$$

In particular, when the population initially occupies state 1 the excited-state population is

$$P_2(t) = |U_{21}(t)|^2 = \frac{\Omega^2}{2(\Delta^2 + \Omega^2)}[1 - \cos(\Upsilon t)], \tag{8.72}$$

and so it oscillates at the generalized Rabi frequency, often termed the *flopping frequency*, Υ. By contrast, the probability amplitudes oscillate at half this frequency.

---

[10] Some authors have omitted the factor $\frac{1}{2}$ by redefining the "Rabi frequency" and the "pulse area".

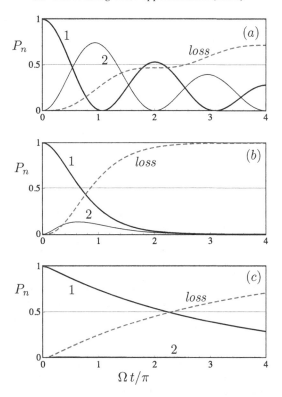

Figure 8.3 Effect of probability loss (say, due to photoionization from excited state) on two-state excitation. Full lines are probabilities P_n, dashed lines are loss, as a function of $\Omega t/\pi$. (a) Top frame shows population histories of states 1 and 2 with weak loss. The behavior is underdamped Rabi oscillations. Each burst of probability into state 2 produces a corresponding burst of loss (e.g. ion signal). (b) Middle frame shows regime of critical damping; the populations undergo no oscillations. (c) Bottom frame shows overdamped behavior; the population flows from state 1 directly into the ionization state. The loss from state 1 proceeds more slowly as the loss rate increases.

When the field is not resonant, $\Delta \neq 0$, the population transfer from state 1 to state 2 produced by a constant RWA Hamiltonian is never complete; always the population is less than $(\Omega/\Upsilon)^2$. With increasing detuning the maximum excitation diminishes and the oscillations become more rapid. Figure 8.4 illustrates this behavior.

8.3.4 Explaining oscillations: Dressed states

Simple analytic solutions to the nonresonant two-state RWA are known for a number of analytic forms of the time-dependent Rabi frequency and detuning. However, it is also possible to find solutions in a form that readily generalizes to multilevel excitation.

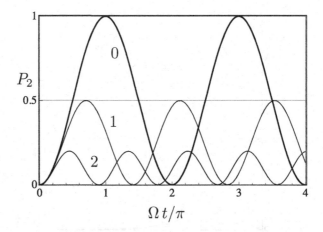

$$\Omega t/\pi$$

Figure 8.4 Excitation probability $P_2(t)$ vs. Rabi cycles $\Omega t/\pi$, showing population oscillations for constant Rabi frequency Ω with detunings $\Delta = 0$, 1 and 2 times Ω. As the detuning increases the oscillations become more rapid and their amplitude diminishes. [Redrawn from Fig. 3.7-1 of B. W. Shore, *The Theory of Coherent Atomic Excitation*. Copyright Wiley-VCH Verlag GmbH & Co. 1990. Reproduced with permission.]

We consider here excitation by a constant real-valued Rabi frequency and nonzero detuning. The constant two-state RWA Hamiltonian matrix is that of eqn. (8.8). We proceed, as suggested in nineteenth century texts on ODEs with constant coefficients, by introducing eigenvectors of the coefficient matrix,

$$\mathbf{W}\mathbf{\Phi}_\pm = \varepsilon_\pm \mathbf{\Phi}_\pm. \tag{8.73}$$

The matrix $\mathbf{W}$ has, for the present discussion, dimension 2, and so there are two of these eigenvectors, labeled here with subscripts $\pm$. Because $\mathbf{W}$ is, apart from conversion to frequency units, the RWA Hamiltonian, the eigenvalues are often termed energy eigenvalues or *quasienergies*, and the eigenvectors are known as energy eigenstates. (They are also known as *dressed states* [Coh92] [Sho90, §10.3]; the original basis states $\boldsymbol{\psi}_n$ are termed *bare states*.)

The energy eigenvalues are readily found to be

$$\varepsilon_\pm = \tfrac{1}{2}[\Delta \pm \Upsilon]. \tag{8.74}$$

Eigenvectors. To find expressions for the eigenvectors of the RWA Hamiltonian matrix $\mathbf{W}$ we introduce an angle θ through the definitions

$$\cos(\theta) = \Delta/\Upsilon, \qquad \sin(\theta) = \Omega/\Upsilon, \tag{8.75}$$

and rewrite the Hamiltonian, using trigonometric identities, as

$$W = \Upsilon \begin{bmatrix} 0 & \frac{1}{2}\sin(\theta)\ e^{-i\varphi} \\ \frac{1}{2}\sin(\theta)\ e^{i\varphi} & \cos(\theta) \end{bmatrix}$$

$$= \Upsilon \begin{bmatrix} 0 & \sin(\theta/2)\cos(\theta/2)\ e^{-i\varphi} \\ \sin(\theta/2)\cos(\theta/2)\ e^{i\varphi} & \cos^2(\theta/2) - \sin^2(\theta/2) \end{bmatrix}. \tag{8.76}$$

It then follows that the normalized eigenvectors and their eigenvalues can be written as[11]

$$\Phi_+ = \sin(\theta/2)\,\psi_1 + \cos(\theta/2)\ e^{i\varphi}\,\psi_2'(t), \qquad \varepsilon_+ = \Upsilon\cos^2(\theta/2),$$
$$\Phi_- = \cos(\theta/2)\,\psi_1 - \sin(\theta/2)\ e^{i\varphi}\,\psi_2'(t), \qquad \varepsilon_- = -\Upsilon\sin^2(\theta/2). \tag{8.77}$$

That is, the eigenvectors of W appear as a rotation, by angle $\theta/2$ away from the bare state basis.

Time dependence. The eigenvectors $\Phi_\pm$ of the time-independent matrix W are stationary states; with advancing time they change only by a phase. If the initial state coincides with one of these, say $\Psi(0) = \Phi_k$, then

$$\Psi(t) = \exp(-i\varepsilon_k t)\,\Phi_k. \tag{8.78}$$

Usually, however, the initial statevector is a superposition of energy eigenstates. For example, when the system is known to be in state 1 at time $t = 0$ the initial state is the superposition

$$\Psi(0) = \psi_1 = \sin(\theta/2)\,\Phi_+ + \cos(\theta/2)\,\Phi_-. \tag{8.79}$$

The energy eigenstates that contribute to this construction evolve in time with simple phases; the effect on $\Psi(t)$ is

$$\Psi(t) = e^{-iWt}\,\Psi(0) = \sin(\theta/2)\ e^{-i\varepsilon_+ t}\,\Phi_+ + \cos(\theta/2)\ e^{-i\varepsilon_- t}\,\Phi_-. \tag{8.80}$$

This equation presents an exact analytic solution to the posed two-state Schrödinger equation with initial condition (8.79). Expressed in (rotating) bare states, through the use of eqn. (8.77), the construction reads

$$\Psi(t) = e^{-i\Delta t/2}\,\Big\{ [\cos(\Upsilon t/2) + i\cos(\theta)\,\sin(\Upsilon t/2)]\,\psi_1$$
$$- i\,\sin(\theta)\sin(\Upsilon t/2)\ e^{i\varphi}\psi_2'(t)\Big\}, \tag{8.81}$$

where Υ is the eigenvalue splitting of eqn. (8.74). The individual bare-state components of the statevector, and the corresponding populations, evidently undergo oscillations. These oscillations of population, at frequency Υ, are here seen to be manifestations of interference between two energy eigenstates formed into a coherent time-dependent superposition, as

---

[11] The overall phase of each of these is arbitrary and is here chosen for convenience.

required by the initial conditions: the initial state is not an eigenstate of the Hamiltonian, and so the statevector cannot remain aligned with the initial state. If, and only if, the initial statevector is in a dressed state will the time variation be a simple overall phase change.

Statevector phase. Although the population returns periodically to the ground state, the statevector acquires a new phase with each return, namely

$$\Psi(t) = e^{-i\Delta t/2} \psi_1. \tag{8.82}$$

This dynamical phase is not directly observable by measurement of populations or dipole moments. However, if the two-state system is linked to other states, then it can be observed as an energy shift, e.g. by probing the two-state system via linkage to a third quantum state. It can also be observed via interferometry when the two states are part of a larger system, see Chap. 20. Measurement of the global statevector phase is analogous to measurement of the electric field phase through interferometric techniques.

8.3.5 Doppler detuning

The detuning that appears in the RWA equations is the difference between a Bohr frequency of an atom and the carrier frequency of radiation, as seen in a reference frame that is fixed with the atomic center of mass. When the atom is moving, the radiation appears Doppler shifted, and hence motional detuning occurs, see Sec. 4.4. The effect can be understood as follows.

The interaction between an electric field $E(r, t)$ and an atom whose center of mass is located at position R is, in the electric dipole approximation, the energy

$$V = -d \cdot E(R, t). \tag{8.83}$$

When the electric field is that of a traveling plane wave, of wavevector k and frequency $\omega = ck$, it has the form

$$E(R, t) = \text{Re}\left\{e\check{\mathcal{E}}\exp(i k \cdot R - i\omega t)\right\}. \tag{8.84}$$

Consider an atom that is steadily moving, with nonrelativistic velocity v, so that the center of mass position is $R = R_0 + vt$. Within this nonrelativistic approximation ($v^2 \ll c^2$), the interaction is proportional to the field

$$E(R_0 + vt, t) = \text{Re}\left\{e\check{\mathcal{E}}\exp[i k \cdot R_0 - i(\omega - k \cdot v)t]\right\}. \tag{8.85}$$

This field is that of a plane wave that is Doppler shifted to the frequency

$$\omega' = \omega - k \cdot v. \tag{8.86}$$

Thus the effect of center-of-mass motion is a detuning

$$\Delta(v) = \Delta(0) - k \cdot v \tag{8.87}$$

shifted by an amount proportional to the component of velocity along the laser propagation axis.

Note that the field amplitude, and hence the Rabi frequency, maintains a constant phase factor $\exp[i\mathbf{k} \cdot \mathbf{R}_0]$, in addition to the Doppler shift. This phase has no effect on population dynamics; it has significance only when we compare induced dipole moments of atoms at different locations, as we do in describing pulse propagation modified by atoms (see App. C.2).

8.3.6 Power broadening

The oscillatory behavior of two-state excitation continues indefinitely, as long as the radiation remains constant and coherent, and no interruptions occur; the populations never approach a steady state. Over many Rabi cycles the time-averaged excitation, obtained by averaging the sinusoid (to zero), is

$$\bar{P}_2 \equiv \lim_{t \to \infty} \frac{1}{T} \int_0^\infty dt \, P_2(t) = \frac{\Omega^2}{2(\Delta^2 + \Omega^2)}. \tag{8.88}$$

On resonance, $\Delta = 0$, this average, $\bar{P}_2 = \frac{1}{2}$, is the value obtained from the rate-equation model for a strong steady field and no degeneracy, in the limit of long times. In the limit of large detuning (i.e. much larger than the Rabi frequency) the excitation is negligible at all times. Figure 8.5 displays the long-time average excitation $\bar{P}_2$ as a function of detuning. As can be seen, the range of detunings where appreciable excitation can occur, on average, is roughly bounded by the Rabi frequency: with larger Rabi frequencies larger detuning can contribute to excitation.

Laser-induced excitation from the ground state to an excited state can always be followed by fluorescence, because both the Rabi frequency and the Einstein A coefficient involve the same dipole transition moment. This fluorescence is strongest when the laser frequency

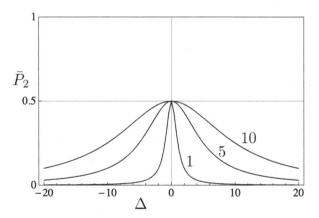

Figure 8.5 Power broadening: the time-averaged excitation $\bar{P}_2$ as a function of the detuning Δ for Rabi frequencies $\Omega = 1, 5$, and 10.

is tuned to the Bohr transition frequency of the two-state system: as a function of this frequency the fluorescence signal of an isolated atom traces out a Lorentzian profile whose central peak coincides with the Bohr frequency. When the Rabi frequency is weak, the width of the Lorentzian is the radiative decay rate, i.e. the inverse of the lifetime of the excited state; see Sec. 10.2.4. As the radiation becomes more intense, so that the Rabi frequency becomes much larger than the spontaneous emission rate, the resonance profile becomes that of the time-averaged excitation, and the profile width is the Rabi frequency. This is an elementary example of *power broadening* of a spectral line.

The resonance condition. The response of a two-state system is greatest – the change to excitation probability is largest – when the detuning remains zero. Applied to a quasiperiodic interaction, such as that of a laser pulse, this implies that the carrier frequency equals the Bohr frequency; the excitation is *resonant*.

In general there will occur such maximal population transfers between two quantum states whenever the relevant diagonal elements of the Hamiltonian are degenerate. For the two-state equations (8.3) this means the equality

$$W_{11}(t) = W_{22}(t). \tag{8.89}$$

With the present choice of phase $\zeta_1(t)$ the resonance condition is that $\Delta(t) = 0$. For multistate excitation the equality $W_{nn}(t) = W_{mm}(t)$ produces a comparable resonance between states n and m (see Sec. 15.3.8). This resonance condition is readily evident from the structure of the matrix $\mathsf{W}(t)$ when we choose the phase $\zeta_n(t)$ such that $W_{nn}(t) = 0$; resonance then occurs with any other state m for which $W_{mm}(t) = 0$.

8.4 Adiabatic time evolution

The opposite extreme of an interaction that begins suddenly is one that grows gradually and for which all changes occur slowly. The meaning of "slow" and "fast" refer, in this context, to the time required for the statevector to make significant changes. This time scale is set, in turn, by the magnitudes of the interaction matrix elements – the detuning and the Rabi frequency. This section presents some examples of such slowly changing interactions. The resulting statevector changes are "adiabatic".

8.4.1 Slow collisions

A traditional example of a slowly varying interaction occurs during the collision of a slowly moving projectile with a target atom or of two structured collision partners. In the simplest approximation we regard the interaction energy as a function of the separation distance R between centers of mass, thereby neglecting the overlap of electronic orbitals. This distance varies with time, either by deliberate manipulation of two collision partners held within traps, or else as a result of uncontrolled encounters in an atomic beam. In the latter case the interaction gains time dependence through the variation of separation distance. Let the relative motion follow a straight line along the z direction with velocity v. Then the separation is

$$R(t) = \sqrt{x^2 + y^2 + (vt)^2} = R_0 - \frac{(vt)^2}{2R_0} + \cdots \qquad (8.90)$$

where $R_0 \equiv R(0) = \sqrt{x^2 + y^2}$ is the distance of closest approach. For such situations we regard the time interval as being of infinite duration, $-\infty < t < +\infty$.

A collision is "slow" if the response time of the atom, typically the inverse of a Rabi frequency, is faster than the rate of change in the interaction, so that the atom has time to adjust to the changing external field.

8.4.2 Chirped frequency; rapid adiabatic passage (RAP)

The production of specific changes in the statevector by resonant excitation requires careful control of the temporal pulse area. Furthermore, such excitation has limited use when the ensemble includes a range of detunings, such as occur with Doppler shifts. An alternative pulsed excitation procedure overcomes such limitations; it can produce equal excitation for a distribution of Doppler-induced detunings, independent of the temporal pulse area. Specifically, the excitation pulse includes not only a variation of the Rabi frequency but a monotonic sweep of the detuning.

The technique, *rapid adiabatic passage* (RAP), requires that statevector changes must be completed during a time interval that is shorter than any incoherence-producing processes, such as spontaneous emission – the overall action is rapid on that time scale – but that within that time interval the detuning should change slowly with time, i.e. adiabatically. The resulting motion of the statevector is an example of *adiabatic following* [Cri73; Ore84a] in which the statevector follows a path in Hilbert space traced by an adiabatic state [Sho90, §3.6], as discussed in Sec. 8.4.3 below.

The desired time variation can be produced by subjecting the atom to a quasistatic electric field, thereby producing Stark shifts, or to a magnetic field, thereby producing Zeeman shifts. When treating laser-induced excitation the required change can be produced by altering the carrier frequency of the laser, as in a chirped-frequency pulse.

The simplest idealization of RAP takes the detuning to vary linearly in time.[12] The RWA equations then read

$$\frac{d}{dt}C_1(t) = -\frac{i}{2}\Omega(t)e^{-i\varphi}C_2(t),$$

$$\frac{d}{dt}C_2(t) = -\frac{i}{2}\Omega(t)e^{i\varphi}C_1(t) - i(\Delta_0 + rt)C_2(t), \qquad (8.91)$$

where r is the rate at which the detuning changes and Δ_0 is a fixed detuning, such as might occur from a single Doppler shift. Such a situation occurs if the laser frequency varies

---

[12] Obviously this cannot continue indefinitely. The infinite limit is a mathematical artifice; it is only necessary that the initial detuning be much larger than the range of static detunings of the ensemble, and that the frequency sweep continue until the desired changes of the statevector are completed.

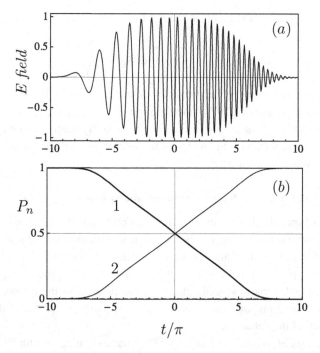

Figure 8.6 Example of chirped rapid adiabatic passage. (*a*) Upper frame shows the electric field vs. time in units of π. (*b*) Lower frame shows the population histories $P_n(t)$. As the detuning sweeps through resonance, here at time $t = 0$, population transfer occurs from the initial state 1 to the final state 2.

linearly with time, i.e. the frequency is *chirped* (basically sweeping from $-\infty$ to $+\infty$), or a Stark shift sweeps the Bohr transition frequency.

Solutions of these coupled equations obtained by direct numerical integration reveal that, if the chirp takes place over a sufficiently long time interval, all population will transfer from the initial state, say state 1, to the excited state, for any value of the static detuning. Indeed, the value of Δ_0 merely sets the time at which resonance occurs, $t = \Delta_0/r$. Figure 8.6 illustrates this behavior. Analytic results confirm this behavior.

The model of chirped RAP finds application in many physical systems. An idealization, the Landau–Zener (or Landau–Zener–Stückelberg) model [Lan32a; Lan32b; Zen32; Stu32], provides an analytic solution. In this model the Rabi frequency is a constant, Ω_0, and the detuning varies linearly with time, at rate r,

$$\Delta(t) = \Delta_0 + rt. \tag{8.92}$$

The solutions to the TDSE for this RWA Hamiltonian are expressible in terms of parabolic cylinder functions. One follows the time evolution from $t \to -\infty$, when the population is entirely in state 1, to a conclusion as $t \to +\infty$. The populations then are

$$\mathcal{P}_1 \equiv P_1(\infty) = \exp(-\pi y), \quad \mathcal{P}_2 \equiv P_2(\infty) = 1 - \exp(-\pi y), \quad y = (\Omega_0)^2/r. \tag{8.93}$$

When y is very large, meaning $(\Omega_0)^2 \gg |\dot{\Delta}|$, the time evolution is *adiabatic*, and population transfers entirely from state 1 to state 2. For smaller values of y the transfer is incomplete, and the final state is a coherent superposition of states 1 and 2. In the limit of weak interaction or rapid chirp, $(\Omega_0)^2 \ll |\dot{\Delta}|$, the time evolution is *diabatic* and the system remains in the initial state 1. This simple model has been extended to allow variation of the Rabi frequency, as is needed for modeling chirped RAP.

8.4.3 Explaining adiabatic passage: Adiabatic states

When the atom is subjected to a swept detuning $\Delta(t)$ the behavior of the population histories, or the statevector underlying them, can best be understood with the aid of an alternative Hilbert-space coordinate system, one in which the needed unit vectors are chosen as instantaneous eigenvectors of the time-varying Hamiltonian $\mathsf{W}(t)$. For the two-state system the defining equation of the eigenvectors $\boldsymbol{\Phi}_{\pm}(t)$ is

$$\mathsf{W}(t)\boldsymbol{\Phi}_{\pm}(t) = \varepsilon_{\pm}(t)\boldsymbol{\Phi}_{\pm}(t). \qquad (8.94)$$

These are known as *adiabatic states*, see App. E, as contrasted with the *diabatic* states $\boldsymbol{\psi}'_n(t)$ that form the original basis states (in a rotating frame). We allow time variation of both the Rabi frequency (taken to be real-valued) and the detuning. The two-state RWA Hamiltonian is that of eqn. (8.8). For the two-state system the two eigenvalues, *adiabatic eigenvalues* or *adiabatic energies*, are

$$\varepsilon_{\pm}(t) = \tfrac{1}{2}[\Delta(t) \pm \Upsilon(t)], \; \Upsilon(t) = \sqrt{\Omega(t)^2 + \Delta(t)^2}. \qquad (8.95)$$

The original diagonal elements of the RWA Hamiltonian, 0 and $\Delta(t)$ in the present two-state example, are known (when multiplied by $\hbar$) as *diabatic energies*. The adiabatic states are as presented in eqn. (8.77) but now with time-varying elements, resulting from a time-dependent angle $\theta(t)$,

$$\cos\theta(t) = \Delta(t)/\Upsilon(t), \qquad \sin\theta(t) = \Omega(t)/\Upsilon(t). \qquad (8.96)$$

As noted below, a slowly swept detuning prepares, and maintains, these adiabatic states. When the statevector initially aligns with one adiabatic state, and the RWA Hamiltonian changes slowly, then the statevector remains aligned with this single adiabatic state. The adiabatic state changes with time, and so the statevector construction changes when expressed with bare-state coordinates. The result is a transfer of population. Figure 8.7 illustrates the behavior.

Connections. The adiabatic states of eqns. (8.77) have the property that, for extremely large positive or negative values of the detuning $\Delta(t)$, they become aligned with the bare (diabatic) states. Specifically, for large positive detuning Δ we have $\cos(\theta) = +1$, so $\Theta(t) = 0$ and

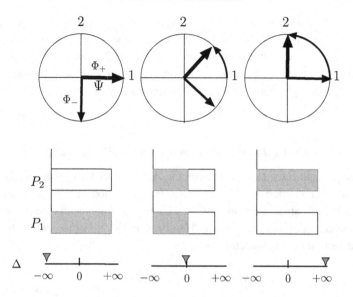

Figure 8.7 Adiabatic population transfer. Top frames show the statevector Ψ and two adiabatic eigenvectors $\Phi_\pm$. The statevector is aligned with Φ_+. Middle frames show histograms of populations P_n for three values of the detuning Δ. Along the base are schematic indicators of the three choices of detuning for the corresponding upper frames. Reading from left to right the detuning scans from negative to positive. [Redrawn from Fig. 18 of B. W. Shore, Acta Physica Slovaka, **58**, 243–486 (2008) with permission.]

the eigenstates are

$$
\begin{aligned}
\Phi_+(t) &= e^{i\varphi}\,\psi_2'(t), & \varepsilon_+ &= +|\Delta|, \\
\Phi_-(t) &= \psi_1, & \varepsilon_- &= 0,
\end{aligned}
\tag{8.97}
$$

and bare state 1 is aligned with the lower-energy dressed state. Conversely, for large negative detuning we have $\cos(\theta) = -1$, so $\Theta(t) = \pi/2$ and the eigenstates are

$$
\begin{aligned}
\Phi_+(t) &= \psi_1, & \varepsilon_+ &= 0, \\
\Phi_-(t) &= -\,e^{i\varphi}\,\psi_2'(t), & \varepsilon_- &= -|\Delta|.
\end{aligned}
\tag{8.98}
$$

Bare state 1 is now aligned with the higher-energy dressed state; the lower-energy state is aligned with bare state 2.

From the connection formulas presented above we can deduce the effect of sweeping the detuning. Suppose that initially $\Psi(t)$ is aligned with ψ_1 and that the detuning is large and negative. Then the state ψ_1 is, in turn, aligned with Φ_+. That adiabatic state will, when $\Delta \to +\infty$, become aligned with $\psi_2'(t)$. If the statevector remains always aligned with this adiabatic state then a transition from 1 to 2 will occur as the detuning sweeps from $-\infty$ to $+\infty$. In this sweep of increasing detuning the system is described by the single adiabatic state, the state having higher energy.

Alternatively, suppose the detuning is initially large and positive, but again $\Psi(t)$ is aligned with ψ_1. This state, for $\Delta \to +\infty$, is aligned with adiabatic state $\Phi_-(t)$. As the detuning sweeps from $+\infty$ to $-\infty$ this adiabatic state will become aligned with $\psi_2'(t)$. Again a transition occurs from 1 to 2, if the statevector remains aligned with an adiabatic state.

Thus it does not matter for population changes whether the detuning sweep is positive or negative; in all cases adiabatic following will produce the complete interchange of population (with a possible phase change)

$$\Psi(t) = \psi_1 \to -e^{i\phi}\psi_2'(t), \qquad \Psi(t) = \psi_2'(t) \to e^{-i\phi}\psi_1. \qquad (8.99)$$

This is (chirped) RAP. The process, though adiabatically slow, so that the statevector remains aligned with an adiabatic state, must nevertheless be completed before decoherence effects disrupt the coherent excitation; it must be rapid compared with any spontaneous emission lifetime of the excited state. Although the linear time dependence of the detuning provides a simple model, adiabatic passage occurs with any detuning variation that changes slowly from very large negative to very large positive or vice versa. It is not necessary that the change be linear or even monotonic with time.

The effects of swept detuning are easy to portray using adiabatic states. There remains a concern: under what conditions can the statevector maintain its alignment with an adiabatic state as the Hamiltonian varies. Appendix E.2.1 offers quantitative guidance. The following paragraphs present qualitative considerations, offering further discussion of adiabatic changes, with specific reference to chirped adiabatic passage of a two-state system.

8.4.4 Energy curves: Crossings and avoided crossings

The constraints on adiabatic passage are often presented by viewing plots of adiabatic energies along with plots of diabatic energies (the diagonal elements of the RWA Hamiltonian). For a two-state system these diabatic energies are 0 and $\Delta(t)$ or, alternatively, $\pm\frac{1}{2}\Delta(t)$. When there is a sweep of detuning, then the (bare) *diabatic* energy curves cross – this occurs at the time when $\Delta(t) = 0$. However, the (dressed) *adiabatic* curves do not share this instantaneous degeneracy: if there is any coupling present, however small the Rabi frequency, the adiabatic curves do not cross; they have an *avoided crossing*; see App. E.

Figure 8.8 illustrates the behavior of these curves, and the corresponding population histories during adiabatic evolution, for two examples of a two-state system subject to a pulsed Rabi frequency. The left-hand pair of frames illustrate the case when the detuning is constant. The diabatic energies remain constant horizontal lines, while the adiabatic energies exhibit reversible changes produced by the Rabi-frequency variation with time. With the choice of parameters here, there occurs CPR.

The right-hand pair of frames show the effect on these curves of a chirped detuning. The diabatic curve for state 2 varies linearly with time, crossing that of state 1 at $t = 0$. For large values of $|t|$, far from $t = 0$, the Rabi frequency is negligible, and the adiabatic curves follow the diabatic curves 0 and $\Delta(t)$. However, as the Rabi frequency grows larger, the two sets of curves differ.

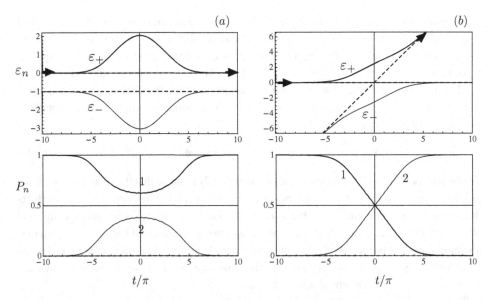

Figure 8.8 Pulsed two-state excitation. Top frames show diabatic eigenvalues ε_k (full lines) and diabatic energies $W_{nn}(t)$ (dashed lines), bottom frames show population histories $P_n(t)$. The arrowheads on the energy curves show those associated with initial and final states; when the motion is adiabatic the system remains in the state associated with eigenvalue ε_+. (*a*) Left frames: adiabatic evolution with constant detuning. The pulse briefly excites the atom but population subsequently returns to the initial state – an example of CPR. (*b*) Right frames: adiabatic evolution with chirped detuning and the pulse of frame (*a*). The composition of the adiabatic state changes as the detuning sweeps from negative to positive. The result is complete population transfer – an example of RAP. [Redrawn from Fig. 4 of Vitanov *et al.*, *Avances in Atomic, Molecular, and Optical Physics*, volume 46, 55 (Academic Press, New York, 2001) with permission.]

To interpret the population histories associated with such curves we begin by considering the system at early times – the left-hand side of the figures. Suppose the statevector is initially a single bare diabatic state, ψ_1. At these times there is no Rabi interaction, and so the diabatic and adiabatic states coincide. The initial statevector is therefore aligned with a single adiabatic state; it is represented by a system point on the coinciding diabatic and adiabatic curves.

As time increases and the energies change, this system point moves across the figure, from left to right, expressing the changes of the energies with time. Its association with a *single* curve can continue only for two extreme idealizations, corresponding to either fast (diabatic) or slow (adiabatic) changes of the RWA Hamiltonian.

During *rapid* change the statevector will remain aligned with the original bare *diabatic* state, and the system point will therefore follow the (dashed) diabatic curve, always associated with the energy of bare diabatic state 1, a horizontal line in this figure. The system point moves steadily along this line, crossing the bare diabatic curve for state 2. At the end the system will still be in bare state 1; it follows a straight-line path on the energy vs. time plot, and no transition occurs.

By contrast, when the changes occur sufficiently slowly (*adiabatically*) the statevector will remain aligned with the *adiabatic* state. The system point will move along the (full) adiabatic curve that initially connects with state 1. During the time when the two sets of curves differ the system point will follow the dressed adiabatic curve, on a path that does not cross any other curve. Initially this path is horizontal, but at later times the path changes direction, eventually heading upward to join the diabatic curve for bare state 2 at large times; a transition will then have occurred.

The choice between the two paths depends on how rapidly the system point moves through the region where the interaction occurs, where the two sets of curves are not the same. If the point moves slowly then it follows the dressed adiabatic curve and a transition occurs. If it moves rapidly then it follows the bare diabatic curve; a *curve crossing* occurs, but no transition. When the changes cannot be clearly classified as fast or slow, the population becomes divided between paths, and at the end the statevector is a coherent superposition of the two diabatic states.

However, if the chirp rate is *too* slow then the adiabatic and diabatic curves will run nearly parallel for a long time interval. It is then not possible to regard the system point as being associated with a single state, either diabatic or adiabatic, and the simple picture of an avoided crossing is not appropriate.

A quantitative description of the relative probabilities for these two extreme possibilities obtains from the Landau–Zener–Stückelberg (LZS) model of a two-state system subject to linearly varying detuning $\Delta(t) = rt$ and constant Rabi frequency Ω, Sec 8.4.2. When the population initially resides in state 1, the probability of finding the system in state 2 at the conclusion of the interaction is

$$\mathcal{P}_2 \equiv P_2(\infty) = \exp(-\pi\,\Omega^2/|r|). \tag{8.100}$$

When $\Omega^2 \gg |r|$, as occurs for small chirp rate, the motion is adiabatic, and a transition occurs. When $\Omega^2 \ll |r|$, as occurs for rapid change, the motion is diabatic, and no transition occurs.

8.4.5 Energy surfaces and adiabatic passage

The adiabatic eigenvalues of the two-state RWA depend on the instantaneous values of two real-valued parameters: the detuning and the magnitude of the Rabi frequency. Therefore plots of these eigenvalues require three dimensions: we deal not with energy curves but with energy surfaces. Rather than follow a system point along a curve, one follows it over a surface. The possible quantum changes depend on the topology of the possible surface paths [Yat02b; Gue03].

Figure 8.9 shows a portion of the two energy surfaces for a two-state system, as a function of detuning and Rabi frequency. Within the front vertical face the Rabi frequency vanishes, and so the surface intersections are straight lines. These correspond to the bare (diabatic)

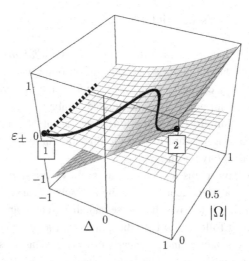

Figure 8.9 Adiabatic energy surfaces as a function of detuning Δ and $|\Omega|$. The full black curve is a parametric plot, on the upper energy surface, of the system point for excitation of state 1 by a chirped Gaussian pulse, as in Fig. 8.8(b). The dashed line is a parametric plot of the system point for static detuning, as in Fig. 8.8(a).

eigenvalues 0 and Δ, and they are associated, respectively, with the bare eigenstates ψ_1 and $\psi'_2(t)$. At the center of this face, where $\Delta = 0$ and $\Omega = 0$, the diabatic curves cross; at that point there occurs a singularity of the RWA Hamiltonian. This *singular point* is the only point at which the two energy surfaces meet.

The behavior of the two-state system in the RWA is governed by initial conditions and by two functions of time, the detuning and the Rabi frequency. When the Rabi frequency maintains a constant phase, it can be treated as a real quantity, and then the system is parametrized by two real-valued functions of time. These fix the two adiabatic eigenvalues and thereby locate a point on the energy surfaces.

Except for the singular point $\Delta = 0, \Omega = 0$ there are always two surfaces on which the point can be placed. If the system is known to be described by a single adiabatic state it can be represented by a system point on one of the two surfaces. Given such a system point, on a single surface, then if the motion is adiabatic, so that the statevector remains aligned with that adiabatic state, then the system point will move on a single energy surface. Figure 8.9 shows the path of such a system point, for a two-state atom that is initially in state 1 and which is acted on by a pulse that has a Gaussian Rabi frequency and a detuning that changes linearly with time, at a constant chirp rate. The initial system point, for negative detuning and zero Rabi frequency, lies on the upper energy surface, i.e. the statevector is approximated by Φ_+, as marked by $\boxed{1}$. With advancing time the detuning increases, as does the Rabi frequency, and the system point moves away from the $\Omega = 0$ plane. If the motion is adiabatic, the system point remains on this surface, passing around the singular point, to arrive at later times back in the $\Omega = 0$ plane, but now on the curve associated with bare state 2, as identified by its slope. The system point is marked $\boxed{2}$.

Directly below the parametric plot of the system point on the upper energy surface is a second curve, not shown, on the lower energy surface. If the motion is not adiabatic, or if the initial state is not a single adiabatic state, then the state of the system will be a coherent superposition of the states represented by these two paths.

Adiabatic passage – inducing a complete transition from the initial state to the other state – can be produced by a variety of pulses. With the aid of the energy-surface picture we can identify the following requirements:

- The system must initially be (almost) in a single adiabatic state. If the system is initially in the ground state, then this means there must be a large detuning prior to the action of the Rabi frequency.
- The subsequent changes should be such that the state vector follows a single adiabatic state. This is the main requirement on changes of $\Delta(t)$ and $|\Omega(t)|$. Note that when $|\Omega| = 0$ the adiabatic states coincide with bare diabatic states.
- The final field should again be such that the adiabatic eigenvalues are well separated, and the system point is identifiable with the desired target state.

8.4.6 Stark-chirped rapid adiabatic passage (SCRAP)

To produce complete population transfer via adiabatic passage the detuning should sweep slowly through resonance. Because the detuning is the difference between the Bohr transition frequency of the atom and the laser carrier frequency, an alteration of either of these two frequencies will produce the desired result. The Bohr frequency, being proportional to the energy difference between two stationary states, can be altered by imposing any slowly varying *nonresonant* electric or magnetic field. Quasistatic electric fields, inducing Stark shifts of the energies, offer one possibility. Pulses of nonresonant laser light offer another means of subjecting the atom to a slowly varying electric field and thereby producing a (dynamic) Stark shift [Has75; Del00; Tow10; Sus11] [Sho90, §4.3]. (Section 13.4.5 discusses the origin of these shifts.)

Figure 8.10 shows a pulse sequence that will produce population transfer by RAP. It involves a Stark-shifting laser pulse (S) that sweeps the Bohr frequency through resonance with the fixed laser carrier while a transition-inducing laser (P) acts. This P pulse terminates before the S pulse ends. The resulting adiabatic passage is, in theory, identical with that produced by a chirped pulse; it has been called *Stark-chirped rapid adiabatic passage* (SCRAP) [Yat99a; Ric00b; San06].

As in all adiabatic processes, the final result does not depend on details of the pulse envelope. It does not depend on the peak value of the Rabi frequency or the temporal pulse area, or on the details of the detuning variation with time; in this sense it is *robust*.

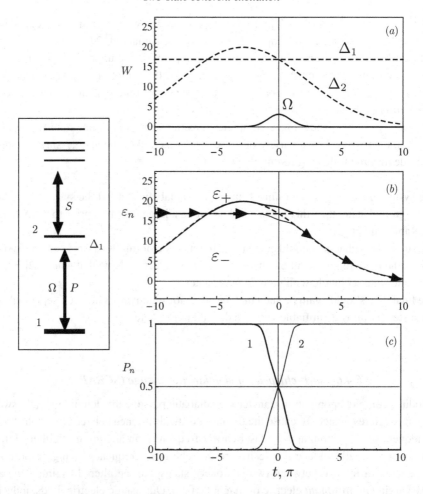

Figure 8.10 Example of Stark-shifted rapid adiabatic passage (SCRAP). The inset at the left shows a schematic diagram of energy levels, with P field, detuned from resonance by Δ_1 responsible for transitions (the Rabi frequency Ω). A nonresonant S field produces a dynamic Stark shift of state 2, Δ_2. The S pulse Δ_2 and the P pulse Ω are offset in time, with P occurring during the falling portion of the S pulse. (a) Elements of the RWA Hamiltonian, the detunings Δ_n and Rabi frequency Ω. The pulsed Rabi frequency is centered at the moment when the Stark-shifted detuning Δ_2 is resonant with the static laser detuning Δ_1. During the P pulse the Stark shift sweeps through resonance. (b) The adiabatic eigenvalues $\varepsilon_\pm$ (full lines) and diabatic energies (dashed lines) for this system. Arrowheads show the progress of the system, first by a curve crossing and then by an avoided crossing. (c) The population P_n vs. time in units of π. Population is transferred adiabatically from state 1 to state 2 (RAP).

8.4.7 Pulse shape effects; population return (CPR)

Unless the carrier frequency ω is very close to resonance with the Bohr frequency (in which case the final population depends only on the temporal pulse area), the pulse-produced excitation $P_2(\infty)$ depends very significantly upon the temporal shape of the pulse. By contrast with the simplicity of resonant excitation, nonresonant excitation depends upon the abruptness with which irradiation commences and ceases. In essence, rapid rise and fall of intensity (meaning significant change during an interval while detuning changes little) leads to rapid small-amplitude population oscillations and to the possibility of leaving a portion of the oscillatory population in an excited state. At the opposite extreme, fields whose amplitudes vary slowly (adiabatically) typically return all population to the initial state.

Figure 8.11 illustrates these effects for three examples of unstructured pulses[13] whose temporal variation is symmetric around a midpoint in time, taken to be $t = 0$. The three examples each show, in the top frames, the time-varying pulse envelope. The middle frames show the populations $P_n(t)$ for resonant excitation. The bottom frames show these populations for constant detuning equal to the peak Rabi frequency, $\Delta = \Omega_0$. For each example the pulse duration is adjusted to give a temporal pulse area of 3π, and so in each of these examples the resonant excitation produces CPR.

Rectangular pulse. The first of the frames, Fig. 8.11(a), shows the behavior of the two-state populations when the initial preparation is in state 1 and the pulse is rectangular: it undergoes an abrupt change from zero to a finite constant value Ω_0 followed by an abrupt pulse cessation after an interval τ,

$$\Omega(t) = \Omega_0 \begin{cases} 1, & |t| \leq \tau/2, \\ 0, & |t| > \tau/2. \end{cases} \tag{8.101}$$

With this pulse shape the RWA Hamiltonian is constant for $|t| < \tau/2$ and the solutions are combination of sines and cosines of the generalized Rabi frequency $\Upsilon = \sqrt{\Delta^2 + \Omega_0^2}$. The final excitation probability is left, upon sudden termination of the pulse, at a value that depends upon the generalized pulse area $\Upsilon\tau$,

$$P_2 \equiv P_2(\infty) = (\Omega_0/\Upsilon)^2 \sin^2(\Upsilon\tau/2). \tag{8.102}$$

As the detuning Δ increases, for given peak Rabi frequency Ω_0, the oscillations become more rapid and the maximum excitation decreases. Thus except for discrete values of the parameter $\Omega_0 = \Delta_0$ (generalized pi pulses) the pulse does not restore the initial conditions, although for large detuning, $|\Delta| \gg \Omega_0$, the final excitation probability tends to zero, independent of the pulse duration,

$$P_2 \approx (\Omega_0/\Delta)^2 \sin^2(\Delta\tau/2), \qquad |\Delta| \gg \Omega_0. \tag{8.103}$$

---

[13] Here "unstructured" means the pulse envelope has only a single maximum.

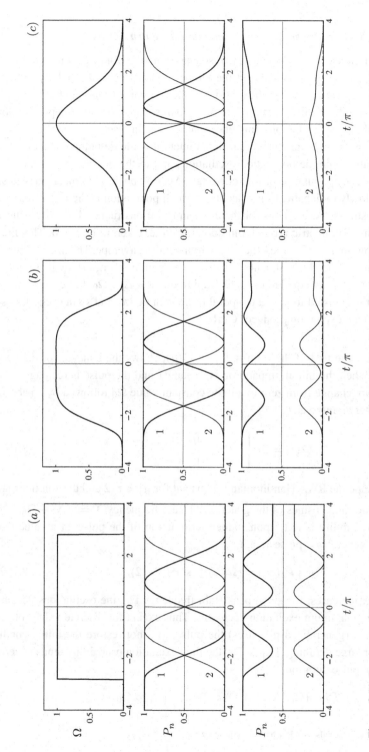

Figure 8.11 Response of two-state atom to rectangular pulse. Top frames show pulse envelope. Middle frames show populations $P_n(t)$ for resonant excitation. The pulse area is 3π and so all population returns to the initial state. Bottom frames show populations for detuned excitation, $\Delta = \Omega$. (a) Rectangular pulse. (b) Gauss-6 pulse. (c) Gaussian pulse. [Redrawn from Figs. 5.1-1–5.1-3 of B. W. Shore, *The Theory of Coherent Atomic Excitation*. Copyright Wiley-VCH Verlag GmbH & Co. 1990. Reproduced with permission.]

130

Smooth pulse. The second frame, Fig. 8.11(*b*), shows resonant and nonresonant excitation for a hypergaussian pulse,

$$\Omega(t) = \Omega_0 \exp\left[-(t-t_0)^6/\tau^6\right]. \tag{8.104}$$

This pulse has a smoother envelope than the rectangular pulse, but it is taken to have the same temporal area, and so the resonant excitation again produces complete population transfer. Again the presence of detuning increases the frequency of population oscillations and diminishes the excitation. However, when this pulse is detuned from resonance, there is less residual population transfer than occurs with the rectangular pulse: there is a more complete restoration of the initial conditions.

Gaussian pulse. The third frame, Fig. 8.11(*c*), shows the behavior produced by a Gaussian pulse,

$$\Omega(t) = \Omega_0 \exp\left[-(t-t_0)^2/\tau^2\right]. \tag{8.105}$$

This pulse is much smoother than either of the other two pulses. Consequently even more population returns to the initial state as the pulse ceases. Nonresonant population oscillations are much less evident in frame (*c*) than in the other frames: the excitation rises and falls along with the pulse envelope.

Hyperbolic secant pulse. An even smoother pulse, the hyperbolic secant pulse, known also as a Rosen–Zener pulse [Ros32; Vit94],

$$\Omega(t) = \Omega_0 \operatorname{sech}(t/\tau) = \frac{2\Omega_0}{\exp(t/\tau) + \exp(-t/\tau)}, \tag{8.106}$$

rises and falls exponentially for $|t| \gg \tau$. Analytic solutions are known, applicable for an interval of infinite duration ($-\infty < t < \infty$). A two-state system prepared in state 1 and excited by a hyperbolic secant pulse for constant detuning Δ undergoes periodic population oscillations, producing the final excitation

$$P_2 = \operatorname{sech}^2(\Delta\tau/2)\sin^2(\Omega_0\tau/2). \tag{8.107}$$

As with the rectangular pulse there occur moments of population transfer and the possibility of complete restoration of the initial state when the pulse area $A_\infty = \pi\Omega_0\tau$ is an exact multiple of 2π. When the detuning is much larger than the inverse of the pulse duration, $|\Delta| \gg \tau$, the final excitation is

$$P_2 \approx 2\exp(-\Delta\tau/2)\sin^2(\Omega_0\tau/2), \qquad |\Delta| \gg \tau. \tag{8.108}$$

This can be made negligibly small, except for resonance $\Delta = 0$, by taking a suitably long pulse.

Population return. Figure 8.12 presents an overview of the pulsed excitation probability $\mathcal{P}_2$, as a function of pulse area and detuning, for the three examples discussed above: Frame (*a*) show results for a rectangular pulse, frame (*b*) is the Gauss-6 pulse, while frame (*c*) shows results for a Gaussian pulse (the hyperbolic secant pulse is similar to this). In these plots the pulse areas $\mathcal{A}_\infty$ are directly proportional to the peak Rabi frequency Ω_0 and the pulse duration τ; an increase in either will increase the area.

In all these examples permanent excitation only occurs for a limited range of detuning. As the pulse becomes smoother, from frame (*a*) to frame (*c*), the pulsed excitation aftermath $\mathcal{P}_2$ becomes appreciable only for small values of the detuning. After a transient excursion to the excited state all population will return to the initially populated ground state (CPR) [Vit95b; Vit95d; Kuh98; Vit01c]. Typically the initial state is unexcited, but CPR will restore initial conditions that include a superposition.

Under CPR conditions fluorescence from the excited state, after pulse cessation, will only occur for small detunings. A typical measure of "small" in this context is the inverse of the Fourier bandwidth of the pulse, typically $\Delta_0 = \mathcal{C}/\tau$, where τ is the pulse duration and $\mathcal{C}$ is a parameter, of order unity, that depends on the temporal shape of the pulse. As the pulse becomes longer lasting the frequency–time uncertainty narrows the range of detunings for which permanent excitation can occur.

Asymmetric pulses. Analysis of various analytic solutions shows that for smooth and unstructured pulses whose envelopes are symmetric in time, so that $\Omega(t) = \Omega(-t)$, adiabatic following in the presence of constant detuning will generally restore the initial conditions (CPR will occur), although there may be a specific combination of pulse area and detuning that leaves some excitation.

By contrast, examples of similar smooth unstructured pulses that are asymmetric in time, $\Omega(t) \neq \Omega(-t)$, generally do not restore the initial conditions. This property, first noted by Bambini and Berman for particular families of pulse shapes, was further extended to more general smooth unstructured pulses [Bam81; Rob81; Rob95; Vit95c]. There may, however, be a particular combination of pulse area and detuning that does leave excitation [Vit94].

If the pulse is very asymmetric, with constant nonzero detuning, such that a slow (adiabatic) increase of Rabi frequency is followed by a rapid (diabatic) decrease, the statevector is frozen into its construction at the peak of the Rabi frequency, Ω_0, and when $\Omega_0 \gg |\Delta|$ the final values are [Vit95c]

$$\mathcal{P}_1 \approx \frac{1}{2}\left(1 + \frac{\Delta}{\Omega_0}\right), \qquad \mathcal{P}_2 \approx \frac{1}{2}\left(1 - \frac{\Delta}{\Omega_0}\right), \qquad \text{when } \Omega_0 \gg |\Delta|. \qquad (8.109)$$

Thus when the evolution is adiabatic and the peak Rabi frequency is much larger than the static detuning, up to half the population can be permanently excited by an asymmetric pulse.

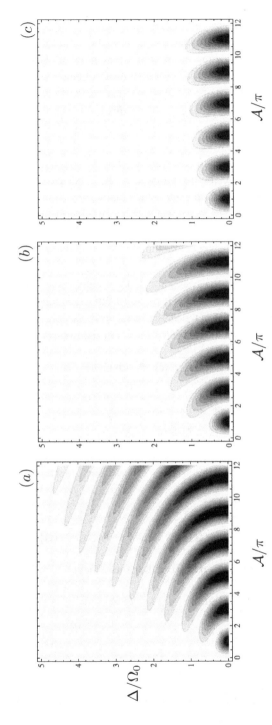

Figure 8.12 Excitation probability $\mathcal{P}_2$ vs. pulse area $\mathcal{A}$ in units of π and detuning Δ/Ω_0 for the pulse shapes of Fig. 8.11: (a) Rectangular pulse; (b) Gauss-6 pulse; (c) Gaussian pulse.

8.4.8 Optimization of two-state RAP

Adiabatic time evolution underlies the technique of chirped RAP or more general techniques based on adiabatic following whereby the statevector remains aligned with an adiabatic eigenvector. Although the combination of pulsed Rabi frequency and swept detuning are capable of producing CPR in the adiabatic limit, practical restrictions (e.g. it is not possible to have pulses of infinite duration) prevent reaching this limit. Thus the experimenter who needs to make the transfer as complete as possible, subject to experimental constraints, faces a problem of optimization. The work of Dykhne, Davis, and Pechukas (DDP) [Dyk60; Dav76b] has provided helpful guidance toward solutions [Gue02], see App. E.3. They provided a formula, eqn. (E.35), for the probability $\mathcal{P}^{non}$ that (from a pulse of infinite duration) there will occur a change of the adiabatic eigenvector with which the statevector is aligned – and a consequent nonadiabaticity – as a result of eigenvalue degeneracy at some complex value of time. By minimizing the effect of such eigenvalue degeneracies, or even eliminating them entirely, one can expect to maintain better adiabatic following for a given pulse duration and chirp rate.

To evaluate the DDP formula it is necessary that the eigenvalue splitting be a single-valued analytic function of complex-valued time subject to a few constraints. The simplest way to minimize the nonadiabaticity of the DDP formula is by designing pulses such that the adiabatic eigenvalues maintain a constant separation. For the two-state model this means

$$\Upsilon(t) \equiv \sqrt{\Omega(t)^2 + \Delta(t)^2} = \text{constant} = \Upsilon_0, \qquad (8.110)$$

subject to the requirement

$$
\begin{aligned}
t &\to -\infty & \Omega &= 0, & \Delta &= +\Upsilon_0, \\
t &= 0 & \Omega &= \Upsilon_0, & \Delta &= 0, \\
t &\to +\infty & \Omega &= 0, & \Delta &= -\Upsilon_0.
\end{aligned}
\qquad (8.111)
$$

A simple choice is the prescription

$$\Omega(t) = \Upsilon_0 \sin[\pi f(t)], \qquad \Delta(t) = \Upsilon_0 \cos[\pi f(t)], \qquad (8.112)$$

where the function $f(t)$ is arbitrary apart from the requirement

$$0 = f(-\infty) \leq f(t) \leq f(\infty) = 1. \qquad (8.113)$$

An example is

$$f(t) = \frac{1}{1 + \exp(-t/T)}. \qquad (8.114)$$

Such pulses can significantly diminish the nonadiabaticity and the infidelity of a quantum-state change [Vas09].

8.5 Comparison of excitation methods

Preceding sections have discussed three classes of two-state excitation: incoherent, impulsive, and adiabatic. Each is associated with different forms of radiation and produces different results; each has application to particular requirements. Figure 8.13 presents illustrative examples of these three mechanisms for producing excitation.

The first mechanism, appropriate for broadband incoherent light (not monochromatic laser light) draws on rate equations. These predict that, at most, half of the population can be transferred to the excited state, and that this saturation value is approached exponentially, at a rate that increases with increasing intensity. Although the transfer is incomplete, the result is not a coherent superposition; it is an incoherent mixture – a *mixed state*, see Sec. 16.2.1.

The second mechanism, resonant coherent excitation, can produce relatively rapid complete population transfer, but the pulse must be terminated with precision if exact inversion

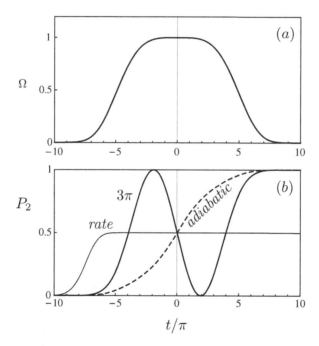

Figure 8.13 Comparison of various excitation schemes. (*a*) A Gauss-6 pulse. (*b*) Examples of excitation probability P_2 vs. time (in units of π) produced by a resonant Gauss-6 pulse: for resonant excitation with area 3π, for adiabatic following (RAP dashed curve) and for rate equation with negligible spontaneous emission. The rate equation saturates the population with half being excited; subsequently population will return to the ground state by spontaneous emission, here neglected. RAP produces complete population transfer. The resonant pulse produces oscillations and the final excitation depends on the pulse area.

is desired. This constraint requires careful control of pulse amplitude and duration (i.e. of integrated Rabi frequency) as well as avoidance of Doppler shifts of detunings. It can produce a coherent superposition of the two states.

The third option, adiabatic passage, requires a larger temporal pulse area than does resonant Rabi oscillation, but in the end it can produce a more robust result, insensitive to details of the pulse. It too can produce a coherent superposition.

9

Weak pulse: Perturbation theory

An important simplification of the theory describing radiation-induced changes occurs when conditions are such that very little excitation occurs. This was the common situation prior to the use of laser radiation, and it therefore was a standard portion of early texts on quantum mechanics.

When changes to the atom are small they can be regarded as small perturbations of existing structure. The resulting simplified mathematics, *time-dependent perturbation theory* [Won76; Lan77; Pal06b] [Sho90, §4.5, 17.1], usually starts from the integral form of the TDSE and approximates the integrand by the initial conditions. The approach is most simply illustrated by considering a two-state system in which all population resides initially, at time t_0, in state 1, so that $C_2(t_0) = 0$. Specifically, we use eqn. (8.15a) of Sec. 8.1.2 with $C_2(t_0) = 0$,

$$C_2(t) = -\frac{\mathrm{i}}{2} \mathrm{e}^{-\mathrm{i}\chi(t)} \int_{t_0}^{t} dt'\, \mathrm{e}^{\mathrm{i}\chi(t')} \Omega(t') \mathrm{e}^{\mathrm{i}\varphi}\, C_1(t'), \qquad (9.1)$$

to obtain an expression for $C_2(t)$. Because the radiative interaction produces, by assumption, little change in the statevector, we approximate the initial-state amplitude, for all time, by its initial value

$$C_1(t) \approx 1. \qquad (9.2)$$

This approximation produces the *first-order perturbation theory* (FOPT) result

$$C_2(t) \approx -\frac{\mathrm{i}}{2} \mathrm{e}^{-\mathrm{i}\chi(t)} \int_{t_0}^{t} dt'\, \Omega(t')\, \mathrm{e}^{\mathrm{i}\chi(t')+\mathrm{i}\varphi}, \qquad \chi(t) = \int_{t_0}^{t} dt'\, \Delta(t'), \qquad (9.3)$$

where φ is the phase of the electric-field envelope. Given a sufficiently simple analytic expression for the time dependence of $\Delta(t)$ and $\Omega(t)$ these integrals can be evaluated to give formulas for the time-varying excitation probability.

9.1 Weak resonant excitation

When the carrier frequency of the radiation is resonant with the Bohr frequency, $\Delta = 0$, the FOPT result is

$$C_1(t) \approx 1, \qquad C_2(t) \approx -\frac{i}{2} \int_{t_0}^{t} dt' \, \Omega(t') \, e^{i\varphi} = -\frac{i}{2} \, \mathcal{A}(t) \, e^{i\varphi}. \tag{9.4}$$

The magnitude of $C_2(t)$ depends only on the Rabi angle $\mathcal{A}(t)$, not on any details of the pulse. The phase of $C_2(t)$ is that of the field φ shifted by $\pi/2$. This formula can only be valid for small values of the integral, such that $|C_2(t)| \ll 1$, so that population change is negligible.

Time-dependent perturbation theory has its most important application to the evaluation of transitions produced by broadband radiative excitation or between a discrete state and a continuum of final states, such as occurs with photoionization. Applied to such situations it produces the traditional transition rates associated with Fermi's Golden Rule; see Sec. 9.6.

9.2 Pulse aftermath and frequency content

The preceding formulas produce particularly useful results when we consider a completed pulse, as expressed by the time interval $-\infty < t < +\infty$, for a pulse with fixed constant detuning Δ. The integral then becomes

$$C_2(\infty) \approx -\frac{i}{2} e^{-i\Delta t} \int_{-\infty}^{+\infty} dt' \, \Omega(t') \, e^{i\Delta t' + i\varphi}. \tag{9.5}$$

This integral is, apart from a numerical factor, the Fourier transform of $\Omega(t)$ evaluated at the detuning frequency,

$$C_2(\infty) = -i\pi \, e^{-i\Delta t + i\varphi} \, \widetilde{\Omega}(\Delta), \tag{9.6}$$

and hence it is proportional to the Fourier transform of the pulse amplitude $\mathcal{E}(t)$. In turn, the excitation probability $|C_2(\infty)|^2$ produced by the completed pulse, to first order and for given detuning Δ, is the absolute square of the Fourier transform of the Rabi frequency,

$$\mathcal{P}^I(\Delta) = |C_2(\infty)|^2 = \left| \frac{1}{2} \int_{-\infty}^{+\infty} dt \, \Omega(t) \, e^{i\Delta t} \right|^2 = |\pi \widetilde{\Omega}(\Delta)|^2. \tag{9.7}$$

In this approximation what matters for producing lasting excitation is the magnitude of the Fourier component of $\Omega(t)$ at the detuning frequency $\Delta = \omega - \omega_0$. Smooth pulses have Fourier transforms that are localized in frequency, whereas pulses with sharp edges tend to have a broader frequency spectrum. Therefore, we expect that for nonresonant excitation sharp pulses will produce greater permanent excitation than smooth pulses. This argument is only suggestive; it takes no account of the possibility that a sharp pulse may leave the population entirely in the initial state at the completion of a flopping cycle.

In this FOPT result the cumulative excitation produced by a pulse depends only on the frequency content, not at all on any details of the temporal pulse shape. An important consequence of this result is that when a pulse has no frequency component at the given detuning, then it produces (in first order) no lasting excitation, $\mathcal{P}^I(\Delta) = 0$. In this first-order approximation the population transfer produced by any resonant pulse ($\Delta = 0$) is proportional to the square of the temporal pulse area,

$$\mathcal{P}^I(0) = (\mathcal{A}_\infty/2)^2, \qquad \text{for } \Delta = 0. \tag{9.8}$$

This result holds only for small values of the temporal pulse area, $\mathcal{A}_\infty \ll 1$. For resonant excitation this absence of frequency content means the pulse has zero temporal pulse area: the electric field has equal positive and negative contributions to this area.

It is not difficult to construct examples in which permanent excitation occurs despite the absence of resonant frequencies. One in which complete population inversion can take place, even though there are no resonant components in the field, occurs with bichromatic light [Ehl90; Una00; Yat03].

Higher-order corrections. The evident lack of excitation predicted (to first order) by eqn. (9.7) does not mean that a pulse has no lasting effect upon the atom. Indeed, exact numerical evaluation of the TDSE may reveal complete population transfer for a pulse for which $\mathcal{P}^I(\Delta) = 0$. To treat such situations within the context of perturbation theory it is necessary to improve the estimate of $C_1(t)$ by using an approximation to the exact expression

$$C_1(t) = 1 - \frac{i}{2} \int_{-\infty}^{t} dt' \, \Omega(t') \, e^{-i\varphi} C_2(t'). \tag{9.9}$$

The second-order estimate requires an integral involving the product of two successive interactions, i.e. two photons. When the radiative interaction is that of an electric dipole moment of a free atom, this vanishes exactly because of parity constraints. Then the next contributory order to the probability amplitude is third-order theory. When the first-order contribution vanishes this gives the prescription

$$\mathcal{P}^{III}(\Delta) = \frac{1}{2^6} \left| \int_{-\infty}^{\infty} dt' \int_{-\infty}^{t'} dt'' \int_{-\infty}^{t''} dt''' \, \Omega(t')\Omega(t'')\Omega(t''') \, e^{i\Delta t' - i\Delta t'' + i\Delta t'''} \right|^2. \tag{9.10}$$

The evaluation of this triple integral is daunting, but is possible in simple cases.

9.3 Example: Excitation despite missing frequencies

To illustrate the limitation of the perturbation-theory result consider a pulse defined by a rectangular Fourier spectrum,

$$\tilde{\Omega}(\Delta) = \begin{cases} \Omega_0/2\Delta_0, & -\Delta_0 \leq \Delta \leq \Delta_0, \\ 0, & |\Delta| > \Delta_0. \end{cases} \tag{9.11}$$

The frequencies present in this pulse are only those that lie within the bandwidth $2\Delta_0$ of being resonant ($\Delta = 0$). Within this range all frequencies are equally likely; beyond it they are absent. The pulse shape in the time domain is a sinc function,

$$\Omega(t) = \Omega_0 \frac{\sin \Delta_0 t}{\Delta_0 t}. \tag{9.12}$$

The peak Rabi frequency, at $t = 0$, is Ω_0 and the temporal pulse area is

$$\mathcal{A}_\infty = \pi \Omega_0 / \Delta_0. \tag{9.13}$$

Although the sinc pulse is not one of the functions for which exact analytic solutions are published, there is no difficulty in obtaining numerical solutions to the TDSE and from these evaluating the population transfer produced by the pulse,

$$\mathcal{P}(\Delta) = P_2(\infty) = |C_2(\Delta; \infty)|^2. \tag{9.14}$$

On resonance ($\Delta = 0$) the detuning is exactly in the middle of the Fourier spectrum. The exact solutions to the Schrödinger equation exhibit the well-known Rabi oscillations,

$$\mathcal{P}(0) = \sin^2(\mathcal{A}_\infty / 2) = \sin^2(\pi \Omega_0 / 2\Delta_0). \tag{9.15}$$

For weak excitation, $\Omega_0 \ll \Delta_0$, this expression gives the approximation $\mathcal{P}(0) \approx (\mathcal{A}_\infty / 2)^2$, predicting a resonant excitation probability proportional to the square of the peak Rabi frequency Ω_0, i.e. directly proportional to the peak intensity.

First-order perturbation theory. The FOPT result for the pulse specified by eqn. (9.11), obtained by carrying out the integral (9.3), is expressible as exponential integrals, from which we obtain the first-order expression for the probability transfer as

$$\mathcal{P}^I(\Delta) = \begin{cases} (\mathcal{A}_\infty / 2)^2, & |\Delta| < \Delta_0, \\ 0, & |\Delta| > \Delta_0. \end{cases} \tag{9.16}$$

Within this approximation any frequency within the allowed bandwidth will behave as though it was resonant and there occurs no population transfer when the detuning Δ lies outside the Fourier bandwidth. For detunings within the regime of allowed transitions ($|\Delta| < \Delta_0$), the population transfer is just the square of the temporal pulse area: it varies in direct proportion to the square of the peak Rabi frequency Ω_0, i.e. to the light intensity, and inversely with the square of the bandwidth Δ_0.

Exact numerical results. Numerical results reveal that population transfer is possible, and indeed may even be complete, for pulses whose detuning lies outside the Fourier bandwidth, i.e. for which there are no resonant frequency components. To obtain a prediction of nonzero population transfer for such frequencies, whose detunings are greater than the bandwidth of the Fourier spectrum, $|\Delta| > \Delta_0$, it is necessary to use higher-order perturbation theory. The next nonzero order, third order, requires evaluation of a triple integral. A tedious calculation yields the third-order transition probability for $\Delta_0 < |\Delta| < 3\Delta_0$ $\mathcal{P}^{III}(\Delta)$ in terms of the dilogarithm function. This third-order approximation vanishes identically for $|\Delta| > 3\Delta_0$,

$$\mathcal{P}^{III}(\Delta) = 0, \qquad (|\Delta| > 3\Delta_0). \tag{9.17}$$

For detunings larger than $3\Delta_0$, a nonzero result can be obtained only from fifth- and higher-order perturbation theory.

The FOPT result of eqn. (9.16) predicts a constant value, equal to 1/4 the square of the pulse area A_∞ [Vit05]. The first-order result accurately describes weak excitation for small values of the pulse area. When the detuning is outside the Fourier bandwidth ($|\Delta| > \Delta_0$), the third-order perturbation theory describes weak excitation well.

Perturbation theory can be expected to be reliable only for weak excitation. In the present context that means small pulse areas and, for given bandwidth, small values of the peak Rabi frequency. When the pulse becomes stronger, one must rely on numerical integration of the Schrödinger equation. From such results one can see that, for sufficiently strong pulses, complete excitation can occur even in the absence of resonant frequency components.

The perturbation theory results give no idea of the details of the time evolution produced by the sinc pulse of eqn. (9.12). This pulse has a long precurser succession of small oscillations, mirrored by a long oscillatory tail. The effects of these oscillations are evident in plots of the time evolution. In particular, although the dominant effect of population transfer occurs during a relatively small time interval, the oscillatory nature of the pulse has influence over a very long time.

Bichromatic field. A simple example of a field in which appreciable population change occurs even though there is no frequency component resonant with the Bohr frequency is the bichromatic field [Ehl90; Una00; Yat03]. In this example the field has two frequency components ω_1 and ω_2 separated by δ. When their average frequency $\bar{\omega} = (\omega_1 - \omega_2)/2$ is resonant, the two-state system undergoes population oscillations, with periodic complete inversion, at the difference frequency δ, even though the field contains no resonant component.

9.4 The Dirac (interaction) picture

The states hitherto labeled 1 and 2 could be any of the discrete quantum states of the system between which the Hamiltonian matrix has links. The following sections take this generalization into account by introducing letter indices n and m for the pairs of states. As in the preceding discussion, the attention remains on weak excitation.

In seeking analytic expressions, as we do here, the mathematics simplifies by choosing a modified Dirac picture for the phases, supplementing the unperturbed energies E_n with any static interaction-induced shifts V_{nn},

$$\hbar\dot{\zeta}_n(t) = E_n + V_{nn}, \tag{9.18}$$

and writing the statevector for the N-state quantum system as in eqn. (7.20),

$$\boldsymbol{\Psi}(t) = \sum_{n=1}^{N} \exp[-i\zeta_n(t)]\, C_n(t)\, \boldsymbol{\psi}_n. \tag{9.19}$$

We consider interaction with monochromatic light of frequency ω in the electric-dipole approximation, taking the field at the center of mass to be

$$\mathbf{E}(t) = \frac{1}{2}\left[\mathbf{e}\check{\mathcal{E}}(t)\,e^{-i\omega t} + \mathbf{e}^*\check{\mathcal{E}}(t)^*\,e^{i\omega t}\right], \tag{9.20}$$

with $\check{\mathcal{E}}(t)$ possibly complex. The equation for the propagator,

$$\frac{d}{dt}\mathsf{U}(\omega;t) = -i\mathsf{W}(\omega;t)\mathsf{U}(\omega;t), \qquad \mathsf{U}(\omega;0) = \mathbf{1}, \tag{9.21}$$

has a coefficient matrix $\mathsf{W}(\omega;t)$ that has null diagonal elements and off-diagonal elements expressible, for $n > m$, as

$$W_{nm}(\omega;t) = \frac{1}{2}\left[\check{\Omega}_{nm}(t) + \check{\Omega}_{nm}(t)^* e^{-2i\omega t}\right] e^{i(\omega_{nm}-\omega)t}, \qquad n > m, \tag{9.22}$$

where the interaction strength is measured by the complex-valued Rabi frequency $\check{\Omega}_{nm}(t)$,

$$\hbar\check{\Omega}_{nm}(t) = -\mathbf{d}_{nm}\cdot\mathbf{e}\check{\mathcal{E}}(t), \tag{9.23}$$

and ω_{nm} is the Bohr transition frequency for the n,m transition, shifted by any static interactions,

$$\hbar\omega_{nm} \equiv E_n + V_{nn} - E_m - V_{mm}. \tag{9.24}$$

The expression (9.22) involves a term that varies with twice the optical frequency. Our interest here lies with more slowly varying amplitudes, for which such rapid variation averages to zero (the rotating-wave approximation). Thus we take the off-diagonal elements of $\mathsf{W}(\omega;t)$ to be

$$W_{nm}(\omega;t) = \frac{1}{2}\check{\Omega}_{nm}(t)\,e^{i(\omega_{nm}-\omega)t}, \qquad n > m. \tag{9.25}$$

There exist nonzero values of this matrix element for every pair of quantum states between which there is a nonzero transition dipole moment along the field polarization direction. With the present choice of phases $\zeta_n(t)$ there occur no detunings as diagonal elements of the matrix $\mathsf{W}(\omega;t)$. Instead the detunings appear in exponentials that multiply the Rabi frequencies; the matrix elements $W_{nm}(\omega;t)$ cannot generally be regarded as slowly varying.

9.5 Weak broadband radiation; transition rates

One of the common models of incoherent excitation treats the interaction as steady (and hence monochromatic) but weak, so that little population transfer occurs and time-dependent perturbation theory provides an adequate description of the system response. To treat general weak excitation of any multistate system we start from the equation for the time-evolution matrix, eqn. (9.21), in which the matrix $\mathsf{W}(\omega;t)$ has no diagonal elements and has off-diagonal elements (transition amplitudes) given by eqn. (9.25), with constant Rabi frequencies, $\check{\Omega}_{nm}(t) = \check{\Omega}_{nm}$.

To treat small changes we first recast the defining equation (9.21) into integral form. For $n > m$ these read

$$U_{nm}(\omega; \tau) = \delta_{nm} - \frac{i}{2} \sum_{m'} \int_0^\tau dt \, \check{\Omega}_{nm'} e^{i(\omega_{nm'} - \omega)t} U_{m'm}(\omega; t'). \qquad (9.26)$$

We then assume that little change occurs, so that under the integral we can make the approximation

$$U_{m'm}(\omega; t') \approx U_{m'm}(\omega; 0) = \delta_{m'm}. \qquad (9.27)$$

The transition amplitude in this perturbative approximation becomes expressible as a definite integral; for $n > m$ this reads

$$U_{nm}(\omega; \tau) = \delta_{nm} - \frac{i}{2} \int_0^\tau dt \, \check{\Omega}_{nm} e^{i(\omega_{nm} - \omega)t}$$

$$= \delta_{nm} - \frac{1}{2} \check{\Omega}_{nm} \frac{e^{i(\omega_{nm} - \omega)\tau} - 1}{\omega_{nm} - \omega}. \qquad (9.28)$$

The transition probability, the square of this transition amplitude, is

$$\mathsf{W}_{nm}(\omega; \tau) \equiv |U_{nm}(\omega; \tau)|^2 = \frac{1}{2} |\check{\Omega}_{nm}|^2 \frac{1 - \cos[(\omega_{nm} - \omega)\tau]}{(\omega_{nm} - \omega)^2}. \qquad (9.29)$$

The *transition rate* is, by definition, the time derivative of the transition probability, suitably averaged.

To evaluate the transition rate we consider the effect of an incoherent sum of frequencies, distributed with probability $\mathsf{p}(\omega - \omega_L)$ about a central laser frequency ω_L. The transition rate for this broadband radiation is

$$\mathcal{R}_{n \to m}(\tau) = \int d\omega \, \mathsf{p}(\omega - \omega_L) \frac{d}{d\tau} |U_{nm}(\omega; \tau)|^2. \qquad (9.30)$$

We rewrite this as

$$\mathcal{R}_{n \to m}(\tau) = \frac{1}{2} |\check{\Omega}_{nm}|^2 \int d\omega \, \mathsf{p}(\omega - \omega_L) f(\omega_{nm} - \omega; \tau), \qquad (9.31)$$

where the time dependence occurs only in the factor

$$f(x; \tau) = \frac{\sin(x\tau)}{x}. \qquad (9.32)$$

The integral (9.31) involves the product of two separate functions of frequency ω. The factor $\mathsf{p}(\omega - \omega_L)$ is independent of time τ and has a frequency spread (a bandwidth) dictated by the radiation. Typically this is a Lorentzian function of ω. The second factor, $f(\omega_{nm} - \omega; \tau)$, is an oscillatory function of τ. It has a peak value τ that increases linearly with time, while the area under the curve remains fixed,

$$\int_{-\infty}^{+\infty} dx \, f(x; \tau) = \pi. \qquad (9.33)$$

The effective width of this function diminishes inversely with increasing time, as it grows steadily more peaked about the frequency $\omega = \omega_{nm}$. At sufficiently long time this function will sample only the values of $p(\omega - \omega_L)$ near $\omega = \omega_{nm}$, and we can remove from the integral the sampled value $p(\omega_{nm} - \omega_L)$. Under these conditions the result is

$$\mathcal{R}_{n \to m}(\tau) \to \mathcal{R}_{n \to m} = \pi p(\omega_{nm} - \omega_L)|\check{\Omega}_{nm}|^2. \tag{9.34}$$

We can apply this formula to slowly varying field amplitudes and introduce slowly varying Rabi frequencies and radiation intensity $I(t)$,

$$\Omega_{nm}(t) = -d_{nm}\mathcal{E}(t)/\hbar, \qquad |\mathcal{E}(t)|^2 = 2I(t)/c\epsilon_0, \tag{9.35}$$

to obtain radiative transition rates as

$$\mathcal{R}_{n \to m}(t) = p(\omega_{nm} - \omega_L)\frac{2\pi e^2}{c\hbar^2\epsilon_0}|r_{nm}|^2 I(t). \tag{9.36}$$

This perturbation-theory result applies for times that are sufficiently long that the effective frequency width of $f(\omega_{nm} - \omega)$ has become much narrower than the radiation frequency distribution $p(\omega - \omega_L)$. This occurs once the time exceeds a few multiples of the inverse of the radiation bandwidth. If the radiation is sufficiently broadband, then eventually a time will arrive when eqn. (9.34) applies. Until that occurs, we are in a transient regime of coherent excitation, in which $\mathcal{R}_{n \to m}(\tau)$ exhibits temporal oscillations (population flopping).

9.6 Fermi's famous Golden Rule

The limiting procedure used to derive eqn. (9.34) from eqn. (9.31) is traditionally presented, in treatments of time-dependent perturbation theory, in the form [Fer50]

$$\mathcal{R}_{n \to m} = \frac{2\pi}{\hbar}|V_{nm}|^2 \delta(E_n - E_m), \tag{9.37}$$

in which the Dirac delta $\delta(x)$ is defined to have the property

$$\int_{-\infty}^{\infty} dx\, F(x)\delta(x - x_0) = F(x_0). \tag{9.38}$$

Formula (9.37), often referred to as "Fermi's famous Golden Rule",[1] must be used with some (incoherent) continuum distribution of energies. The Dirac delta picks from this continuum those values for which the transition conserves energy.

For example, when applied to the (incoherent) photoionization reaction

$$A + \hbar\omega \to A^+ + e \tag{9.39}$$

---

[1] This is actually the *second* of two "Golden Rules" of Fermi; the first is the rule for obtaining the second-order perturbation-theory value for the energy shift of state n as $\sum_m V_{nm}V_{mn}/(E_n - E_m)$.

the Dirac delta selects from the final states of photoelectrons those whose kinetic energy E_{kin} meets the requirement[2]

$$E_{atom} + \hbar\omega = E_{ion} + E_{kin}. \tag{9.40}$$

In this situation the energies of atom and ion are discrete, but those of the unbound photo-electron are not *a priori* restricted to discrete values; the Dirac delta then selects a discrete energy-conserving value from the available continuum. The dipole matrix element describes a *bound–free* transition,

$$r_{n,E} = \int d\mathbf{r}\,\psi_n(\mathbf{r})^*\,\mathbf{r}\cdot\mathbf{e}\,\psi_E(\mathbf{r}). \tag{9.41}$$

As can be recognized from the derivation of the Golden Rule for transition rates, a number of restrictions apply. In particular, the transition rates are not applicable to the coherent excitation scenarios that produce Rabi oscillations and adiabatic following.

---

[2] As mentioned in Sec. 15.1 the kinetic energy of the photoelectron includes a ponderomotive component that becomes significant for intense fields.

10
The vector model

The statevector for a two-state system involves two complex-valued probability amplitudes $C_n(t)$. With rotating coordinate 2 and $E_1 = 0$ the construction is, as in eqn. (8.47),

$$\Psi(t) = C_1(t)\, \psi_1 + C_2(t)\, \psi_2'(t). \qquad (10.1)$$

The two complex amplitudes are constrained by the normalization requirement $\sum_n |C_n(t)|^2 = 1$. When the two-state system remains isolated from links to other states the overall phase is not of interest. In this situation we need only two real-valued time-varying parameters to characterize the statevector. As noted in Sec. 5.6.1, a simple description presents the statevector as a point on a two-dimensional sphere, the Bloch sphere, parametrized by two angles. This description, analogous to the Poincaré-sphere representation of the polarization state of the electric field of a ray, has several useful applications, and allows a generalization in which incoherent processes are treated. The resulting formalism, in which both the system properties and the Hamiltonian are represented by vectors in three-dimensional abstract spaces, was first described by Feynman, Vernon, and Hellwarth[1] [Fey57] [Sho90, §8.5]. The following paragraphs discuss the model, with its simple means of depicting the dynamics of a two-state atom. Section 16.10.2 discusses an N-state generalization.

10.1 The Feynman–Vernon–Hellwarth equations

The parametrization of the two-state RWA Hamiltonian requires a real-valued detuning and a real-valued Rabi frequency ω with phase φ (or a complex-valued Rabi frequency $\check{\Omega} = \Omega e^{i\varphi}$). Because the absolute phase of the field is uncontrollable, it is convenient to set this to zero at the start of a pulse, but subsequent manipulation of the laser field may require a later nonzero value. The description of the quantum states then requires three real-valued functions for complete characterization.

The Bloch variables. A very satisfactory choice of variables for the two-state atom are the three real-valued quantities now known as *Bloch variables*, see Sec. 5.6.1 [Sho90, §8.1].

---

[1] The model results rather simply from the two-state density matrix, see Sec. 16.6.2.

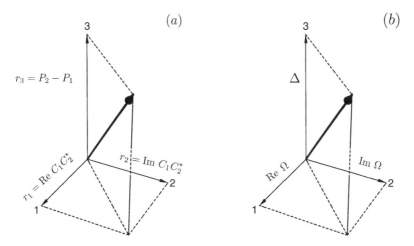

Figure 10.1 (*a*) The Bloch vector. The 1, 2 axes are set by the initial phase. (*b*) The angular velocity vector (torque vector). The projection into the 1, 2 plane is determined by the field phase φ.

These are often denoted u, v, w, but to emphasize their interpretation as components of a vector $\mathbf{r}$ (the Bloch vector) in an abstract three-dimensional space I use the notation r_1, r_2, r_3 for the Cartesian components of this vector, and denote the angle in spherical coordinates (the *Bloch angles*) as θ and φ of eqns. (5.34) and (5.35),[2]

$$u(t) \equiv r_1(t) = 2\text{Re}\ [C_1(t)C_2(t)^*] = \sin\theta(t)\cos\varphi(t),$$
$$v(t) \equiv r_2(t) = 2\text{Im}\ [C_1(t)C_2(t)^*] = \sin\theta(t)\sin\varphi(t), \qquad (10.2)$$
$$w(t) \equiv r_3(t) = |C_2(t)|^2 - |C_1(t)|^2 = \cos\theta(t).$$

In the absence of probability loss this Bloch vector maintains unit length,

$$|\mathbf{r}(t)|^2 = r_1(t)^2 + r_2(t)^2 + r_3(t)^2 = 1, \qquad (10.3)$$

and hence it maps the Hilbert-space statevector onto a point on the surface of a unit sphere (the Bloch sphere). Figure 10.1(*a*) illustrates the coordinates of the Bloch vector. The 1, 2 axes are set by the initial E-field phase. It is customary, and convenient, to take this as zero, thereby fixing the relative orientation of the coordinate systems. Missing from the Bloch vector parametrization of the statevector is any information about the overall phase. Thus the motion of the Bloch vector does not reveal the sign changes of probability amplitudes that accompany a single Rabi cycle.

From the components of the Bloch vector we evaluate the populations as

$$P_2(t) = \tfrac{1}{2}[1 + r_3(t)], \qquad P_1(t) = \tfrac{1}{2}[1 - r_3(t)], \qquad (10.4)$$

---

[2] The Bloch variables are here defined with respect to a coordinate system that rotates steadily at the field carrier frequency ω.

and the expectation value of the dipole moment as

$$\langle \mathbf{d}(t) \rangle = 2\mathrm{Re}\left\{ \mathbf{d}_{12} C_1(t)^* C_2(t) \exp(-\mathrm{i}\omega t) \right\}$$

$$= d_{12}\left\{ r_1(t)\cos(\omega t) - r_2(t)\sin(\omega t) \right\}. \tag{10.5}$$

When the RWA applies, as here assumed, the variables $C_n(t)$ and $r_j(t)$ vary more slowly than the carrier phase ωt. The Bloch-vector component $r_3(t)$ is the population inversion; components $r_1(t)$ and $r_2(t)$ are termed *coherences*. Component $r_1(t)$ is said to be *in phase* with the electric field of the laser, while component $r_2(t)$ is *in quadrature* with this field.

Phase. The Bloch vector, like the Stokes vector, provides no information about the overall phase of a statevector or of an adiabatic state. Initially this phase can be chosen arbitrarily, but later alterations can be produced. Section 20.7.2 discusses measurement of this phase.

The vector equations. For the TDSE with the RWA Hamiltonian of eqn. (8.8) of Sec. 8.1,

$$W(t) = \begin{bmatrix} 0 & \tfrac{1}{2}\Omega(t)\,\mathrm{e}^{-\mathrm{i}\varphi} \\ \tfrac{1}{2}\Omega(t)\,\mathrm{e}^{\mathrm{i}\varphi} & \Delta \end{bmatrix}, \tag{10.6}$$

the Bloch variables satisfy the equation

$$\frac{d}{dt}\begin{bmatrix} r_1 \\ r_2 \\ r_3 \end{bmatrix} = \begin{bmatrix} 0 & -\Delta & -\Omega\sin\varphi \\ \Delta & 0 & -\Omega\cos\varphi \\ \Omega\sin\varphi & \Omega\cos\varphi & 0 \end{bmatrix}\begin{bmatrix} r_1 \\ r_2 \\ r_3 \end{bmatrix}. \tag{10.7}$$

These are a special case (absent relaxation) of the RWA *optical Bloch equations* (OBE),[3] in a Hilbert-space reference frame rotating with angular velocity ω [Sho90, §8.4]. Although not shown explicitly, the Bloch variables and the elements of the RWA Hamiltonian can all vary with time.

These three coupled equations can be cast into a torque equation for the three-dimensional vector $\mathbf{r}$,

$$\frac{d}{dt}\,\mathbf{r}(t) = \boldsymbol{\Upsilon}(t) \times \mathbf{r}(t), \tag{10.8}$$

in which the elements of the RWA Hamiltonian appear organized into an *angular velocity vector* $\boldsymbol{\Upsilon}$ (often termed the *torque vector*), whose components are

$$\Upsilon_1(t) = \Omega(t)\cos\varphi(t), \qquad \Upsilon_2(t) = -\Omega(t)\sin\varphi(t), \qquad \Upsilon_3(t) = \Delta(t). \tag{10.9}$$

The length of this vector is the rms of the Rabi frequency (i.e. the interaction energy) and the detuning,

$$\Upsilon(t) = |\boldsymbol{\Upsilon}(t)| = \sqrt{\Delta(t)^2 + \Omega(t)^2}. \tag{10.10}$$

---

[3] Often the term "Bloch equations" is reserved for equations that include relaxation, as in Sec 10.2.1. The application here is to pulses of laser radiation, as contrasted with the original introduction of equations applicable to radio-frequency excitation, hence the term optical Bloch equations (OBE).

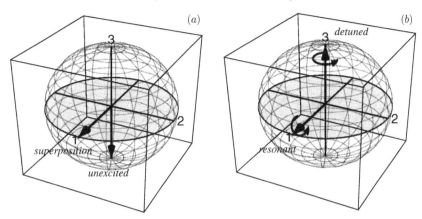

Figure 10.2 (*a*) The initial Bloch vector points to the south pole of the Bloch sphere. (*b*) Possible torque vectors: for resonant excitation the vector is in the equatorial plane; for detuning without any field it lies on the polar axis.

The azimuthal angle φ of the vector is the field phase while the polar angle is defined by the relationship

$$\cot\theta(t) = \Delta(t)/\Omega(t). \qquad (10.11)$$

Figure 10.1(*b*) shows this vector.

Picturing the dynamics. From the torque equation we recognize that the Bloch vector rotates at the instantaneous rate $|\Upsilon(t)|$ about an axis defined by the components of the angular velocity vector $\Upsilon(t)$. Figure 10.1(*b*) illustrates the angular velocity vector, parametrizing the RWA Hamiltonian. The projections onto the 1, 2 axes are set by the E-field phase φ.

By regarding the action of the angular velocity vector as producing a torque that turns the Bloch vector it is straightforward to visualize the dynamics resulting from a simple pulse. Initially the atom is unexcited, and the Bloch vector points to the south pole of the Bloch sphere, as shown in Fig. 10.2(*a*). For resonant excitation the torque vector Υ lies in the equatorial plane; typically we take the Rabi frequency to be real-valued, and then the torque vector lies along the 1 axis of the Bloch sphere. When the laser is detuned, but the field is absent (as occurs before or after the arrival of a pulse), the torque vector lies along the polar axis. Figure 10.2(*b*) illustrates these situations.[4]

Figure 10.3 illustrates two examples of excitation by a constant-amplitude pulse. Frame (*a*) shows the motion of the Bloch vector **r** for resonant excitation. Because the pulse is resonant, with real amplitude, the torque vector Υ lies along the 1 axis, and the Bloch vector traces a disk in the 2, 3 plane. Frame (*b*) shows the Bloch vector motion for excitation with constant detuning and small initial inversion but no field, as would occur after the pulse ceases. The torque vector lies along the polar axis, and the Bloch vector traces out a cone

---

[4] In the absence of any pulsed radiation the carrier frequency ω used to define the detuning is arbitrary. It should be chosen in anticipation of the pulse that will arrive later.

(a)   (b)

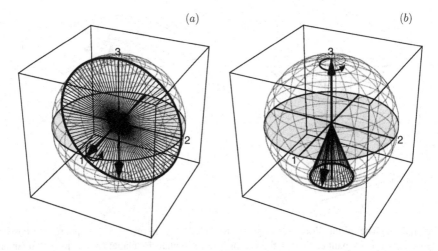

Figure 10.3 Two-state vector model, showing the action of an angular velocity vector on the Bloch vector. (*a*) Resonant excitation ($\Delta = 0$). The torque vector lies along the 1 axis and the Bloch vector follows a circular path in the 2, 3 plane. (*b*) Detuned excitation without Rabi frequency. The torque vector lies along the 3 axis, and moves the Bloch vector in a cone about this axis.

around this axis. For this illustration the initial Bloch vector is offset from the south pole; some previous excitation must have occurred.

10.1.1 Steady field

Although the torque equation for the Bloch vector holds for any pulsed excitation (within the RWA), it is most useful when applied to situations when the RWA Hamiltonian has constant or slowly varying elements. Some examples follow.

Resonant. When the excitation is resonant and the initial state is that of population entirely in state 1 the tip of the Bloch vector moves along a great-circle path, along a longitude in the plane normal to the angular velocity vector. The projections of this motion onto the 3 axis are the Rabi oscillations of population inversion. Figure 10.4 illustrates this behavior.

Nonresonant. When the excitation is nonresonant the angular velocity vector lies out of the equatorial plane of the Bloch sphere. It produces motion of the Bloch vector on a conical surface, as shown in Fig. 10.5.

10.2 Coherence loss; relaxation

The two-state coherences r_1 and r_2 are responsible for the effects of matter upon a resonant (or near-resonant) field. During times that are shorter than a relaxation time for the coherence, the coherences will retain phase properties and will therefore be capable of producing

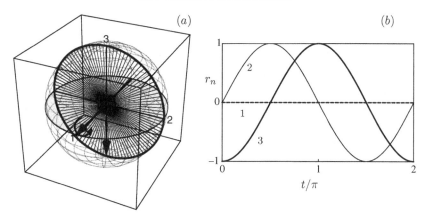

Figure 10.4 (*a*) Depiction of rotating Bloch vector for resonant excitation. The Bloch vector traces a great circle. (*b*) Plots of the time dependence of the three components of the Bloch vector r_n. [Redrawn from Fig. 26 of B. W. Shore, Acta Physica Slovaka, **58**, 243–486 (2008) with permission.]

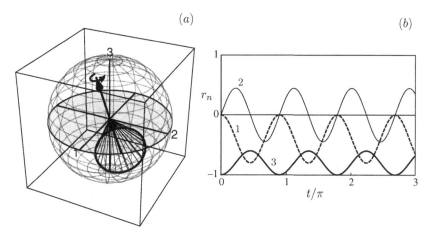

Figure 10.5 As in the previous figure, 10.4, but with detuning. The Bloch vector traces a circular path that has components along all three axes.

coherent modifications of the electric field envelopes. Two classes of relaxation phenomena contribute to the loss of phase for the coherences:

Inhomogeneous: The sample contains atoms that have different velocities (and therefore different Doppler shifts of detunings) or different static surroundings (and therefore different energy shifts and different detunings).
Homogeneous: All atoms of a particular species undergo spontaneous emission at the same rate, and hence all have the same probability loss.

Subsequent sections describe methods for treating each of these forms of relaxation.

10.2.1 Bloch equations with relaxation

The TDSE is able to treat spontaneous emission as a probability loss from the excited state but it cannot properly deal with the arrival of this population in the ground state: the change does not preserve phases and so it cannot be described by a changing statevector. The Bloch variables allow treatment of spontaneous emission as well as other processes that fail to maintain phase coherence.

Spontaneous emission at rate A_{21} from state 2 to state 1 will change the two populations in a manner that can be described by rate equations. In the absence of coherent radiation these read

$$\frac{d}{dt}P_1 = A_{21}P_2, \quad \frac{d}{dt}P_2(t) = -A_{21}P_2, \tag{10.12}$$

and thus the change in the inversion is expressed by the equation

$$\frac{d}{dt}r_3 = -2A_{21}r_3. \tag{10.13}$$

The emission does not preserve statevector phases – it is a random process – and so it will tend to diminish the phase coherence. The affect of spontaneous emission on the coherences obtains from the equations

$$\frac{d}{dt}(r_1 + ir_2) = -A_{21}(r_1 + ir_2). \tag{10.14}$$

Other processes, such as elastic collisions, also produce a coherence loss. To account for such effects the needed revisions of the Bloch equations are often written as

$$\frac{d}{dt}r_1(t) = -\frac{1}{T_2}r_1(t) - \Delta\, r_2(t) + \Omega_I\, r_3(t),$$

$$\frac{d}{dt}r_2(t) = -\frac{1}{T_2}r_2(t) + \Delta\, r_1(t) + \Omega_R\, r_3(t), \tag{10.15}$$

$$\frac{d}{dt}r_3(t) = -\frac{1}{T_1}[r_3(t) + 1] - \Omega_I\, r_1(t) - \Omega_R\, r_2(t),$$

where the complex-valued Rabi frequency has been separated into real and imaginary parts,

$$\check{\Omega} = \Omega_R + i\Omega_R = \Omega e^{i\varphi} = \Omega\cos\varphi + i\Omega\sin\varphi. \tag{10.16}$$

Here the incoherent processes are parametrized by two times, traditionally denoted T_1 and T_2. The *longitudinal* relaxation time T_1 parametrizes the effect of population-changing events: spontaneous emission and inelastic (energy-changing) collisions; it is the average time required for the atom to change states (e.g. decay from the excited state to the ground state or collisional excitation). The *transverse* relaxation time T_2 parametrizes the effect of events that alter phases; it is the average time required for spontaneous emission or elastic collisions to alter phases.

The effect of T_2 is to pull the Bloch vector toward the vertical axis, towards $r_1 = r_2 = 0$. The effect of T_1 is to pull the Bloch vector toward the south pole of the Bloch sphere, towards $r_3 = -1$. In the absence of coherent excitation, $\Omega = 0$, this point is the ultimate resting point of the Bloch vector – all population resides in the ground state.

These equations model a closed system: there is no radiative decay from state 2 except to state 1. When spontaneous emission is the only relaxation process the two relaxation times have the connection $2T_1 = T_2 = 1/A_{21}$. More generally there exists the constraint

$$T_2 \leq 2T_1. \tag{10.17}$$

10.2.2 Adiabatic elimination: The rate equation limit

Over times that are longer than the coherence relaxation time T_2 the coherences equilibrate with the more slowly changing populations. We obtain these equilibrium values by setting the time derivatives of r_1 and r_2 to zero. The result of this adiabatic elimination is

$$r_1(t) = T_2 \frac{\Omega_R T_2 \Delta - \Omega_I}{1 + (T_2\Delta)^2} r_3(t), \qquad r_2(t) = -T_2 \frac{\Omega_I T_2 \Delta + \Omega_R}{1 + (T_2\Delta)^2} r_3(t). \tag{10.18}$$

These formulas produce for the inversion the equation

$$\frac{d}{dt} r_3(t) = -\frac{1}{T_1} - \frac{1}{T_1} \frac{(1/T_2)^2 + \Delta^2 + (T_1/T_2)|\check{\Omega}|^2}{\Delta^2 + (1/T_2)^2} r_3(t). \tag{10.19}$$

This has the form of a rate equation; upon introducing rate coefficients $\mathcal{R}_{n \to m}(t)$ we rewrite it as

$$\frac{d}{dt} r_3(t) = [\mathcal{R}_{1 \to 2}(t) - \mathcal{R}_{2 \to 1}(t)] - [\mathcal{R}_{1 \to 2}(t) + \mathcal{R}_{2 \to 1}(t)] r_3(t). \tag{10.20}$$

The rate coefficients are expressible in terms of Rabi frequencies as

$$\mathcal{R}_{2 \to 1}(t) = \frac{1}{T_1}\left[1 + \frac{(T_1/2T_2)|\check{\Omega}|^2}{\Delta^2 + (1/T_2)^2}\right], \qquad \mathcal{R}_{1 \to 2}(t) = \frac{1}{2T_2}\left[\frac{|\check{\Omega}|^2}{\Delta^2 + (1/T_2)^2}\right]. \tag{10.21}$$

10.2.3 The excitation cross section

It is useful to express the laser-induced excitation rate as the product

$$\mathcal{R}_{1 \to 2}(t) = \sigma_{12}(\Delta, \mathbf{e}) \mathcal{F}(t) \tag{10.22}$$

of the photon fluence (photons per unit area per unit time)

$$\mathcal{F}(t) = I(t)/\hbar\omega_{12} \tag{10.23}$$

and a frequency-dependent cross section for excitation produced by the field of direction $\mathbf{e}$, in an atom whose transverse relaxation time is T_2,

$$\sigma_{12}(\Delta, \mathbf{e}) = \sigma_{12}(0, \mathbf{e}) \frac{(1/T_2)^2}{\Delta^2 + (1/T_2)^2}. \tag{10.24}$$

The frequency distribution is a Lorentzian function, here presented in terms of the resonant value of this cross section, expressible as

$$\sigma_{12}(0, \mathbf{e}) \equiv \sigma_{12} = \frac{|d_{12}|^2 \omega_{12}}{\hbar c \epsilon_0 (1/T_2)} = 4\pi\alpha \, |\mathbf{r}_{12} \cdot \mathbf{e}|^2 \, \omega_{12} T_2. \tag{10.25}$$

The latter formula uses the Sommerfeld fine-structure constant

$$\alpha = \frac{e^2}{4\pi\epsilon_0 \hbar c} \approx \frac{1}{137}. \tag{10.26}$$

As is evident, the cross section varies directly as an effective area of the transition moment $\pi |\mathbf{r}_{12} \cdot \mathbf{e}|^2$ and with the decoherence time T_2, i.e. inversely with the coherence-relaxation rate. The frequency-integrated cross section is independent of this rate,

$$\int_{-\infty}^{+\infty} d\Delta \, \sigma_{12}(\Delta, \mathbf{e}) = \frac{\pi}{T_2} \sigma_{12}. \tag{10.27}$$

Comments. Radiative rate equations made their first appearance as phenomenological descriptions of radiative processes, expressing the insight of Einstein (see [Sho90, §2.2]). Derivations of the Einstein A and B coefficients from time-dependent perturbation theory can be found in many textbooks on radiation theory. More general rate equations, allowing for incoherent changes produced by collisions, are widely used in chemical kinetics [Lev69; Nik74].

10.2.4 Steady-state excitation; power broadening

When the illumination of the atom is steady then spontaneous emission provides a balance to the excitation, as recognized by Einstein, and the various elements of the Bloch vector reach a steady state. More generally, when the relaxation time T_2 is short the variables r_1 and r_2 equilibrate with the instantaneous value of the inversion. When the rate is constant the equilibrium solution to the rate equation, obtained by setting $\dot{r}_3 = 0$, is

$$\bar{r}_3 = -\frac{\Delta^2 + (1/T_2)^2}{\Delta^2 + (1/T_2)^2 + (T_1/T_2)|\check{\Omega}|^2}. \tag{10.28}$$

When the Rabi frequency is small, or when the detuning is very large, all population resides in the ground state:

$$\text{If } \Delta^2 + (1/T_2)^2 \gg (T_1/T_2)|\check{\Omega}|^2 \text{ then } \bar{r}_3 = -1. \tag{10.29a}$$

By contrast, when the Rabi frequency is large the equilibrium inversion vanishes, $\bar{r}_3 = 0$, meaning the two populations are equal:

$$\text{If } |\check{\Omega}|^2 \gg \Delta^2 + (1/T_2)^2 \text{ then } \bar{r}_3 = 0. \tag{10.29b}$$

However, this is not a coherent superposition of the two states. Within the steady-state approximation it is not possible to obtain positive equilibrium inversion; the equilibrium populations must satisfy the inequality $\bar{P}_2 < \bar{P}_1$.

Power broadening. When spontaneous emission, at rate A_{21}, provides the only incoherent process the relaxation times are $T_2 = 2/A_{21}$ and $T_1 = 1/A_{21}$. Then the equilibrium inversion is

$$\bar{r}_3 = -\frac{(A_{21}/2)^2 + \Delta^2}{\Delta^2 + (A_{21}/2)^2 + (1/2)\Omega^2}. \tag{10.30}$$

Under these conditions the steady-state excited-state population is

$$\bar{P}_2 = \frac{(\Omega/2)^2}{\Delta^2 + (A_{21}/2)^2 + (1/2)\Omega^2}. \tag{10.31}$$

That is, the excitation is greatest, as one expects, for resonance, $\Delta = 0$, where it can be no larger than $\bar{P}_2 = 0.5$. The dependence of this equilibrium population on detuning Δ, the *excitation spectrum*, is a *Lorentz profile* whose FWHM is

$$\Delta_{1/2} = \sqrt{(A_{21})^2 + 2\Omega^2}. \tag{10.32}$$

In the absence of the laser radiation this is the *natural line width* $\Delta_{1/2} = A_{21}/2$. But as the excitation becomes stronger, as parametrized by the radiation power, the width increases; it exhibits *power broadening*. In the limit of strong excitation the width is directly proportional to the Rabi frequency, i.e. to the square root of the intensity.

The occurrence of power broadening depends upon the conditions of excitation, here taken to be steady illumination with damping. When the excitation is pulsed, and observations take place after the pulse, then power broadening does not occur [Vit01c].

Steady state; linear optics. In the absence of relaxation the solutions to the Bloch equation, like those of the Schrödinger equation, oscillate interminably with ongoing time; the variables never approach constant values. However, when phase incoherence is present, as parametrized by T_2, then Bloch components eventually approach steady values $\bar{r}_j$. The steady coherences relate to the steady inversion $\bar{r}_3$ by the formulas

$$\bar{r}_1 = \frac{[\Omega_I - (\Delta T_2)\Omega_R]}{1 + (\Delta T_2)^2}\bar{r}_3, \qquad \bar{r}_2 = \frac{[\Omega_R - (\Delta T_2)\Omega_I]}{1 + (\Delta T_2)^2}\bar{r}_3. \tag{10.33}$$

Each of these components is directly proportional to a Rabi frequency and hence the dipole moment of eqn. (10.5) is proportional to the amplitude of the E field. The resulting linear response is the physics that underlies the linear approximation for the polarization field discussed in App. C.4 as the foundation for conventional linear optics (as contrasted with nonlinear optics).

Fluorescence spectrum. This steady-state model provides a description of excitation that is useful for understanding the frequency variation of excited-state photoionization – the *excitation spectrum.* By contrast, the fluorescence radiation has its own distribution of frequencies. A spectroscopist, collecting the fluorescence, passes this radiation through a spectrometer to determine the traditional *fluorescence spectrum.* The totality of this radiation, into all solid angles and integrated over all frequencies, is proportional to the product of the emission rate A_{21} and the excited-state population, a quantity available from the Bloch equation. However, the simple Bloch-vector model presented here does not give a description of the fluorescence spectrum. When the laser field becomes sufficiently strong that Rabi oscillations dominate the atom dynamics – coherent excitation – the periodic variation of the dipole moment creates sidebands to the fluorescent signal. These are offset from the carrier frequency and are separated by the Rabi frequency. The spectrum appears as a *Mollow triplet* [Mol69] [Sho90, §11.6]. The treatment of such observations requires a more elaborate treatment than the simple average excitation probability.

10.2.5 Inhomogeneous relaxation; Ramsey pulses

There are simple techniques that can overcome inhomogeneous relaxation. These can be understood from examination of the distribution of Bloch vectors for two atoms, each subject to a different detuning Δ because of a different environment. Figure 10.6 illustrates the evolution of the two Bloch vectors, labeled here as positive Δ (top row) and negative Δ

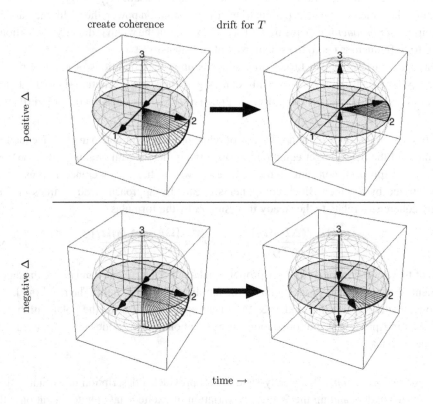

Figure 10.6 Column 1: The preparation of a coherence of two Bloch vectors, for atoms that have different environments and hence different detunings. Column 2: the different detunings (here positive and negative) produce different rotations of the Bloch vectors in the equatorial plane. The top row shows an example for positive detuning, the bottom row has negative detuning.

(bottom row). Both atoms are first prepared, using $\pi/2$ pulses, as coherences – the Bloch vector is rotated by a resonant pulse onto the equator of the Bloch sphere. Following this brief pulse the atoms are left undisturbed. Because they have different detunings, the torque vector rotates the Bloch vectors differently. In the figure the two detunings have opposite signs.

After an interval T of this Bloch vector rotation the two vectors will be as shown in the second column of Fig. 10.6. One must imagine a distribution of such different Bloch vectors describing the medium; as T grows the average of the coherences will diminish.

The first step of restoration of the initial coherence takes place at some time, say T, when rotation of the Bloch vectors has proceeded substantially. At that time we apply a π pulse. This has the effect of rotating all the Bloch vectors from one side of the equatorial plane to the other, as shown in the first column of Fig. 10.7. We now wait for a second time interval T. During that time the Bloch vectors continue their rotation in the equatorial plane, and become aligned along the -2 axis: they are reversed from their initial preparation.

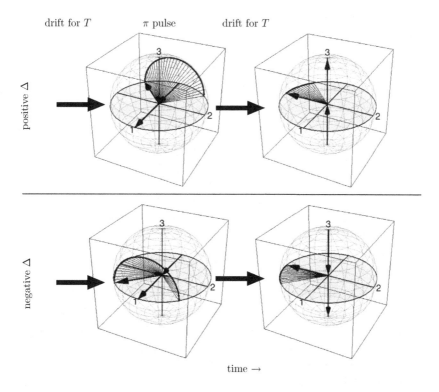

Figure 10.7 A pi pulse applied after time T will produce a coherence that will, after a second interval T, be opposite the original Bloch vector; it is phased for emission rather than absorption (it produces a photon echo). A pi pulse applied at that moment will reproduce the initial coherence. The top row shows an example for positive detuning, the bottom row has negative detuning.

Because the initial coherence resulted from absorption, the new coherence is phased for emission, and it will therefore produce a signal (a *photon echo*). A second π pulse applied at this time, $2T$, will restore the initial Bloch vector. Because all Bloch vectors were originally aligned, this pulse rotates all of them back to the initial position: The pair of π pulses has restored the original coherence.

The total time of the process can be lengthened by applying the second pulse at time $t = 3T$ rather than $t = 2T$. The restoration then occurs after a total time interval of $4T$. This succession of pulse pairs can be repeated. Eventually, however, homogeneous relaxation processes will diminish the coherence.

11

Sequential pulses

The statevector rotated by a single pulse can provide the initial state for a second pulse; the effect of a sequence of pulses can be treated as a succession of statevector rotations. The cumulative effect is best presented with the aid of the time-evolution matrix $U(t, t_0)$, a solution to the Schrödinger equation that reduces to the unit matrix at the initial time $t = t_0$. In particular, we consider a succession of interactions, during each of which the RWA Hamiltonian remains constant. For a constant matrix $W(t) = W$ only a single time argument is needed for the time-evolution matrix and the relevant defining equations are

$$\mathbf{C}(t + \tau) = \mathsf{U}(\tau)\mathbf{C}(t), \tag{11.1}$$

where

$$\frac{d}{d\tau}\mathsf{U}(\tau) = -i\mathsf{W}\mathsf{U}(\tau), \qquad \mathsf{U}(0) = \mathbf{1}. \tag{11.2}$$

11.1 Contiguous pulses

A succession of such matrices offers a valuable tool for treating more general pulses as a succession of constant interactions, thereby providing an algorithm for numerical integration. Let a constant interaction, described by RWA Hamiltonian $W^{(1)}$, persist for an interval T_1 after time t_1. Let this interval be followed by a succession of contiguous intervals T_m, during each of which the interaction is treated as constant. During the mth interval the RWA Hamiltonian is a constant matrix $W^{(m)}$ and the time-evolution operator is the associated matrix $U^{(m)}(t)$. Figure 11.1 illustrates a succession of four such intervals.

The effect of these successive interactions upon the probability amplitudes is expressible as a succession of matrix multiplications that carry the statevector through a succession of Hilbert-space rotations. For the four intervals of Fig. 11.1 the result, at the end of the pulse sequence, has the form

$$\mathbf{C}(t) = \mathsf{U}^{(4)}(T_4)\mathsf{U}^{(3)}(T_3)\mathsf{U}^{(2)}(T_2)\mathsf{U}^{(1)}(T_1)\mathbf{C}(t_1), \quad t = t_1 + T_1 + T_2 + T_3 + T_4. \tag{11.3}$$

The intervals may be of arbitrary duration consistent with the requirement that the system remain unaffected by interruptions. In particular, by breaking the description of a general pulse into a succession of small contiguous increments, one can describe its effect by

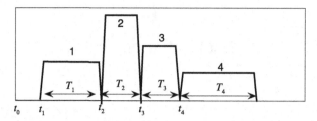

Figure 11.1 A succession of four contiguous constant interactions. To clarify the extent of the separate pulses they are shown as turning on and off during small intervals. [Redrawn from Fig. 22 of B. W. Shore, Acta Physica Slovaka, **58**, 243–486 (2008) with permission.]

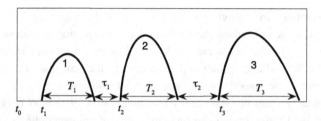

Figure 11.2 A succession of three nonoverlapping pulses, of duration T_n and separated by intervals τ_n. [Redrawn from Fig. 23 of B. W. Shore, Acta Physica Slovaka, **58**, 243–486 (2008) with permission.]

means of a succession of evolutions with constant interactions.[1] This construction, when automated, provides a very effective algorithm for evaluating numerically the effect of an arbitrary pulse. The intervals T_n need not be of uniform duration, although that constraint simplifies the construction of numerical procedures. The mathematics is analogous to that of the Jones calculus of Sec. 3.6.1 applied to a succession of transmissive plates.

11.2 Pulse trains

The description of time evolution by means of independent interactions applies to any succession of pulses [Mly81; Mil92; Tem93; Vit95a]. In particular, there can be time intervals between the interactions, i.e. the system is subjected to a train of pulses, separated by interaction-free intervals τ_n. Figure 11.2 illustrates an example of such a sequence, for three pulses.

Although there is no interaction during the pauses between pulses, it is essential to account for these intervals. Let t_0 be the reference time for expressing the sinusoidal time variation of the field as $\mathcal{E}(t)\cos(\omega t - \varphi)$. That is, the field phase is $\varphi = \omega t_0$. This reference time t_0 also serves to define a rotating coordinate system, earlier written as

$$\psi'_1(t) = \psi_1, \qquad \psi'_2(t) = \exp(-i\omega t)\psi_2, \qquad (11.4)$$

---

[1] The motion of the tip of the statevector can be regarded as a *quantum walk*.

and to express the RWA interaction energy as $(\hbar/2)\Omega(t)\exp(i\varphi)$. In this reference frame the statevector has the construction

$$\Psi(t) = \sum_n C_n(t)\psi'_n(t). \tag{11.5}$$

We now wish to consider a succession of independent interactions, each described by a slowly varying RWA Hamiltonian matrix $\mathsf{W}^{(m)}(t)$. Traditionally, and conveniently, one takes the start of a pulse as the reference time for the rotating Hilbert-space coordinates during a pulse, rather than maintaining an ongoing reference system. This means that during the mth pulse we write the statevector as

$$\Psi(t) = \sum_n C'_n(t)\psi_n^{(m)}(t), \tag{11.6}$$

where the time-dependent basis states

$$\psi_1^{(m)}(t) = \psi_1, \qquad \psi_2^{(m)}(t) = \exp[-i\omega(t - t_m)]\psi_2, \tag{11.7}$$

provide the reference system appropriate to the interval that begins at time t_m. That is, at time $t = t_m$ the rotating coordinates $\psi_n^{(m)}(t)$ coincide with the static basis ψ_n. The probability amplitudes in the pulse-centric basis of eqn. (11.6) incorporate phases from the ongoing carrier frequency

$$\mathbf{C}'(t) = \mathsf{T}(t)\mathbf{C}(t), \tag{11.8}$$

where

$$\mathsf{T}(\tau) = \begin{bmatrix} 1 & \exp(-i\omega\tau) \\ \exp(+i\omega\tau) & 1 \end{bmatrix}. \tag{11.9}$$

The RWA Hamiltonian $\mathsf{W}^{(m)}(t)$ must incorporate the rotation of the coordinates in defining the Rabi frequency. Typically this is taken to be a real-valued quantity at some reference time, say t_0. The ongoing rotation of the Hilbert-space coordinates then introduces phases that must be incorporated into the Hamiltonian at the start of each pulse.

During interactionless intervals the RWA Hamiltonian is a simple diagonal matrix W^0. When detuning is present this can introduce important phase factors in the probability amplitudes.

Given these properties of the successive time evolutions, we can write the time evolution as the product of pulse interactions $\mathsf{U}^{(m)}(t)$, free evolution $\exp[-i\mathsf{W}^0\tau_m]$, and revisions of the coordinate system phases $\mathsf{T}(\tau)$. For the three pulses of Fig. 11.2 the construction is (the operations read from bottom to top)

$$
\begin{aligned}
\mathbf{C}(t) = {}& \exp[-i\mathbf{W}^0(t - t_3)] && \textit{interval 3} \\
& \times \mathbf{U}^{(3)}(T_3) && \textit{pulse 3} \\
& \times \mathbf{T}(t_3 - t_2) && \textit{update coordinates} \\
& \times \exp[-i\mathbf{W}^0\tau_2] && \textit{interval 2} \\
& \times \mathbf{U}^{(2)}(T_2) && \textit{pulse 2} \\
& \times \mathbf{T}(t_2 - t_1) && \textit{update coordinates} \\
& \times \exp[-i\mathbf{W}^0\tau_1] && \textit{interval 1} \\
& \times \mathbf{U}^{(1)}(T_1) && \textit{pulse 1} \\
& \times \mathbf{T}(t_1 - t_0)\mathbf{C}(t_0). && \textit{preliminary}
\end{aligned}
$$

The individual evolution matrices $\mathbf{U}^{(n)}(t)$ appearing here may be composites, constructed as in eqn. (11.3), or they may be analytic constructions appropriate to one of the many interactions for which there exist analytic solutions to the Schrödinger equation [Kyo06].

11.3 Examples

An interesting application of pulse sequences occurs with two pulses, each of which produces a Bloch-vector rotation of $\pi/2$, i.e. a 50:50 superposition. The overall effect depends explicitly on the phase change during the pause (see Sec. 11.4). The resulting variation of excitation, between constructive and destructive interference, is analogous to the bright and dark intensity fringes viewed in the traditional two-slit interference experiment.

The preceding discussion assumes that there is no external field present during the pauses between pulses. If, instead, there occurs a static energy-shifting field, perhaps differing from pulse to pulse, then this must be regarded as a detuning and treated as in eqn. (11.3).

Another interesting example occurs when the pulse train consists of contiguous identical pulses (i.e. the pauses are regarded as part of the pulse). For this analysis we write the RWA Hamiltonian as

$$
\mathbf{W} = \frac{1}{2} \begin{bmatrix} -\Delta(t) & \Omega(t)^* \\ \Omega(t) & \Delta(t) \end{bmatrix}. \tag{11.10}
$$

This form corresponds to the choice of phases

$$
\hbar\dot{\zeta}_1 = \tfrac{1}{2}(E_1 + E_2 - \hbar\omega), \qquad \hbar\dot{\zeta}_2 = \tfrac{1}{2}(E_1 + E_2 + \hbar\omega), \tag{11.11}
$$

rather than the choice made earlier with eqn. (8.45) for eqn. (8.8) of Sec. 8.1. Let the effect of a single pulse be expressed by the unitary matrix

$$
\mathbf{U}^{(1)} = \begin{bmatrix} (a_1 + ib_1) & (c_1 + id_1) \\ -(c_1 - id_1) & (a_1 + ib_1) \end{bmatrix}. \tag{11.12}
$$

Then the effect of N pulses is obtained from the matrix [Vit95a]

$$\mathsf{U}^{(N)} = (\mathsf{U}^{(1)})^N = \begin{bmatrix} \cos(N\vartheta) + ib_1 f_N(\vartheta) & (c_1 + id_1) f_N(\vartheta) \\ -(c_1 - id_1) f_N(\vartheta) & \cos(N\vartheta) - ib_1 f_N(\vartheta), \end{bmatrix}, \qquad (11.13)$$

where

$$f_N(\vartheta) \equiv \frac{\sin(N\vartheta)}{\sin\vartheta}, \qquad \cos\vartheta = a_1, \qquad \sin\vartheta = \sqrt{1 - a_1^2}. \qquad (11.14)$$

Thus if population is in state 1 prior to the arrival of the pulse train, the excited state population after N pulses is

$$P_2^{(N)} = P_2^{(1)} \frac{\sin^2(N\vartheta)}{\sin^2\vartheta}. \qquad (11.15)$$

This is the quantum analog of the pattern of optical intensity fringes produced by a diffraction grating.

Trains of short pulses, each different and each producing only a small change (a "kick" in system space), have been suggested as a means of crafting specified Hilbert-space rotations [Sha07].

11.4 Pulse pairs

The Bloch vector and the torque equation of motion for it provide a very simple interpretation of sequential-pulse excitation effects. One of these occurs when a first pulse, resonant and with constant Rabi frequency, ceases for some time interval before resuming. During the illumination halt there occurs no change of the Bloch vector, but there does occur an ongoing accumulation of phase, originating with the rotating coordinate system; see Sec. 11.2. This affects the orientation of the angular velocity vector when it next acts, and hence it affects the changes induced by the second pulse.

To analyze the effect of successive pulses we write the initial pulsed field as

$$\mathbf{E}^{(1)}(t) = \mathrm{Re}\left\{ \mathbf{e}\mathcal{E}(t - t_1) e^{-i\omega(t - t_1) + i\varphi_1} \right\}, \qquad (11.16)$$

with t_1 as the time when the pulse envelope starts. That is, the envelope function appearing in the Bloch equation is a function of the time interval $\tau = t - t_1$ measured from the pulse start; it vanishes for $t - t_1 < 0$. The rotating Hilbert-space coordinate associated with this field is

$$\psi_2'(t) = e^{-i\omega(t - t_1)} \psi_2. \qquad (11.17)$$

The second pulse begins at a later time t_2, and we write the resulting field as

$$\mathbf{E}^{(2)}(t) = \mathrm{Re}\left\{ \mathbf{e}\mathcal{E}(t - t_2) e^{-i\omega(t - t_2) + i\varphi_2} \right\}. \qquad (11.18)$$

To place these two expressions on a common time scale, one associated with the rotating coordinate of eqn. (11.17), we rewrite eqn. (11.18) as

$$\mathbf{E}^{(2)}(t) = \mathrm{Re}\left\{ \mathbf{e}\mathcal{E}(t - t_2) e^{+i\Delta\varphi} e^{-i\omega(t - t_1) + i\varphi_1} \right\}. \qquad (11.19)$$

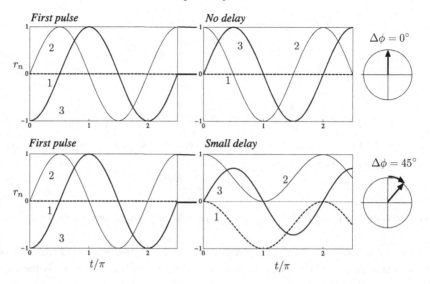

Figure 11.3 Two possible continuations of the excitation; here the pause introduces a phase change $\Delta\varphi = 0$ (top row) or 45° (lower row). To the right are clock faces indicating the phase change during the pulse-free interval. [Redrawn from Fig. 29 of B. W. Shore, Acta Physica Slovaka, **58**, 243–486 (2008) with permission.]

The phase increment appearing here,

$$\Delta\varphi = \omega(t_2 - t_1) - \varphi_1 + \varphi_2, \tag{11.20}$$

acts as a modifier of the field envelope $\mathcal{E}(t)$. We see that, in addition to any specific phase change $\varphi_1 - \varphi_2$ between pulses there occurs an inevitable change $\omega(t_2 - t_1)$ associated with the translation of the time scale used for the rotating coordinate system. This dynamic phase is directly proportional to the steady carrier frequency ω and to the time interval $t_2 - t_1$ between the pulses. Thus the effect of a second pulse depends on the phase of the field envelope, as parametrized here by the phase shift $\Delta\varphi$.

Figure 11.3 shows two examples. In the first (top row) there occurs no pause between pulses, $\Delta\varphi = 0$. In the lower row, a phase delay of $\Delta\varphi = 45°$ occurs. As can be seen, the result is a smaller modulation of the population oscillations (component 3) and component 2. Instead, Bloch-vector component 1 begins to oscillate.

Figure 11.4 shows two more examples of possible phase increments. In the first case (top row), for phase increment 80°, there is little excitation produced by the second pulse. As the phase increment changes, to 100°, there is also little variation of the Bloch vector, and the signs of components 1 and 3 are reversed.

An interesting situation occurs when the phase increment is exactly 90°, midway between the two cases of Fig. 11.4. As can be anticipated from the previous figure, the three components of the Bloch vector remain constant: the second pulse has no effect, even though a pulsed field is present. Such a pulse has been termed a "do nothing" pulse. Although the Bloch vector does not change, the statevector acquires a time-varying phase associated with

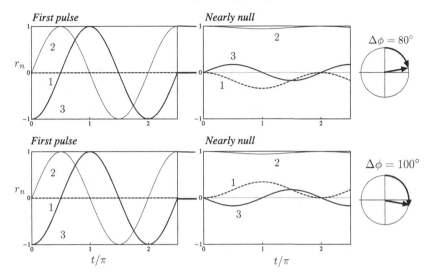

Figure 11.4 Other possible continuations of the excitation; here the pause introduces a phase change $\Delta\varphi = 80°$ (top row) or $100°$ (lower row). In each case the changes of the Bloch vector are slight, but with different signs of components 1 and 3.

an energy shift. This is observable by probing with a second, weak, field. When the delay is $180°$ the effect of the delay interval is to reverse the excitation process – essentially to reverse time.

11.5 Vector picture of pulse pairs

The behavior just described can be understood by considering the combination of Bloch vector and torque vector. The first pulse rotates the Bloch vector in the 2, 3 plane. Figure 11.5 shows two examples in which the motion of the Bloch vector, tracing a curve on the Bloch sphere, leaves a series of radii. In frame (*a*) the pulse has temporal area $\pi/2$, moving the Bloch vector from the south pole into the equatorial plane. In frame (*b*) the pulse has temporal area π, moving the Bloch vector to the north pole and thereby producing complete inversion.

Figure 11.6 shows the effect of a second $\pi/2$ pulse on the two Bloch vectors of Fig. 11.5, under the assumption that there has been a phase change of the field by $90°$, and the torque vector therefore lies long the 2 axis. In frame (*a*) this torque vector is parallel to the Bloch vector, and so no motion occurs. In frame *b* the torque vector produces a coherent superposition. If it continues to act, it will cause the Bloch vector to trace a disk in the 1, 3 plane, with periodic inversions.

In summary, the effect of the pause prior to resumption of the pulse is as follows:

- The envelope phase shift $\Delta\varphi$ shifts the phase of the complex-valued Rabi frequency, and hence the orientation of the angular velocity vector Υ with components $\Upsilon_1 = \mathrm{Re}\,\check{\Omega} = \Omega\cos\varphi$, $\Upsilon_2 = \mathrm{Im}\,\check{\Omega} = \Omega\sin\varphi$, $\Omega_3 = \Delta$.

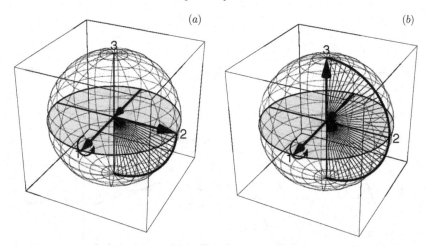

Figure 11.5 Vector picture of resonant excitation. The torque vector lies along the 1 axis, the Bloch vector, initially pointing to the south pole of the Bloch sphere, traces a curve marked by a succession of radii. (*a*) A $\pi/2$ pulse, producing a coherent superposition (to equatorial plane). (*b*) A π puse, producing complete inversion (to north pole).

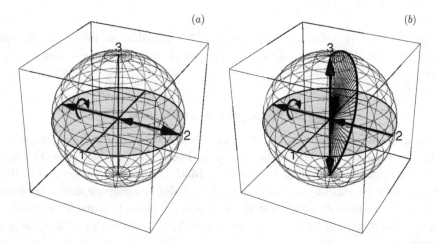

Figure 11.6 The effect of a second pulse, phase shifted from the first one by 90°, so that the torque vector lies along the 2 axis. (*a*) Acting on the coherent superposition of Fig. 11.5(*a*), no change occurs. (*b*) Acting on an inverted system, Fig. 11.5(*b*), observable change occurs to the population, here undergoing return to state 1 (the south pole of the Bloch sphere).

- Subsequent population change may be zero, if the angular velocity vector aligns with the Bloch vector. When resonant, $\Delta = 0$, this occurs if $\Delta\varphi = \pi/2$.
- A phase change $\Delta\varphi = \pi$ is equivalent to time reversal.

Such effects occur only if the laser field remains coherently phased. When the delay interval is longer than the coherence time, then the phase becomes random. The resulting effect on the atom can only be evaluated statistically: The Bloch vector components are averages of those for a distribution of phase changes. Rabi oscillations will not then be seen at long times.

11.6 Creating dressed states

As has been noted, a $\pi/2$ pulse, acting on a two-state system in a single quantum state, will produce a 50:50 superposition of two states. If the laser field is then altered by $\pi/2$, there will occur no further motion of the Bloch vector: The system, in that particular coherent superposition state, is immune to that radiation. The torque vector produces no change because it is either parallel to, or antiparallel to, the Bloch vector.

Let us denote the Bloch vector for such an immune state by $\mathbf{r}^{im}$. It is parallel (or antiparallel) to the torque vector Υ. The apparent lack of temporal change in $\mathbf{r}^{im}$ implies that the system is in an eigenstate of the RWA Hamiltonian [Wol05]. The connection between the Bloch vector of the immune state $\mathbf{r}^{im}$ and the Hamiltonian eigenstates is as follows.

A statevector, parametrized by two Bloch angles, a polar angle θ and an azimuthal angle ϕ (not necessarily the field phase φ of eqn. (11.21)),

$$\Psi = \cos(\theta/2)\,\psi_1 + e^{-i\phi}\sin(\theta/2)\,\psi_2, \tag{11.21}$$

corresponds to the Bloch vector whose Cartesian components are

$$r_1 = \cos\phi\,\sin\theta, \quad r_2 = \sin\phi\,\sin\theta, \quad r_3 = \cos\theta. \tag{11.22}$$

We compare the corresponding dressed (adiabatic) eigenstates of the RWA Hamiltonian of eqn. (10.6) by describing them with parameters appropriate to the Bloch sphere. The lower-energy eigenstate of that Hamiltonian is

$$\Phi_- = \cos(\theta/2)\,\psi_1 + e^{i\varphi}\sin(\theta/2)\,\psi_2, \tag{11.23}$$

where φ is the laser phase[2] and eqn. (10.11) defines the angle θ. When represented on the Bloch sphere this eigenvector has the Cartesian components

$$\begin{aligned}
s_1 &= \cos(-\varphi)\sin\theta = \cos\varphi\,\sin\theta, \\
s_2 &= \sin(-\varphi)\sin\theta = -\sin\varphi\,\sin\theta, \\
s_3 &= \cos\theta.
\end{aligned} \tag{11.24}$$

The second eigenstate is

$$\begin{aligned}
\Phi_+ &= \sin(\theta/2)\,\psi_1 + e^{i\varphi}\cos(\theta/2)\,\psi_2, \\
&= \cos(\theta/2 - \pi/2)\,\psi_1 + e^{i\varphi+i\pi}\sin(\theta/2 - \pi/2)\,\psi_2.
\end{aligned} \tag{11.25}$$

---

[2] Note that the factor $e^{i\varphi}$ here negates the $e^{-i\varphi}$ of the RWW Hamiltonian.

When represented on the Bloch sphere this has the components

$$
\begin{aligned}
s_1 &= \cos(-\phi + \pi)\sin(\theta - \pi) = -\cos\phi\,\sin\theta, \\
s_2 &= \sin(-\phi + \pi)\sin(\theta - \pi) = \sin\phi\,\sin\theta, \\
s_3 &= \cos(\theta - \pi) \qquad\qquad\quad = -\cos\theta.
\end{aligned}
\tag{11.26}
$$

These components of Φ_+ are the reverse of the components of the other eigenvector, Φ_-: the two eigenvectors are on opposite sides of the Bloch sphere.

Written in spherical coordinates the torque vector reads

$$
\begin{aligned}
\Upsilon_1 &= \Omega\cos\varphi \;\;= |\Upsilon|\cos\varphi\,\sin\theta, \\
\Upsilon_2 &= -\Omega\sin\varphi = -|\Upsilon|\sin\varphi\,\sin\theta, \\
\Upsilon_3 &= \Delta \qquad\quad = |\Upsilon|\cos\theta.
\end{aligned}
\tag{11.27}
$$

This vector is aligned with the direction of the lower-energy eigenstate. Thus if we place the system into the lower-energy eigenstate, with ϕ of eqn. (11.21) equal to the field phase φ, the Bloch vector will be parallel to the torque vector, and will produce no change of the Bloch vector. Conversely, a technique that revises the torque vector to make it parallel to the Bloch vector will place the system into the lower-energy eigenstate. Similarly, if the torque vector is antiparallel to the Bloch vector the system is in the higher-energy eigenstate, and again no change will be seen. With either of these two alignments of the Bloch vector a pulse produces no change: it is a "do nothing" pulse.

Adiabatic following. Once the torque vector and the Bloch vector are aligned (and hence the system is in one adiabatic eigenstate), a slow controlled motion of the torque vector will retain that alignment – the system remains in an adiabatic state. By steering the torque vector an experimenter thereby steers the Bloch vector. This is adiabatic following. When the torque vector does not move sufficiently slowly the Bloch vector shows additional motion to the adiabatic following.

11.7 Zero-area pulses

As long as the phase $\varphi(t)$ remains constant, it is possible to regard $\mathcal{E}_k(t)$, and hence the associated Rabi frequency $\Omega_k(t)$, as real and positive. However, various techniques allow one to alter the field phase during the course of a pulse. Such changes will alter the Rabi frequency; it may become negative or even complex-valued. Under suitable conditions there may occur equal contributions from positive and negative envelope functions, i.e. it has a temporal pulse area of zero [Cri70; Vas06; Sho10; Ran09]. To parametrize such pulses it is useful to introduce the *absolute area*, the integral over the pulse of the absolute magnitude of the Rabi frequency (with superscript *ab* denoting absolute value and subscript λ identifying a particular pulse field) [Ran09],

$$
\mathcal{A}_\lambda^{ab} = \int_{-\infty}^{\infty} dt'\, |\Omega_\lambda(t')|.
\tag{11.28}
$$

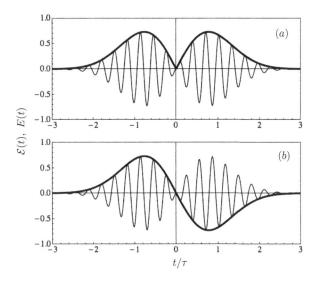

Figure 11.7 Two examples of relative values of electric field components $E(t) \equiv \hat{\mathbf{e}} \cdot \mathbf{E}(t)$ (light lines) and two concomitant envelopes $\mathcal{E}(t)$ (thick lines), differing by a phase change at the pulse midpoint, $t = 0$. (a) A pulse whose envelope $\mathcal{E}(t) = |f(t + .5\tau) - f(t - .5\tau)|$ is positive at all times, producing a positive pulse area. (b) A pulse in which the envelope $\mathcal{E}(t) = f(t + .5\tau) - f(t - .5\tau)$ changes sign at $t = 0$, from positive at $t < 0$ to negative for $t > 0$, producing a pulse area of zero. The time scale is in units of the pulse width parameter τ used in constructing the envelope from $f(t) \equiv \exp[-(t/\tau)^2]$. [Redrawn from Fig. 1 of B. W. Shore *et al.*, Opt. Comm., **283**, 730–736 (2009) with permission.]

When a zero-area pulse is combined with a time-varying detuning it is possible to produce complete population transfer without a crossing of the diabatic energy curves. That is, the detuning does not change sign. It is only necessary that the adiabatic curves touch. (By definition, the larger of the two adiabatic curves can never lie below the other curve, but the two curves will touch if there is a degeneracy.) Examples of population transfer produced by pulse segments (pulselets) that combine to zero temporal area have been presented by Rangelov *et al.* [Ran09]. Both complete population inversion and complete population return occur for different choices of structured pulses.

The concept of a zero-area laser pulse may be somewhat puzzling, and so I offer here some explanatory comments. Figure 11.7 shows examples of two pulses, each with the same carrier frequency and the same intensity centered at time $t = 0$, but with different envelope phases. In one of the pulses (solid lines) the phase remains constant, and the envelope is positive at all times. In the other pulse (dashed lines for $t > 0$) there occurs a phase change of π at $t = 0$. The subsequent envelope is negative, and its integral exactly reverses the accumulation of pulse area that occurred prior to $t = 0$. The result, at the conclusion of the pulse, is a pulse area of zero. Such a resonant pulse, acting on a two-state system, will produce temporary excitation but will, upon conclusion of the pulse, return all population to the initial state.

Zero-area pulses can be produced by a variety of techniques, including the following.

Ultrashort shaped pulses: Nowadays various pulse-shaping devices, acting on the Fourier components of a temporal pulse of picosecond or femtosecond duration, permit the creation of pulses that have predetermined phase and amplitude variations [Ass98; Bri04; Wol07]. From the Fourier transform of the pulse,

$$\widetilde{\Omega}(\Delta) = \frac{1}{2\pi} \int_{-\infty}^{\infty} dt \, e^{i\Delta t} \, \Omega(t), \tag{11.29}$$

one evaluates the pulse area from the value on resonance ($\Delta = 0$),

$$\mathcal{A}_{\infty} = 2\pi \widetilde{\Omega}(0). \tag{11.30}$$

Hence, a zero-area pulse can be produced by modifying the Fourier spectrum to nullify resonant frequency component $\widetilde{\Omega}(0)$. Pulse shapers can also be used to produced specified detunings $\Delta(t)$.

Beam splitting and recombination: A zero-area pulse can also be produced by passing a laser beam through a beam splitter to produce two pulses, delaying one of these by τ, and then recombining the two split pulses with a phase shift of π. The recombined pulse, with Rabi frequency

$$\Omega(t) = \Omega_0 [f(t - \tau) - f(t)], \tag{11.31}$$

is a zero-area pulse. Such an implementation requires an interferometric stability within a wavelength λ.

Self-induced transparency: When a laser pulse that has a temporal area between 0 and 2π undergoes self-induced transparency in passing through a resonant medium it will, after sufficient distance, be reshaped into a zero-area pulse. This is the assertion of the McCall–Hahn area theorem, see Sec. 21.2.7 [McC69; All87]. Once this reshaping has occurred, the pulse propagates without producing any permanent excitation of the medium: the pulse itself alters the absorption properties, a phenomenon termed *self-induced transparency* (SIT), see Sec. 21.2.7.

Quasistatic magnetic fields: The dynamics of nuclear magnetic resonance of a spin-$\frac{1}{2}$ particle is that of eqn. (9.25), but with a Rabi frequency that originates with the product of a magnetic dipole and a slowly varying magnetic field. It is this field whose temporal integral provides the pulse area. A zero-area pulse can be produced by suitably reversing this field.

12

Degeneracy

The discrete energies E_n of bound states are not the only indicators of quantum-mechanical properties and quantization. Rotational motion of atoms or molecules, associated with angular momentum, is also quantized, both in magnitude and in direction; only discrete orientations are allowed with respect to any selected (but arbitrary) axis of quantization [Sho90, §18.1]. In the absence of external fields, the energy of a free atom or molecule does not depend on this orientation, and the energy states are degenerate. This chapter discusses that degeneracy, and the theory of coherent excitation of such degenerate quantum states.

Angular momentum degeneracy. The theoretical building blocks for describing rotational motion are angular momentum states $|J, M\rangle$, discussed in App. A, associated with a dimensionless vector operator $\hat{\mathbf{J}}$ and its component $\hat{J}_z$ along a quantization axis, taken as defining the z axis.[1] The label J, the *angular momentum quantum number*, derives from the eigenvalue $J(J+1)$ of the operator $\hat{\mathbf{J}}^2$ and therefore quantifies the magnitude of the angular momentum. It may be an integer or half integer. The label M, the *magnetic quantum number*, is the eigenvalue of $\hat{J}_z$; it quantifies the projection of angular momentum along a reference axis. The values of M differ by integers and range from $-J$ to $+J$ in integer steps. The total number of such values, $2J+1$, is an integer, the degeneracy of the quantum state $|J, M\rangle$.

Angular momentum of an isolated quantum system, such as an atom or molecule, refers always to the center of mass (which may be moving). Because such a system comprises constituent particles (electrons and nuclei) there are several sources of angular momentum that add vectorially to produce a total angular momentum. Although the magnitude of each contributory angular momentum is fixed, it is only the projection of this total angular momentum that quantum theory constrains to discrete values.

12.1 Zeeman sublevels

When an atom in a quantum state of definite angular momentum, as labeled by J, is subjected to an external magnetic field there occurs an energy shift directly proportional to the

---

[1] As discussed in App. A, the angular momentum vector associated with the dimensionless operator $\hat{\mathbf{J}}$ is $\hbar\mathbf{J}$.

magnetic quantum number M. These are the *Zeeman energy shifts* associated with static magnetic fields. A static electric field produces an energy shift proportional to the absolute value $|M|$. These are known as *Stark shifts*. When the electric field is that of radiation the shifts are termed *dynamic Stark shifts* or *AC Stark shifts*. In the absence of such external fields the energy of a confined quantum-mechanical particle is unaffected by orientation; all $2J + 1$ orientations have the same energy, and the associated quantum states are degenerate. It is common practice, based upon terminology from atomic spectroscopy [Mar06b], to refer to a set of degenerate quantum states as an *energy level*. The individual constituents, distinguishable by orientation, are magnetic or Zeeman *sublevels*.

When degeneracy occurs, the laser-field polarization provides a means of more specifically controlling the laser-induced excitation. Because the states of free atoms can be taken as states with well-defined angular momentum, simple selection rules on the magnetic quantum number M are associated with particular polarizations.

Coordinate changes. The interpretation of the population occurring in any Zeeman sublevel depends on the orientation of the reference system. The connection between reference frames differing by a rotation $\hat{\mathbf{R}}$ is provided by the rotation matrix of order J, through the relationship [Zar88; Lou06; Auz10]

$$|J, M\rangle_{rot} = \sum_{M'} \mathcal{D}^{(J)}_{M,M'}(\hat{\mathbf{R}}) |J, M'\rangle. \tag{12.1}$$

The rotations between the two reference frames are typically parametrized using the Euler angles α, β, γ that connect the original and the new reference frames. With this parametrization the change of reference frame is expressed by the formula

$$|J, M\rangle_{rot} = e^{-i\alpha M} \sum_{M'} d^{(J)}_{M,M'}(\beta)\, e^{-i\gamma M'} |J, M'\rangle, \tag{12.2}$$

in which $d^{(J)}_{M,M'}(\beta)$ is the *reduced rotation matrix* (see App. A.1.4).

12.2 Radiation polarization and selection rules

Laser-induced transitions originate, in the electric-dipole approximation, with transition dipole moments within the atom. These can either emit or absorb electric-dipole radiation. Typically one regards the free atom as existing in an angular momentum state defined by the pair of quantum numbers J, M. The formula for evaluating the dipole moment between angular momentum states is given by the Wigner–Eckart theorem, discussed in treatises on angular momentum [Zar88; Lou06; Bie09; Auz10] [Sho90, §20.5]:

$$\langle J'M'|\mathbf{e}_q \cdot \mathbf{d}|JM\rangle = -ea_0\sqrt{S(J, J')}\,(-1)^{J'-M'}\begin{pmatrix} J' & 1 & J \\ -M' & q & M \end{pmatrix}, \tag{12.3}$$

where $ea_0 = 2.542$ debye is the atomic unit of dipole moment [Sho90, §2.9], $S(J, J')$ is the dimensionless transition strength of eqn. (H.1) and $(:::)$ is a three-j symbol (see App. A.2).

The unit vectors $\mathbf{e}_q$, for $q = -1, 0, +1$, used to define the directions of a three-dimensional vector in a spherical basis, have the following relationship to the three unit vectors $\mathbf{e}_x$, $\mathbf{e}_y$, and $\mathbf{e}_z$ of Cartesian coordinates:

$$\mathbf{e}_{\pm 1} = \mp \frac{1}{\sqrt{2}} [\mathbf{e}_x \pm i \mathbf{e}_y], \qquad \mathbf{e}_0 = \mathbf{e}_z. \tag{12.4}$$

Selection rules. The six arguments of the three-j symbol incorporate basic selection rules for dipole radiation (electric or magetic). First, the two atomic angular momentum quantum numbers J and J' must form a triangle with unity as the third side. This constraint is because the dipole field has unit angular momentum, and this must equal the change in atomic angular momentum.[2] Thus it is not possible to have a radiative transition, via a single photon, between two states that each have angular momentum $J = 0$. Nor is it possible to have a transition, via dipole radiation, in which $|J - J'| > 1$. Such transitions occur with quadrupole or higher multipole radiation.

The nonzero linkages amongst sublevels are restricted by the requirement

$$M - M' + q = 0. \tag{12.5}$$

Thus the magnetic quantum numbers M and M' can similarly differ by no more than 1; the choice $q = 0$ requires $M = M'$. This constraint expressed the fact that the projection of angular momentum onto the propagation axis (the helicity q) cannot exceed unity. When the two angular momenta are the same, $J = J'$, an additional restriction occurs: no transitions occur from $M = 0$ or $M' = 0$ without a change in magnetic quantum number, i.e. there is no interaction link $M = 0 \leftrightarrow M' = 0$. To summarize, these geometrically based selection rules are

$$|J - J'| = 0, \pm 1, \quad \text{but not } J = 0 \leftrightarrow J' = 0$$

$$|M - M'| = 0, \pm 1, \quad \text{but not } M = 0 \leftrightarrow M' = 0 \text{ when } J = J'. \tag{12.6}$$

A further selection rule applies to transitions between states of free atoms, though not to transitions in atoms held within a crystalline environment: the electric dipole transition moment is nonzero only between states of opposite parity. This attribute refers to symmetry under coordinate inversion: odd-parity states change sign whereas even-parity states remain unchanged.

12.2.1 Emission: Angular momentum fields

The radiation spontaneously emitted during a transition between atomic states of well-defined angular momentum quantum numbers J, M also carries angular momentum; the relevant fields are the angular momentum multipole fields of App. D.2 [Sho90, §19.4].

---

[2] The electromagnetic field, being a vector field in three-dimensional Euclidean space, has intrinsic angular momentum (i.e. spin) of one. Particular fields may also carry orbital angular momentum about a reference axis.

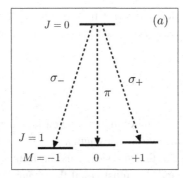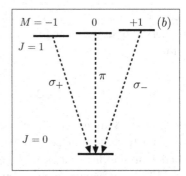

Figure 12.1 Spontaneous emission transitions between magnetic sublevels create one of three electric dipole fields, denoted σ_+, σ_-, and π. (*a*) For transitions from an excited state $J = 0$ to sublevels of $J = 1$. (*b*) For transitions from excited sublevels of $J = 1$ to ground state $J = 0$. The energy levels of $J = 1$ are shown with Zeeman shifts, so that each transition has a different frequency. [Redrawn from Fig. 33 of B. W. Shore, Acta Physica Slovaka, **58**, 243–486 (2008) with permission.]

There are three basic angular momentum dipole fields that can be created, conventionally denoted σ_+, σ_-, and π, each identified with a specific change of magnetic quantum number, ΔM. The association is

$$\Delta M = 0 \quad \text{for } \pi,$$
$$\Delta M = +1 \quad \text{for } \sigma_+, \quad\quad\quad (12.7)$$
$$\Delta M = -1 \quad \text{for } \sigma_-.$$

Figure 12.1 shows examples of the transitions that produce these emission fields. Frame (*a*) shows the three fields that are generated by a transition from a single initial sublevel; frame (*b*) shows the three fields that produce a given final sublevel. Because the sublevels differ in energy (by the Zeeman shift) the three spectral lines have different frequencies: These are *Zeeman components* of a spectral line.

The three Zeeman components shown in the figure differ not only in frequency but also in polarization characteristics; the fields are examples of three basic dipole fields, shown in Fig. 12.2. The π-field vectors remain fixed in direction, although oscillating in length. The $\sigma_\pm$ fields have vectors that rotate about the z axis, with frequency ω.

12.2.2 Absorption: Linear momentum fields

To excite an atom with laser radiation it is necessary to associate these emission fields, which are multipole fields characterized by unit angular momentum, with laser fields, which are characterized by *linear momentum*, see App. D.2. The laser fields typically are idealized as plane waves, meaning spatial and temporal properties governed by the phase function of eqn. (3.17).

A general field vector can be written, using Cartesian coordinates, as

$$\mathbf{e} = a_z \mathbf{e}_z + a_x \mathbf{e}_x + a_y \mathbf{e}_y. \quad\quad\quad (12.8)$$

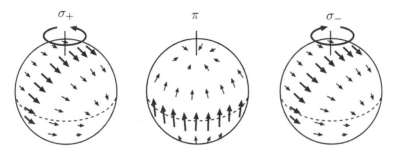

Figure 12.2 Basic electric-dipole fields with definite angular momentum. Left: the σ_+ field appears along the polar axis as circularly polarized, while when viewed in the equatorial plane it is horizontally linearly polarized; center: the π field vanishes in polar directions, and is vertically polarized in the equatorial plane; right: the σ_- field is similar to the σ_+ field but has opposite sense of polarization rotation. [Redrawn from Fig. 2.7 of B. W. Shore and D. H. Menzel, *Principles of Atomic Spectra* (Wiley, New York, 1968) with permission.]

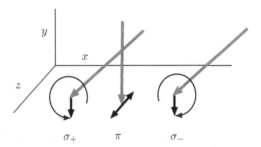

Figure 12.3 Examples of polarizations based on helicity, eqn. (12.9). Long arrows show propagation direction, shorter arrows indicate polarization plane for linear (π) or circular ($\sigma_\pm$) polarizations. [Redrawn from Fig. 35 of B. W. Shore, Acta Physica Slovaka, **58**, 243–486 (2008) with permission.]

Because the electric and magnetic fields of radiation must be transverse to the propagation direction it proves useful to identify these unit vectors in a coordinate system aligned with the propagation direction. For radiation propagating along the z axis (the quantization axis) we use orthogonal complex spherical unit vectors (helicity states of the field), as given by eqn. (12.4), and instead of eqn. (12.8) we have the corresponding expansion

$$\mathbf{e} = a_0\mathbf{e}_0 + a_{+1}\mathbf{e}_{+1} + a_{-1}\mathbf{e}_{-1}. \tag{12.9}$$

Figure 12.3 sketches the three polarization choices available with this basis.

The appearance of the three components of the polarization vector $\mathbf{e}$ depends explicitly upon the choice of coordinate orientation, e.g. whether the z axis lies along the propagation direction or along the direction of linear polarization. The choice of orientation is typically described by the three Euler angles α, β, and γ [Zar88; Lou06; Auz10],

$$\mathbf{e}_{q'}(\alpha, \beta, \gamma) = e^{i\gamma q'} \sum_{q=-1,0,+1} \mathbf{e}_q \, e^{-i\gamma q} \, d^{(1)}_{q,q'}(\beta). \tag{12.10}$$

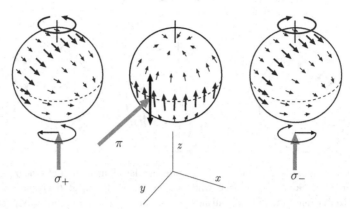

Figure 12.4 Electric dipole fields in absorption. The $\sigma_\pm$ fields at the left and right are most strongly coupled to circularly polarized plane waves propagating vertically. The π field at the center has strongest coupling to a vertically polarized field incident in the equatorial plane. [Redrawn from Fig. 36 of B. W. Shore, Acta Physica Slovaka, **58**, 243–486 (2008) with permission.]

Here $d^{(1)}_{M,M'}(\beta)$ is the reduced rotation matrix of order 1 [Sho90, §18.5], tabulated in App. A.1.4. The most general polarization, elliptical polarization, can be expressed as a superposition of any two independent unit vectors that are perpendicular to the propagation direction; see Sec. 3.6 and Fig. 3.2.

12.2.3 Connection, linear and angular momentum fields

The combination of energy selectivity, associated with the monochromatic nature of laser light, and sublevel selectivity, associated with polarization, provides controls with which to select specific pairs of quantum states for pulsed manipulation. Figure 12.4 shows possible connections between the polarization of a directed beam and the multipole fields needed for inducing transitions between angular momentum states. By using all three independent polarization fields, each specified by a complex amplitude, we have at most six control parameters with which to create, and detect, degenerate superpositions.

It is important to recognize that any rotation of the coordinates alters the apparent identification of individual magnetic sublevels. Under rotation, parametrized by Euler angles α, β, and γ, a single angular momentum state appears as a coherent superposition of states having the same J [Zar88; Lou06; Auz10] [Sho90, §18.4],

$$|J, M'\rangle_{rot} = \sum_M \mathcal{D}^{(J)}_{M',M}(\alpha, \beta, \gamma)|J, M\rangle. \tag{12.11}$$

Similarly, the complex-valued unit vectors that describe the direction of the electric field undergo the transformation

$$\mathbf{e}_{q'}(\alpha, \beta, \gamma) = \sum_q \mathcal{D}^{(1)}_{q',q}(\alpha, \beta, \gamma)\mathbf{e}_q, \qquad q, q' = 0, \pm 1. \tag{12.12}$$

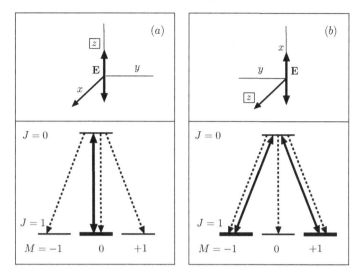

Figure 12.5 Comparison of linkages associated with coordinate choices for linearly polarized light. Upper frames show the Cartesian coordinates x, y, z, propagation axis (light arrow) and electric field axis (thick double-headed arrow). Lower frames show the linkages: thick arrows for laser–field coupling and dashed arrows for spontaneous emission. (*a*) The z axis (the quantization axis) is aligned with the **E** field (i.e. the polarization axis); propagation is along the x axis. (*b*) The z axis is aligned with the propagation direction, the x axis is aligned with the polarization. Linkages are expressed in a helicity basis. [Redrawn from Fig. 37 of B. W. Shore, Acta Physica Slovaka, **58**, 243–486 (2008) with permission.]

As an example, the rotation of $J = 1$ from x to z produces the result

$$|1, 0\rangle_{rot} = \frac{1}{\sqrt{2}} \Big[|1, -1\rangle - |1, +1\rangle \Big]. \tag{12.13}$$

A single state becomes a superposition, and vice versa. Figure 12.5 shows an example of how the coordinate choice affects the linkage pattern.

12.3 The RWA with degeneracy

The RWA for circularly (or elliptically) polarized light requires consideration of linkages involving all magnetic sublevels. As an example, the interaction Hamiltonian for a single field of positive helicity is

$$\mathsf{V}(t) = -\,\mathrm{Re}\,[d_{+1}\check{\mathcal{E}}(t)\,\mathrm{e}^{-\mathrm{i}\omega t}] = -\tfrac{1}{2}d_{+1}\check{\mathcal{E}}(t)\,\mathrm{e}^{-\mathrm{i}\omega t} + \tfrac{1}{2}d_{-1}\check{\mathcal{E}}^*\,\mathrm{e}^{+\mathrm{i}\omega t}. \tag{12.14}$$

Consider a degenerate ground level having angular momentum $J = 1$ linked by this Hamiltonian to an excited state having $J = 0$, as in Fig. 12.5. Three states are coupled by this interaction; let us label them by their magnetic quantum numbers. Set the zero of energy to be $E_{-1} = E_{+1} = 0$ and express the statevector as[3]

$$\Psi(t) = C_{-1}(t)\,\psi_{-1} + C_0(t)\,e^{-i\omega t}\,\psi_0 + C_{+1}(t)\,\psi_{+1}. \qquad (12.15)$$

Using angular momentum selection rules we obtain from the TDSE the three coupled equations

$$\frac{d}{dt}C_{-1} = -\frac{i}{2}\check{\Omega}_P^* C_0,$$

$$\frac{d}{dt}C_0 = -i\,\Delta C_0 - \frac{i}{2}\check{\Omega}_P C_{-1} - \frac{i}{2}\check{\Omega}_S\,e^{+2i\omega t}C_{+1}, \qquad (12.16)$$

$$\frac{d}{dt}C_{+1} = -\frac{i}{2}\check{\Omega}_S^*\,e^{-2i\omega t}C_0,$$

where the complex-valued Rabi frequencies are

$$\hbar\check{\Omega}_P^* \equiv -\langle -1|d_{-1}|0\rangle\check{\mathcal{E}}^*, \qquad \hbar\check{\Omega}_S \equiv -\langle +1|d_{+1}|0\rangle\check{\mathcal{E}}. \qquad (12.17)$$

The RWA neglects terms that vary as twice the carrier frequency. When we omit these terms we obtain the two-state RWA equations

$$\frac{d}{dt}C_{-1} = -\frac{i}{2}\check{\Omega}_P^* C_0,$$

$$\frac{d}{dt}C_0 = -i\,\Delta C_0 - \frac{i}{2}\check{\Omega}_P C_{-1}. \qquad (12.18)$$

More generally, when the radiation is elliptically polarized, meaning two fields, then the interaction Hamiltonian will have the form

$$V(t) = -\text{Re}\left\{d_{+1}\check{\mathcal{E}}_P(t)\,e^{-i\omega t} + d_{-1}\check{\mathcal{E}}_S(t)\,e^{-i\omega t}\right\}. \qquad (12.19)$$

There will now be linkages to C_0 involving both C_{-1} and C_{+1}. Each linkage will have counter-rotating terms involving $\exp(\pm i 2\omega t)$. The RWA neglects these, and provides a three-state equation, for the lambda linkage, of the form

$$\frac{d}{dt}C_{-1} = -\frac{i}{2}\check{\Omega}_P^* C_0,$$

$$\frac{d}{dt}C_0 = -i\,\Delta C_0 - \frac{i}{2}\check{\Omega}_P C_{-1} - \frac{i}{2}\check{\Omega}_S C_{+1}, \qquad (12.20)$$

$$\frac{d}{dt}C_{+1} = -\frac{i}{2}\check{\Omega}_S^* C_0.$$

These equations, based on the TDSE, do not account for spontaneous emission; they only treat the laser-induced transitions. To account for spontaneous emission, an incoherent process, it is necessary to employ a density matrix, as discussed in Chap. 16. Nevertheless, even without displaying and solving the required equations of motion it is possible to make qualitative observations about some effects of spontaneous emission. The following sections present examples.

---

[3] In App. A.1.3 the basis vectors ψ_μ are identified with spin-one eigenvectors and denoted χ_μ.

12.4 Optical pumping

Under conditions of thermal equilibrium populations are evenly distributed amongst magnetic sublevels; each sublevel of a given J has population $1/(2J+1)$. Various laser-induced processes alter this uniform distribution. Distributions of population amongst magnetic sublevels, whether coherent or incoherent, are commonly classed as being either *oriented* or *aligned* [Sho90, §21.9] [Auz10]. *Orientation* means that there is a nonzero averaged magnetic moment. *Alignment* means that the averaged magnetic moment is zero but the population is not uniformly distributed amongst magnetic sublevels – although sublevels of $+M$ and $-M$ have equal population.

The redistribution often takes place by a combination of coherent and incoherent processes. Illumination of a resonant transition will transfer population into an excited state, from which it will decay by spontaneous emission. When the ground level is degenerate (say, angular momentum $J = 1$) then spontaneous emission from an excited quantum state (a sublevel) will proceed by all three polarization fields, in accord with selection rules $\Delta M = 0, \pm 1$, each with equal probability. When the illumination is steady and polarized, then there will occur cycling of population: each absorption from a ground sublevel will be followed by spontaneous return to another low-lying sublevel. The result, after many excitation–deexcitation cycles, will be a redistribution of population amongst the ground sublevels – an example of *optical pumping* [Hap72; Kni88; Hap10; Auz10], illustrated by the following figures.

Population concentration. As a first example, Fig. 12.6 illustrates how optical pumping with circularly polarized light produces complete population transfer into a single quantum state, the sublevel having maximum magnetic quantum number – in this case $M = +1$. The laser-induced excitation obeys the selection rule $\Delta M = +1$, while spontaneous emission returns population to all three sublevels. Population eventually accumulates in sublevel $M = J$, on which the illumination has no effect. Complete transfer will occur to this sublevel if, and only if, there are no spontaneous emissions other than those that return population to the initially populated Zeeman sublevels – such transitions introduce probability losses.

Population removal. As a second example, Fig. 12.7 illustrates the redistribution that occurs when linearly polarized light excites atoms from angular momentum $J = 1$ to $J = 0$. The quantization axis here is taken along the direction of linear polarization, so that the selection rule $\Delta M = 0$ holds for excitation. Radiative decay, by contrast, returns population to all three sublevels. Population eventually accumulates into sublevels $M = \pm 1$, which do not undergo laser-induced excitation. After many optical cycles the population is entirely removed from $M = 0$. However, the final population distribution, in $M = \pm 1$, is not a coherent superposition; it is an incoherent mixture of two quantum states (a mixed state).

Population trapping. The redistribution of population from optical pumping appears differently if we choose the quantization axis differently. Figure 12.8 shows the linkages when

Degeneracy

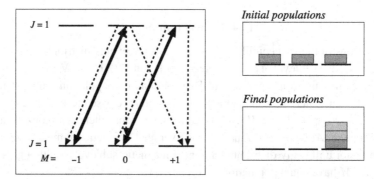

Figure 12.6 Optical pumping can produce complete transfer into a single quantum state. Left frame: Linkage pattern for radiative transitions for $J = 1 \to J = 1$ using circularly polarized light. Solid lines: linkage of circularly polarized laser radiation. Dashed lines: allowed linkages of spontaneous emission. Right frames: Population distributions before (above) and after (below) many optical pumping cycles of excitation and spontaneous emission. Initially all sublevels are equally populated. Light fill marks the portion affected by laser radiation, dark fill marks the portion unaffected by laser radiation. After optical pumping (below) all population is placed into sublevel $M = +1$. [Redrawn from Fig. 38 of B. W. Shore, Acta Physica Slovaka, **58**, 243–486 (2008) with permission.]

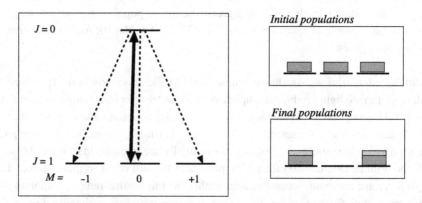

Figure 12.7 Optical pumping can produce complete depletion. Left frame: Linkage pattern for radiative transitions for $J = 1 \to J = 0$. Solid line: linkage of linearly polarized laser radiation, in a reference system aligned with the polarization axis. Dashed lines: allowed linkages of spontaneous emission. Right frames: As in Fig. 12.6, population distributions before (above) and after (below) many cycles of optical pumping. After optical pumping (below) all population is removed from sublevel $M = 0$. [Redrawn from Fig. 39 of B. W. Shore, Acta Physica Slovaka, **58**, 243–486 (2008) with permission.]

the polarization is linear but the quantization axis is chosen to be perpendicular to the polarization direction, say in the direction of propagation. It is then necessary to express linear polarization as a superposition of right- and left-circularly polarized light, for which the excitation selection rules are $\Delta M = \pm 1$. It might appear that, with this coordinate system, population will accumulate entirely in the $M = 0$ sublevel. This is not correct, as will be

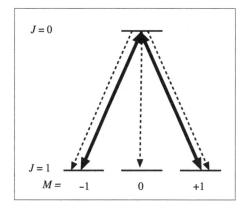
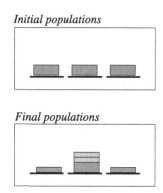

Figure 12.8 Optical pumping can produce trapped population. Left frame: As in Fig. 12.7, linkage patterns for laser radiation (solid lines) and spontaneous emission (dashed lines) for linearly polarized light, in a reference system aligned perpendicular to the polarization direction. The linear polarization is expressed as a (coherent) superposition of two circular polarizations. Right frame: As in Fig. 12.7, population distributions before (above) and after (below) many cycles of optical pumping. Dark fill marks the portion unaffected by laser radiation, light fill marks the portion affected by laser radiation. After optical pumping (below) population is only partially removed from sublevels $M = \pm 1$, leaving a trapped coherent superposition of quantum states. Section 14.3 offers further insight into population trapping states. [Redrawn from Fig. 40 of B. W. Shore, Acta Physica Slovaka, **58**, 243–486 (2008) with permission.]

explained in Secs. 13.4.2, 15.4.6 and App. F: Owing to coherence, only a portion of the initial population is affected by the laser radiation. The remainder is unaffected by the radiation; it remains in a *population trapping* state that does not become excited and therefore does not fluoresce – a *dark state* coherent superposition of $M = \pm 1$. The portion affected by the radiation, indicated by dark fill in Fig. 12.7, becomes an incoherent mixture of two quantum states after optical pumping.

The preceding examples illustrate situations in which radiative excitation produces measurable changes of quantum-state probabilities. However, because these rely on spontaneous emission, they do not exhibit the phase coherence that is of central concern in the present monograph. For such purposes it is necessary to rely solely upon laser-induced transitions, and to deal with coherent excitation.

12.5 General angular momentum

As angular momentum J increases, so too do the number of linkages possible between sublevels. A few simple examples will illustrate some of the possibilities, both for pairs of degenerate sublevels, as discussed above, and for more general chains of degenerate sublevels, as will be discussed in Chap. 15.

12.5.1 Two levels

Figure 12.9 illustrates linkage patterns for a $J = 2 \leftrightarrow J = 1$ transition (in the absence of spontaneous emission) for two degenerate levels, as viewed with two choices of coordinates.

Frame (*a*), appropriate to linear polarization aligned with the quantization axis, shows that there are three independent two-state transitions, associated with states $M = 0, \pm 1$. The states with $M = \pm 2$ are unlinked, and do not undergo excitation; they are dark states, or population-trapping states, meaning that they are unaffected by the laser radiation that would otherwise produce excitation and subsequent fluorescence.

Frame (*b*) shows the same system but with a helicity basis for the polarizations, i.e. a combination of right- and left-circular polarization. The linkages now appear as two independent systems: a five-state letter-M linkage pattern (starting with $J, M = 2, -2$) and a three-state lambda linkage (starting with $J, M = 2, -1$). All magnetic sublevels have linkages to excited states. This appearance is deceiving; the number of dark states cannot depend on the (arbitrary) choice of coordinate alignment. As discussed in App. F, each of the

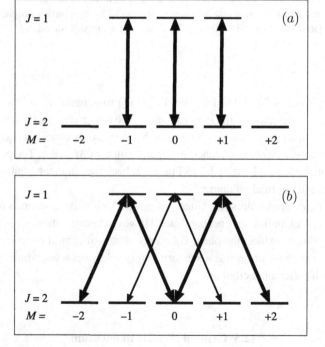

Figure 12.9 Linkage patterns for $J = 2 \leftrightarrow J = 1$, in the absence of spontaneous emission. (*a*) Linear polarization, in direction of quantization axis. Excitation occurs through three independent linkages; two (dark) states are unaffected. (*b*) Linear polarization, in helicity basis. There are two independent excitation linkages: a five-state "letter-M" (heavy lines) and a three-state lambda (lighter lines). [From Fig. 41 of B. W. Shore, *Acta Physica Slovaka*, **58**, 243–486 (2008) with permission.]

independent linkage patterns of frame (*b*) has a single dark state. Thus, as in the portrayal of frame (*a*), the full seven-state system has two dark states.

12.5.2 Multiple levels: Parallel chains

Using the angular momentum selection rules for a single pair of degenerate sublevels it is straightforward to evaluate selection rules for a chain of excitations between degenerate levels. Figures 12.10 and 12.11 illustrate examples, for particular simple choices of the polarization at each step [Sho90, §20.12]. These display the linkages as they would occur in a ladder, from least to greatest excitation energy, but the conclusions apply to any resonant sequence, regardless of the relative energies.

Linear polarization. Figure 12.10 presents examples of linear polarization at each excitation step. Frame (*a*) shows a three-step sequence in which each step involves a larger value of J, starting from $J = 0$. The single initial state has a single complete linkage path to a state of highest excitation ($J = 3$). All of the states have some field-induced links with other states, though not all are connected to the initial ground state.

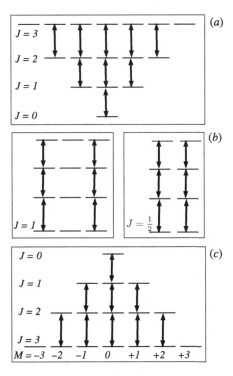

Figure 12.10 Parallel angular momentum chains, linear polarization. (*a*) Increasing J, (*b*) uniform J, (*c*) decreasing J. [Redrawn from Fig. 20.12-1 of B. W. Shore, *The Theory of Coherent Atomic Excitation*. Copyright Wiley-VCH Verlag GmbH & Co. 1990. Reproduced with permission.]

184 Degeneracy

Frame (*b*) shows two examples in which angular momentum remains constant along the chain, either $J = 1$ (left hand) or $J = 1/2$ (right hand). For integer J there occurs no transition $M = 0 \leftrightarrow M = 0$, and so one set of sublevels is unlinked: there will be one initial sublevel that is unaffected by the radiation (a dark state). With half-integer J there are complete linkages between each of the lowest and highest excitation states; there is no dark state.

Frame (*c*) shows an example in which angular momentum decreases with excitation. At each step there are two fewer sublevels linked to lower excitation. Thus with the three excitation stages shown there are six ground-state sublevels (of $J = 3$) that have no link to the most excited sublevel ($J = 0$). Of these, two are completely dark; all but these two states have some field-induced link to other states.

Circular polarization. Figure 12.11 illustrates the same sets of J sequences, but with circular polarizations. Frame (*a*) shows that, as with linear polarization, there exists a single linkage chain between the ground state ($J = 0$) and a fully excited state ($J = 3$). In this

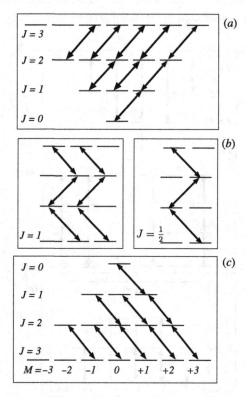

Figure 12.11 Parallel angular momentum chains, circular polarization. (*a*) Increasing J, (*b*) uniform J, alternating left- and right-circular polarization, (*c*) decreasing J. [Redrawn from Figs. 20.12-2 and 20.12-3 of B. W. Shore, *The Theory of Coherent Atomic Excitation*. Copyright Wiley-VCH Verlag GmbH & Co. 1990. Reproduced with permission.]

illustration each excitation step has the same circular polarization; a comparable connection occurs with alternating right- and left-circular polarization.

Frame (*b*) illustrates two sequences that provide complete linkages between initial states and states of highest excitation, when all levels have the same J. In these cases it is necessary to employ alternating right- and left-circular polarizations in order to have a complete path between ground sublevels and those at the end of the chain. At each step of excitation there is a single unlinked (dark) sublevel.

Frame (*c*) shows circular polarization linkages in a sequence in which J decreases at each step. As with linear polarization, at each step two sublevels are unlinked to higher excitation. This disconnection pattern also occurs when the sense of polarization changes at each step, as in frame (*b*).

Comment. When spontaneous emission can be neglected the effect of degeneracy is often expressible as a set of independent systems, corresponding to different values of the magnetic quantum number of the initial level. From each of these states there arises a chain of excitation. The linkages are those of independent chains, and the overall excitation probability is the sum of probabilities for the separate chains (i.e. one must sum probabilities, not probability amplitudes). Thus if J_0 denotes the angular momentum of the initial level, and the sublevels have equal initial probability, then the population in level J at time t is the sum of contributing sublevel populations,

$$P_J(t) = \frac{1}{2J_0 + 1} \sum_M |C_{JM}(t)|^2, \tag{12.21}$$

where $C_{JM}(t)$ is the solution for an RWA Hamiltonian appropriate to a particular set of M values. The average over magnetic sublevels, each with a distinct orientation and hence a distinct Rabi frequency, acts to diminish the amplitude of population oscillations.

It is important to recognize that this sum of individual probabilities cannot, in general, be reproduced as the square of any single probability amplitude for an averaged interaction – there is generally no "average atom" that will exhibit the averaged dynamics.

13

Three states

Quantum changes of three-state systems have some similarities with those of two-state systems. When subject to steady illumination the populations may undergo oscillations similar to the Rabi cycling of two-state systems, and various forms of adiabatic following are possible. Analytic solutions to the relevant TDSE exist [Sho90, Chap. 23]. The additional degree of freedom, typically allowing controllable parameters of a second laser pulse, allows a wider variety of controlled excitation. The resulting differences and similarities to two-state systems have been discussed at length [Whi76; Sho77; Rad82b; Yoo85; Car87].

13.1 Three-state linkages

Two-field linkages. The simplest extension of two-state excitation allows two laser fields, here identified by letters P (for pump) and S (for Stokes), as befits the stimulated Raman process discussed in Chap. 14. The carrier frequencies of the two fields, ω_P and ω_S, are each assumed to be close to resonance with one, and only one, Bohr frequency, so that each field can be uniquely identified with a particular transition (failure of this restriction, and the resulting linkage ambiguity, is discussed in [Una00]). I will assume that the P field is (near) resonance only with the 1–2 transition, while the S field is (near) resonant only with the 2–3 transition; these interactions thereby form a two-step linkage chain. This system has three possible linkage patterns, shown in Fig. 13.1.

The linkage patterns (sometimes called configurations) differ by the ordering of the energies of the linked states. With the assumption that population initially occupies state 1, as in Fig. 13.1, the definitions are:

Ladder: The ladder system has the energy ordering $E_1 < E_2 < E_3$.
Lambda: In the lambda-linkage system the middle state has largest energy, $E_1 < E_2$ and $E_3 < E_2$, and the links, when displayed on an energy diagram such as Fig. 13.1, appear as the two legs of a greek letter "lambda". This is the arrangement found in the traditional Raman process, in which an applied pump field (P here) places population into an excited state, from which it decays by spontaneous emission of a Stokes field (S here).

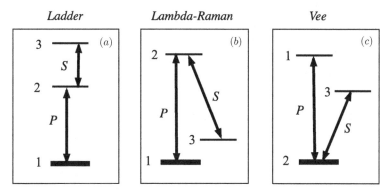

Figure 13.1 Types of three-state two-field excitation linkages when initial population is in a single state, shown as a thick line: (*a*) the ladder, (*b*) lambda and (*c*) vee linkages. The ground state is the initial state of the three-state chain for the ladder and lambda, but is the middle state of the vee.

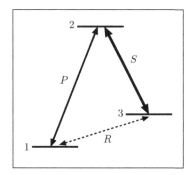

Figure 13.2 Three-state loop linkage, showing states 1, 2, 3 and links *P*, *S*, and *R*. The links *P* and *S* form a Raman linkage, in which the Raman transition, link *R*, completes a loop.

Vee: The "letter vee" system is an inverted lambda: the middle level has lowest energy, $E_2 < E_1$ and $E_2 < E_3$. In this arrangement the initial state (i.e. the ground state) lies in the middle of the chain.

Three-field loop linkage. The presence of a third field, labeled *R* (for Raman) and assumed to be (near) resonance with the 1–3 transition, can complete a loop of linkages, as shown in Fig. 13.2.

Selection rules prevent such loops for electric-dipole transitions in free space – such linked states should have opposite parity. Thus it is not possible to have a closed loop of excitation linkages with only electric-dipole radiation in a free atom. However, there exist alternative interactions, and alternative systems (e.g. atoms in a crystal), for which closed loops are possible. Section 15.5 discusses examples.

13.2 The three-state RWA

The most general three-state system has three possible interstate linkages, driven by three fields, as shown in the loop linkage pattern of Fig. 13.2. These links involve the electric field, at the center of mass,

$$\mathbf{E}(t) = \mathcal{E}_P(t)\mathrm{Re}\left\{\mathbf{e}_P\,e^{-i\omega_P t + i\varphi_P}\right\} + \mathcal{E}_S(t)\mathrm{Re}\left\{\mathbf{e}_S\,e^{-i\omega_S t + i\varphi_S}\right\}$$
$$+ \mathcal{E}_R(t)\mathrm{Re}\left\{\mathbf{e}_R\,e^{-i\omega_R t + i\varphi_R}\right\}, \tag{13.1}$$

characterized by carrier frequencies ω_λ, phases φ_λ, real-valued pulse envelopes $\mathcal{E}_\lambda$ and complex unit vectors $\mathbf{e}_\lambda$, each identified by a label $\lambda = P, S, R$. All of these may vary with time; the time variation of the phase φ_λ contributes to the instantaneous frequency,

$$\omega_\lambda(t) = \omega_\lambda - \dot{\varphi}_\lambda. \tag{13.2}$$

As indicated in Fig. 13.2, the three linkages form a loop, although typically we deal with only two fields, taken to be P and S. These are the pump and Stokes fields of a Raman transition (or probe and strong field for electromagnetically induced transparency (EIT); see Sec. 21.4.3). The instantaneous frequencies of these fields differ from their respective time-dependent Bohr frequencies,

$$\omega_{nm}(t) = |E_n + V_{nn}(t) - E_m - V_{mm}(t)|/\hbar, \tag{13.3}$$

by the detunings

$$\Delta_P(t) = \omega_{12}(t) - \omega_P + \dot{\varphi}_P, \qquad \Delta_S(t) = \omega_{23}(t) - \omega_S + \dot{\varphi}_S. \tag{13.4}$$

The third link, the R field, has a carrier frequency ω_R close to the *Raman frequency*, i.e. the Bohr frequency of the 1–3 transition; it has the detuning

$$\Delta_R(t) = \omega_{13}(t) - \omega_R + \dot{\varphi}_R. \tag{13.5}$$

In the absence of this field the system is that of stimulated Raman scattering (SRS); see Chap. 14.

As with two-state coherent excitation, our interest lies with near-resonant fields. For such interactions it is desirable to introduce Hilbert-space reference axes that rotate in keeping with carrier frequencies and to make a RWA. For that purpose we express the statevector as the superposition of three quantum states, 1, 2, and 3, whose bare energies (eigenvalues of the atom Hamiltonian H^{at} are E_1, E_2, and E_3,

$$\Psi(t) = \sum_{n=1,3} C_n(t)\,\boldsymbol{\psi}'_n(t), \qquad \boldsymbol{\psi}'_n(t) = e^{-i\zeta_1(t)}\boldsymbol{\psi}_1, \tag{13.6}$$

with time-dependent phases $\zeta_n(t)$ and rotating coordinates $\boldsymbol{\psi}'_n(t) \equiv \exp[-\zeta_n(t)]\boldsymbol{\psi}_n$ chosen, as discussed below, to eliminate explicit appearance of rapidly varying terms in the equations of motion. The specific choice depends on the linkage pattern – on the ordering of the energies along the chain – as will be noted below. With these constructions for the statevector and the field the Schrödinger equation becomes a set of coupled ODEs,

$$\frac{d}{dt}\mathbf{C}(t) = -i\mathbf{W}(t)\mathbf{C}(t). \tag{13.7}$$

Diagonal elements. The diagonal elements of the lossless Hamiltonian matrix $\hbar\mathbf{W}$ are, by construction, the energies of the system shifted by amounts that depend on the choice of phases $\zeta_n(t)$,

$$\hbar W_{nn}(t) \equiv \hbar\Delta_n(t) = E_n + V_{nn}(t) - \hbar\dot{\zeta}_n(t). \tag{13.8}$$

States n and m are said to be resonant if $W_{nn}(t) = W_{mm}(t)$.

Off-diagonal elements. For definiteness I will consider electric-dipole transitions, for which the interaction Hamiltonian is that of an atomic dipole moment $\mathbf{d}$ affected by a classical electric field,

$$V(t) = -\mathbf{d} \cdot \mathbf{E}(t). \tag{13.9}$$

All that matters is that the interaction is expressible as the product of an atomic operator, here the dipole transition moment $\mathbf{d}$, and a field, here the electric field $\mathbf{E}$ evaluated at the stationary center of mass (taken as the coordinate origin), over which the experimenter has control.

We assume, with justification discussed below, that each of the links shown in Fig. 13.2 is near resonant with one and only one of the three fields P, S, R. With the assumption that each field can be uniquely and unambiguously associated with a single transition, the off-diagonal elements of $\hbar\mathbf{W}(t)$ are, with omission of explicit time dependence,

$$\hbar W_{12} = -\frac{1}{2}\mathcal{E}_P\left[d_{12}\,e^{-i\omega_P t + i\varphi_P} + d_{21}\,e^{i\omega_P t - i\varphi_P}\right]e^{-i(\zeta_2 - \zeta_1)},$$

$$\hbar W_{23} = -\frac{1}{2}\mathcal{E}_S\left[d_{23}\,e^{-i\omega_S t + i\varphi_S} + d_{32}\,e^{i\omega_S t - i\varphi_S}\right]e^{-i(\zeta_3 - \zeta_2)}, \tag{13.10}$$

$$\hbar W_{13} = -\frac{1}{2}\mathcal{E}_R\left[d_{13}\,e^{-i\omega_R t + i\varphi_R} + d_{13}\,e^{i\omega_R t - i\varphi_R}\right]e^{-i(\zeta_3 - \zeta_1)},$$

where the dipole moments are

$$d_{12} = d_{21}^* = \langle 1|\mathbf{d}\cdot\mathbf{e}_P|2\rangle, \quad d_{23} = d_{32}^* = \langle 2|\mathbf{d}\cdot\mathbf{e}_S|3\rangle, \quad d_{13} = d_{31}^* = \langle 1|\mathbf{d}\cdot\mathbf{e}_R|3\rangle. \tag{13.11}$$

Any two of these can be made real valued by appropriate choice of wavefunction phases. Here, and in the following, the slowly varying time dependence of pulse envelopes $\mathcal{E}_\lambda$ and the matrix $\mathbf{W}$ is not explicitly shown.

13.2.1 The RWA

We choose the (arbitrary) time-dependent phases $\zeta_n(t)$ to cancel the exponential carrier frequencies and phases from some of the exponentials that appear in the off-diagonal elements of $W(t)$. This choice defines the rotating-wave picture. We then treat the other exponentials as rapidly varying and therefore negligible, thereby making a three-state RWA. This means that we take the phase differences $\zeta_n(t) - \zeta_m(t)$ to be equal to field phases $\omega_\lambda t - \varphi_\lambda$. The specific choices depend on the ordering of the energy levels, as discussed below.

The ladder system. For the ladder system we typically make the phase choices

$$\hbar\dot\zeta_1 = E_1 + V_{11}, \qquad \hbar\dot\zeta_2 = \hbar\dot\zeta_1 + \omega_P, \qquad \hbar\dot\zeta_3 = \hbar\dot\zeta_2 + \omega_S, \tag{13.12}$$

to give the detunings

$$\Delta_1 = 0, \qquad \Delta_2 = \Delta_P, \qquad \Delta_3 = \Delta_P + \Delta_S. \tag{13.13}$$

The choice of null first detuning, $\Delta_1 = 0$, suits the usual initial condition $C_1(0) = 1$. States 1 and n are resonant if $\Delta_n = 0$.

The lambda system. In the lambda system we make the phase choices

$$\hbar\dot\zeta_1 = E_1 + V_{11}, \qquad \hbar\dot\zeta_2 = \hbar\dot\zeta_1 + \omega_P, \qquad \hbar\dot\zeta_3 = \hbar\dot\zeta_2 - \omega_S. \tag{13.14}$$

The resulting detunings are

$$\Delta_1 = 0, \qquad \Delta_2 = \Delta_P, \qquad \Delta_3 = \Delta_P - \Delta_S. \tag{13.15}$$

The third detuning for the lambda linkage is the *difference* of two detunings, whereas for the ladder system the third detuning is the *sum* of two single-step detunings.

The vee system. The vee system is distinguishable from the other linkage patterns by the initial conditions:[1] Whereas the initial population of the lambda and ladder systems resides in one end of the linkage chain, for the vee system the population initially resides in the *middle* state of a chain, $C_2(0) = 1$. When the interaction occurs, population proceeds towards each end of the chain, from which it returns with constructive or destructive interference of amplitudes. We choose the phases to reference state 2 as the energy zero,

$$\hbar\dot\zeta_1 = \hbar\dot\zeta_2 - \hbar\omega_P, \qquad \hbar\dot\zeta_2 = E_2 + V_{22}, \qquad \hbar\dot\zeta_3 = \hbar\dot\zeta_2 - \omega_S t. \tag{13.16}$$

The resulting detunings are

$$\Delta_1 = -\Delta_P, \qquad \Delta_2 = 0, \qquad \Delta_3 = -\Delta_S. \tag{13.17}$$

Neglecting counter-rotating terms. With these phases we make the RWA by neglecting terms that vary as $\exp[i(\omega_n + \omega_m)t]$ – these are *counter-rotating* terms:

$$e^{\pm i 2\omega_P t} \to 0, \qquad e^{\pm i 2\omega_S t} \to 0, \qquad e^{\pm i(\omega_S + \omega_P)t} \to 0. \tag{13.18}$$

---

[1] This distinction fails for a system that has already been subjected to coherent excitation, thereby producing a distribution of population amongst the three states.

As with the two-state RWA these approximations require that the carrier frequencies be much larger than the Rabi frequencies that are responsible for excitation changes.

Selectivity. We must also neglect terms that oscillate at difference frequencies,

$$e^{\pm i(\omega_S - \omega_P)t} \rightarrow 0. \tag{13.19}$$

This approximation requires that the separation between carrier frequencies be much larger than the Rabi frequencies. However, it is not necessary that the two fields have different frequencies; if the frequencies are equal, then the fields must differ in their polarization, **e**. The needed selectivity then obtains through selection rules.

The RWA Hamiltonian. Each of the linkage patterns shown in Fig. 13.1 leads to a similar RWA Hamiltonian; the only significant difference is in the diagonal elements – the detunings. The following discussion will assume that, as indicated in Fig. 13.2, the numbering of states makes E_2 the largest of the energies: $E_2 > E_1$ and $E_2 > E_3$.

The first of the phases we define to nullify $\Delta_1(t)$,

$$\hbar\zeta_1(t) = E_1 t + \int_0^t dt' \, V_{11}(t'). \tag{13.20a}$$

The others we take to be, with omission of explicit time dependence,

$$\zeta_2 = \zeta_1 + \omega_P t + \varphi_2, \tag{13.20b}$$

$$\zeta_3 = \zeta_2 - \omega_S t + \varphi_3 = \zeta_1 + \omega_P t - \omega_S t + \varphi_2 + \varphi_3, \tag{13.20c}$$

where the phases φ_2 and φ_3 are still to be assigned. The result of the RWA is the off-diagonal elements

$$\hbar W_{12} = -\frac{1}{2} d_{21} \mathcal{E}_P \, e^{-i(\varphi_P + \varphi_2)} \equiv \hbar W_P^*,$$

$$\hbar W_{23} = -\frac{1}{2} d_{23} \mathcal{E}_S \, e^{+i(\varphi_S - \varphi_3)} \equiv \hbar W_S, \tag{13.21}$$

$$\hbar W_{13} = -\frac{1}{2} d_{13} \mathcal{E}_R \, e^{i(\omega_R - \omega_P - \omega_S)t} \, e^{-i(\varphi_R + \varphi_2 + \varphi_3)} \equiv \hbar W_R^*,$$

and the matrix W has the structure (the numbers to the right show the ordering of the states upon which the matrix acts)

$$\mathsf{W} = \begin{bmatrix} 0 & W_P^* & W_R^* \\ W_P & \Delta_P & W_S \\ W_R & W_S^* & \Delta_P - \Delta_S \end{bmatrix}. \tag{13.22}$$

Phase convention. The choice of phase for the fields at a given time is arbitrary. I have here followed the convention that when $E_n > E_m$ the matrix element of W_{mn} is shown as an explicit complex conjugate. For the lambda pattern and the linkage of Fig. 13.2, in which state 2 has highest energy, this means $W_{12} = W_P^*$ and $W_{32} = W_S^*$. Alternatively, for the chain pattern $W_{23} = W_S^*$.

Real-valued interactions. The choice

$$\varphi_2 = -\varphi_P, \qquad \varphi_3 = \varphi_S \tag{13.23}$$

removes the field phase from the elements W_R and W_S. When the dipole matrix elements are real-valued, $d_{nm} = d_{mn}$, the resulting elements are real valued,

$$\hbar W_P = -\frac{1}{2}d_{12}\mathcal{E}_P = \hbar W_P^*, \qquad \hbar W_S = -\frac{1}{2}d_{23}\mathcal{E}_S = \hbar W_S^*, \tag{13.24a}$$

while the third element may be complex and time dependent,

$$\hbar W_R^* = -\frac{1}{2}d_{13}\mathcal{E}_R \exp[i \Delta_{psr}t - i\varphi_{psr}]. \tag{13.24b}$$

Here Δ_{psr} is the beat frequency of the three carrier frequencies and the phase φ_{psr} is a static phase if the three fields retain static phases:

$$\Delta_{psr} = \omega_R + \omega_S - \omega_P, \qquad \varphi_{psr} = \varphi_R + \varphi_S - \varphi_P. \tag{13.25}$$

Rabi frequencies. It is traditional to parametrize the strength of the atom–laser interaction, here embodied in the off-diagonal elements of the RWA Hamiltonian matrix $\hbar\mathsf{W}$, by means of Rabi frequencies. As shown above, it is always possible to employ basis states whose phases (perhaps time dependent) make elements of the RWA Hamiltonian real, and hence it is usually possible to treat the Rabi frequencies as real-valued functions of time, here taken to be

$$\Omega_S = -d_{23}\mathcal{E}_S/\hbar, \qquad \Omega_P = -d_{12}\mathcal{E}_P/\hbar, \qquad \Omega_R = -d_{13}\mathcal{E}_R/\hbar. \tag{13.26}$$

Various choices of phases and signs, and factors of 2, appear in the literature of symbols for the RWA Hamiltonian and its parametrization by means of frequencies. In the present work, and most of the literature, a Rabi frequency refers to a frequency at which two-state oscillations occur in probabilities, not probability amplitudes. The choice of minus sign retains the structure of the Schrödinger equation when Ω is used in place of W.

The use of real-valued Rabi frequencies, and the placement of field phase into basis states, may obscure some changes of interest when the field phases are controllably changed. To make explicit such time dependence it is often preferable to use complex-valued Rabi frequencies that exhibit field phases:

$$\check{\Omega}_S = -d_{23}\mathcal{E}_S e^{-i\varphi_S}/\hbar, \quad \check{\Omega}_P = -d_{12}\mathcal{E}_P e^{-i\varphi_P}/\hbar, \quad \check{\Omega}_R = -d_{13}\mathcal{E}_R e^{-i\varphi_R}/\hbar. \tag{13.27}$$

Frequency matching. An important special case occurs when the three photon frequencies obey the condition

$$\omega_P = \omega_S + \omega_R, \qquad \Delta_{psr} = 0. \qquad (13.28)$$

Then the element $W_R(t)$ becomes slowly varying, and the RWA Hamiltonian matrix takes the form (possibly slowly varying)

$$W = \frac{1}{2} \begin{bmatrix} 0 & \Omega_P\, e^{-i\varphi_P} & \Omega_R\, e^{i\varphi_R} \\ \Omega_P\, e^{i\varphi_P} & 2\Delta_P & \Omega_S\, e^{i\varphi_S} \\ \Omega_R\, e^{-i\varphi_R} & \Omega_S\, e^{-i\varphi_S} & 2\Delta_R \end{bmatrix}. \qquad (13.29)$$

The RWA detunings are those of the P field and the R field from their respective Bohr frequencies ω_{12} and ω_{13}.

Two-photon resonance, between states 1 and 3, occurs when the frequency matching of eqn. (13.28) holds and, in addition, the detuning Δ_3 vanishes. This means

$$\Delta_R = 0, \qquad \text{or} \qquad \Delta_S = \Delta_P, \qquad \Delta_{psr} = 0. \qquad (13.30)$$

No direct coupling between states 1 and 3 occurs in this description – the linkage pattern does not form a closed loop. This constraint usually follows from parity selection rules, but there are situations where such couplings must be considered; see Sec. 15.5. Furthermore, field P is assumed to make no contribution to the 2–3 coupling, nor does field S contribute to the 1–2 coupling. Such couplings are present, as eqn. (13.29) shows, but they are assumed here to have the rapid time variation that is neglected in the RWA.

Quasistatic fields. The presentation above dealt with laser fields as the source of interaction energy. The basic RWA equations, and the various solutions presented in the sections that follow, require only that there be slowly varying or constant terms in the matrix $W(t)$. Thus one can consider treating quasistatic or DC fields by taking the carrier frequencies to be zero.

13.2.2 Three-state analytic solutions, constant intensity

As with the two-state atom, the three-state RWA Schrödinger equation has exact analytic solutions when the RWA Hamiltonian matrix $\hbar W$ is constant. Although the general analytic expressions are cumbersome, special cases produce simple results.

The eigenvalues. The analytic solutions are most easily presented in terms of eigenvectors Φ and eigenvalues ε of W (the dressed states and quasienergies),

$$W\Phi_\nu = \varepsilon_\nu \Phi_\nu. \qquad (13.31)$$

The eigenvalues ε_ν are the roots of the determinantal equation

$$\text{Det } [W - \varepsilon 1] = 0. \qquad (13.32)$$

For the Raman Hamiltonian,

$$\mathsf{W}(t) = \begin{bmatrix} \Delta_1 & \frac{1}{2}\Omega_P e^{-i\varphi_P} & 0 \\ \frac{1}{2}\Omega_P e^{i\varphi_P} & \Delta_2 & \frac{1}{2}\Omega_S e^{-i\varphi_S} \\ 0 & \frac{1}{2}\Omega_S e^{i\varphi_S} & \Delta_3 \end{bmatrix}, \tag{13.33}$$

this equation produces the cubic *characteristic equation*

$$\varepsilon^3 + a\varepsilon^2 + b\varepsilon + c = 0, \tag{13.34}$$

where the coefficients are

$$\begin{aligned} a &= -(\Delta_1 + \Delta_2 + \Delta_3), \\ b &= \Delta_1\Delta_2 + \Delta_2\Delta_3 + \Delta_3\Delta_1 - \frac{1}{4}\left[|\Omega_P|^2 + |\Omega_S|^2\right], \\ c &= \Delta_1\Delta_2\Delta_3 - \frac{1}{4}\Delta_3|\Omega_S|^2 - \frac{1}{4}\Delta_1|\Omega_P|^2. \end{aligned} \tag{13.35}$$

The three roots to this equation are the quasienergies of the three-state atom. Explicit analytic expressions for the general solutions to cubic equations may be found in many reference works. In practice it is often more convenient to employ a computer routine that determines roots of equations or, more generally, that obtains matrix eigenvalues and eigenvectors.

13.2.3 Three-state eigenvectors

Given the eigenvalues, we can construct the eigenvectors in various ways. In one approach we rewrite the defining equation for an eigenvector $\mathbf{\Phi}$ of the time-independent Hamiltonian W, eqn. (13.31), as

$$\mathsf{M}(\varepsilon_\nu)\mathbf{\Phi} = 0 \quad\text{or}\quad \sum_i M_{ij}(\varepsilon_\nu)\Phi_j = 0, \tag{13.36}$$

where

$$\mathsf{M}(\varepsilon) = \mathsf{W} - \varepsilon\mathbf{1}. \tag{13.37}$$

For the Raman Hamiltonian this means

$$\mathsf{M}(\varepsilon) = \begin{bmatrix} \Delta_1 - \varepsilon & W_P^* & 0 \\ W_P & \Delta_2 - \varepsilon & W_S \\ 0 & W_S^* & \Delta_3 - \varepsilon \end{bmatrix}. \tag{13.38}$$

The eigenvalues ε are solutions to the determinantal equation $\operatorname{Det}\mathsf{M}(\varepsilon_\nu) = 0$. In turn, the determinant of any matrix M is expressible in terms of cofactors: for any row i the determinant is the sum of the elements times the cofactor of the element,

$$\operatorname{Det}\mathsf{M} = \sum_j M_{ij} M^{ij}. \tag{13.39}$$

For the matrix $\mathsf{M}(\varepsilon) = \mathsf{W} - \varepsilon\mathbf{1}$ the determinant vanishes when ε is an eigenvalue. This means that for any row i

$$\sum_j M_{ij}(\varepsilon_\nu)M^{ij}(\varepsilon_\nu) = 0. \tag{13.40}$$

We can therefore obtain an unnormalized eigenvector by taking the components $V_j(\varepsilon_\nu) = M^{ij}(\varepsilon_\nu)$ from a set of cofactors. For example, on taking the first row we have

$$V_1(\varepsilon_\nu) = (-1)^0[(\Delta_2 - \varepsilon_\nu)(\Delta_3 - \varepsilon_\nu) - |W_S|^2],$$
$$V_2(\varepsilon_\nu) = (-1)^1(\Delta_3 - \varepsilon_\nu)W_P, \tag{13.41}$$
$$V_3(\varepsilon_\nu) = (-1)^0 W_S^* W_P.$$

From the middle row we obtain the components

$$V_1(\varepsilon_\nu) = (-1)^1(\Delta_3 - \varepsilon_\nu)W_P^*,$$
$$V_2(\varepsilon_\nu) = (-1)^0(\Delta_1 - \varepsilon_\nu)(\Delta_3 - \varepsilon_\nu), \tag{13.42}$$
$$V_3(\varepsilon_\nu) = (-1)^1(\Delta_1 - \varepsilon_\nu)W_S^*.$$

When the eigenvalues are expressed in analytic form, then such a formula provides an analytic expression for the components of the eigenvectors.

Although it is sometimes useful to have such closed-form expressions for eigenvalues and eigenvectors, it is usually more satisfactory to order the eigenvalues by size, as discussed in App. E. With three states this leads to the labels $-, 0, +$ used below. With that convention, curves of labeled eigenvalues do not cross as changes occur in the underlying Hamiltonian.

The probability amplitudes. From eigenvalues and eigenvectors of the RWA Hamiltonian matrix $\hbar\mathsf{W}$ we can construct solutions to the Schrödinger equation that satisfy any required initial condition: we write the statevector as the superposition

$$\Psi(t) = \sum_\nu A_\nu \exp(-\mathrm{i}\varepsilon_\nu t)\,\Phi_\nu \tag{13.43}$$

and determine the constants A_ν such that the initial statevector is

$$\Psi(0) = \begin{bmatrix} C_1(0) \\ C_2(0) \\ C_3(0) \end{bmatrix}. \tag{13.44}$$

The desired solution is expressible as a sum of three exponentials,

$$C_n(t) = \sum_{\nu m} \Phi_{\nu n} \exp(-\mathrm{i}\varepsilon_\nu t)v_{\nu m}^* C_m(0), \tag{13.45}$$

where $\Phi_{\nu n}$ is the nth component of eigenvector Φ_ν. We have here an exact solution to the three-state RWA for any detunings and any Rabi frequencies. To construct this solution we need only evaluate the three eigenvalues ε_ν and obtain the three eigenvectors.

Although presented here for a three-state system, the formulas readily extend to multistate systems. The sums over ν and n or m then run over N values for N states.

13.2.4 Eigenvectors, two-photon resonance

An important special case is when there occurs a two-photon resonance, so that $\Delta_1 = \Delta_3 = 0$. The RWA Hamiltonian for this two-photon resonant excitation reads

$$\mathsf{W} = \frac{1}{2} \begin{bmatrix} 0 & \Omega_P e^{-i\varphi_P} & 0 \\ \Omega_P e^{i\varphi_P} & 2\Delta & \Omega_S e^{i\varphi_S} \\ 0 & \Omega_S e^{-i\varphi_S} & 0 \end{bmatrix}. \tag{13.46}$$

The eigenvalues are

$$\varepsilon_0 = 0, \qquad \varepsilon_\pm = \frac{1}{2}(\Delta \pm \Upsilon), \tag{13.47}$$

and the eigenvectors can be taken as

$$\begin{aligned} \Phi_0 &= \left[\Omega_S e^{i\varphi_S} \psi'_1(t) & & -\Omega_P e^{i\varphi_P} \psi'_3(t)\right]/\mathcal{N}_0, \\ \Phi_\pm &= \left[\Omega_P e^{-i\varphi_P} \psi'_1(t) & +(\Delta \pm \Upsilon)\psi'_2(t) & +\Omega_S e^{-i\varphi_S} \psi'_3(t)\right]/\mathcal{N}_\pm, \end{aligned} \tag{13.48}$$

where the $\mathcal{N}$ are normalization constants and the eigenvalue splitting is

$$\Upsilon = \sqrt{\Delta^2 + |\Omega_S|^2 + |\Omega_P|^2}. \tag{13.49}$$

The primes on $\psi'_m(t)$ indicate that the basis vectors are in a rotating coordinate system. Alternatively, the eigenvectors can be rewritten as

$$\Phi_0 = \cos\Theta\, e^{i\varphi_S}\, \psi'_1(t) \qquad - \sin\Theta\, e^{i\varphi_P} \psi'_3(t), \tag{13.50a}$$

$$\Phi_\pm = \sin\Theta \cos\varphi\, e^{-i\varphi_P} \psi'_1(t) \pm \sin\varphi\, \psi'_2(t) + \cos\Theta \cos\varphi\, e^{-i\varphi_S} \psi'_3(t), \tag{13.50b}$$

where the defining angles are

$$\tan\Theta = \frac{\Omega_P}{\Omega_S}, \qquad \tan\varphi = \frac{\Upsilon - \Delta}{\sqrt{|\Omega_P|^2 + |\Omega_S|^2}}. \tag{13.51}$$

When the linkages are fully resonant, so $\Delta = 0$, then $\varphi = \pi/2$ and the eigenvectors are

$$\begin{aligned} \Phi_0 &= \cos\Theta\, e^{i\varphi_S}\, \psi'_1(t) & -\sin\Theta\, e^{i\varphi_P}\psi'_3(t), \\ \Phi_\pm &= \left[\sin\Theta\, e^{-i\varphi_P}\psi'_1(t) & \pm\psi'_2(t) & +\cos\Theta\, e^{-i\varphi_S}\psi'_3(t)\right]/\sqrt{2}. \end{aligned} \tag{13.52}$$

When $\Omega_s = 0$ then $\Theta = 0$ and the eigenvectors are

$$\begin{aligned} \Phi_0 &= e^{i\varphi_S}\, \psi'_1(t), \\ \Phi_\pm &= \pm\psi'_2(t) & +\cos\varphi\, e^{-i\varphi_S}\psi'_3(t), \end{aligned} \tag{13.53}$$

and so state $\psi'_1(t)$ is aligned with the single adiabatic state Φ_0.

13.2.5 Pulsed analytic solutions

Analytic solutions for the three-state TDSE have been published for a variety of pulse shapes [Car87; Car88a; Car88b; Car90]. Many of these involve the generalized hypergeometric function

$$_3F_2(a,b,c;d,e;z) = 1 + \frac{a}{1!}\frac{b}{d}\frac{c}{e}z^1 + \frac{a(a+1)}{2!}\frac{b(b+1)}{d(d+1)}\frac{c(c+1)}{e(e+1)}z^2 + \cdots, \qquad (13.54)$$

with various choices for the five parameters a, b, c, d, e and the variable $z(t)$. Some of these are for special cases of more general N-state systems whose RWA Hamiltonians have special symmetries, for example

$$\mathbf{W}(t) = c_1(t)\mathbf{S}_x + c_2(t)\mathbf{S}_y + c_3(t)\mathbf{S}_z + d(t), \qquad (13.55)$$

where the $\mathbf{S}_j$ are N-dimensional spin matrices (see App. A.1.3). For this Hamiltonian the TDSE solutions are expressible in terms of those for a two-state system [Maj32; Hio87].

13.3 Resonant chains

The next few paragraphs discuss the population histories for resonant three-state chains with constant Rabi frequencies, with particular attention to cases when the two Rabi frequencies are equal. When there are no probability losses the RWA Hamiltonian is the same for the ladder, the lambda, and the vee configurations of a three-state system. However, the differing initial conditions lead to different population histories. Population starting at one end of the chain will undergo the same behavior for each configuration. However, when population starts from the middle of the chain, as it does for the vee linkage, then the subsequent time dependence differs.

13.3.1 Equal Rabi frequencies

A simple theoretical situation occurs when both Rabi frequencies are constant and equal, and the two transitions are each resonant. The RWA Hamiltonian then reads[2]

$$\mathbf{W} = \frac{1}{2}\begin{bmatrix} 0 & \Omega & 0 \\ \Omega & 0 & \Omega \\ 0 & \Omega & 0 \end{bmatrix}. \qquad (13.56)$$

In preparation for the discussion of the analytic solutions for this system it is helpful to have in mind examples of the population histories. These depend upon the initial conditions, i.e. on which state initially has the population. Figure 13.3 illustrates the basic cases of (a) the ladder or lambda, and (b) the "letter vee" linkage patterns.

---

[2] This matrix is a multiple of the spin-one matrix $\mathbf{S}_x$ of Appendix A.1.3. The solutions are therefore special cases of those described in Sec. 15.3.5.

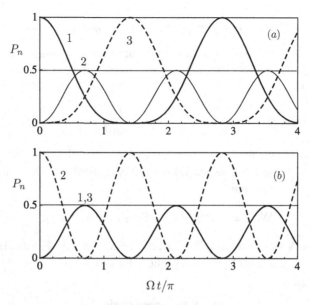

Figure 13.3 Excitation of three-state chain: populations P_n vs. Rabi cycles $\Omega t/\pi$. (*a*) The ladder or lambda pattern, initial population in state 1. (*b*) The vee pattern, initial population in state 2.

Ladder populations. In Fig. 13.3(*a*) the population initially resides in a state at one end of the chain, $C_1(0) = 1$, as in the ladder or lambda linkage. Starting from state 1 population flows into state 2 and then arrives, completely, in state 3. From there it returns through state 2 to the initial state for further periodic cycling. Population arrives in state 2 both in transit to state 3 and on returning. Thus the population oscillations of this state are twice as fast as those for states 1 and 3. Population never concentrates entirely in state 2; it holds at most half the population.

Vee-linkage populations. When the initial population resides in the quantum state lying at one end of a chain, the initial changes can only move population toward a single neighboring state. However, if the initially populated state is linked to two states, as it is with the vee linkage, then the initial changes can occur toward each of these two states. Figure 13.3(*b*) illustrates this behavior, for initial condition $C_2(0) = 1$. It is apparent that the populations of states 1 and 3 follow identical patterns. What is not evident is that these two states form a coherent superposition, as discussed in Sec. 13.3.2 below.

Eigenvalues and eigenvectors. The analytic solutions for this system are readily obtained from the eigenvalues and eigenvectors of the RWA Hamiltonian matrix. These are

$$
\begin{aligned}
\varepsilon_0 &= 0, & \Phi_0(t) &= \frac{1}{2}\Big[\psi_1'(t) & & -\psi_3'(t)\Big], \\
\varepsilon_\pm &= \pm\Omega/\sqrt{2}, & \Phi_\pm(t) &= \frac{1}{2}\Big[\psi_1'(t) & \pm\sqrt{2}\psi_2'(t) & +\psi_3'(t)\Big].
\end{aligned}
\tag{13.57}
$$

For the initial condition $C_1(0) = 1$ the statevector, expressed in terms of these eigenvectors, reads

$$\Psi(t) = \frac{1}{2}\left[\exp(-i\varepsilon_- t)\Phi_-(t) + \sqrt{2}\Phi_0(t) + \exp(-i\varepsilon_+ t)\Phi_+(t)\right]. \tag{13.58}$$

Reexpressed in terms of the original basis states the construction reads

$$\Psi(t) = \frac{1}{2}\left[\psi_1'(t) - \psi_3'(t)\right] + \frac{1}{2}\left[\psi_1'(t) + \psi_3'(t)\right]\cos(\sqrt{2}\Omega t/2)$$

$$- \frac{i}{\sqrt{2}}\psi_2'(t)\sin(\sqrt{2}\Omega t/2). \tag{13.59}$$

From this expression we see that all of the populations vary periodically with time, and that the periodicity of the middle state, 2, is half that of the first and last states:

$$P_1(t) = \frac{1}{8}[2 + \cos 2x + 4\cos x],$$

$$P_2(t) = \frac{1}{4}[1 - 2\cos 2x], \tag{13.60}$$

$$P_3(t) = \frac{1}{8}[2 + \cos 2x - 4\cos x],$$

where time variation occurs through the argument $x \equiv \sqrt{2}\Omega t$.

Time-dependent Rabi frequencies. These solutions readily apply to resonant pulsed excitation, so long as all Rabi frequencies share the same time dependence, say $\Omega(t) = f(t)\Omega_0$. The RWA Hamiltonian then factors as $W(t) = f(t)W$. By introducing the time scale

$$\tau(t) = \int_{-\infty}^{t} dt'\, f(t') \tag{13.61}$$

we obtain the equation

$$\frac{d}{d\tau}C(\tau) = -iWC(\tau), \tag{13.62}$$

in which the RWA Hamiltonian matrix $\hbar W$ is constant. The solutions are those discussed above, but with the scaled time variable $\tau(t)$. This rescaling technique allows the solution to any resonant equations in terms of solutions for a constant RWA Hamiltonian, as in Sec. 8.3.1.

13.3.2 Resonant "letter vee": Bright state

The difference between the population histories of the ladder and the "letter-vee" linkage follow from the initial conditions and the construction of the eigenvectors of the RWA Hamiltonian matrix. The initial conditions for the vee,

$$C_2(0) = 1, \qquad C_1(0) = C_3(0) = 0, \tag{13.63}$$

mean that the statevector has no component of the eigenvector $\Phi_0(t)$; it is

$$\Psi(t) = \frac{1}{2}\left[\exp(-i\varepsilon_-t)\Phi_-(t) - \exp(-i\varepsilon_+t)\Phi_+(t)\right]. \tag{13.64}$$

This combination always involves a fixed superposition of states 1 and 3,

$$\Psi(t) = \frac{1}{\sqrt{2}}\psi_2'(t)\cos(\Upsilon t/\sqrt{2}) - \frac{i}{2}\left[\psi_1'(t) + \psi_3'(t)\right]\sin(\Upsilon t/\sqrt{2}), \tag{13.65}$$

where Υ is defined in eqn. (13.49). The three states have the same periodicity, and population regularly returns entirely to the initial state, state 2.

The bright state. The mathematical description of the vee linkage takes particularly simple form when we introduce a combination of probability amplitudes that incorporate all of the interaction with state 2. Consider a three-state chain, resonant for the two-photon transition, with constant Rabi frequencies

$$\frac{d}{dt}C_1(t) = -\frac{i}{2}\Omega_P e^{-i\varphi_P}C_2(t),$$

$$\frac{d}{dt}C_2(t) = -i\,\Delta C_2(t) - \frac{i}{2}\Omega_P e^{i\varphi_P}C_1(t) - \frac{i}{2}\Omega_S e^{i\varphi_S}C_3(t), \tag{13.66}$$

$$\frac{d}{dt}C_3(t) = -\frac{i}{2}\Omega_S e^{-i\varphi_S}C_2(t).$$

Let the initial conditions, at $t = 0$, be that all population resides in the middle level of the chain – a "letter vee" pattern:

$$C_2(0) = 1, \qquad C_1(0) = 0, \qquad C_3(0) = 0. \tag{13.67}$$

We place all of the change to state 2 into a "bright" amplitude $C_B(t)$, complementing this with a "dark" amplitude $C_D(t)$,

$$C_B(t) = \left[\Omega_P e^{i\varphi_P}C_1(t) + \Omega_S e^{-i\varphi_S}C_3(t))\right]/\Omega_T, \tag{13.68a}$$

$$C_D(t) = \left[\Omega_S e^{i\varphi_S}C_1(t) - \Omega_P e^{-i\varphi_P}C_3(t)\right]/\Omega_T, \tag{13.68b}$$

where

$$\Omega_T = \sqrt{|\Omega_P|^2 + |\Omega_S|^2} \tag{13.69}$$

is the rms Rabi frequency. The amplitude $C_D(t)$ is uncoupled from the other two amplitudes and remains constant,

$$\frac{d}{dt}C_D(t) = 0, \tag{13.70}$$

while the other two amplitudes obey the equations

$$\frac{d}{dt}C_2(t) = -i\,\Delta C_2(t) - \frac{i}{2}\Omega_T C_B(t),$$

$$\frac{d}{dt}C_B(t) = -\frac{i}{2}\Omega_T C_2(t).$$

(13.71)

From these we deduce that the solutions oscillate at half the mean-square Rabi frequency.

These amplitudes are subject to the initial conditions $C_B(0) = C_D(0) = 0$. When the excitation is resonant, $\Delta = 0$, the desired solution for state 2 is

$$C_2(t) = \cos[\Omega_T t/2],$$

(13.72)

while that for the bright amplitude is

$$C_B(t) = \frac{i}{2}\Omega_T \frac{d}{dt}C_2(t).$$

(13.73)

From that expression we find the solutions for states 1 and 3 to be

$$C_1(t) = i\,(\Omega_P/\Omega_T)\sin[\Omega_T t/2], \qquad C_3(t) = i\,(\Omega_S/\Omega_T)\sin[\Omega_T t/2].$$

(13.74)

The population periodically leaves state 2 into a fixed superposition of states 1 and 3. The populations of these states are weighted by the squares of the relevant Rabi frequencies Ω_P and Ω_S.

13.4 Detuning

Nonzero detuning may originate from several sources. An experimenter may deliberately adjust a carrier frequency to be offset from the relevant Bohr frequency. Quasistatic fields may produce Stark or Zeeman shifts V_{nn} that alter the Bohr frequencies. Or motion of the atoms may introduce Doppler shifts, as discussed in the following section. Each of these effects appears as a detuning, possibly time dependent.

13.4.1 Doppler shifts in ladders; Doppler-free excitation

To a moving atom all traveling-wave laser frequencies appear Doppler shifted. When an atom is irradiated by several beams each different laser wavevector $\mathbf{k}$ produces a different Doppler shift, dependent on the component of $\mathbf{k}$ along the velocity vector. For the three-state ladder the RWA detunings associated with velocity v are

$$\hbar\Delta_1(v) = 0,$$

$$\hbar\Delta_2(v) = E_2 - E_1 - \hbar\omega_1',$$

(13.75)

$$\hbar\Delta_3(v) = E_3 - E_1 - \hbar\omega_1' - \hbar\omega_2',$$

where primes denote Doppler-shifted frequencies. Expressed in terms of stationary-atom detunings $\Delta_n(0)$ these formulas read

$$\Delta_1(v) = 0,$$

$$\Delta_2(v) = \Delta_2(0) - s_1, \tag{13.76}$$

$$\Delta_3(v) = \Delta_3(0) - s_1 - s_2,$$

where

$$s_n = \mathbf{v} \cdot \mathbf{k}_n \tag{13.77}$$

is the Doppler shift of the nth traveling wave. For optical frequencies the magnitude of each shift differs only slightly from the shift Δ appropriate to a mean optical frequency $\bar{\omega}$:

$$|s_n| = |\Delta| \times [1 + (\bar{\omega} - \omega_n)/\bar{\omega}] \simeq |\Delta|. \tag{13.78}$$

However the sign of s_n (i.e. whether the frequency is red or blue shifted) depends upon the relative direction of atom velocity and photon direction. We must therefore distinguish *collinear* propagation, wherein all laser photons travel in the same direction, from *counterpropagating* lasers, wherein alternate lasers propagate in opposite directions. (Other orientations of laser axes are also possible; the choices of co- and counterpropagation represent instructive extremes.) When the lasers are all resonant for stationary atoms, $\Delta_n(0) = 0$, the collinear case yields cumulative detunings that increase linearly with excitation state n

$$\Delta_n(v) = \{0, -\Delta, -2\Delta, -3\Delta, \ldots\}. \tag{13.79}$$

By contrast, the counterpropagating beams yield the alternating elements

$$\Delta_n(v) = \{0, -\Delta, 0, -\Delta, \ldots\}. \tag{13.80}$$

Figure 13.4 shows these two cases. In either case we construct Doppler-averaged probabilities with the formula

$$\bar{P}_n(t) = \int_0^\infty d\Delta \, \mathrm{p}(\Delta) \, P_n(\Delta; t), \tag{13.81}$$

where $\mathrm{p}(\Delta)$ is the probability of observing Doppler shift Δ. Let us here examine the behavior of the integrand $P_n(\Delta; t)$, the population histories that occur when the single-step detuning is Δ.

The two cases of Fig. 13.4 exhibit very different behavior, as is readily seen from a model in which a three-state ladder, undergoing coherent excitation by two steady and equal Rabi frequencies, is subject to a constant ionization loss from state 3, as modeled by including an imaginary part of the RWA Hamiltonian element W_{33}, as in Sec. 8.3.2. Figures 13.5 and 13.6 show examples of the excitation histories $P_n(\Delta; t)$ of such a system, as a function of time t and of first-step Doppler shift Δ. The separate frames show histories appropriate to each of the three states, as well as the population that is lost to ionization. A vertical plane parallel to the time axis at the center of the detuning axis, $\Delta = 0$, provides the population history appropriate to resonant excitation.

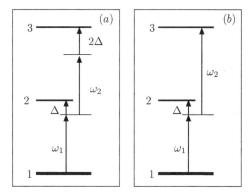

Figure 13.4 Three-state ladder configuration with Doppler detuning. (*a*) Collinear lasers for the two steps. (*b*) Counterpropagating lasers for the two steps.

Collinear lasers. Figure 13.5 shows the results when the two detunings are those of collinear lasers, so that the second-step detuning is twice the first-step detuning. As is typical of ladder linkages, the landscape patterns of the depicted surfaces are symmetric with respect to detuning; the excitation does not depend upon whether the detuning Δ is positive or negative. For any fixed detuning we observe regular patterns of population pulsations as a function of time. These oscillate more regularly and more rapidly with increasing Doppler shift Δ. Finally, we observe that excitation, and ionization, becomes less complete for large detunings; only those atoms whose Doppler shifts are less than a few Rabi frequencies will ionize. (Unit detuning here means detuning equal to the Rabi frequency.) Conversely, all atoms whose Doppler shift is a small fraction of a Rabi frequency can be treated as stationary and hence resonant; they all undergo resonant excitation.

Counterpropagating lasers. The population histories associated with counterpropagating laser beams, shown in Fig. 13.6, differ significantly from those obtained with collinear laser beams. For these calculations the second-step Doppler shift is opposite to that of the first step, and so the second-step detuning is zero. As with collinear beams, the patterns are symmetric about resonant detuning, $\Delta = 0$. The histories along a vertical slice with $\Delta = 0$ are identical with the resonant excitation slices of Fig. 13.5. As in that figure we observe low-amplitude high-frequency oscillations of excited-state population. However, large first-step Doppler shifts have much smaller effects on ionization than with collinear lasers. The cancellation of Doppler shifts, apparent as null cumulative detuning for state 3, here maintains conditions needed for resonant two-photon transitions. The resulting behavior shown in the figure is relatively insensitive to Doppler shift.

Doppler-free excitation. From the set of detunings of eqn. (13.80) we recognize that each atom, regardless of its velocity, will be able to maintain exact resonance conditions for two-photon absorption. This property holds even in the presence of radiative decay. When relaxation processes are present the two-photon resonance match must be within a tolerance

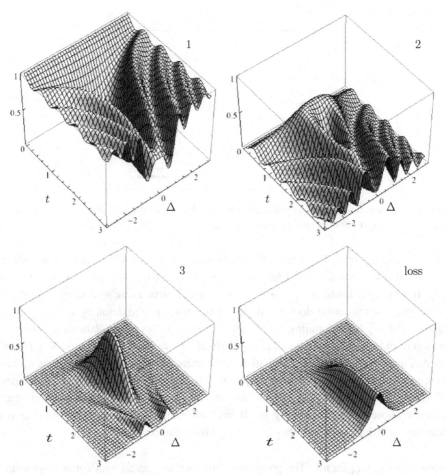

Figure 13.5 Perspective view of population histories $P_n(\Delta; t)$ and ionized fraction as a function of time and the first-step detuning Δ (in units of the equal Rabi frequencies) for a three-state atom excited by collinear lasers. The two Rabi frequencies are $\Omega_1 = \Omega_2 = 1$. [Redrawn from Fig. 9 of B. W. Shore and J. Ackerhalt, Phys. Rev. A, **15**, 1640 (1977) with permission.]

set by the inverse of the upper-state lifetime. Consequently a frequency scan of excitation reveals a strong signal when the two-photon energy is within the natural (radiative) width rather than the Doppler width. This principle forms the basis for *Doppler-free spectroscopy* [Gry77; Gia80; Dem03].

Exact cancellation of Doppler shifts, as in eqn. (8.83), occurs only for lasers having equal wavelength. Even then relativistic effects introduce residual second-order Doppler shifts proportional to $(v/c)^2$, and these prevent exact cancellation. Nevertheless Doppler-free excitation has numerous practical applications and has been widely studied.

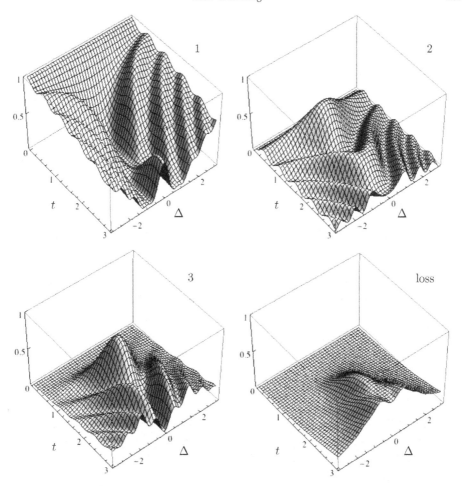

Figure 13.6 As Fig. 13.5 but for counterpropagating laser beams. [Redrawn from Fig. 10 of B. W. Shore and J. Ackerhalt, Phys. Rev. A, **15**, 1640 (1977) with permission.]

13.4.2 Detuned lambda system; population trapping

When there are no probability losses the RWA Hamiltonian is the same for the ladder, the lambda, and the vee configurations of a three-state system. However, when losses are present (e.g. photoionization or spontaneous emission) then the linkages are no longer equivalent. Typically the losses of a ladder occur at the end of the chain, whereas losses for the lambda linkage occur from the middle of the chain. This difference produces a qualitative difference in population histories. Figure 13.7 compares the difference of population histories for population that starts from state 1. In frame (*a*) the loss occurs at the end of the chain, from state 3. The population undergoes damped oscillations, eventually becoming completely lost. By contrast, frame (*b*) shows the population histories when the loss occurs from the

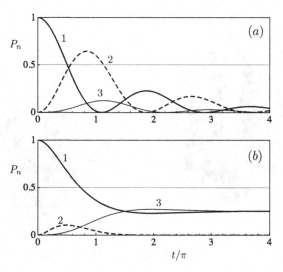

Figure 13.7 Population histories $P_n(t)$ for resonant three-state chain. (*a*) Loss from state 3. All population is eventually lost. (*b*) Loss from state 2. Population becomes trapped in a coherent superposition of states 1 and 3.

middle level, state 2. Here the population does not become completely lost; it becomes trapped in a coherent superposition of states 1 and 3 (a trapped state).

Bright and dark states. The occurrence of trapped population can be readily understood by introducing the amplitudes $C_B(t)$ and $C_D(t)$ of eqn. (13.68b). When population starts in state 1 these amplitudes have the initial values

$$C_D(0) = e^{i\varphi_S}\Omega_S/\Omega_T, \qquad C_B(0) = e^{i\varphi_P}\Omega_P/\Omega_T, \qquad (13.82)$$

where $\Omega_T^2 = \Omega_S^2 + \Omega_P^2$, and hence for the calculations of Fig. 13.7(*b*) these have equal values. When loss is present from state 2 but detunings are absent these amplitudes obey the equations

$$\frac{d}{dt}C_D(t) = 0,$$

$$\frac{d}{dt}C_B(t) = -\frac{i}{2}\Omega_T C_2(t),$$

$$\frac{d}{dt}C_2(t) = -\frac{i}{2}\Omega_T C_B(t) - \frac{\Gamma}{2}C_2(t). \qquad (13.83)$$

The amplitude $C_D(t)$ remains fixed at its initial value, while amplitude $C_B(t)$ decays with exponential damping. After a sufficiently long time there remains no population in state 2;

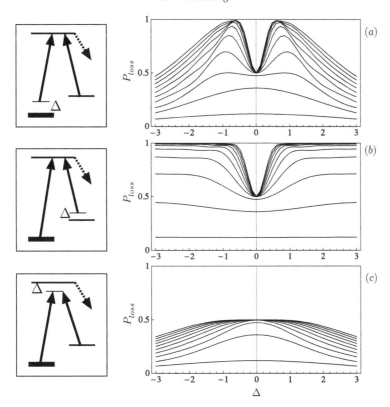

Figure 13.8 Snapshots at successive times of the lost portion $P_{loss}(t) = 1 - \sum_n P_n(t)(t)$ versus detuning Δ for the lossy lambda system. (a) Detuning of state 1. (b) Detuning of state 3. (c) Detuning of state 2. In each case resonant detuning $\Delta = 0$ leaves half of the population in a coherent superposition "dark" state. [Redrawn from Figs. 13.7-7 and 13.7-8 of B. W. Shore, *The Theory of Coherent Atomic Excitation*. Copyright Wiley-VCH Verlag GmbH & Co. 1990. Reproduced with permission.]

states 1 and 2 approach steady values such that

$$\Omega_S C_1(\infty) = \Omega_P C_3(\infty), \tag{13.84}$$

as seen in Fig. 13.7(b). This superposition is a trapped state.

The occurrence of trapped population, and a trapped state, requires resonant excitation. The presence of detuning, along with loss, prevents the creation of nonzero stationary states. Figure 13.8 illustrates this alteration. The frames show the lost portion $P_{loss}(t) = 1 - \sum_n P_n(t)$ at successively later times, as a function of detuning. Unless the detuning vanishes, the probabilities eventually become zero.

13.4.3 Large intermediate detuning

When intermediate-state detuning is present, the excitation dynamics can exhibit a variety of characteristics, depending on how the detuning compares with the Rabi frequencies. A

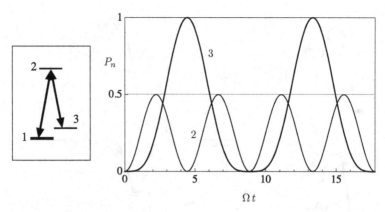

Figure 13.9 Inset at left: three-state linkage pattern for fully resonant excitation. Main frame at right: Population histories of intermediate state 2 and final state 3 for resonant excitation, $\Delta_1 = \Delta_2 = \Delta_3$, and equal Rabi frequencies, $\Omega_P = \Omega_S \equiv \Omega$. [Redrawn from Fig. 43 of B. W. Shore, Acta Physica Slovaka, **58**, 243–486 (2008) with permission.]

particularly simple, and useful, example occurs when the detuning is much larger than the Rabi frequencies.

Consider the three-state Hamiltonian for a lambda linkage induced by simultaneous P and S fields (here, for typographical simplicity, the Rabi frequencies are taken as complex, denoted $\check{\Omega}$, incorporating the phases of the fields)

$$\frac{d}{dt}C_1 = -i\Delta_1 C_1 - \frac{i}{2}\check{\Omega}_P^* C_2,$$
$$\frac{d}{dt}C_2 = -i\Delta_2 C_2 - \frac{i}{2}\check{\Omega}_P C_1 - \frac{i}{2}\check{\Omega}_S C_3,$$
$$\frac{d}{dt}C_3 = -i\Delta_3 C_3 - \frac{i}{2}\check{\Omega}_S^* C_2. \qquad (13.85)$$

When the excitation is completely resonant at each step then $\Delta_1 = \Delta_2 = \Delta_3$, and the two Rabi frequencies have equal magnitudes, $|\check{\Omega}_P| = |\check{\Omega}_S| \equiv \Omega$, then there occurs periodic excitation from state 1, through state 2, to a complete transfer into state 3, followed by a return. Figure 13.9 illustrates this periodic behavior.

When the intermediate-state detuning Δ_2 becomes large, less population reaches state 2. However, as long as there is two-photon resonance, meaning $\Delta_1 = \Delta_3$ with the present description, population will continue to reach state 3. Figure 13.10 illustrates the population histories that result when intermediate-state detuning is large but two-photon resonance prevails.

In the limit of very large intermediate-state detuning Δ_2, negligible population occurs in state 2. However, there occurs a periodic population transfer – Rabi oscillation – between states 1 and 3. The description of this phenomenon involves the reduction of the three-state system to an effective two-state system using adiabatic elimination of state 2.

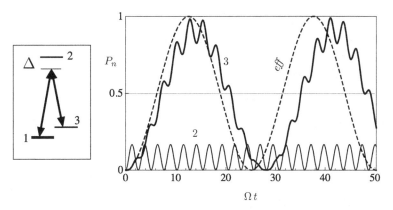

Figure 13.10 Inset at left: three-state linkage pattern for detuned intermediate state, but two-photon resonance. Main frame at right: Population histories of intermediate state 2 and final state 3 for steady intensity with large detuning ($\Delta = 20 \times \Omega$). Intermediate-state population is small and rapidly oscillating; the population cycles between states 1 and 3. The dashed curve shows prediction of effective two-state system, an approximation that becomes increasingly accurate as the detuning grows larger and the population reaching state 2 diminishes. [Redrawn from Fig. 44 of B. W. Shore, Acta Physica Slovaka, **58**, 243–486 (2008) with permission.]

13.4.4 Adiabatic elimination

When the detuning Δ_2 of the intermediate state 2 is very large, meaning $\Delta_2| \gg |\Omega|$ as we now assume, then the derivative of C_2 varies rapidly. We are not concerned with such rapid variations, and we average them out by taking an average over many cycles. The average of the derivative vanishes, and we can solve the resulting equation for the average amplitude $\bar{C}_2$,

$$\bar{C}_2 = -\frac{1}{2\Delta_2}\left[\check{\Omega}_P C_1 + \check{\Omega}_S C_3\right], \tag{13.86}$$

which thereby becomes linked (adiabatically) to states 1 and 2. Using this result we can *adiabatically eliminate* the occurrence of state 2 in the equations of motion [Sho90, §4.3]. The result is

$$\begin{aligned}
\frac{d}{dt}C_1 &= -i\,\Delta_1 C_1 + i\,\frac{|\check{\Omega}_P|^2}{4\Delta_2}C_1 + i\,\frac{\check{\Omega}_P^* \check{\Omega}_S}{4\Delta_2}C_3, \\
\frac{d}{dt}C_3 &= -i\,\Delta_3 C_3 + i\,\frac{\check{\Omega}_S^* \check{\Omega}_P}{4\Delta_2}C_1 + i\,\frac{|\check{\Omega}_S|^2}{4\Delta_2}C_3.
\end{aligned} \tag{13.87}$$

The result is expressible as an effective two-state RWA Hamiltonian

$$\mathsf{W}^{\text{eff}} = \begin{bmatrix} \Delta_1 + M_{11} & M_{12} \\ M_{12}^* & \Delta_3 + M_{22} \end{bmatrix}, \tag{13.88}$$

involving a two-photon interaction M, i.e. an interaction involving the product of two electric-field amplitudes, expressed here as the product of two Rabi frequencies.

The effective Hamiltonian produces two effects: The diagonal elements of M alter the detunings, while the off-diagonal elements are responsible for transitions, as an effective Rabi frequency

$$\check{\Omega}^{eff} = \frac{\check{\Omega}_S^* \check{\Omega}_P}{2\Delta_2}. \tag{13.89}$$

13.4.5 The polarizability

The altered diagonal elements of the effective Hamiltonian express energy shifts induced by the electric field of the laser, commonly termed dynamic Stark shifts, by contrast to the *static* shifts produced by slowly varying (DC) electric fields. These shifts, in elements M_{11} and M_{22}, are proportional to the intensity of the P and S fields respectively, a proportionality that we can write as

$$M_{11} = |\check{\mathcal{E}}_P|^2 X(1, 1), \qquad M_{22} = |\check{\mathcal{E}}_S|^2 X(2, 2). \tag{13.90}$$

The coefficient $X(n, n)$ is the *polarizability*[3] of state n [Van65; Auz95; Mar06a]. We can similarly express the off-diagonal element of M in terms of E-field envelopes. For the lambda system the expression reads

$$M_{12} = \check{\mathcal{E}}_P^* \check{\mathcal{E}}_S X(1, 2). \tag{13.91}$$

In any real atom there are many nonresonant states, and each of them can be regarded as an appropriate intermediate state in the three-state chain considered here. Thus the effective Hamiltonian should sum over all possible intermediate states. A further alteration of the formula is needed to account properly for the counter-rotating terms that the RWA neglects. With these two corrections one obtains the following expression for the (generalized) polarizability

$$X(n, m) = \sum_k \frac{d_{nk} d_{km}}{4\hbar} \left[\frac{1}{E_k - E_n - \hbar\omega} + \frac{1}{E_k - E_n + \hbar\omega} \right]. \tag{13.92}$$

The first of the bracketed fractions dominates for a frequency near a resonance; the second term originates in the counter-rotating terms of the Hamiltonian.

13.4.6 Multiphoton transitions

The off-diagonal elements of M produce transitions that bear interpretation as *multiphoton* transitions. In the present example these are two-photon transitions between states of common parity. For the lambda linkage considered here this is a transition involving the

---

[3] The polarizability is a tensor usually shown with subscripts that identify the Cartesian components of the two electric fields that appear in its construction.

replacement of a P-field photon with an S-field photon; the Hamiltonian is that of a stimulated Raman transition. For a ladder linkage the transition involves simultaneous absorption or emission of the two fields. These situations require the two-photon resonance condition

$$E_3 - E_1 = \hbar(\omega_P \pm \omega_S), \tag{13.93}$$

where the positive sign accompanies the ladder linkage, the negative sign the lambda linkage.

More generally the effective Hamiltonian produces N-photon transitions, e.g. three-photon (between states of opposite parity). The calculation of multiphoton transition rates occupied much effort during the early decades of laser usage [Fai87; Mai91; Del00].

These off-diagonal elements of the effective Hamiltonian replace the Rabi frequency used in all of the preceding discussions. It is important to recognize that any interaction that produces a multiphoton transition will also produce a dynamic Stark shift.

13.5 Unequal Rabi frequencies

When the two Rabi frequencies of a three-state chain are constant and equal, $\Omega_S = \Omega_P$, and both fields are on resonance with the relevant Bohr frequency, so that $\Delta_1 = \Delta_2 = \Delta_3 = 0$, then the population cycles periodically completely out of the initial state. However, when the Rabi frequencies are not equal the time dependence is more complicated. Figure 13.11 illustrates examples of such behavior, for population that initially resides in state 1. In frame (a) the two Rabi frequencies are equal, $\Omega_S = \Omega_P$; population cycles periodically into state 3, the final state of the chain. In frame (b) the S field is larger, $\Omega_S = 2\Omega_P$. The oscillations are more rapid and population never entirely leaves state 1. In frame (c) the S field is larger still, as are the oscillations.

This behavior is reminiscent of the effect of detuning: it appears that as the S-field Rabi frequency Ω_S becomes larger than that of the P field the first step of the excitation chain is detuned from resonance. This is indeed an accurate description of the system, as we can verify by evaluating the time-averaged populations $\bar{P}_n$. Figure 13.12 shows the average initial-state population $\bar{P}_1$ for the three cases of Ω_S shown in Fig. 13.11, as a function of the P-field detuning Δ_1, with $\Delta_2 = \Delta_3 = 0$ as before. We see that as Ω_S becomes larger than Ω_P it is necessary to introduce a detuning in order to retain the maximum removal of population from state 1. The choice of detuning Δ_{res} that maximizes the population removal is approximately $\Delta_{res} \approx \sqrt{|\Omega_S|^2 - |\Omega_P|^2}$, an approximation that improves as Ω_S becomes larger.

Figure 13.13 shows the population histories of Fig. 13.12 but with nonzero P-field detunings chosen to approximate the values that maximize the average transfer out of state 1. We observe that population periodically leaves state 1 completely, moving into a superposition of states 2 and 3. As the S-field Rabi frequency increases, the system becomes more like a simple two-state system.

The following section offers an interpretation of these observations.

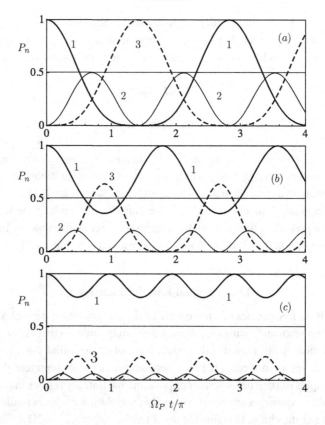

Figure 13.11 Effect of unequal Rabi frequencies. Populations P_n as a function of Rabi cycles $\Omega_P t/\mathbf{P}$ for three-state system with null detunings $\Delta_1 = \Delta_2 = \Delta_3 = 0$ and constant Rabi frequencies for three choices of S-field Rabi frequency. (a) $\Omega_S = \Omega_P$. (b) $\Omega_S = 2\Omega_P$. (c) $\Omega_S = 4\Omega_P$.

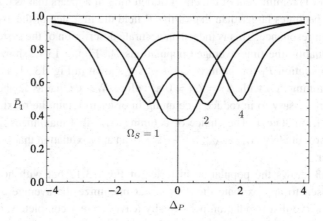

Figure 13.12 Time-averaged ground-state populations $\bar{P}_1$ vs. P-field detuning Δ_P for the three cases of Ω_S shown in the previous figure, $\Omega_S/\Omega_P = 1, 2$ and 4 times Ω_P. Minima occur near the values $\Delta_{res} = 0, 0.9$ and 1.9.

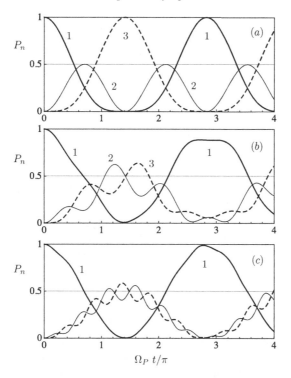

Figure 13.13 Population histories as in Fig. 13.12 but with P field detuned. (*a*) $\Omega_S = \Omega_P$, $\Delta_P = 0$. (*b*) $\Omega_S = 2\Omega_P$, $\Delta_P = 1$. (*c*) $\Omega_S = 4\Omega_P$, $\Delta_P = 2$.

13.5.1 Weak probe field: Autler–Townes splitting

When laser-induced excitation from the ground state (via P field) occurs in the presence of a second laser-induced excitation (the S field) we must enlarge our mathematical description from a two-state system to a three-state system. The fluorescence from the first excited state now can be modified by the S field, as a result of population dynamics within a three-state system. When the S field is sufficiently strong (i.e. with Rabi frequency much larger than that of the first-step P transition) then the excitation behavior changes dramatically: Once the system is in an excited state, it undergoes many S-field induced Rabi oscillations between the two strongly coupled states 2 and 3 before returning to the ground state via the P-field coupling. These Rabi oscillations affect the fluorescence signal: instead of observing the single Lorentz profile of the two-state system, one observes a splitting of the peak into two components, the *Autler–Townes doublet* separated by the Autler–Townes splitting [Aut55; Gra78a; Zub96]. The following paragraphs explain the origin of this effect.[4]

---

[4] Often the Autler–Townes doublet is explained by introducing a photon-number basis for the fields, as would be appropriate for atoms within a cavity [Sho90, §10.5]. But with laser excitation the fields are readily described as classical fields; the granularity of quantized fields is not evident. Thus it is useful to have a description that makes no explicit mention of photon-number states. The use of adiabatic states (also called dressed states), as in the present discussion, avoids the need to quantize the radiation field, and offers a straightforward way to treat slowly varying fields.

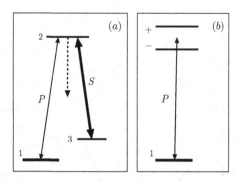

Figure 13.14 (*a*) A weak probe field P couples the initially populated state 1 with an excited state 2, which in turn is coupled by a strong S transition to state 3. The linkages are shown as resonant. (*b*) The couplings in the adiabatic (dressed) basis of the two strongly coupled states 2 and 3. The strong S field acts to split the resonance transition for the weak P field, thereby preventing population transfer from state 1 to the excited state 2 without altering the P-field detuning.

Consider the situation when the population initially resides in state 1, the ground state, which couples via P field to excited state 2, which in turn has linkage via S field to low-lying state 3. Our interest here is in situations where the S field (a *dressing field*) is much stronger than the P field, which acts as a weak *probe field* to the fluorescing excited state. Figure 13.14(*a*) shows the assumed linkage pattern.

We take the detunings to be those appropriate to the lambda linkage,

$$\Delta_1 = 0, \qquad \Delta_2 = \Delta_P, \qquad \Delta_3 = \Delta_P - \Delta_S, \tag{13.94}$$

so that the RWA Hamiltonian matrix, for the lambda linkage, has the following structure (the labels to the right show the states upon which the matrix acts)

$$\mathsf{W} = \begin{bmatrix} 0 & \frac{1}{2}\check{\Omega}_P^* & 0 \\ \frac{1}{2}\check{\Omega}_P & \Delta_P & \frac{1}{2}\check{\Omega}_S \\ 0 & \frac{1}{2}\check{\Omega}_S^* & \Delta_P - \Delta_S \end{bmatrix} \begin{matrix} \psi_1' \\ \psi_2' \\ \psi_3' \end{matrix}. \tag{13.95}$$

Here the fluorescence loss is neglected and the Rabi frequencies $\check{\Omega}$ are regarded as complex valued. The box encloses the portion of the RWA Hamiltonian matrix that involves the strongly coupled states 2 and 3.

13.5.2 Subsystem dressed states

To treat this system we regard states 2 and 3 as a single strongly coupled two-state system and we introduce a change of basis states that describe this strongly coupled pair of states. That is, we introduce eigenstates of the RWA Hamiltonian of the two strongly coupled states (examples of dressed states or, when the RWA Hamiltonian has some slow time dependence, adiabatic states),

$$\mathsf{W}^{(2)} = \begin{bmatrix} \Delta_P & \frac{1}{2}\check{\Omega}_S \\ \frac{1}{2}\check{\Omega}_S^* & \Delta_P - \Delta_S \end{bmatrix} \begin{matrix} \psi_2' \\ \psi_3' \end{matrix}. \tag{13.96}$$

The two eigenvalues are

$$\varepsilon_\pm = \Delta_P + \frac{1}{2}\left[-\Delta_S \pm \sqrt{(\Delta_S)^2 + |\check{\Omega}_S|^2} \right]. \tag{13.97}$$

The diagonalization of this 2×2 submatrix of the full Hamiltonian is accomplished by a unitary transformation U that produces the result

$$\mathsf{U}^\dagger \mathsf{W}^{(2)} \mathsf{U} = \begin{bmatrix} \varepsilon_- & 0 \\ 0 & \varepsilon_+ \end{bmatrix}. \tag{13.98}$$

Using this transformation matrix we introduce, in place of the original amplitudes C_2 and C_3 of the strongly coupled states, the pair of dressed-state amplitudes A_+ and A_-,

$$\begin{bmatrix} C_2(t) \\ C_3(t) \end{bmatrix} = \mathsf{U} \begin{bmatrix} A_+(t) \\ A_-(t) \end{bmatrix}. \tag{13.99}$$

Using these, and assuming that the RWA Hamiltonian varies only slowly,[5] we write the equation of motion as

$$\frac{d}{dt}\begin{bmatrix} C_1(t) \\ A_+(t) \\ A_-(t) \end{bmatrix} = -\frac{i}{2}\begin{bmatrix} 0 & \check{\Omega}_+^* & \check{\Omega}_- \\ \check{\Omega}_+ & 2(\Delta_P + \Delta_S + \delta) & 0 \\ \check{\Omega}_-^* & 0 & 2(\Delta_P + \Delta_S - \delta) \end{bmatrix}\begin{bmatrix} C_1(t) \\ A_+(t) \\ A_-(t) \end{bmatrix}, \tag{13.100}$$

where δ is the generalized (nonresonant) Rabi frequency of the strongly coupled transition,

$$\delta \equiv \sqrt{(\Delta_S)^2 + |\check{\Omega}_S|^2}, \tag{13.101}$$

and there are two new Rabi frequencies, $\check{\Omega}_\pm$. These are proportional to the original probe-field Rabi frequency and to relevant elements of the matrix that transforms from bare to dressed states,

$$\check{\Omega}_\pm = \check{\Omega}_P U_{2,\pm}. \tag{13.102}$$

---

[5] Appendix E.2 discusses the nonadiabatic couplings that occur between amplitudes A_k when there is variation of the RWA Hamiltonian.

The ground state, where population initially resides, is now coupled to two states – the two dressed states that have replaced the original bare states. Each of these dressed states is a linear superposition of the original bare states. In particular, each contains some component of state 2, which has the coupling to the ground state, state 1. Thus each one of the dressed states couples (by an amount that depends on S-field detuning) to the initially populated state, although each one receives population at a different rate (depending on the P-field detuning).

Figure 13.14(b) illustrates the couplings in the new, adiabatic, basis. The Hamiltonian in this basis has couplings between state 1 and each of the adiabatic states,

$$
\mathsf{W}^A = \begin{bmatrix} 0 & \frac{1}{2}\check{\Omega}^* _- & \frac{1}{2}\check{\Omega} _+ \\ \frac{1}{2}\check{\Omega} _- & \boxed{\begin{matrix} \varepsilon _- & 0 \\ 0 & \varepsilon _+ \end{matrix}} \\ \frac{1}{2}\check{\Omega}^* _+ & \end{bmatrix} \begin{matrix} \psi_1 \\ \Phi_- \\ \Phi_+ \end{matrix} \,.
\tag{13.103}
$$

In general, there will occur resonance transitions whenever a diagonal element of the Hamiltonian is equal to the element associated with the initially populated state. In the present situation, with population starting in state 1, that is the first element on the diagonal. For the matrix of eqn. (13.95) this resonance condition reads either $0 = \Delta_P$, meaning single-photon resonance between states 1 and 2, or else $0 = \Delta_P - \Delta_S$, corresponding to two-photon resonance between states 1 and 3. For the matrix of eqn. (13.103) the two possible resonance conditions are $\varepsilon_\pm = 0$, meaning

$$
\Delta_P = \mp \frac{1}{2} [\delta - \Delta_S].
\tag{13.104}
$$

In the absence of the S field the two values coincide, requiring $\Delta_P = 0$: resonance occurs when the P-field carrier matches the Bohr frequency. However, when the S field is present there are two distinct possibilities for the resonance condition. What was a single resonance in the absence of the 2–3 coupling will now appear as two resonances, the Autler–Townes doublet, separated by δ, the Autler–Townes splitting of eqn. (13.101) [Sho90, §10.4].

The simplest situation is when the strong (dressing) laser is resonant with the original Bohr frequency of the excited-state transition, so that $\Delta_S = 0$. That is,

$$
\hbar\omega_S = E_3 - E_2.
\tag{13.105}
$$

The two probe-field detunings for resonance are $2\Delta_P = \pm|\check{\Omega}_S|$. These occur at probe-field frequencies such that

$$
\hbar\omega_P = E_2 - E_1 \pm \frac{1}{2}\hbar|\check{\Omega}_S|.
\tag{13.106}
$$

That is, when $\Delta_S = 0$ the two resonances are separated by the strong-field Rabi frequency.

The amplitudes of the two components of the doublet depend on the effective Rabi frequency that couples the resonance to the ground state. For example, when the coupling is into the minus component, then the relevant Rabi frequency is $\check{\Omega}_-$.

Although I have not included fluorescence loss from either of the excited states, such effects can be included if one uses a density matrix rather than a statevector as the fundamental entity of interest (see Chap. 16). There will still be two resonances in the coupling out of the ground state and into excited states. The fluorescence will be strongest when the resonance condition holds. Thus a measurement of fluorescence, as a function of probe-field detuning Δ_P, will reveal an Autler–Townes doublet, separated by the strong-field Rabi frequency $|\check{\Omega}_S|$. Such measurements offer a means of determining Rabi frequencies. When so doing, the S-field resonance condition $\Delta_S = 0$ is obtained by adjusting the S-field frequency to minimize the Autler–Townes splitting δ for fixed value of $\check{\Omega}_S$.

13.5.3 Dressed state preparation and probing

Section 11.6 discussed the preparation of dressed states of the two-state system by means of phase-shifted pulses. The individual dressed states discussed there are those of the strongly coupled pair in the three-state system discussed here. Another procedure for placing the system into a single dark state is to implement RAP by a chirped pulse [Wol05]. Starting from large negative detuning, and the system in the ground state, the system begins in the positive-energy eigenstate: $\psi_1'(t)$ is aligned with $\Phi_+(t)$. If we proceed by inducing adiabatic following, then as the detuning increases the statevector will remain aligned with this adiabatic state: the system will remain in the dressed state of higher energy. It is the energy of this adiabatic state that will be found by a weak link to a third state. Because the slow chirp retains alignment of the statevector and adiabatic state, the weak probe field will only reveal a single energy-shifted resonance, a single one of the Autler–Townes doublets.

The Bloch sphere portrayal of this adiabatic passage by increasing detuning is simple. The motion of the Bloch vector is produced by a torque vector that initially points toward the south pole, parallel to the Bloch vector. The Bloch vector will remain aligned with (follow) this torque vector as it moves toward the equator. At the moment of null detuning the torque vector lies in the equatorial plane and the separation of the two adiabatic states is equal to the Rabi frequency.

Alternatively we can carry out chirped RAP starting with large positive detuning. The alignment of bare state 1 is then with the lower-energy dressed state; the torque vector points toward the north pole, antiparallel to the Bloch vector. Again adiabatic passage will retain the alignment between the statevector and an adiabatic state; the system remains in a single adiabatic state.

Photoelectron probing. The discussion above, with Fig. 13.14, considers an electromagnetic field as the probe of the dressed states. Photoelectrons offer an alternative probe, as indicated in Fig. 13.15 [Wol10]. Here a strong field S produces an Autler–Townes splitting of two dressed states. A weak probe field P photoionizes electrons into the continuum,[6] where they are distinguishable by their kinetic energy $E_\pm^e$. When the dressed states are

---

[6] In the experiement, a two-photon transition.

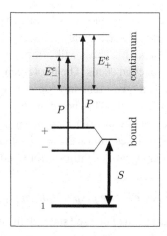

Figure 13.15 Probing dressed states with photoelectron spectroscopy. Peaks in the energy spectrum of the electrons reveal Autler–Townes splitting, here greatly exaggerated, produced by the strong field *S*. A weak probe field *P* photoionizes the excited state. When only a single dressed state is populated there is only a single photoelectron peak.

equally populated there occur two peaks in the photoelectron energy spectrum. A single peak occurs when only a single dressed state is populated [Wol10].

13.6 Laser-induced continuum structure (LICS)

Linkages into a continuum are responsible for such incoherent processes as photoionization and photodissociation. However, the presence of a continuum does not necessarily eliminate coherent behavior. In particular, coherent population transfer is possible even though the intermediate state of a three-state chain is embedded in a continuum. A widely studied example of such a situation is depicted in Fig. 13.16(*a*). Here a strong (dressing) field *S*, of amplitude $\mathcal{E}_S(t)$ and carrier frequency ω_S, connects state 2 to a continuum of energy *E*. A weak (probe) field *P*, with amplitude $\mathcal{E}_P(t)$ and carrier frequency ω_P, links this continuum state to an initially populated state 1. The presence of the strong *S* field endows an otherwise unstructured continuum with a spectral feature centered around energy $E_2 + \hbar\omega_S$, termed *laser-induced continuum structure* (LICS) [Kni90; Nak94; Yat99b], produced by a "pseudo-autoionizing state" [Arm75].

Fano profile. The effect of LICS is often expressed as a photoionization cross section, using one of two forms

$$\sigma(\omega) = \frac{Ax + B}{x^2 + 1} + C(\omega) = \sigma_a \frac{(x+q)^2}{x^2 + 1} + \sigma_b(\omega), \qquad x = (\omega - \omega_0)/\gamma. \quad (13.107)$$

The first of these, with parameters A, B, C, is the Breit–Wigner formula found with nuclear reaction cross sections [Bre36; Sho67; Sho68]. The second is the Fano profile [Fan61; Fan65], with parameters σ_a, σ_b, q. Figure 13.16(*b*) shows examples of this profile.

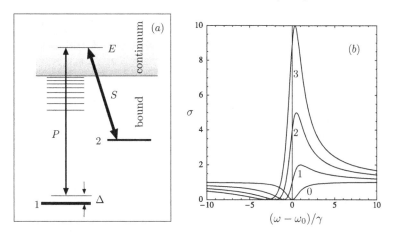

Figure 13.16 (*a*) Schematic linkage through a continuum. The strong field *S* creates structure in the continuum, probed by weak field *P*. (*b*) Examples of Fano profiles $\sigma(\omega)$ for $\sigma_b = 1$, $\sigma_a = 0$ and Fano parameter $q = 0, 1, 2$, and 3.

When $q = 0$ the Fano profile has a minimum at $x = 0$. When $\sigma_b = 0$ this is a transparency window. As q becomes positive the profile becomes asymmetric: the minimum remains σ_b, at $x = -q$, while the maximum, at $x = 1/q$, increases to $\sigma_b + (1 + q^2)\sigma_a$.

Formalism. The total electric field is

$$\mathbf{E}(t) = \mathrm{Re}\left\{ \sum_{\lambda = P, S} \mathbf{e}_\lambda \mathcal{E}_\lambda(t) \exp(-\mathrm{i}\omega_\lambda t) \right\}. \tag{13.108}$$

The *P* and *S* lasers are tuned close to the two-photon Raman resonance between states 1 and 2, with energies E_1 and E_2 respectively; their carrier frequencies differ from exact resonance by the two-photon detuning Δ,

$$\hbar\Delta \equiv \hbar(\omega_P - \omega_S) - (E_2 - E_1). \tag{13.109}$$

We express the statevector of the atom, using the Dirac picture, as a combination of states 1 and 2 together with a sum over all other bound states and an integral over continuum states:

$$\Psi(t) = C_1(t)\exp(-\mathrm{i}E_1 t/\hbar)\,\psi_1 + C_2(t)\exp(-\mathrm{i}E_2 t/\hbar)\,\psi_2 \tag{13.110}$$

$$+ \sum_{n \neq 1,2} C_n(t)\exp(-\mathrm{i}E_n t/\hbar)\,\psi_n + \sum_j \int dE\, C_{E,j}(t)\exp(-\mathrm{i}Et/\hbar)\,\psi_{E,j}.$$

Here $\psi_{E,j}$ is the continuum state with energy E (label j identifies different possible continua).

As is customary [Kni90; Nak94; Yat99b] we adiabatically eliminate all amplitudes except those of states 1 and 2. In so doing, we obtain effective interactions, and dynamic energy shifts, expressible in terms of a polarizability tensor, see Sec. 13.4.5 [Sho90, §14.9],

$$X(m,n;\omega) = \sum_k \frac{\langle m|\mathbf{d}|k\rangle\langle k|\mathbf{d}|n\rangle}{E_k - E_m - \hbar\omega} + \int dE \sum_j \frac{\langle m|\mathbf{d}|E, j\rangle\langle E, j|\mathbf{d}|n\rangle}{E - E_m - \hbar\omega}, \tag{13.111}$$

where $\mathbf{d}$ is the dipole-moment operator. The continuum integral includes the resonance energy $E = E_m + \hbar\omega$, which is dealt with by breaking the integration into a nonsingular principal value part (denoted by $\mathcal{P}$).

$$\mathrm{Re}\,X(m,n;\omega) = \sum_k \frac{\langle m|\mathbf{d}|k\rangle\langle k|\mathbf{d}|n\rangle}{E_k - E_m - \hbar\omega} + \mathcal{P}\int dE \sum_j \frac{\langle m|\mathbf{d}|E, j\rangle\langle E, j|\mathbf{d}|n\rangle}{E - E_m - \hbar\omega}, \tag{13.112}$$

and a resonant part (yielding the transition rate of a Fermi Golden Rule),

$$\mathrm{Im}\,X(m,n;\omega) = \sum_j \pi \langle m|\mathbf{d}|E = E_m + \hbar\omega, j\rangle\langle E = E_m + \hbar\omega, j|\mathbf{d}|n\rangle. \tag{13.113}$$

The adiabatic elimination yields a two-state time-evolution equation, for which the RWA Hamiltonian is [Kni90; Yat99a]

$$W = -\frac{1}{2}\begin{bmatrix} 2S_1(t) - i\Gamma_1(t) - i\gamma_1 & -(i+q)\Omega(t) \\ -(i+q)\Omega(t) & 2S_2(t) - i\Gamma_2(t) - i\gamma_2 - 2\Delta \end{bmatrix}. \tag{13.114}$$

The theoretical ionization rates $\Gamma_n(t)$ and Stark shifts $S_n(t)$ are obtainable by summing partial rates:

$$\Gamma_n(t) = \sum_{j,\lambda} \Gamma_{n\lambda}^{(j)}(t), \qquad S_n(t) = \sum_\lambda S_{n\lambda}(t), \tag{13.115}$$

where $\Gamma_{n\lambda}^{(j)}(t)$ is the ionization rate from state n to continuum j caused by the P field ($\lambda = P$) or S field ($\lambda = S$),

$$\Gamma_{n\lambda}^{(j)}(t) = \frac{\pi}{2\hbar}|\mathcal{E}_\lambda(t)|^2 \sum_{\lambda,j} |\langle n|\mathbf{e}_\lambda\cdot\mathbf{d}|E = \hbar\omega_\lambda - E_i, j\rangle|^2, \tag{13.116}$$

and $\hbar S_{n\lambda}(t)$ is the dynamic Stark shift of the energy of state n produced by laser λ:

$$S_{n\lambda}(t) = -\frac{1}{4\hbar}|\mathcal{E}_\lambda(t)|^2\,\mathbf{e}_\lambda\cdot[X(n,n;\omega_\lambda) + X(n,n;-\omega_\lambda)]\cdot\mathbf{e}_\lambda^*. \tag{13.117}$$

The quantity

$$\Omega(t) = \sum_j \sqrt{\Gamma_{1P}^{(j)}(t)\Gamma_{2S}^{(j)}(t)} \tag{13.118}$$

is an effective Rabi frequency for the two-step transition from the state 1 to state 2 via linkage through the continuum states j with energy $E \simeq \hbar\omega_P - E_2 + E_1 \simeq \hbar\omega_S - E_1 + E_2$. The quantity q, known as the Fano q parameter [Fan61; Fan65], can be evaluated from the expression [Kni90; Yat99a]

$$q\,\Omega(t) = \frac{1}{2}|\mathcal{E}_P(t)\mathcal{E}_d(t)|\,\mathbf{e}_P\cdot[X_{12}(\omega_p) + X_{12}(-\omega_d)]\cdot\mathbf{e}_d^*. \tag{13.119}$$

The Fano parameter q appearing here is basically the ratio of polarizability (such as occurs in a Raman transition) to the product of two dipole transition moments into the continuum. In systems with $|q| \gg 1$ the Raman-type transitions dominate. The ionization rate is enhanced, when tuning the probe laser across the two-photon resonance, because Raman-type transitions open additional (multiphoton) ionization channels from the initial state to the ionization continuum. If $q = 0$ no Raman-type transitions are present and no enhancement of ionization occurs.

14

Raman processes

The traditional Raman process alluded to in Chap. 13 is a three-state sequence of transitions in which radiative excitation (induced by a *pump* field) is followed by spontaneous emission that produces a final state differing from the initial state [Her50b] [Sho90, §17.5]. When the final state of the sequence is more energetic than the initial state the resulting emission line (to the red of the pump wavelength) is known as a *Stokes* spectral line. The difference between the pump frequency and the Stokes frequency, the *Raman frequency*, defines the excitation energy of the final state relative to the initial state. When, instead, the final state has lower energy than the initial state, as can occur when the initial quantum state is already excited, then the emission is an *anti-Stokes* line, at a bluer wavelength than the pump field. The overall Raman scattering is a two-photon process.

Typically Raman spectroscopy deals with molecules; the two-photon transitions are then between vibrational-rotational states, through electronically excited intermediate states, that are characterized in part by vibrational quantum number v and rotational angular momentum quantum numbers J, M. From any given excited electronic state there are many fluorescing transitions, corresponding to various vibrational and rotational quantum numbers of the final state. The wavelengths of the various Stokes and anti-Stokes lines (i.e. the Raman frequencies) characterize the particular molecular species, and so they have provided a valuable diagnostic tool for spectroscopists.

The Raman processes of interest here, as producers of coherent quantum-state manipulation, are those in which both the pump field and the Stokes (or anti-Stokes) field are experimenter-controlled laser fields, i.e. both transitions are stimulated by existing fields, a process that has been termed stimulated Raman scattering.

14.1 The Raman Hamiltonian

The coherent dynamics of stimulated Raman processes – the coherent flow of population amongst three states – is the same for any three-state chain, whether in lambda or ladder configuration. For definiteness let us consider the lambda linkage pattern, as is appropriate for a stimulated Raman process. Figure 14.1 shows the energies of the states and the two fields: the pump field P, with frequency ω_P, produces excitation into state 2 from the initially populated state 1. The *Stokes field* S, with frequency ω_S, produces deexcitation

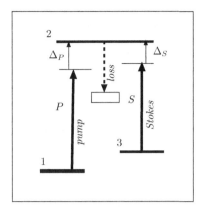

Figure 14.1 The stimulated Raman linkage. The P field (pump) links initially populated state 1 with the excited state 2, from which spontaneous emission loss can occur. The S field (Stokes) links the excited state with the metastable state 3. The single-field detunings Δ_P and Δ_S are shown. [Redrawn from Fig. 49 of B. W. Shore, Acta Physica Slovaka, **58**, 243–486 (2008) with permission.]

from this intermediate state to the final state 3. The figure shows explicitly the spontaneous emission loss from state 2; the other two states are assumed to be stable.

The RWA Hamiltonian for the stimulated Raman processes is basically that of eqn. (13.33), but with phases suited to the lambda linkage and the allowance for loss from the excited state (time dependence is not shown explicitly),

$$
\mathsf{W} = \begin{bmatrix} 0 & \frac{1}{2}\Omega_P\,\mathrm{e}^{-\mathrm{i}\varphi_P} & 0 \\ \frac{1}{2}\Omega_P\,\mathrm{e}^{+\mathrm{i}\varphi_P} & \Delta_P - \mathrm{i}\Gamma/2 & \frac{1}{2}\Omega_S\,\mathrm{e}^{+\mathrm{i}\varphi_S} \\ 0 & \frac{1}{2}\Omega_S\,\mathrm{e}^{\mathrm{i}\varphi_S} & \Delta_P - \Delta_S \end{bmatrix}. \tag{14.1}
$$

Here Γ is the rate at which probability is lost from state 2 and the two detunings are as defined earlier,

$$
\hbar\Delta_P = E_2 + V_{22} - E_1 - V_{11} - \hbar\omega_P, \qquad \hbar\Delta_S = E_2 + V_{22} - E_3 - V_{33} - \hbar\omega_S. \tag{14.2}
$$

14.2 Population transfer

One of the goals of laser excitation has been to transfer population from an initial state to a chosen final state. Raman processes provide a mechanism for transferring population via two-photon transitions, into final states that cannot be reached by electric-dipole radiation at optical frequencies. These include states that are degenerate with the initial state.

14.2.1 Steady fields

The simplest use of Raman scattering to produce population transfer, shown in Fig. 14.2(*a*), employs a pump field of preselected frequency to produce a first step of excitation, but

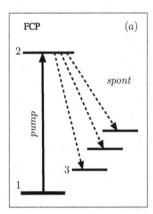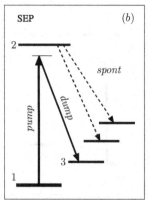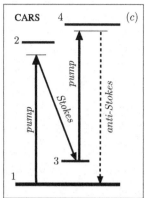

Figure 14.2 Raman processes that produce population transfer. (*a*) Franck–Condon pumping (FCP) converts pump radiation into Stokes radiation via spontaneous emission. Some of this connects with state 3. (*b*) Stimulated emission pumping (SEP) selects a particular Stokes transition to stimulate by a dump field. The pump field need not be resonant, but the pump and dump fields must together satisfy a two-photon resonance condition. (*c*) Coherent anti-Stokes Raman scattering (CARS) uses a three-photon transition, pump–Stokes–pump, to create a coherent dipole moment (between states 4 and 1) that then produces an anti-Stokes field. [Redrawn from Fig. 46 of B. W. Shore, Acta Physica Slovaka, **58**, 243–486 (2008) with permission.]

relies on spontaneous emission to produce the final transition (the Stokes field). From the excited state there occur many possible emission routes, each with its own Stokes field. The relative probability of a specific one (the *branching ratio*) depends on wavefunctions of the two states. For molecular transitions between vibrational states the overlap of the wavefunctions is expressed as Franck–Condon factors, and hence the population transfer process is termed *Franck–Condon pumping* (FCP) [Ber80]. Because the process relies on spontaneous emission, it does not allow creation of a coherent superposition. And because there are numerous possible final states, it is not possible to achieve appreciable transfer into any one state.

Rather than rely on nature to produce the Stokes field, one can impose a laser field to select one of the deexcitation choices, via stimulated emission, a process often termed *stimulated Raman scattering* (SRS) [Ray90]. By using a second laser field the two-step Raman process, of excitation and decay, can be made rapid and more selective. The second field is typically termed the Stokes field, independent of its wavelength. Figure 14.2(*b*) illustrates this population transfer scheme, often termed *stimulated emission pumping* (SEP) [Dai94]. The use of two simultaneous laser fields, as in SEP, allows selective transfer to a chosen final state, diluted only by any competing spontaneous emission. When used in this way the pump field need not be resonant, but the pump and dump fields must together satisfy a two-photon resonance condition.

Another class of Raman processes, used more for spectroscopic purposes or microscopy than for population transfer, is the coherent anti-Stokes Raman scattering (or spectroscopy)

(CARS) sketched in Fig. 14.2(*c*) [Yur77]. This is an example of nonlinear optics (specifically four-wave mixing) in which a three-photon process, involving pump–Stokes–pump transitions, creates a dipole moment that acts as a source of anti-Stokes radiation; see Sec 15.5. As with SEP, the pump and Stokes fields need not be resonant; an overall frequency-matching constraint picks out a particular anti-Stokes frequency; see Sec. 15.5.

The lambda linkage pattern associated with the Raman process is basically a three-state excitation chain. Conceptually the simplest scheme for transferring population along a chain is to use resonant excitation with equal Rabi frequencies that share the same time dependence. As shown in Sec. 13.3.1 population then flows from the initial state, 1, to a succession of states along the chain. For the three-state system this can produce Rabi oscillations that will periodically place all population into state 3. Figure 13.3(*a*) illustrates the population flow for such a situation.[1]

Such schemes are possible for pulsed excitation, but they require that all Rabi frequencies must rise and fall together, with carefully controlled relative values (typically best results occur when all Rabi frequencies are equal). The technique also suffers from the temporary placement of population in the excited state, from which it may be lost via spontaneous emission to other states. (This loss can be reduced by using lasers which, while satisfying two-photon resonance conditions, are not in single-photon resonance.)

Intuitive pulse sequence. One might expect that an effective and robust way to transfer population between state 1 and state 3 would be by placing population first into intermediate state 2 (by means of a π pulse or by RAP) and then transferring this to state 3. This *intuitive* pulse sequence exposes the atom first to the *P* field and then to the *S* field. Figure 14.3 illustrates such a sequence. In the first step, shown in frame (*a*), the *P* field induces complete transfer into excited state 2. In the second step, shown in frame (*b*), the *S* field transfers this excited-state population to the desired target state 3.

The pulse sequence of Fig. 14.3, with *P* preceding *S*, fits the intuitive understanding of how excitation proceeds when it is described by incoherent rate equations. When such equations apply, only a portion of the population can be transferred at each step (one-half at most), because the populations equilibrate under the influence of the excitation.

Sequential pulse transfer via coherent excitation has a major potential drawback: the pulses must place all population into an intermediate state from which spontaneous emission can occur. Thus undesirable population losses occur. It turns out that coherent excitation provides an alternative procedure, one in which (almost) no population resides in state 2, yet (almost) complete population transfer can occur. To understand the possibility we return to an examination of the appropriate TDSE, as following from the Raman Hamiltonian in the RWA.

---

[1] Note that although excitation with simultaneous and equal Rabi frequencies can produce complete population transfer, it cannot produce a superposition that does not include state 2.

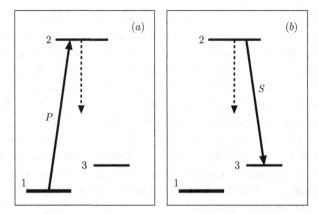

Figure 14.3 Population transfer via stimulated Raman transition with intuitive ordering of pulses. (*a*) The pump pulse places population into the excited state. (*b*) The Stokes (or dump) pulse moves population into the final state. During residence in the excited state some population is lost by spontaneous emission, as indicated by the dashed line.

14.2.2 Pulse-pair sequences; STIRAP

The S and P fields of the stimulated Raman linkage are independent, because they differ in either frequency or polarization, and in principle they may have different pulse shapes. In particular, the centers of the pulses may differ: one may precede the other. The results depend both on the timing and on the coherence properties of the radiation.

Incoherent pulses. When the radiation is incoherent then rate equations describe its effect. Let us assume that population initially resides entirely in state 1. When the S field acts first, it produces no alteration of the atom; only the P field can move population from the initial state, placing it into state 2. Figure 14.4 shows an example of a pulse pair in which the P field precedes, and does not overlap, the S field. During the P pulse population equilibrates between states 1 and 2. In the illustration the transfer of population is as complete as possible for incoherent radiation: half of the initial population transfers into state 2. A subsequent S pulse transfers half of this into state 3, for an overall transfer of 25% into that state.

Coherent pulses. With coherent radiation, excitation is governed by the TDSE, and more complete population transfer becomes possible. Figure 14.5 shows a sequence of resonant pulses, again with P preceding S. The population transfer produced by resonant radiation depends only on the temporal pulse area. In this example each pulse has area π. The P pulse therefore transfers all population into state 2. The subsequent S pulse transfers all of this into state 3. The result is 100% population transfer into state 3.

If the goal of the excitation procedure is maximum population transfer, then clearly the pair of offset π pulses is superior. However, often it is not possible to craft the pulses such that the temporal areas are exactly π. Then the resulting transfer is smaller, perhaps even zero.

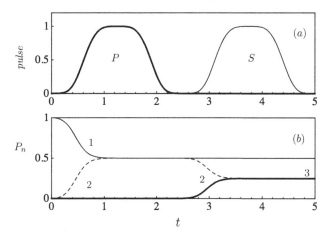

Figure 14.4 Population transfer predicted by rate equations for a P pulse followed by an S pulse, for times much shorter than the lifetime for spontaneous emission. (a) The pulses. (b) The populations $P_n(t)$. The two pulses successively equilibrate pairs of levels, and, in the absence of spontaneous emission, the final population in state 3 is $P_3(\infty) = 0.25$. (Pulse envelopes are Gauss-4.)

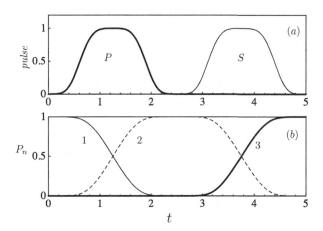

Figure 14.5 Pulse shapes and populations as in Fig. 14.4 for coherent excitation by resonant pulse pairs, each with temporal pulse area π, with P preceding S. There occurs first a complete transfer of population into state 2, and then complete transfer into state 3. The final population in state 3 is $P_3(\infty) = 1$. (Pulse envelopes are Gauss-4.)

To illustrate the sensitivity of the overall process to the first pulse, Fig. 14.6 shows an example of a pulse pair in which the first pulse has temporal area 2π. This pulse produces a complete Rabi cycle, returning all population into the initial state. The subsequent S pulse has no effect.

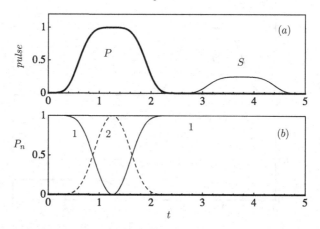

Figure 14.6 Coherent excitation as in Fig. 14.5 but with a P pulse of temporal area 2π. The first pulse produces complete population return, and so the second pulse has no effect. The final population in state 3 is $P_3(\infty) = 0$.

The three examples of the above figures all deal with pulse ordering that can be expected to move population from state 1 into state 2 and then into state 3, a sequence that seems intuitively obvious. However, any population arriving in state 2 will be subject to loss by spontaneous emission, and thus, unless the overall pulse sequence is faster than the radiative lifetime, the overall population transfer will not be complete.

Counterintuitive pulse sequence: STIRAP. One might expect that a pulse sequence in which S precedes P would not be as successful, because the S field has no linkage with the initially populated state – it is a "counterintuitive" pulse sequence [Sho95b]. This supposition is not correct. Figure 14.7 shows an example in which the S field precedes, and overlaps, the P field. The pulse area is significantly larger than 2π. Essentially all population is transferred into state 3, $P_3(\infty) = 1$. Moreover, state 2 never acquires appreciable population. This behavior is an example of *stimulated Raman adiabatic passage* (STIRAP), a procedure developed in the research group of Bergmann [Gau90].

14.2.3 The STIRAP mechanism

The mechanism of STIRAP defined in the previous section is a simple example of adiabatic time evolution that has many applications and has been extensively discussed; for reviews see [Ber98; Vit01a; Vit01b]. As first developed in nondegenerate three-state systems it requires the following four conditions to hold:

1. Pulses: Two pulsed interactions, S and P, in a Raman (lambda) linkage

$$1 \leftarrow (P) \rightarrow 2 \leftarrow (S) \rightarrow 3. \tag{14.3}$$

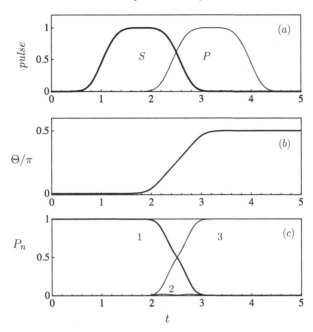

Figure 14.7 Coherent excitation by a counterintuitive pulse sequence in which S precedes but overlaps P. The final population in state 3 is $P_3(\infty) = 1$. (a) The pulses. (b) The mixing angle $\Theta(t) = \arctan\Omega_S / \Omega_P$ in units of π. (c) The populations $P_n(t)$. The changes are adiabatic, and the populations follow the change in mixing angle.

2. Delay: For population initially in state 1 the S field should precede but overlap with the P field (a *counterintuitive pulse sequence*).
3. Resonant: The P and S fields should maintain two-photon resonant detuning, $\Delta_P = \Delta_S$.
4. Adiabatic: The time evolution should be adiabatic.

When these four conditions hold the consequences are (the STIRAP mechanism)

1. Population is transferred completely from state 1 to state 3.
2. Negligible population is ever in the intermediate state 2.

These consequences do not depend on having specific pulse shapes or precisely controlled delay between the pulses. They do not require particular values of peak Rabi frequencies (so long as these are not too large), pulse duration (as long as these are shorter than the coherence time) or temporal pulse areas (so long as these are sufficiently large). Thus STIRAP is said to be *robust*. Furthermore they are independent of the single-photon detunings Δ_P and Δ_S so long as these satisfy the two-photon resonance condition.

14.3 Explaining STIRAP

The STIRAP dynamics can be understood as a five-stage process (see [Vit01a]) with time intervals indicated by roman numerals in Fig. 14.8.

I. In the first stage the presence of the S field alone establishes alignment of the statevector Ψ with the dark adiabatic state $\Phi_0(t)$.

II. During the second stage the strong S field acts to produce a dynamic Stark shift (an Autler–Townes splitting) such that the weak P field has no effect (an example of electromagnetically induced transparency, EIT), see Fig. 13.14.

III. As the P field becomes stronger, and the S field weaker, the evolution is by means of adiabatic passage. The following paragraphs discuss this regime.

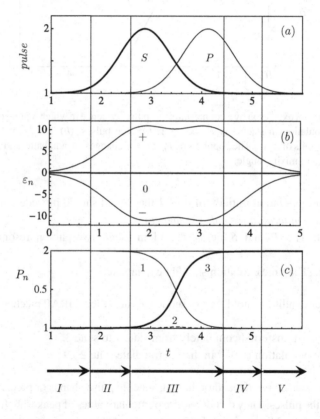

Figure 14.8 The STIRAP process. Top frame: the S and P pulse sequence, with S preceding but overlapping with P. Middle frame: the adiabatic eigenvalues vs. time. Bottom frame: population histories. Along the bottom appear roman-numeral labels of the five regimes discussed in the following section. In this figure there occurs a dip at the center of the adiabatic curves. When the pulse separation is optimized there occurs no dip. [Redrawn from Fig. 3 of K. Bergmann *et al.*, Rev. Mod. Phys., **70**, 1003 (1998) with permission.]

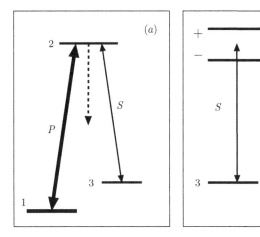

Figure 14.9 Final stage of STIRAP. (*a*) Strong *P* field, weak *S* field, after population has been transferred adiabatically to state 3. (*b*) Dressed-state picture: the *P* field acts to split the resonance transition for the weak *S* field, thereby preventing population transfer from state 3 to the excited state 2.

IV. In the fourth stage the strong *P* field produces a dynamic Stark shift such that the weak *S* field has no effect, see Fig. 14.9.

V. In the final stage only the *P* field is present, and the dark adiabatic state, along with the statevector, becomes aligned with the target state $\boldsymbol{\psi}_3$.

14.3.1 The dark state

During the intermediate stage of the STIRAP dynamics the time evolution is best described using the three adiabatic states of the RWA Raman Hamiltonian, eqn. (13.50b). When the two-photon detuning vanishes ($\Delta_P = \Delta_S$) the eigenvalues are

$$\varepsilon_0 = 0, \qquad \varepsilon_\pm = \frac{1}{2}\left[\Delta_P \pm \sqrt{\Delta_P^2 + \Omega_P^2 + \Omega_S^2}\right], \tag{14.4}$$

and the eigenstates are

$$\boldsymbol{\Phi}_\pm(t) = \frac{e^{-i\varphi_S}}{\sqrt{2}}\begin{bmatrix} \sin\Theta(t) \\ \pm e^{-i\alpha} \\ \cos\Theta(t)\,e^{-i\beta} \end{bmatrix},$$

$$\boldsymbol{\Phi}_0(t) = \begin{bmatrix} \cos\Theta(t) \\ 0 \\ -\sin\Theta(t)\,e^{-i\beta} \end{bmatrix}, \qquad \text{no component of state 2.}$$

$$\tag{14.5}$$

Here $\Theta(t)$ is the (Raman) *mixing angle*, defined through eqn. (13.51), $\tan\Theta = \Omega_P/\Omega_S$ and the phases are those of the fields,

$$\alpha \equiv \varphi_P, \qquad \beta \equiv \varphi_P - \varphi_S. \tag{14.6}$$

It is generally possible to set $\varphi_S = 0$, because absolute phases are uncontrollable. Only if both pulses derive from the same laser field (i.e. states 1 and 3 are degenerate) through optical elements is the phase difference $\varphi_P - \varphi_S$ controllable.

Notably the null-eigenvalue adiabatic state has no component of the excited state ψ_2. Therefore it cannot fluoresce; it is a *dark* state [Alz79; Ari96; Mil98; Wyn99; Kis01b]. It has the construction (in the rotating coordinate basis)

$$\Phi_0(t) = \frac{\Omega_S(t)\psi_1 - \Omega_P(t)e^{-i\beta}\psi_3'(t)}{\sqrt{|\Omega_P|^2 + |\Omega_S|^2}} = \cos\Theta(t)\psi_1 - \sin\Theta(t)e^{-i\beta}\psi_3'(t). \tag{14.7}$$

For the STIRAP pulse sequence, of Stokes preceding pump, this adiabatic state has the following properties

Initially only S field	Finally only P field
$\Omega_P(t) = 0$	$\Omega_S(t) = 0$
$\Phi_0(t) = \psi_1$	$\Phi_0(t) = -e^{-i\beta}\psi_3'$
initial state	*target state*

We see that if we can ensure that the time evolution is adiabatic, then the statevector $\Psi(t)$ follows the adiabatic state $\Phi_0(t)$ and population transfers $1 \to 3$. The final state acquires a phase factor $e^{-i\beta}$ that depends on the difference between P and S field phases. Unless the two fields derive from a common laser field (so $\omega_P = \omega_S$), this is a value fixed by our arbitrary choice of initial phases, and can be taken as zero; any other choice merely implies a redefinition of Hilbert-space coordinates.

Picturing the Hilbert space. Figure 14.10 illustrates the various Hilbert-space vectors associated with STIRAP. Frame (*a*) shows the three adiabatic states initially, when only the S field is present. The adiabatic vector Φ_0 lies along the 1 axis, while the other two vectors lie in the 2,3 plane. Frame (*b*) shows the statevector Ψ at this time, aligned along the 1 axis.

As the S pulse weakens and the P pulse grows, the framework of adiabatic vectors rotates about the 2 axis. After the S pulse vanishes these vectors remain fixed, with Φ_0 along the 3 axis and $\Phi_\pm$ in the equatorial plane. Frame (*a*) of Fig. 14.11 illustrates this change. If the motion of the statevector is adiabatic, then it remains aligned with Φ_0. The result is the change shown in frame (*b*).

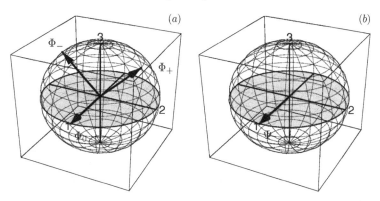

Figure 14.10 (*a*) The initial adiabatic vectors. (*b*) The initial statevector, aligned along the 1 axis.

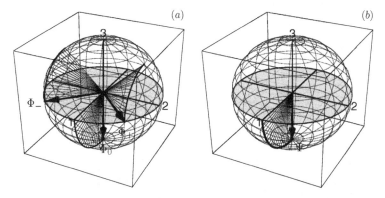

Figure 14.11 (*a*) The motion of the adiabatic vectors as the pulses proceed from *S* to *P*. (*b*) When the time evolution is adiabatic the statevector follows the motion of adiabatic state $\boldsymbol{\Phi}_0(t)$. The final statevector is aligned along the 3 axis.

14.3.2 The adiabatic basis; adiabatic conditions

To determine the conditions needed for the adiabatic evolution of the STIRAP process we express the statevector as a superposition of adiabatic states,

$$\boldsymbol{\Psi}(t) = \sum_k A_k(t) \boldsymbol{\Phi}_k(t). \tag{14.8}$$

The connection between expansion coefficients is a unitary transformation

$$\mathbf{C}(t) = \mathsf{U}(t)\mathbf{A}(t), \tag{14.9}$$

where (here the right-hand labels identify the ordering of the states)

$$\mathsf{U}(t) = \begin{bmatrix} \sin\Theta(t)\sin\varphi(t) & \cos\Theta(t) & \sin\Theta(t)\cos\varphi(t) \\ \cos\varphi(t) & 0 & -\sin\varphi(t) \\ \cos\Theta(t)\sin\varphi(t) & -\sin\Theta(t) & \cos\Theta(t)\cos\varphi(t) \end{bmatrix} \begin{matrix} 1 \\ 2 \\ 3 \end{matrix} \tag{14.10}$$

with

$$\tan \Theta(t) = \Omega_P(t)/\Omega_S(t), \qquad \tan \varphi(t) = \Omega_T(t)/\Delta, \qquad \Omega_T(t) = \sqrt{\Omega_P(t)^2 + \Omega_S(t)^2}. \tag{14.11}$$

The equation of motion for the amplitudes $A_k(t)$ reads

$$\frac{d}{dt}\mathbf{A}(t) = -i\mathbf{W}^A(t)\mathbf{A}(t), \tag{14.12}$$

where (again right-hand labels identify the states)

$$\mathbf{W}^A = \mathbf{U}^{-1}\mathbf{W}\mathbf{U} - i\mathbf{U}^{-1}\dot{\mathbf{U}} = \begin{bmatrix} \varepsilon_+(t) & i\dot{\Theta}\sin\varphi(t) & i\dot{\varphi} \\ -i\dot{\Theta}\sin\varphi(t) & 0 & -i\dot{\Theta}\cos\varphi(t) \\ -i\dot{\varphi} & i\dot{\Theta}\cos\varphi & \varepsilon_-(t) \end{bmatrix} \begin{matrix} + \\ 0 \\ - \end{matrix}. \tag{14.13}$$

Here the adiabatic eigenvalues are

$$\varepsilon_+(t) = \Omega_T(t)\cot[\varphi(t)/2], \qquad \varepsilon_-(t) = \Omega_T(t)\tan[\varphi(t)/2], \tag{14.14}$$

and the derivatives are

$$\dot{\Theta}(t) = \frac{\Omega_S(t)\frac{d}{dt}\Omega_P(t) - \Omega_P(t)\frac{d}{dt}\Omega_S(t)}{\Omega_P(t)^2 + \Omega_S(t)^2}, \tag{14.15}$$

$$\dot{\varphi}(t) = \frac{\Omega_T(t)\frac{d}{dt}\Delta(t) - \Omega_T(t)\frac{d}{dt}\Delta}{\Omega_P(t)^2 + \Omega_S(t)^2 + \Delta(t)^2}. \tag{14.16}$$

The statevector will remain aligned with a particular adiabatic state, or with a fixed super-position of adiabatic states, whenever the off-diagonal elements of the matrix $\mathbf{W}^A$ can be neglected. To ensure this condition, in which the statevector undergoes adiabatic following, we require that the off-diagonal elements be much smaller than the separation of diagonal elements (note that this criterion cannot be fulfilled when the eigenvalues are degenerate). This is least when $\Delta = 0$, when it is

$$|\varepsilon_\pm(t) - \varepsilon_0| = \frac{1}{2}\sqrt{\Omega_P(t)^2 + \Omega_S(t)^2} \equiv \frac{1}{2}\Omega_T(t). \tag{14.17}$$

From this expression we deduce the instantaneous (or local) requirement

$$\left|\Omega_S(t)\frac{d}{dt}\Omega_P(t) - \Omega_P(t)\frac{d}{dt}\Omega_S\right| \ll \frac{1}{2}\Omega_T(t)^3. \tag{14.18}$$

For STIRAP the local condition for adiabatic following is that the rate of change in the mixing angle should remain much smaller than the separation of eigenvalues. When $\Delta = 0$ this condition is

$$|\dot{\Theta}(t)| \ll \frac{1}{2}\sqrt{\Omega_P(t)^2 + \Omega_S(t)^2}. \tag{14.19}$$

From this we obtain a *global* adiabatic condition by integrating the local condition,

$$\pi \ll \int_{-\infty}^{\infty} dt\sqrt{\Omega_P(t)^2 + \Omega_S(t)^2}. \tag{14.20}$$

That is, the temporal pulse area must be large – typically more than 10π. When this condition applies the statevector remains aligned with the dark adiabatic state,

$$\Psi(t) = \Phi_0(t), \qquad (14.21)$$

and will adiabatically follow the Hilbert-space motion of this vector.

14.4 Demonstrating STIRAP

Several types of experiments demonstrate the STIRAP mechanism [Ber98; Vit01a; Vit01b]. These rely on observing some indicator of population transfer while varying such parameters as the time delay between the S and P pulses, or the detuning of one of the fields.

14.4.1 Vary pulse delay

A particularly clear demonstration makes use of controlled temporal separation of the P and S pulses while holding fixed the frequencies such that the two-photon resonance condition holds. Figure 14.12 presents results obtained with a molecular beam; the needed temporal delay was obtained by altering the physical position of the S and P laser beams through which the molecular beam passed at a right angle. In this experiment a subsidiary laser field serves to probe the final population transfer. When the S field precedes the P field, as occurs toward the left-hand side of the figure, it has no effect on the dynamics. Population transfer

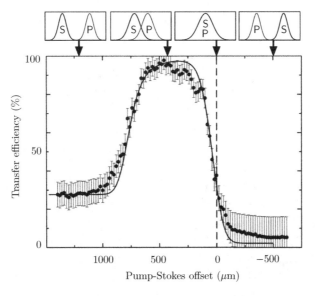

Figure 14.12 Population transfer via stimulated Raman process, as a function of spatial separation between S and P laser beams (and hence delay between S and P pulses, as indicated with small inserts along the top). Positive values of separation correspond to S preceding P. [Redrawn from Fig. 9 of K. Bergmann *et al.* Rev. Mod. Phys., **70**, 1003 (1998) with permission.]

occurs via excitation followed by spontaneous emission (an example of FCP). When the P field precedes the S, as at the right-hand side, then any population transfer produced by the P field is reversed by the S field. The maximum transfer occurs, not when the pulses coincide, but when the S field precedes, and overlaps, the P field. This type of plot is a clear indicator of STIRAP.

14.4.2 The dark resonance

By monitoring the fluorescence from the excited state 2 we obtain a direct measure of population placed there by the P field. In the absence of the S field this fluorescence signal, as a function of P-field detuning, exhibits a Lorentz profile whose width originates with the lifetime of the excited state. When the S field is present and two-photon detuning occurs the population transfer takes place through the dark adiabatic state: no population enters the excited state. The result is a "dark" resonance, as seen in Fig. 14.13, The spectral width of this narrow feature is the two-photon line width.

14.4.3 The bright resonance

The STIRAP process transfers population only when the P and S frequencies satisfy the two-photon resonance condition. One can monitor the success of this transfer by inducing a transition from state 3 (the final state of the STIRAP process) into a fourth state, using a

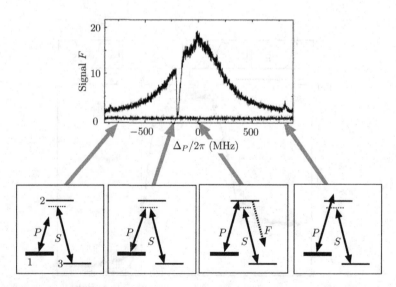

Figure 14.13 The STIRAP dark resonance: fluorescence signal F from state 2 vs. detuning of P field. The P field takes population into state 2, from which it produces a fluorescence signal F. When the P and S fields satisfy the two-photon resonance condition population transfers directly to state 3 without passing through state 2: there is then no fluorescence signal. Frames along the bottom show the changing detuning of the P laser. [Redrawn from Fig. 8 of K. Bergmann *et al.* Rev. Mod. Phys., **70**, 1003 (1998) with permission.]

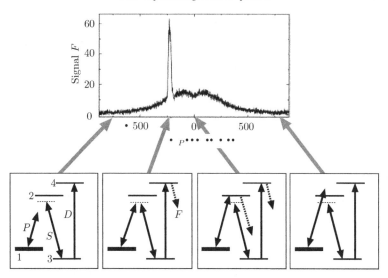

Figure 14.14 The STIRAP bright resonance: fluorescence signal F from state 4 vs. detuning of P field. A probe field D moves population into state 4 from state 3. Population can arrive into state 3 either when the P and S fields satisfy the two-photon resonance condition, or when the P field places population resonantly into state 2, from which it decays to state 3. Frames along the bottom show the changing detuning of the P laser. [Redrawn from Fig. 2 of K. Bergmann *et al.*, Rev. Mod. Phys., **70**, 1003 (1998) with permission.]

D field, from which fluorescence produces a signal. This signal will be present even in the absence of the S field, because the P field produces excitation whose spontaneous emission populates state 3, and hence leads to a fluorescence signal. However, the population transfer to state 3 is much larger when a stimulated Raman transition assists that transfer, as occurs when there is a two-photon resonance.

Figure 14.14 illustrates this effect. This demonstration monitored the fluorescence from state 4 of an excitation chain that begins with initially populated state 1, and relies on the sequence of P, S, and D fields to produce this fluorescence. In the absence of any resonance with the S field, the P field excites population into state 2, which spontaneously decays to state 3, from which the D field carries the population into state 4, whose fluorescence produces the signal. The dependence of this signal on pump detuning traces out the natural width of the 1–2 transition. Only a portion of the decays place population into state 3, from which the subsequent fluorescence signal derives.

When the S field is present there will occur, when the P and S fields together satisfy the two-photon resonance condition (within the narrow limits of the two-photon linewidth), a more complete transfer of population into state 3. This produces a narrow bright line in the fluorescence signal as a function of P-field frequency.

14.5 Optimizing STIRAP pulses

Early demonstrations of STIRAP excitation were produced by passing atoms or molecules through Gaussian-profile laser beams. For such conditions the experimenter can adjust the peak values of the two Rabi frequencies and, for given atom velocity, the width of the Gaussian. The population transfer is most complete if the two peak Rabi frequencies and the two pulse widths are equal. The only adjustable parameter is then the separation between the two pulses. The Gaussian pulse pair

$$\Omega_P(t) = \Omega_0 \exp[-(t - \tau/2)^2/T^2], \qquad \Omega_S(t) = \Omega_0 \exp[-(t + \tau/2)^2/T^2], \qquad \tau = 1.2T \tag{14.22}$$

have temporal area $\mathcal{A} = \Omega T \sqrt{\pi}$ and are often used in simulations. The choice $\tau = 1.2T$ produces the most complete transfer.

When an experimenter has more control over pulse shapes it is reasonable to ask if, for given pulse fluence or given temporal pulse area, pulse shapes can be found that will produce more complete transfer than these Gaussians. As shown by Vasilev *et al.* [Vas09], it is possible to design such pulses.

The STIRAP procedure relies for success upon adiabatic following. It achieves complete population transfer – fidelity of unity – only in the adiabatic limit in which the temporal pulse areas become infinitely large. Because practical constraints limit the pulse areas it is often desirable to optimize the pulse shapes to maximize the population transfer – the fidelity.

Constant rms Rabi frequency. One successful approach, based on insights gained from the Dykne–Davis–Pechukas (DDP) formula (see App. E.3) applied to STIRAP [Vas09], constructs pulses that maintain constant separation of the adiabatic eigenvalues by imposing the constraint

$$\Omega_P(t)^2 + \Omega_S(t)^2 = \text{constant}, \tag{14.23}$$

along with the STIRAP condition

$$\lim_{t \to -\infty} \frac{\Omega_P(t)^2}{\Omega_S(t)^2} = 0, \qquad \lim_{t \to -\infty} \frac{\Omega_S(t)^2}{\Omega_P(t)^2} = 0. \tag{14.24}$$

One such pulse pair, used for simulations restricted to the finite interval $0 \geq t \leq T$ are the "shark-fin" pulses

$$\Omega_P(t) = \Omega_0 \sin(\pi t/2T), \qquad \Omega_S(t) = \Omega_0 \cos(\pi t/2T). \tag{14.25}$$

These have temporal pulse area $\mathcal{A}_\infty = \Omega_0 T \, 2/\pi$. More generally one could use the form

$$\Omega_P(t) = \Omega_0 \sin[\pi f(t)/2], \qquad \Omega_S(t) = \Omega_0 \cos[\pi f(t)/2], \tag{14.26}$$

where the function $f(t)$ is arbitrary apart from the requirement that it be monotonic and have the limits

$$0 = f(-\infty) \leq f(t) \leq f(\infty) = 1. \tag{14.27}$$

Such forms are mathematically convenient but their experimental implementation would require either abrupt start and stop (with possible consequent nonadiabatic change) or would require that pulses continue indefinitely. However, it is only important that the condition (14.23) hold within the time interval during which the adiabatic eigenvector undergoes almost all of its Hilbert-space realignment, and that the onset of the S pulse maintain adiabatic following. For this purpose Vasilev *et al.* [Vas09] suggested the construction

$$\Omega_P(t) = \Omega_0 F(t) \sin[\pi f(t)/2], \qquad \Omega_S(t) = \Omega_0 F(t) \cos[\pi f(t)/2], \qquad (14.28)$$

in which the filtering function $F(t)$ terminates the pulses smoothly and remains relatively constant during the population change. To maintain adiabaticity the two functions should satisfy the condition

$$|\dot{f}(t)| \ll \Omega_0 |F(t)|. \qquad (14.29)$$

Specifically they used the functions

$$f(t) = \frac{1}{1 + \exp(-\lambda t/T)}, \qquad \lambda = 4, \qquad (14.30)$$

$$F(t) = \exp[-(t/T_0)^{2n}], \qquad n = 3, \qquad T_0 = 2T. \qquad (14.31)$$

These pulses gave appreciably lower infidelity than the optimal Gaussian pulses.

Pulse smoothness. Simulations of STIRAP often use Gaussian pulses, as is appropriate for atoms passing through a Gaussian laser beam. It is noteworthy that a Gaussian pulse has infinite support. Therefore when one models a Gaussian pulse over any finite interval, as must occur with numerical simulations, there will inevitably occur an abrupt change of the field. Although this change may be small, nonetheless it can have observable effects on the population histories, typically introducing oscillations when there is detuning.

To ensure accurate modeling of a pulse one should use a pulse of finite support, such as a sine squared pulse, taken to be zero beyond the range of a single half cycle,

$$f(t) = \begin{cases} 0, & t < 0, \\ \sin^2(t/T), & 0 \le t \le \pi T, \\ 0, & \pi T < t. \end{cases} \qquad (14.32)$$

14.6 Two-state versions of STIRAP

The Raman linkage (or the ladder) in which two-photon detuning is maintained allows several analogies with two-state behavior that can assist the design of pulse pairs that will produce a desired final state. The equation for the fully resonant Raman linkage,

$$\frac{d}{dt} \begin{bmatrix} C_1(t) \\ C_2(t) \\ C_3(t) \end{bmatrix} = -\frac{i}{2} \begin{bmatrix} 0 & \Omega_P(t) & 0 \\ \Omega_P(t) & 0 & \Omega_S(t) \\ 0 & \Omega_S(t) & 0 \end{bmatrix} \begin{bmatrix} C_1(t) \\ C_2(t) \\ C_3(t) \end{bmatrix}, \qquad (14.33)$$

allows two reformulations that lead to connections with two-state behavior and analytic solutions that follow from that analogy. The following sections discuss these.

Raman processes

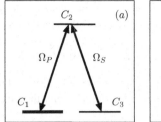
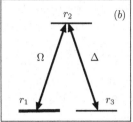

Figure 14.15 (*a*) The three-state system with linkages between the variables C_n by the Rabi frequencies Ω_P and Ω_S. (*b*) The analogous equations for the two-state system are the Bloch variables r_n coupled by the Rabi frequency Ω and the detuning Δ.

14.6.1 Three real variables

Whenever the RWA equations have null detunings it is possible to introduce phases such that all probability amplitudes are real. With a reordering of variables the RWA equations for the fully resonant three-state system can be presented as the equation

$$\frac{d}{dt}\begin{bmatrix} -C_3 \\ -iC_2 \\ C_1 \end{bmatrix} = \frac{1}{2}\begin{bmatrix} 0 & -\Omega_S & 0 \\ \Omega_S & 0 & -\Omega_P \\ 0 & \Omega_P & 0 \end{bmatrix}\begin{bmatrix} -C_3 \\ -iC_2 \\ C_1 \end{bmatrix}. \tag{14.34}$$

As has been noted [Vit06], this set of equations has the same structure as the Bloch equations for a coherently driven two-state system,

$$\frac{d}{dt}\begin{bmatrix} r_1 \\ r_2 \\ r_3 \end{bmatrix} = \begin{bmatrix} 0 & -\Delta & 0 \\ \Delta & 0 & -\Omega \\ 0 & \Omega & 0 \end{bmatrix}\begin{bmatrix} r_1 \\ r_2 \\ r_3 \end{bmatrix}. \tag{14.35}$$

The real-valued two-state Bloch variables r_1, r_2, and r_3 correspond to the real-valued three-state probability amplitudes $-C_3$, $-iC_2$, and C_1. The two-state detuning Δ corresponds to half the three-state S-field Rabi frequency, while the two-state Rabi frequency Ω corresponds to half the P-field Rabi frequency. Figure 14.15 illustrates this connection.

The initial condition for the unexcited two-state system, $r_3(\infty) = w(-\infty) = -1$, corresponds to $C_1(-\infty) = -1$ in the three-state system. The STIRAP process uses adiabatic following to move population into state 3, meaning $|C_3| = 1$. This corresponds, in the two-state system, to creating the conditions $r_1 = 1, r_2 = r_3 = 0$, or $u = 1$, $v = w = 0$.

States for which $r_1^2 + r_2^2 = 1$ and $r_3 = 0$ are states of *maximal coherence* in the two-state system. In the three-state system these correspond to a depleted initial state, $C_1 = 0$. Hence a process that depletes the initial state in the three-state system, e.g. STIRAP, will create a maximally coherent superposition state in the two-state system. Both systems can describe population loss. Table 14.1 summarizes the connections between the two-state system and the fully resonant three-state system.

Table 14.1. *Correspondence between three- and two-state systems*

Three-state system	Two-state system
$\frac{1}{2}\Omega_S(t)$	$\Delta(t)$
$\frac{1}{2}\Omega_P(t)$	$\Omega(t)$
$C_1(t)$	$r_3(t) = w(t)$
$-iC_2(t)$	$r_2(t) = v(t)$
$-C_3(t)$	$r_1(t) = u(t)$
$C_1(-\infty) = -1$	$r_3(-\infty) = -1$
$-C_3(+\infty) = 1$	$r_1(+\infty) = 1$
depleted initial state	maximal coherence
$C_1 = 0$	$r_3 = 0$

We can exploit this analogy to devise procedures, analogous to the STIRAP adiabatic following of three-state systems, that will create two-state superposition states. The counterpart of the STIRAP dark state is a combination of Bloch variables $r_3(t)$ and $r_1(t)$,

$$d(t) = r_3(t)\cos\vartheta(t) - r_1(t)\sin\vartheta(t), \quad \text{with} \quad \vartheta(t) = \arctan\frac{\Omega(t)}{\Delta(t)}. \quad (14.36)$$

When the detuning pulse $\Delta(t)$ precedes the pump pulse $\Omega(t)$ the mixing angle $\vartheta(t)$ has the same asymptotic values as in STIRAP; hence

$$d(-\infty) = r_3(-\infty) \quad \text{and} \quad d(\infty) = r_1(\infty). \quad (14.37)$$

We see that in a two-state system we can move the Bloch vector from alignment with r_3 to alignment with r_1 by applying first a detuning pulse and then an excitation pulse, while maintaining adiabatic conditions. The result will be that the Bloch vector moves on the Bloch sphere from the south pole to the equator, where it is aligned with r_1, i.e. the system ends in a coherent superposition of the two states.

Because the adiabatic passage is robust, this procedure is robust: it depends only weakly on the overlap of the two pulses and the peak values of $\Delta(t)$ and $\Omega(t)$. Applied to the two-state dynamics the local condition for adiabatic following, eqn. (14.19) becomes

$$\left|\dot{\vartheta}(t)\right| \ll \sqrt{\Delta(t)^2 + \Omega(t)^2}. \quad (14.38)$$

The global adiabatic condition is obtained after integration of eqn. (14.38),

$$\frac{\pi}{2} \ll \int_{-\infty}^{\infty} dt\sqrt{\Delta(t)^2 + \Omega(t)^2}. \quad (14.39)$$

The STIRAP pulse sequence, in which the S field precedes the P field, is the counterpart of a two-state procedure in which $\Delta(t)$ precedes $\Omega(t)$ and which produces the final super-position state described by $r_1 = 1$. The opposite pulse sequence, in which $\Omega(t)$ precedes $\Delta(t)$, will produce a superposition in which $|r_1|^2 + |r_2|^2 = 1$, $r_3 = 0$.

14.6.2 Two complex variables

The three-state fully resonant equations that follow from eqn. (14.33) can be written in terms of two complex amplitudes $B_1(t)$ and $B_2(t)$, as suggested by Carroll and Hioe [Car90], through the definitions

$$
\begin{aligned}
C_1 &= -[B_1^* B_2 + B_2^* B_1] = -2 \operatorname{Re} B_1^* B_2, \\
C_2 &= -[B_1^* B_2 - B_2^* B_1] = -2\mathrm{i} \operatorname{Im} B_1^* B_2, \\
C_3 &= |B_2|^2 - |B_1|^2.
\end{aligned}
\tag{14.40}
$$

The equations then read

$$
\frac{d}{dt}
\begin{bmatrix} B_1(t) \\ B_2(t) \end{bmatrix}
= -\frac{\mathrm{i}}{4}
\begin{bmatrix} -\Omega_S(t) & \Omega_P(t) \\ \Omega_P(t) & \Omega_S(t) \end{bmatrix}
\begin{bmatrix} B_1(t) \\ B_2(t) \end{bmatrix}.
\tag{14.41}
$$

This is the basic two-state TDSE but with the replacement of interaction variables shown in Table 14.1. With this replacement any analytic solutions for detuned two-state equations pro-vide solutions to resonant Raman excitation. The connection with STIRAP comes through the initial and final conditions, which require that initially the system is described by a coherent superposition, $B_1 - B_2$,

Time	C_1	C_2	C_3	B_1	B_2
Initial	1	0	0	$1/\sqrt{2}$	$-1/\sqrt{2}$
Final	0	0	-1	1	0

14.6.3 STIRAP with large single-photon detuning

Another class of two-state behavior in three-state systems occurs when, in the TDSE

$$
\frac{d}{dt}
\begin{bmatrix} C_1(t) \\ C_2(t) \\ C_3(t) \end{bmatrix}
= -\frac{\mathrm{i}}{2}
\begin{bmatrix} 0 & \Omega_P(t) & 0 \\ \Omega_P(t) & 2\Delta(t) & \Omega_S(t) \\ 0 & \Omega_S(t) & 0 \end{bmatrix}
\begin{bmatrix} C_1(t) \\ C_2(t) \\ C_3(t) \end{bmatrix},
\tag{14.42}
$$

the single-photon detuning $\Delta(t)$ is very large. Then the equation can be rewritten, after adiabatically eliminating state 2, as the two-state equations

$$\frac{d}{dt}\begin{bmatrix} C_1(t) \\ C_3(t) \end{bmatrix} = -\frac{i}{2}\begin{bmatrix} 0 & \Omega_{\mathrm{eff}}(t) \\ \Omega_{\mathrm{eff}}(t) & 2\Delta_{\mathrm{eff}}(t) \end{bmatrix}\begin{bmatrix} C_1(t) \\ C_3(t) \end{bmatrix}, \tag{14.43}$$

where the effective Rabi frequency and effective detuning are, respectively,

$$\Omega_{\mathrm{eff}}(t) = \frac{\Omega_P(t)\Omega_S(t)}{2\Delta(t)}, \qquad \Delta_{\mathrm{eff}}(t) = \frac{\Omega_P(t)^2 - \Omega_S(t)^2}{4\Delta(t)}. \tag{14.44}$$

The adiabatic eigenvalues are

$$\varepsilon_{\pm}(t) = \pm\left|\frac{\Omega_P(t)^2 + \Omega_S(t)^2}{4\Delta(t)}\right|. \tag{14.45}$$

With constant $\Delta(t)$ the diabatic energy $\hbar\Delta_{\mathrm{eff}}(t)$ is directly proportional to the difference between S- and P-field intensities; there will occur a crossing of the diabatic energy curves as these change. However, the effective detuning $\Delta_{\mathrm{eff}}(t)$ does not become infinitely large, as it does with the LZS model of chirped adiabatic passage.

14.7 Extending STIRAP

The basic three-state STIRAP has been extended in many ways, both theoretically and experimentally, to systems that involve more than three states [Ber98; Vit01a; Vit01b]. In all of these generalizations there occurs a pulse sequence that induces adiabatic transfer. The following sections describe a few of these.

14.7.1 STIRAP and B-STIRAP

The STIRAP process, with the S field preceding but overlapping the P field, produces complete population transfer from state 1 to state 3. It is also possible to obtain complete population transfer in the reverse direction, using the same pair of pulses. That is, population starts in state 3, and the S pulse precedes the P pulse. To accomplish this it is necessary that there be appreciable single-photon detuning, but still null two-photon detuning. Figure 14.16 shows examples of such population transfer, using pulses whose shapes are sine squared. This reversal of population transfer, using detuned pulses intuitively ordered, is termed backward (or bright-state) STIRAP, abbreviated as B-STIRAP. Unlike normal (forward) STIRAP, the reverse necessarily places population temporarily into the excited state; though the process is adiabatic, the statevector is not in a dark state.

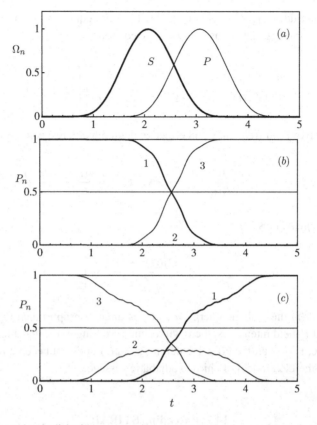

Figure 14.16 Example of adiabatic passage using sine squared pulses and one-photon detuning. (*a*) The two pulses. (*b*) The populations P_n when state 1 is the initial state: the STIRAP process. (*c*) The populations when state 3 is the initial state: this is the B-STIRAP process.

14.7.2 Fractional or F-STIRAP

The STIRAP procedure produces, via adiabatic following, complete transfer of population from state 1 to state 3. However, during the pulse the system is in a coherent superposition

$$C_1(t) = \cos \Theta(t), \qquad C_3(t) = -e^{-i\beta} \sin \Theta(t), \tag{14.46}$$

determined by the mixing angle $\Theta(t)$ of eqn. (14.47),

$$\tan \Theta(t) = \Omega_P(t)/\Omega_S(t). \tag{14.47}$$

Thus if we are able to "freeze" the mixing angle at some selected value, say $\Theta(t) = \alpha$, then the final state will be the coherent superposition defined by the probability amplitudes

$$C_1(\infty) = \cos \alpha, \qquad C_3(\infty) = -e^{-i\beta} \sin \alpha, \tag{14.48}$$

a process termed fractional STIRAP (or F-STIRAP) [Mar91; Vit99a; Kis05]. The required conditions on the pulses now read, instead of eqn. (14.24),

$$\lim_{t \to -\infty} \frac{\Omega_P(t)}{\Omega_S(t)} = 0, \qquad \lim_{t \to -\infty} \frac{\Omega_S(t)}{\Omega_P(t)} = \tan \alpha. \tag{14.49}$$

There are several ways to produce this result. We might abruptly terminate the two pulses at the moment they have the desired ratio. Alternatively we can use pulses that have the same time dependence at late times, for example

$$\begin{aligned} \Omega_P(t) &= \Omega_0 \sin \alpha \exp[-(t - \tau/2)^2/T^2], \\ \Omega_S(t) &= \Omega_0 \left\{ \exp[-(t + \tau/2)^2/T^2] + \cos \alpha \exp[-(t - \tau/2)^2/T^2] \right\}. \end{aligned} \tag{14.50}$$

Alternatively we might consider a version of eqn. (14.28),

$$\Omega_P(t) = \Omega_0 F(t) \sin[\pi f(t)/2], \qquad \Omega_S(t) = \Omega_0 F(t) \cos[\pi f(t)/2]. \tag{14.51}$$

Both of these choices have been examined by Vasilev *et al.* [Vas09].

14.7.3 Vacuum or V-STIRAP

Stimulated processes, by definition, require pre-existing fields to induce transitions. Laser fields provide readily manipulated fields for this purpose, and most of this monograph treats these as the fields responsible for, say, stimulated Raman processes. However, vacuum fields can also produce measurable effects [Mil94]. An important example occurs when an atom is confined within a cavity. The vacuum field of the cavity is always present, and together with the dipole moment of the 2–3 transition it provides a constant *S*-field *vacuum Rabi frequency*.[2] By introducing a pulsed-laser *P* field it is possible to produce a STIRAP-like transition as the mixing angle changes in response to the growing *P* field. Section 22.3 discusses this procedure, used to place a single photon into a cavity.

14.7.4 Multiple intermediate states

The linkages from initial state 1 to final state 3 need not be through a single intermediate state. The transfer can occur as well when a set of $N - 2$ discrete states, each with slightly different detuning, all have the same linkage pattern to these initial and final states – a *multiple-lambda linkage* involving pairs of Rabi frequencies $\Omega_P^{(k)}(t)$ and $\Omega_S^{(k)}(t)$. Figure 14.17 illustrates the linkages.

In order for there to be a null-eigenvalue dark state, as needed for the STIRAP mechanism to operate, it is necessary that the ratio $\Omega_P^{(k)}(t)/\Omega_S^{(k)}(t)$ be the same for each link [Vit00a]. When this condition holds it is possible to achieve STIRAP-like complete population transfer with no transient population in the intermediate states.

---

[2] This is evaluated with the single-photon electric field of the cavity, so it may also be termed the single-photon Rabi frequency.

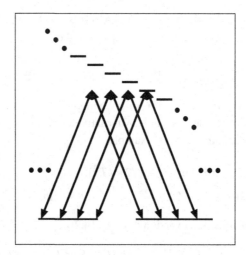

Figure 14.17 Stimulated Raman linkage through a set of intermediate states. In the simplest quasicontinuum model the excited levels are equidistant and infinite in number.

Under less restrictive conditions on the paired Rabi frequencies it is sometimes possible to find an adiabatic transfer (AT) state which, although allowing transient population in intermediate states, can produce complete population transfer [Vit01a]. Sec. 15.6.2 discusses STIRAP in chains.

Quasicontinuum. The number of intermediate states $N - 2$ used in a model need not be finite; they have been taken as an infinite set of equidistant energy states, a *quasicontinuum* of states, each with the same Rabi frequency [Car93]. However, although that simple model has analytic solutions, the needed simplification of intermediate-state structure is unrealistically restrictive [Vit01b].

Continuum. As the energy levels of a quasicontinuum become more closely spaced they approach a true continuum. It should be noted that this need not be the "flat" structureless continuum used in simple estimates of "golden-rule" transition rates. The consequences of continuum structure have been discussed in several papers [Kni90; Yat97; Yat99b].

14.7.5 Hyper-Raman STIHRAP

Hitherto I have assumed that the P and S Rabi frequencies originate with single-photon interactions, proportional to the square root of the intensity. More generally, each of these interactions can originate with some multiphoton effective Hamiltonian, the simplest of which is a two-photon interaction whereby the Rabi frequency scales directly as the intensity. Such interactions generalize the Raman process and have been termed hyper-Raman processes. Their use in STIRAP-type processes has been termed stimulated hyper-Raman adiabatic passage (STIHRAP) [Yat98; Gue98]. Figure 14.18 illustrates some examples in which one or both of the two links are by means of multiphoton Rabi frequencies.

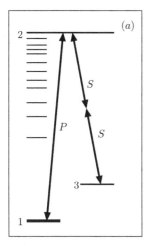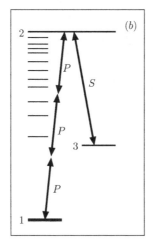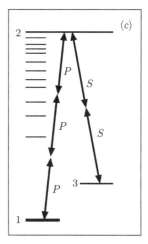

Figure 14.18 Examples of STIHRAP linkages: the two legs of the conventional lambda linkage are here multiphoton links. (*a*) The 2–3 transition is via two *S* photons. (*b*) The 1–2 transition is via three *P* photons. (*c*) Both transitions are via multiphoton Rabi frequencies.

14.7.6 STIRAP with sublevels

Adiabatic passage in a three-state system, as STIRAP, readily generalizes in several ways to treat the degeneracy, or near degeneracy, that occurs with magnetic sublevels [Kis04; Tha04; Kis05; Boo08]. Figure 14.19 shows an example of the most general linkage pattern that occurs with a three-level system in which the linkages begin in a state $J = 0$, proceed via *P* field into sublevels of $J = 1$, and then by *S* field to sublevels of $J = 2$. Such linkage patterns have been studied extensively in excitation of metastable neon atoms in atomic beams [Sho95a; Mar95; Mar96; Vew09].

Preparing superpositions of sublevels. In the absence of any static magnetic field the sublevels are degenerate, and sublevel selectivity must be obtained by adjustment of laser polarizations. Figure 14.20 shows examples of obtainable linkage patterns for the neon system of Fig. 14.19. For clarity the initial state $^3P_0$ is placed at the top of the linkage pattern; the STIRAP process moves population from the top to the bottom of the diagram. As can be seen, by using STIRAP it is possible to prepare not only selected sublevels but particular coherent superpositions – Zeeman coherences. By adjusting the relative directions of the two laser beams (collinear or perpendicular) and the individual field polarizations (linear, circular, or elliptical) an experimenter selects various linkage patterns. The upper-left frame shows the most general linkage pattern, a repetition of the linkage of Fig. 14.19. The other frames bear two-letter labels denoting, respectively, the polarization of *P* and *S* fields: *E* (elliptical), *L* (linear), and *C* (circular). These linkage patterns have been implemented in studies of metastable neon beams [Vew09].

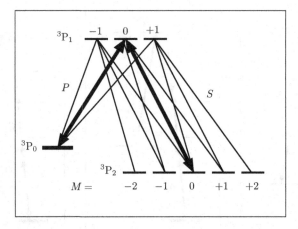

Figure 14.19 Linkage patterns available with suitable polarization choices in the $J = 1 \leftrightarrow J = 0 \leftrightarrow J = 1$ of metastable neon. When there is no magnetic field the Zeeman sublevels are degenerate; the choice of polarization then selects the transitions or superpositions of transitions that will be active. [Redrawn from Fig. 56 of B. W. Shore, Acta Physica Slovaka, **58**, 243–486 (2008) with permission.]

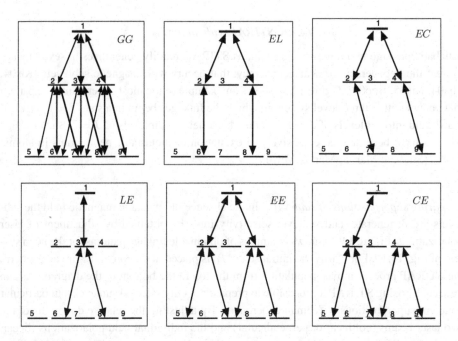

Figure 14.20 Linkage patterns accessible amongst degenerate sublevels by suitable choice of P and S polarizations. Frame GG shows the most general linkage and defines the numbering scheme. The linkages of frames EL and LE produce a superposition of states 5, 8; linkage EC produces a superposition of states 7, 9; linkages EE and CE produce superpositions of 5, 7, 9.

Complete transfer into a sublevel. When a static magnetic field is present the natural coordinate system is one in which the quantization axis lies along the direction of that field. By suitably choosing the orientation of the laser propagation axis, and polarization, in this reference frame it is possible to have each of the linkages of Fig. 14.19 present. The magnetic field produces a Zeeman shift of the sublevels, so that only particular linkages are resonant. Thus with fixed laser polarization and beam axes it is possible, by altering only the frequencies of the light, to place all population into any single selected sublevel of $J = 2$ via STIRAP-like adiabatic passage [Sho95a; Mar95; Mar96].

In a degenerate system whose degeneracies do not decrease along the sequence 1–2–3 it is possible to reduce the TDSE to a set of independent three-state chains, using a Morris–Shore transformation (see App. F) within each of which STIRAP can be implemented [Kis04]. Thus complete population transfer is possible, in such systems, between degenerate levels.

Transfer within a level. The STIRAP mechanism allows population transfer within a set of degenerate sublevels. A simple extension is the linkage pattern available with a $J = 1$ to $J = 0$ excitation transition with arbitrary polarization, see Fig. 14.21(a). This has a tripod linkage pattern, discussed in Sec. 15.4.1.

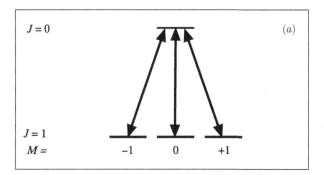

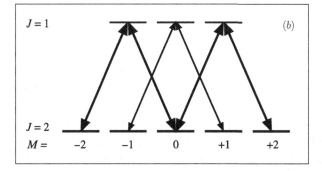

Figure 14.21 (a) Linkage pattern for $J = 1$ to $J = 0$ with general polarization – the propagation directions of the three fields cannot all be parallel. The links form a tripod. (b) Linkage pattern for transition $J = 2$ to $J = 1$ available with elliptically polarized light, with quantization axis along propagation direction. There are two independent systems: a three-state lambda (light lines) and a five-state "letter-M" (heavy lines).

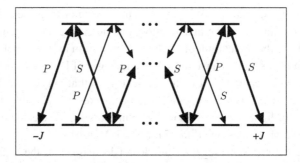

Figure 14.22 A chain of P and S linkages in a degenerate system of Zeeman sublevels for integer J, starting from $M = -J$. A STIRAP-like $S-P$ pulse sequence will transfer all population from $M = -J$ to $M = +J$.

The transition $J = 2 \leftrightarrow 1$ has, for elliptically polarized light and quantization axis along the propagation direction, see Fig. 12.9(b) of Sec. 12.5, two sets of linkages: a three-state lambda (involving $M = \pm 1$ of $J = 2$) and a five-state "letter-M" pattern (involving $M = 0, \pm 2$ of $J = 2$). When spontaneous emission can be neglected these two sets are independent.

The lambda and "letter-M" linkages generalize to a linkage pattern in which a single S field alternates with a single P field to produce a chain involving degenerate ground states and degenerate excited states, as shown in Fig. 14.22. Preliminary optical pumping with circularly polarized light can place all the population into the state at one end of the chain, say the $M = -J$ state. With this as the single initial state, the usual $S-P$ pulse sequence will transfer all population to the other end of the chain. There will never occur population in the excited state, although at intermediate times the population will be distributed amongst the ground sublevels. The underlying theory, for chains of arbitrary length, has been discussed in several papers [Vit01b; Vit01c].

14.7.7 Loop STIRAP

The STIRAP process requires large temporal pulse areas, meaning a combination of large peak Rabi frequency Ω_0 and long pulse duration T: the deviation from adiabatic following scales as $1/\Omega_0 T$. As T becomes long, incoherent processes may intervene; as the field becomes more intense various dynamic Stark shifts may become detrimental.

As recognized by Unanyan et al. [Una97; Fle99; Sho08], one approach to mediating these difficulties is by means of a loop linkage: in addition to the P field linking states 1, 2 and the S field linking states 2, 3, we introduce an overlapping R (for Raman, also termed D for detuning) field between states 1, 3, as shown in Fig. 13.2. Such a link, though not occurring with three electric-dipole transitions of a free atom, is possible with a magnetic-dipole transition (e.g. a quasistatic magnetic field), via quadrupole transition, or with transitions of an atom in a solid environment. The presence of a suitable R pulse makes possible almost complete population transfer, $1 \to 3$, with relatively small temporal pulse areas.

The analysis is simplest when the three fields are each resonant with their respective Bohr frequencies, so that the three RWA equations read

$$\frac{d}{dt}C_1 = -\frac{i}{2}\Omega_P C_2 - \frac{i}{2}\Omega_R e^{i\phi} C_3,$$
$$\frac{d}{dt}C_2 = -\frac{i}{2}\Omega_P C_1 - \frac{i}{2}\Omega_S C_3, \tag{14.52}$$
$$\frac{d}{dt}C_3 = -\frac{i}{2}\Omega_R e^{-i\phi} C_1 - \frac{i}{2}\Omega_S C_2,$$

where the phase on Ω_R is $\phi = \phi_P - \phi_S + \phi_R$.

Bright and dark basis. To treat this problem we introduce bright and dark states with amplitudes

$$\begin{aligned} C_B &= \sin\Theta C_1 + \cos\Theta C_3, & C_1 &= \sin\Theta C_B + \cos\Theta C_D, \\ C_D &= \cos\Theta C_1 - \sin\Theta C_3, & C_3 &= \cos\Theta C_B - \sin\Theta C_D, \end{aligned} \tag{14.53}$$

where

$$\tan\Theta = \Omega_P / \Omega_S. \tag{14.54}$$

The equations satisfied by these variables can be written in matrix form as

$$\frac{d}{dt}\begin{bmatrix} C_D \\ C_B \\ C_2 \end{bmatrix} = -i \begin{bmatrix} -\Delta_D & \frac{1}{2}\check{\Omega}_{BD}^* & 0 \\ \frac{1}{2}\check{\Omega}_{BD} & +\Delta_D & \frac{1}{2}\Omega_{SP} \\ 0 & \frac{1}{2}\Omega_{SP} & 0 \end{bmatrix} \begin{bmatrix} C_D \\ C_B \\ C_2 \end{bmatrix}. \tag{14.55}$$

The redefined variables provide linkage in a chain $D - B - 2$, from dark state to bright state to excited state, rather than a closed loop as in eqn. (14.53). The bright-state amplitude C_B embodies all of the interaction between the excited state 2 and the other two states, via the rms Rabi frequency Ω_{SP},

$$\Omega_{SP} = \sqrt{|\Omega_S|^2 + |\Omega_P|^2}. \tag{14.56}$$

The coupling between dark and bright states is through the complex-valued Rabi frequency

$$\check{\Omega}_{BD} = 2i\dot{\Theta} + \Omega_R(\cos 2\Theta \cos\phi - i\sin\phi), \tag{14.57}$$

in which the usual nonadiabatic coupling $\dot{\Theta}$ is augmented by a term that depends on the R pulse. This pulse also produces a detuning,

$$\Delta_D = \Omega_R \sin 2\Theta \cos\phi. \tag{14.58}$$

As with traditional STIRAP we assume that the S pulse precedes the P pulse, so that initially (at $t = 0$) the mixing angle Θ is zero and the amplitudes are

$$C_1(0) = C_D(0) = 1, \qquad C_B(0) = C_2(0) = 0. \tag{14.59}$$

Also as with STIRAP we assume the pulse sequence concludes with a mixing angle $\Theta = \pi/2$, so that the dark-state amplitude C_D changes from $C_1 = 1$ to $-C_3$, maintaining always unit magnitude.

Adiabatic condition. To avoid diluting this amplitude we need to maintain isolation of the dark state, i.e. maintain adiabatic evolution. To ensure the evolution is adiabatic, and that the statevector follows the dark state, we require that the bright–dark coupling Ω_{BD} should at all times be much smaller than the Rabi frequency Ω_{SP} that controls the separation of the eigenvalues. This condition holds separately for the real and imaginary parts of Ω_{BD} and gives the constraints

$$|2\dot{\Theta} + \Omega_R \sin\phi)| \ll |\Omega_{SP}|, \qquad |\Omega_R \cos 2\Theta \cos\phi| \ll |\Omega_{SP}|. \tag{14.60}$$

In the absence of the R field the condition for adiabatic following is

$$|2\dot{\Theta}| \ll |\Omega_{SP}|, \tag{14.61}$$

meaning that the pulse envelope should change very little during one rms Rabi period. The presence of the R field lessens this constraint. In particular, if $\phi = \pi/2$ and the R-field envelope is crafted to follow the change of the rms $P - S$ field,

$$\Omega_R = -2\dot{\Theta}, \tag{14.62}$$

then population remains exactly trapped at all times. Because the mixing angle changes by $\pi/2$ the temporal area of the R pulse should be π. Unlike a traditional two-state π pulse, however, this area need not be precisely π to obtain complete population transfer. Numerical simulations demonstrate that complete 1–3 population transfer can take place with P and S pulses whose temporal areas are much less than what would be required for traditional STIRAP, and that this successful transfer is not sensitive to any of the temporal pulse areas [Vit01b; Vit01c].

15

Multilevel excitation

A variety of extensions and generalizations of two- and three-state excitation hold interest, either theoretically or for applications. This chapter discusses some examples of those extension for systems having N quantum states, first for incoherent rate-equation processes (Sec. 15.1 below) and then for coherent excitation (in the remainder of this chapter).

15.1 Multiphoton and multiple-photon ionization

Amongst the earliest applications of laser-induced excitation was to successions of radiation-mediated transitions that ultimately produce ionization. In the simplest situations the laser-supplied energy adds to the initial internal energy E_1 enough to produce a free electron of kinetic energy E_{kin} and an ion having internal energy E_{ion}. The energy required for this photoionization is greater than can be supplied by a single photon, $\hbar\omega$, and so some minimum number N_{min} of monochromatic photons are needed, as set by the requirement that N_{min} be the smallest integer such that

$$E_{ion} < E_1 + N_{min}\hbar\omega. \tag{15.1}$$

The kinetic energy E_{kin} carried by the photoelectron after deposition of energy $N\hbar\omega$ in the atom is

$$E_{kin} = N\hbar\omega - (E_{ion} - E_1). \tag{15.2}$$

When N frequencies supply the energy, in single-photon increments, the requirement is

$$E_{ion} < E_1 + \hbar\omega_1 + \hbar\omega_2 + \cdots + \hbar\omega_N, \tag{15.3}$$

and the kinetic energy is

$$E_{kin} = \sum_n \hbar\omega_n - (E_{ion} - E_1). \tag{15.4}$$

Excess-photon ionization. The inequality of eqn. (15.1) merely establishes a minimum for the number of photons required to produce ionization, based on the requirement that the electron kinetic energy be positive. As experiments have demonstrated, it is possible to produce photoelectrons with successively larger energies, differing by increments $\hbar\omega$. These photoelectrons are the result of excess photon absorption; the phenomenon has been termed *above-threshold ionization* (ATI) [Gon80; Del89; Ebe91] or *excess-photon ioniza-tion* (EPI) [Sho87]. Typically such photoionization processes take place with cw radiation, with all frequencies present simultaneously.

Multiple-photon ionization. A variety of mechanisms lead to photoionization. One such is a succession of independent single-photon transitions, each near resonant with one step of an overall N-step ionization process

$$1 \rightarrow 2 \rightarrow \cdots \rightarrow N \rightarrow \text{ions}.$$

The individual steps of this *multiple-photon* process need not be coherent, and each one is independent of other steps. Typically one describes each step by a rate equation, using single-photon golden-rule transition rates. The equations have the form [Ack77]

$$\frac{d}{dt} P_n = P_{n-1} \mathcal{R}_{n-1 \rightarrow n} - P_n [\mathcal{R}_{n \rightarrow n-1} + \mathcal{R}_{n \rightarrow n+1}] + P_{n+1} \mathcal{R}_{n+1 \rightarrow n}, \qquad (15.5)$$

where the rate coefficients $\mathcal{R}_{n \rightarrow m}$ are the single-step transitions rates. With such a description there is a definite sense of time ordering of the transitions: population proceeds, one step at a time, through a sequence of N steps. Each excitation step requires that population has completed a succession of preliminary steps. If the overall process uses a set of distinguishable frequencies, then these are absorbed in a definite sequence.

Multiphoton ionization. When some of the steps are not resonance then a portion of the sequence proceeds by *multiphoton* processes, in which the rate coefficients require the simul-taneous presence of several photons. The rate coefficients, calculated using the golden rule of Sec. 9.6, require off-diagonal elements of a multiphoton effective Hamiltonian, as dis-cussed in Sec. 13.4.6. The simplest of these processes is the direct N-photon transition from state 1 to the photoionization continuum that lies just beyond energy E_N. The multiphoton transition amplitude must be evaluated by summing all possible products of N dipole transi-tion moments paired with field envelopes, and divided by $N - 1$ associated detunings. Only for relatively simple systems can this summation be carried out reliably. However, empiri-cal values for the multiphoton rate coefficient (or multiphoton Rabi frequency) suffice for simulations and optimization studies.

   It is not possible to assign a time ordering of the photon interactions for a multipho-ton process: A multiphoton excitation step requires the simultaneous presence of several photons, possibly of different frequencies.

Resonance-enhanced multiphoton ionization (REMPI). When one (and only one) state of the chain, say state n, is resonant with the sum of several photon frequencies, so that

$$E_n = E_1 + \hbar\omega_1 + \cdots + \hbar\omega_{n-1}, \qquad (15.6)$$

then the overall photoionization takes place in two steps. The first step involves $n - 1$ photons, present simultaneously, while the second involves $N - n - 1$ photons. The overall

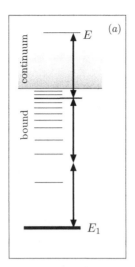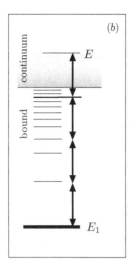

Figure 15.1 Resonance-enhanced multiphoton ionization. (*a*) 2 + 1 REMPI, involving a two-photon transition. (*b*) 3 + 1 REMPI, involving a three-photon transition.

process is an example of *resonance-enhanced multiphoton ionization* or REMPI, revealed by scanning one of the frequencies while monitoring the ionization signal. Figure 15.1 illustrates the excitation patterns for examples of REMPI.

Ponderomotive energy. In a strong field, such as is typically required to produce multiphoton ionization, the kinetic energy of the photoelectron comprises two parts. One is a rapid jittering motion at the field carrier frequency. This has a cycle-averaged kinetic energy equal to the ponderomotive energy E^{pond} and proportional to the field intensity. The other component of the kinetic energy of the photoelectron is a slower drift away from the atom, and out of the laser field. The N photons must supply both contributions to the photoelectron energy [Ebe91].

15.2 Coherent excitation of N-state systems

Coherent-excitation dynamics follows the dictates of the TDSE. We here consider those equations for an arbitrary set of N states. As will be seen, the applicable theory retains many of the idealizations and approximations used for $N = 2$ and 3.

As with those earlier discussions we introduce an N-state rotating Hilbert-space coordinate system by expressing the statevector as the superposition

$$\Psi(t) = \sum_{n=1}^{N} C_n(t) \exp[-i\zeta_n(t)] \, \psi_n. \tag{15.7}$$

The probability amplitudes $C_n(t)$ are solutions to the TDSE, expressed in vector-matrix form as

$$\frac{d}{dt}\mathbf{C}(t) = -i\mathbf{W}(t)\mathbf{C}(t), \tag{15.8}$$

where the elements of the $N \times N$ Hamiltonian matrix are

$$\hbar W_{nm}(t) = \begin{cases} E_n + V_{nn}(t) & \text{if } n = m, \\ V_{nm}(t)\exp[i\,\zeta_n(t) - i\,\zeta_m(t)] & \text{if } n \neq m. \end{cases} \tag{15.9}$$

Typically the phases $\zeta_n(t)$ can be chosen such that, with neglect of rapidly varying exponentials (the RWA), the elements of the matrix $\mathbf{W}(t)$ are constant or slowly varying – any appreciable changes occur during many optical cycles, for example. When the number of states N is modest, then the evaluation of numerical solutions to these coupled ODEs is usually straightforward.

As the number of quantum states increases, so too does the variety of possible linkage patterns – the nonzero elements of the matrix $\mathbf{W}(t)$. In many cases it is possible to introduce a generalization of the RWA such that the Hamiltonian has detunings as diagonal elements and slowly varying Rabi frequencies as the off-diagonal elements [Ein76] [Sho90, §14.2]. The slow variation allows procedures based on generalizations of either Rabi oscillations (e.g. generalizations of π pulses) or adiabatic evolution (generalizations of RAP or STIRAP). The following sections provide an overview of several multistate coherent excitation models, treated with a generalized RWA [Sho90, Chap. 14].

15.2.1 Multilevel linkages

Linkage patterns depend upon two characteristics of the system. First, the intrinsic set of transition moments – usually those of electric dipole moments – that provide a network of connections between states, a framework within which particular Hamiltonian matrix elements may be activated. Second, the fields that the experimenter chooses to apply. The choice of polarizations selects certain dipole moments, eliminating others. The choice of carrier frequencies further restricts the active linkages to those that are near resonance with one of the frequencies – or which are accessed by a multiphoton-transition moment. The combination of intrinsic moments and laser fields provides the nonzero elements $V_{nm}(t)$ of the interaction Hamiltonian and allows identification of the N essential states – those quantum states whose dynamics must be followed with the TDSE.

The elements $V_{nm}(t)$ need not be present simultaneously. Sequences of pulses may activate only portions of the full $N \times N$ Hamiltonian matrix. Often elaborate linkage patterns can be reduced to a succession of single-pulse two-state linkages, in which full or partial population transfer takes place by a succession of nonoverlapping pulses. These require careful control of pulse area (for generalization of π pulses), or of relatively slow changes (for adiabatic following). Thus such procedures have not been widely used for coherent excitation at optical frequencies. Instead, it is desirable to consider the overall pattern of available linkages. Three types of linkages are discussed in this chapter:

Chain: Every state is connected to at most two states, with two "end" or "terminal" states that have only one link. Conventionally one terminal state is regarded as state 1, and the numbering proceeds sequentially along the links to state N. The resulting Hamiltonian matrix is tridiagonal. When the states are displayed on an energy diagram the chain may appear as kinked, but this geometry is not relevant to the dynamics. Section 15.3 discusses such linkages.

Branch: A branched state has more than two links to other states. Section 15.4 discusses examples in which there is a single branch.

Loop: A loop occurs when a succession of links in a chain leads back on itself. In such a situation there are two pathways between any two states and consequently the corresponding transition amplitude is the sum of two terms. Interference, constructive or destructive, takes place to make the transition sensitive to phase differences. Section 14.7.7 discussed the simplest example, that of a three-state system. Section 15.5 discusses some four-state examples. The loop linkage underlies the interferometric measurements discussed in Chap. 20.

For special combinations of dipole moments and field amplitudes the Hamiltonian matrix may have symmetry properties that facilitate analytic solutions [Hio85b; Hio87]. Such is the case when the Hamiltonian is expressible as the sum of three N-dimensional spin matrices, as discussed in Sec. 15.3.5.

In all these linkages it is the topology of the linkage, rather than the geometry exhibited on an energy-level diagram, that dictates the dynamics. Thus, as noted below, various energy-level structures have equivalent linkages (e.g. a tridiagonal Hamiltonian matrix) and the same dynamics. The following sections provide an overview of several multistate coherent excitation models, treated with a generalized RWA [Sho90, Chap. 14].

15.2.2 Multilevel analytic solutions

The availability of computer programs that can produce solutions to coupled ODEs involving large numbers of variables and complicated time variations of the coefficient matrices allows the simulation of ever-more realistic models of pulsed excitation. Nevertheless, closed-form analytic solutions still have value. They allow one to examine various asymptotic limits of parameters and to establish various bounds upon probabilities, to identify ranges of parameters of interest and to examine the sensitivity to errors and fluctuations of test cases. Thus the past investigations of analytic solutions retain value, and interest continues in the properties of these solutions, embodied in the various special functions of mathematical physics.

Chapters 8 and 13 have presented closed-form solutions for two- and three-state systems, for various forms of the RWA Hamiltonian. It is natural to ask [Ebe77]: for what N, and what patterns of Rabi frequencies and detunings, are analytic solutions available? Analytic solutions for chain linkages and constant intensity are known for $N \to \infty$ as well as for large but finite N [Bia77; Let78; Sav96] [Sho90, Chap. 15]. These make use of dressed states and properties of tridiagonal matrices. The solutions serve as useful reference templates with which to interpret phenomena in other situations.

Common time dependence. Numerous closed-form solutions are known for the multistate RWA Schrödinger equation. The simplest solutions occur when all of the elements of the RWA Hamiltonian matrix have the same time dependence, say $f(t)$,

$$W_{nm}(t) = f(t) W_{nm}(0), \qquad f(0) = 1. \tag{15.10}$$

This condition requires that any nonzero diagonal elements (the detunings) should share the time dependence of the off-diagonal elements (the Rabi frequencies), as occurs when all the transitions are resonant and all of the Rabi frequencies are obtained from a single pulse envelope. When eqn. (15.10) applies then we define a scaled time variable τ as the definite integral of the function $f(t)$, as in Sec. 13.3.1,

$$\tau(t) = \int_0^t dt' \, f(t'), \tag{15.11}$$

and obtain the coupled equations

$$\frac{d}{d\tau} C_n(\tau) = -i \sum_m W_{nm}(0) C_m(\tau), \tag{15.12}$$

in which there occurs a constant matrix of coefficients $W_{nm}(0)$.

The solutions to the TDSE for constant Hamiltonian have been extensively studied and reported, most notably for multistate chains [Bia77; Let78; Sav96] [Sho90, Chap. 15].

15.2.3 Multilevel adiabatic states

Although direct numerical integration of the TDSE offers useful results, evaluations of slowly varying eigenvectors $\boldsymbol{\Phi}_\nu(t)$ and eigenvalues $\varepsilon_\nu(t)$ of the matrix $\mathsf{W}(t)$ – the dressed or adiabatic states – can provide valuable insights as well as useful numerical tools. For a constant RWA Hamiltonian W we can, in principle, always construct the solutions to the N-state time-dependent Schrödinger equation as the superposition

$$C_n(t) = \sum_{\nu m} \Phi_{\nu n} \exp(-i\varepsilon_\nu t) \Phi_{\nu m}^* C_m(0) = \sum_m U_{nm}(t, 0) C_m(0), \tag{15.13}$$

where $\Phi_{\nu n}$ is the nth component of eigenvector $\boldsymbol{\Phi}_\nu$ and $C_m(0)$ are the initial probability amplitudes. This construction can always be produced numerically. In a number of cases it is possible to obtain closed-form analytic expressions for the $\Phi_{\nu m}$ and ε_ν, thereby facilitating the evaluation of asymptotic limits.

When the initial population is entirely concentrated in state m the subsequent probability amplitudes have the construction

$$P_n(t) = \sum_{\nu\nu'} \exp[i(\varepsilon_{\nu'} - \varepsilon_\nu)t] \, \Phi_{\nu n} \Phi^*_{\nu'n} \, \Phi^*_{\nu m} \Phi_{\nu'm}, \qquad (15.14)$$

and thus they generally exhibit oscillatory behavior at *beat frequencies* – differences of the adiabatic eigenvalues. These are absent if the needed eigenvector is aligned with the initial state, as occurs when adiabatic following takes place. Under those conditions only a single eigenfrequency occurs.

Periodicities. When the RWA Hamiltonian is constant then the formula of eqn. (15.14) exhibits explicitly the frequency content of each probability. The solutions will be periodic if each exponential is periodic, a condition that requires the eigenvalue differences to be multiples of a single value – they must be *commensurable*. This condition always holds for $N = 2$ and $N = 3$, but it generally does not hold for larger N. Section 15.3.5 discusses an exception, when solutions for any N are periodic. Periodic solutions will, after one period, exactly reproduce the initial condition. They may also allow periodic placement of all population into an excited state. However, periodicity is not a necessary condition for such population transfer; various forms of adiabatic passage (e.g. STIRAP) can also accomplish this change.

15.3 Chains

The simplest extensions of two- and three-state RWA equations are those for a single chain of interactions, in which each state links to no more than two other states (nearest neighbors) and no loops occur [Sho90, Chap. 15], i.e. linkages of the type

$$1 \leftrightarrow 2 \leftrightarrow 3 \leftrightarrow \cdots \leftrightarrow N.$$

For such linkages the TDSE comprises a set of equations having the form

$$\frac{d}{dt} C_n(t) = -i \Delta_n(t) C_n(t) - \frac{i}{2} \Omega_{n-1}(t) C_{n-1} - \frac{i}{2} \Omega_n(t) C_{n+1}. \qquad (15.15)$$

Under suitable conditions discussed below it is possible to generalize the RWA such that the Rabi frequencies $\Omega_n(t)$ and detunings $\Delta_n(t)$ become slowly varying functions of time, each associated with a single carrier frequency.

15.3.1 The multilevel RWA

We introduce a rotating-wave picture by introducing explicit time-dependent phases $\zeta_n(t)$ in the construction of the statevector,

$$\Psi(t) = \sum_n C_n(t) \exp[-i\zeta_n(t)] \, \psi_n. \qquad (15.16)$$

The Hamiltonian matrix then has as diagonal elements the detunings

$$\hbar W_{nn}(t) = E_n + V_{nn}(t) - \hbar \dot{\zeta}_n(t) \equiv \hbar \Delta_n(t), \tag{15.17}$$

while the off-diagonal elements are

$$\hbar W_{n-1,n}(t) = V_{n-1,n}(t) \exp[i\zeta_{n-1}(t) - i\zeta_n(t)]. \tag{15.18}$$

We make the RWA by assuming that when $W_{n-1,n}(t)$ is averaged over rapid variations, each pair of exponentiated phases select a single carrier frequency. The details of this procedure depend on the ordering of the energy levels and on whether there are multiple frequencies or only a single frequency.

Multiple frequencies. Consider the electric-dipole interaction $H_{n,n+1}(t) = -\mathbf{d}_{n,n+1} \cdot \mathbf{E}(t)$ and let the field at the center of mass have the structure (see App. C.2.3)

$$\mathbf{E}(t) = \mathrm{Re}\left\{ \sum_\lambda \mathbf{e}_\lambda \mathcal{E}_\lambda(t)\, e^{-i\omega_\lambda t - i\varphi_\lambda} \right\}. \tag{15.19}$$

Let the carrier frequency closest to resonance with the link between states $n-1$ and n be ω_n. That is, we assume that, for each pair of states, there is a single readily identifiable carrier frequency ω_n such that, in the absence of field-induced shifts,

$$|E_n - E_{n-1}| \approx \hbar \omega_n. \tag{15.20}$$

To make the RWA for a chain we first choose some reference state, typically the initially populated state, say i. We set the detuning for that reference state to zero by choosing the phase derivative to be

$$\hbar \dot{\zeta}_i(t) = E_i + V_{ii}(t), \quad \text{so } \Delta_i(t) = 0. \tag{15.21}$$

We proceed along the chain, defining successive phase derivatives to remove the exponential of the near-resonant frequency,

$$\dot{\zeta}_n(t) = \begin{cases} \dot{\zeta}_{n-1} + \omega_n & \text{if } E_n > E_{n-1}, \\ \dot{\zeta}_{n-1} - \omega_n & \text{if } E_n < E_{n-1}. \end{cases} \tag{15.22}$$

We can, if we wish, incorporate field phase φ_n into the rotating Hilbert-space coordinate $\psi_n'(t)$ by writing, when $t = 0$ is the initial time and the carrier frequencies remain constant,

$$\zeta_n(t) = \zeta_{n-1}(t) \pm \omega_n t \pm \varphi_n. \tag{15.23}$$

The signs here are those of eqn. (15.22). This choice, which I will adopt below, produces real-valued elements of the matrix $\mathbf{W}(t)$ when the transition moments $\mathbf{d}_{n,n+1} \cdot \mathbf{e}_\lambda$ are real and the laser phases remain constant. We then neglect all exponentials that vary at the *sum* frequencies $\omega_n + \omega_\lambda$. This produces the off-diagonal matrix elements

$$W_{n,n+1}(t) = \tfrac{1}{2}\Omega_n^{(n)}(t) + \tfrac{1}{2}\sum_{\lambda \neq n}\Omega_n^{(\lambda)}\exp[i(\omega_n - \omega_\lambda)t], \tag{15.24}$$

where

$$\hbar\Omega_n^{(\lambda)}(t) = -\mathbf{d}_{n,n+1} \cdot \mathbf{e}_\lambda \mathcal{E}_\lambda(t). \tag{15.25}$$

As the final step we require that the *difference* frequencies be much larger than the characteristic frequencies appropriate to changes of interest, i.e. larger than the Rabi frequencies,

$$|\omega_n - \omega_\lambda| \gg |\Omega_n(t)|. \tag{15.26}$$

By making this final approximation we neglect the contributions to $W(t)$ from the difference frequencies $\omega_n - \omega_\lambda$ of eqn. (15.24), and obtain as the interaction terms of the RWA Hamiltonian the usual Rabi frequencies,

$$W_{n,n+1}(t) \equiv \tfrac{1}{2}\Omega_n(t) = \tfrac{1}{2}\Omega_n^{(0)}(t). \tag{15.27}$$

The field phases φ_n are here incorporated into the definitions of the Hilbert-space unit vectors rather than appearing explicitly as multipliers of the Rabi frequencies.

Typically the initial population resides in state 1, at one end of the chain. Then for the ladder system, in which the energies increase steadily, $E_n > E_{n-1}$, the detunings are

$$\hbar\Delta_n(t) = E_n + V_{nn}(t) - [E_1 + V_{11}(t) + \hbar\omega_1 + \cdots + \hbar\omega_{n-1}], \tag{15.28}$$

with $\Delta_1(t) = 0$. The element $\Delta_n(t)$ is the *cumulative detuning* from state 1 after $n-1$ excitation steps. Resonance occurs between state 1 and state n when $\Delta_n = 0$, implying an $(n-1)$-photon transition. More generally, states n and m are resonant when $\Delta_n(t) = \Delta_m(t)$. This equality may hold only briefly, or may be more permanent during pulsed excitation.

Single frequency. When the field has only a single carrier frequency,

$$\mathbf{E}(t) = \mathrm{Re}\left\{\mathbf{e}\mathcal{E}(t)\, e^{-i\omega t + i\varphi}\right\}, \tag{15.29}$$

then we choose the phase derivatives to be

$$\dot\zeta_n(t) = \begin{cases} \dot\zeta_{n-1} + \omega & \text{if } E_n > E_{n-1}, \\ \dot\zeta_{n-1} - \omega & \text{if } E_n < E_{n-1}. \end{cases} \tag{15.30}$$

The result, for the ladder system with $E_n > E_{n-1}$, is the sequence of phase derivatives

$$\hbar\dot\zeta_n(t) = E_1 + V_{11}(t) + (n-1)\hbar\omega \tag{15.31}$$

and the detuning energies

$$\hbar\Delta_n(t) = E_n + V_{nn}(t) - \hbar\dot\zeta_n = E_n + V_{nn}(t) - E_1 - V_{11}(t) - (n-1)\hbar\omega. \tag{15.32}$$

The requirement for resonance between state 1 and state n, in the absence of dynamic shifts $V_{nn}(t)$, is that the frequency satisfy the multiphoton resonance condition for $n-1$ photons,

$$E_n = E_1 + (n-1)\hbar\omega. \tag{15.33}$$

The chain RWA Hamiltonian. Whatever the nature of the excitation, the RWA Hamiltonian of a multistate chain is a tridiagonal matrix that has the diabatic energies (the detunings) on its diagonal, and the laser-induced couplings (half the Rabi frequencies) as off-diagonal elements. When these are real-valued the matrix is *tridiagonal*, i.e. it has the form

$$
\mathsf{W}(t) = \frac{1}{2}
\begin{bmatrix}
2\Delta_1 & \Omega_1 & 0 & \cdots & 0 & 0 \\
\Omega_1 & 2\Delta_2 & \Omega_2 & \cdots & 0 & 0 \\
0 & \Omega_2 & 2\Delta_3 & \cdots & 0 & 0 \\
\vdots & \vdots & \vdots & \ddots & \vdots & \vdots \\
0 & 0 & 0 & \cdots & 2\Delta_{N-1} & \Omega_{N-1} \\
0 & 0 & 0 & \cdots & \Omega_{N-1} & 2\Delta_N
\end{bmatrix}.
\tag{15.34}
$$

In general all elements are time dependent, although that is not shown explicitly here. A (multiphoton) resonance occurs between states n and m whenever $\Delta_n = \Delta_m$. Typically we choose phases such that $\Delta_1 = 0$; then resonance occurs between state 1 and state n when $\Delta_n = 0$.

Ladders and bent chains. As recognized in the phase choices of eqns. (15.22) and (15.30) the energies of the chain need not be ordered as in a ladder, with $E_n > E_{n-1}$; in the RWA the ordering of the energy levels is irrelevant. Figure 15.2 shows examples of some excitation patterns that are possible with a four-state chain: a simple ladder and two "bent" patterns. Within the RWA the linkages of this figure are equivalent.[1] For five states the choices include the "letter-M" or "letter-W" patterns as well as generalized lambda and vee patterns.

15.3.2 Analytic solutions

When the RWA Hamiltonian is constant there exist known analytic solutions for a number of Rabi-frequency sequences, both fully resonant and with detuning. The connection with conventional special functions and classical polynomials occurs through the three-term recurrence relationship that follows from the tridiagonal form of the RWA Hamiltonian when one seeks the eigenstates and eigenvalues [Bia77] [Sho90, §15.3].

The solutions for a constant RWA Hamiltonian can be used for pulsed excitation whenever all of the elements of the matrix $\mathsf{W}(t)$ have the same time dependence, as in eqns. (15.10)–(15.12). These solutions allow extension, to N-state chains, of the two-state pi pulses.

Adiabatic population transfer along a multistate chain has also been studied. Interesting differences occur between chains having an even number of states, e.g. the $N = 4$ linkages shown in Fig. 15.2, and those with an odd number of states, e.g. the $N = 5$ linkage of the "letter-M" pattern of Fig. 15.5(a). In the odd-N chains it is possible to implement a variant of the STIRAP procedure to produce complete population transfer between terminal states of the chain [Sho91a; Mar91; Vit98b; Vit99b].

---

[1] What matters, in addition to the values of the RWA Hamiltonian elements, is the graph structure.

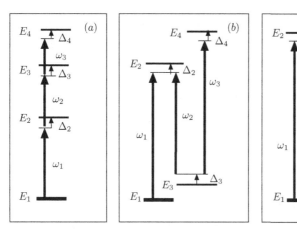

Figure 15.2 Some possible orderings of energies along a four-state chain. (*a*) A ladder, in which $E_n > E_{n-1}$ for all n. (*b*) A "letter-N" pattern, in which E_3 is less than both E_2 and E_4. (*c*) A "lambda" pattern in which E_2 is the largest energy.

15.3.3 Population flow

The simplest examples of multistate excitation are for a fully resonant chain of N states, i.e. $\Delta_n = 0$ for all n. Figure 15.3 presents a comparison of the three types of Rabi-frequency sequences discussed in the present section, for an $N = 10$-state chain: (*a*) uniform Rabi, (*b*) pseudospin (see Sec. 15.3.5), and (*c*) harmonic oscillator (see Sec. 15.3.4). Of these cases, only the pseudospin model is exactly periodic, allowing complete population transfer into the last state of the chain.

Plots of the population histories $P_n(t)$ vs. n at fixed times t resemble, during the first stages of excitation, fluid flow (a wavepacket) along the chain [Sho90, §15.5]. Figure 15.4 shows two examples, for the uniform Rabi-frequency sequence, frame (*a*), and for the harmonic oscillator, frame (*b*).

Although during the early stages of excitation the probabilities are localized in the initial state and its immediate linkage neighbors, the time-varying population distribution along a chain does not always appear as a localized wavepacket. The precise behavior depends upon the sequences of Rabi frequencies. For some choices the wavepacket may undergo such severe alteration as to be unrecognizable after a short time. Other sequences of Rabi frequencies, the pseudospin, produce exactly periodic behavior, in which the initial state repeatedly receives all the population, albeit with a phase change; see Sec. 15.3.5.

Although the RWA Hamiltonian has the same tridiagonal structure for any chain, the dynamics will differ significantly if the population starts in one of the intermediate states of the chain rather than a terminal state (1 or N). As an example, the five-state "letter M" linkage pattern of Fig. 15.5(*a*) places the initial population into state 1, whereas the "letter-W" linkage of Fig. 15.5(*b*) places initial population into one of the intermediate states of

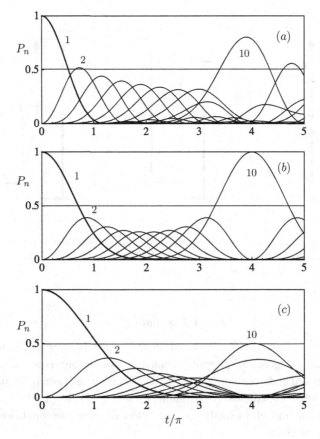

Figure 15.3 Population histories P_n vs. t/π for fully resonant $N = 10$-state ladder with constant Rabi frequencies. (*a*) Uniform Rabi frequencies, $\Omega_n = 1$. (*b*) Pseudospin model, $\Omega_n = \sqrt{n(N-n)}/4$. (*c*) Harmonic oscillator, $\Omega_n = \sqrt{n}/2$.

the chain, from which it departs along two distinct paths; subsequent interference occurs when population returns.

Initial state is at chain end. When the detunings are all resonant, population flows from the initial state 1 along the chain to the end of the chain, whereupon it returns (but only with specific choices for Rabi frequencies will this be complete and hence periodic). Figure 15.6(*a*) presents an example of such resonant behavior, for excitation in which all Rabi frequencies are equal. The two frames show different initial conditions: state 1, in frame (*a*), and state 2, in frame (*b*), hold all initial population.

Initial state is not the chain end. As mentioned in the discussion of the "vee" linkage, when population starts in an intermediate state of the chain it can flow initially in two directions.

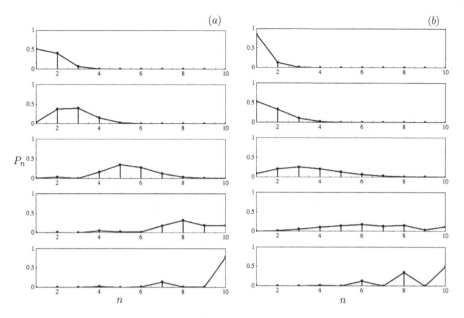

Figure 15.4 Snapshots of probability histograms, $P_n(t)$ vs. n at five equidistant times. (a) Uniform Rabi frequencies, $\Omega_n = 1$. (b) Harmonic oscillator, $\Omega_n = \sqrt{n}/2$.

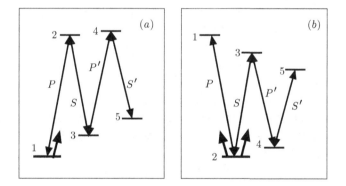

Figure 15.5 Five-state chains. (a) Arranged in "letter-M" pattern, showing links from fields P, P', S, and S'. (b) The "letter-W" linkage. Here population initially resides in one of the intermediate states, and proceeds initially toward two states, as indicated by short arrows; subsequently interference occurs. In both cases the vertical positions of the energy levels in the diagram are irrelevant for the RWA linkage.

Figure 15.6(b) illustrates an example, for the "letter-W" system. Note that, just as in frame (a), the population in state 3 varies sinusoidally. In general the N-state solutions (with constant RWA Hamiltonian) are not periodic; Sec. 15.3.5 discusses an exception.

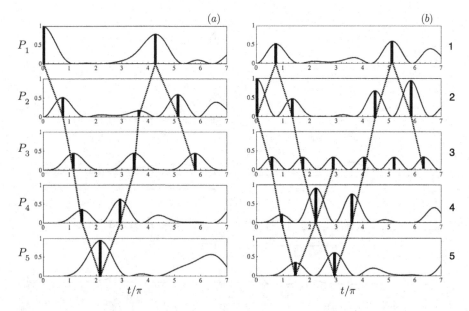

Figure 15.6 Population histories $P_n(t)$ vs. t/π for the five-state ladder, fully resonant with constant uniform Rabi frequencies $\Omega_n = 1$, for two different initial conditions. (*a*) Initial population in state 1. Dashed lines, connecting vertical bars, show time variation of peak populations; these increase nearly linearly with time to the end of the chain, whereupon they reverse. Note that population does not return completely to the initial state 1; the population changes are not periodic (except for state $n = 3$ of $N = 5$). (*b*) Initial population in state 2. The population initially flows into states 1 and 3. [Redrawn from Figs. 62 and 63 of B. W. Shore, Acta Physica Slovaka, **58**, 243–486 (2008) with permission.]

15.3.4 The harmonic oscillator

The harmonic oscillator, driven by dipole coupling of arbitrary time dependence, is one of the soluble quantum systems widely studied [Sho90, §15.1]. It has an infinite number of evenly spaced nondegenerate energy levels, and so a single laser field will have an RWA Hamiltonian with detunings

$$\Delta_n = n\Delta_1. \tag{15.35}$$

The linkage pattern is tridiagonal (i.e. nearest-neighbor couplings), with Rabi frequencies that increase monotonically,

$$\Omega_n(t) = \sqrt{n}\,\Omega_1(t). \tag{15.36}$$

When the field is steady, population placed into state 1 will pass successively toward states of higher excitation. Eventually the detuning, if present, will become so large that the next-step Rabi frequency will not produce further excitation; states that are further along the chain will receive small or negligible excitation. The turning point for the excitation occurs when the cumulative detuning Δ_n exceeds the Rabi frequency Ω_n [Sho90, §15.10]. The population return to the initial state is not complete; the behavior is not periodic.

When the excitation is resonant in the first step, it will be resonant for all subsequent steps, and population will pass toward ever-higher excitation. This progress must eventually end for any real system. Typically this occurs because the system has anharmonicity, so that the energy levels become closer together with increasing excitation. Then the cumulative detuning will eventually overcome the Rabi frequency, and excitation will progress no further. Alternatively, the excitation will lead to photoionization or dissociation and consequent probability loss from the discrete states.

15.3.5 The pseudospin model

An interesting multistate excitation chain, sometimes termed the *Cook–Shore model* or *pseudospin model*, occurs when there are N states, coupled only between adjacent states so that the RWA Hamiltonian is tridiagonal, and with elements [Maj32; Coo79; Hio87] [Sho90, §18.6]

$$W_{nn} = n\Delta_1 + \Delta_0, \qquad W_{n+1,n} = \tfrac{1}{2}\check{\Omega}_0\sqrt{n(N-n)}. \tag{15.37}$$

Here $\check{\Omega}_0$ is allowed to be complex valued. The sequences of Rabi frequencies are slightly larger at the center of the chain than at the ends, for example

$$
\begin{aligned}
N &= 2: & &\check{\Omega}_0\{1\}, \\
N &= 3: & &\check{\Omega}_0\{\sqrt{2},\ \sqrt{2}\}, \\
N &= 4: & &\check{\Omega}_0\{\sqrt{3},\ \sqrt{4},\ \sqrt{3}\}, \\
N &= 5: & &\check{\Omega}_0\{\sqrt{4},\ \sqrt{6},\ \sqrt{6}\ \sqrt{4}\}.
\end{aligned}
$$

The structure of these matrix elements is identical to what occurs when expressing the interaction of a magnetic moment (proportional to spin angular momentum $\mathbf{S}$) with a steady magnetic field $\mathbf{B}$. That is, the interaction is expressible in the form of a scalar product,

$$\mathsf{W} = \Upsilon_x \mathsf{S}_x + \Upsilon_y \mathsf{S}_y + \Upsilon_z \mathsf{S}_z \equiv \boldsymbol{\Upsilon}\cdot\mathbf{S}. \tag{15.38}$$

The three matrices S_k are three $N \times N$ matrix representations of the group $SU(N)$, i.e. spin matrices in N dimensions, see App. A.1.3. The part of the magnetic vector is here taken by the vector $\boldsymbol{\Upsilon}$, defined by the three Cartesian components

$$\Upsilon_x = \mathrm{Re}\,(\check{\Omega}_0) \equiv \Upsilon_R, \quad \Upsilon_y = \mathrm{Im}\,(\check{\Omega}_0) \equiv \Upsilon_I, \quad \Upsilon_z \equiv \Delta_1, \tag{15.39}$$

in an abstract three-dimensional space. This form of interaction produces a rotation of the statevector, an N-dimensional analog of the rotation of the Bloch vector. The magnitude of the angular velocity vector,

$$\Upsilon = |\boldsymbol{\Upsilon}| = \sqrt{\Delta_1^2 + |\check{\Omega}_0|^2}, \tag{15.40}$$

defines a rotation rate: it is the rms value of the detuning Δ_1 and the Rabi frequency Ω_0. That is, the system is equivalent to a spin angular momentum s, such that the number of states is $N = (2s+1)$. We identify the nth state (with $n = 1, 2, \ldots, N$) as being a particular magnetic substate, labeled by the eigenvalue μ (with $\mu = -s, \ldots, +s$) of the matrix S_z. The correspondence is

$$N = 2s + 1, \qquad\qquad s = \frac{1}{2}(N - 1), \tag{15.41}$$

$$n = \mu + s + 1, \qquad\qquad \mu = n - \frac{1}{2}(N + 1). \tag{15.42}$$

Eigenvalues. The RWA Hamiltonian can be diagonalized by rotating the vector Υ onto the vertical axis, using the two Euler angles α and β,

$$\tan(\alpha) = \mathrm{Im}(\check{\Omega}_0) / \mathrm{Re}(\check{\Omega}_0), \qquad\qquad \tan(\beta) = |\check{\Omega}_0| / \Delta_1. \tag{15.43}$$

In this coordinate system the Hamiltonian appears as a constant multiple of the angular momentum operator $\mathsf{S}_{z'}$

$$\mathsf{W}' = \Upsilon \mathsf{S}_z. \tag{15.44}$$

Because the spin matrix has eigenvalues that are evenly spaced, separated by unity, it follows that the eigenvalues of W have the form

$$\varepsilon_\mu = \mu \Upsilon, \tag{15.45}$$

where μ takes integer (or half-integer) values ranging from $-s$ to $+s$ in unit steps. Thus all the eigenvalues of W are multiples of the basic frequency unit Υ, and so the eigenstates are completely periodic.

Eigenstates. The eigenstates of W are angular momentum states in the rotated coordinate system. Referred back to the original basis they are, with Dirac notation,

$$|\alpha, \beta; s, \mu\rangle = \sum_{\mu'} |s, \mu'\rangle \, \mathcal{D}^{(s)}_{\mu'\mu}(\alpha, \beta, 0). \tag{15.46}$$

Here $\mathcal{D}^{(s)}_{\mu'\mu}(\alpha, \beta, 0)$ is the rotation matrix of order s parametrized by two Euler angles α, β, see App. A.1.4. The time evolution is evaluated by expressing the state in rotated coordinates. The result is

$$C_n(t) = \sum_{\mu''n'} \mathcal{D}^{(s)}_{\mu\mu''}(\alpha, \beta, \Upsilon t) \, \mathcal{D}^{(s)}_{\mu'\mu''}(\alpha, \beta, 0)^* C_{n'}(0), \tag{15.47}$$

where

$$s = 2N + 1, \qquad \mu \equiv n - \frac{1}{2}(N + 1), \qquad \mu' \equiv n' - \frac{1}{2}(N + 1). \tag{15.48}$$

This equation presents an exact analytic expression for the time dependence of the probability amplitude $C_n(t)$ for fixed parameters Ω_0 (Rabi frequencies) and Δ_1 (detunings). It involves a rotation operator $\mathcal{D}^{(S)}(\alpha, \beta, \Upsilon t)$ descriptive of a coordinate system turning steadily at the rate $\Upsilon = \sqrt{\Delta_1^2 + |\Omega_0|^2}$. When detuning is absent, the vector Υ lies in the x, y plane. When the Rabi frequency Ω_0 is real, Υ lies in the x, z plane. In particular, when Ω_0 is real and there is no detuning, the vector lies along the x axis, and time evolution

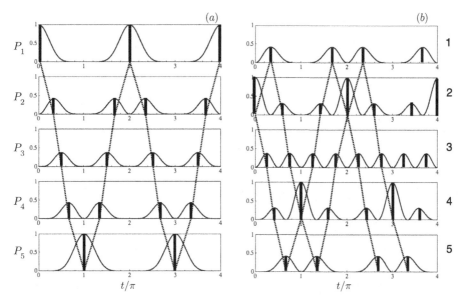

Figure 15.7 Resonant population flow for five-state pseudospin model, $\Omega_n = \sqrt{n(N-n)}$. Plots are of $P_n(t)$ vs. t. Vertical bars mark the times at which populations peak; the population histories are periodic. (*a*) Starting in state 1. (*b*) Starting in state 2. [Redrawn from Fig. 64 of B. W. Shore, Acta Physica Slovaka, **58**, 243–486 (2008) with permission.]

amounts to rotation about this axis at the usual Rabi frequency. In all cases the behavior is periodic – as can be recognized from the fact that the eigenvalues are all multiples of a common frequency.

Figure 15.7 illustrates an example of the perfect periodicity observed with the pseudospin model, for $N = 5$ and no detuning. The two frames illustrate different initial conditions: frame (*a*) shows the population histories starting from state 1. In frame (*b*) the population is initially in state 2.

Pulsed excitation. The treatment of a static RWA Hamiltonian presented above generalizes to pulsed excitation in which the three coefficients Υ_i of eqn. (15.38) become time dependent [Hio87]:

$$W(t) = \sum_{i=x,y,z} \Upsilon_i(t) S_i. \tag{15.49}$$

The fundamental example is that of $N = 2$, or spin $s = 1/2$, for which the TDSE reads, using the phase choice of eqn. (11.11),

$$\frac{d}{dt} \begin{bmatrix} C_{-1/2} \\ C_{1/2} \end{bmatrix} = -\frac{i}{2} \begin{bmatrix} -\Delta_0(t) & \check{\Omega}_0(t) \\ \check{\Omega}_0(t)^* & \Delta_0(t) \end{bmatrix} \begin{bmatrix} C_{-1/2} \\ C_{1/2} \end{bmatrix}, \tag{15.50}$$

with arbitrary time dependence for the detuning $\Delta_0(t)$ and complex-valued Rabi frequency $\check{\Omega}_0(t)$. The two-state (spin $s = \frac{1}{2}$) time-evolution matrix can be expressed as

$$\mathsf{U}^{(2)} = \begin{bmatrix} a(t) & b(t) \\ -b(t)^* & a(t)^* \end{bmatrix}, \qquad |a(t)|^2 + |b(t)|^2 = 1, \tag{15.51}$$

where $a(t)$ and $b(t)$ depend on the particular choice of detuning and Rabi-frequency pulses. Given functions $a(t)$ and $b(t)$ for the two-state system, we obtain the evolution matrix for the N-state system as

$$U^{(N)}_{m,m'} = D^{(s)}_{m,m'}[a(t), b(t)], \tag{15.52}$$

where $D^{(s)}_{m,m'}[a(t), b(t)]$ is the $N = 2s + 1$ dimensional representation of the unitary group $U(N)$, expressible as powers (of order $N - 1$) of $a(t)$ and $b(t)$ and their complex conjugates. This procedure, presented by Hioe [Hio87], provides an analytic solution for the N-state pseudospin system in terms of solutions to the two-state system. In particular, if the system is initially in state $m = -s$ at one end of the chain, then the probability of later finding state m is

$$|C_m(t)|^2 = \binom{2s}{s-m} |a(t)|^{2(s-m)} |b(t)|^{2(s+m)}. \tag{15.53}$$

Thus the condition for complete population transfer in N levels coincides with the condition for complete transfer in the two-state problem.

15.3.6 Time-averaged populations

Observations that take place during many Rabi cycles, yet still within a time interval that allows little loss of probability or coherence, sample time-averaged populations. Such a situation also occurs when we have an ensemble of atoms, each of which undergoes many Rabi cycles, but with an appreciable variation in the Rabi angle – as happens when the different atoms are illuminated for different times. We calculate the appropriate probabilities from the averaging integral

$$\bar{P}_n(T) = \frac{1}{T} \int_{t_0}^{T} dt\, |C_n(t)|^2 \tag{15.54}$$

as the limit, over many Rabi cycles,

$$\bar{P}_n = \lim_{T \to \infty} \bar{P}_n(T). \tag{15.55}$$

When the illumination is steady the desired values can be obtained from the eigenvalues and eigenvectors of the RWA Hamiltonian that governs the time evolution [Sho90, §15.9]. From eqn. (15.14) we find that, for population initially in state m and nondegenerate eigenvalues, the time-averaged populations are

$$\bar{P}_n = \sum_{\nu} |\Phi_{\nu n}|^2 |\Phi_{\nu m}|^2. \tag{15.56}$$

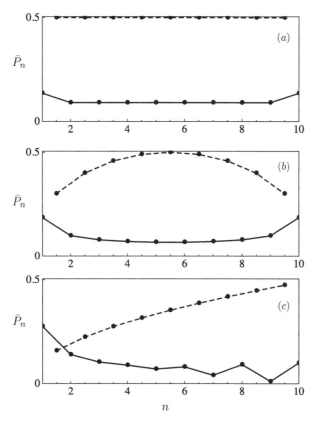

Figure 15.8 Time-averaged populations $\bar{P}_n$ (thick lines) and relative Rabi frequencies (dashed lines) vs. n for the cases of Fig. 15.3. The detunings are all zero. The Rabi frequencies are plotted between the states they connect.

A simple case is a fully resonant chain (a tridiagonal RWA Hamiltonian with null diagonal elements) in which all the Rabi frequencies are constant and equal. When the initial state is m the average probability for state n is [Sho90, §15.9]

$$\bar{P}_n = \frac{1}{N+1}\left[1 + \frac{1}{2}\delta_{n,m} + \frac{1}{2}\delta_{n,N+1-m}\right]. \tag{15.57}$$

The probability is uniformly distributed amongst all states, as befits chaotic behavior, apart from an enhancement (by a factor $3/2$) of the initial state m and its "mirror image" $N+1-m$. When the system is initially at one end of the excitation chain, then it is the upper end of the chain that has this enhancement.

Figure 15.8 shows calculated time-averaged populations for the three cases of Fig. 15.3. In each case, because the initial population is in state 1, the averages are largest at the ends of the chain. The figure also shows the relative values of the Rabi frequencies. The population tends to remain longer where the Rabi frequencies are small. The harmonic oscillator is notable for the very low probability in the penultimate state. Further examples of the relationship between Rabi frequencies and time averages will be found in [Sho90].

When plotted as a function of detuning, such time-averaged populations reveal the presence of resonances: particular choices of detuning for which two states have averaged probabilities of $\bar{P}_n \approx \frac{1}{2}$ [Sho90, §15.11]. These may be adjacent states of the chain or, more generally, they may be multiphoton resonances, as discussed in Sec. 15.3.8.

15.3.7 Detuned chains

In general there is no simple pattern to the energy levels and consequently no simple pattern to the choice of carrier frequencies. However, several models do have simple energy-level sequences in chains. A simple example is the energy levels of a harmonic oscillator, whose evenly spaced energies are parametrized by the integer vibrational quantum number v,

$$E_v^{vib} = \hbar\omega_e(v + \tfrac{1}{2}), \qquad v = 0, 1, 2, \dots. \tag{15.58}$$

The vibrations of real molecules do not follow this simple energy pattern for large v. Instead, the potential for large motions departs from quadratic form; the motion becomes anharmonic. The vibrational energies of an *anharmonic* oscillator are typically written as a power series in $(v + \tfrac{1}{2})$ [Her50a],

$$E_v^{vib} = \hbar\omega_e\left(v + \tfrac{1}{2}\right) - \hbar\omega_e x_e\left(v + \tfrac{1}{2}\right)^2 + \hbar\omega_e y_e\left(v + \tfrac{1}{2}\right)^3 + \cdots, \tag{15.59}$$

where x_e and y_e are small dimensionless constants. Reexpressed in terms of the numbering convention used in the present work the energies, within a quadratic approximation, follow the pattern

$$E_n = E_1 + (n-1)\hbar\omega_B - (n-1)(n-2)\hbar a, \tag{15.60}$$

where $a = \omega_e x_e$ is an anharmonicity parameter and

$$\omega_B \equiv \omega_e - 2\omega_e x_e \tag{15.61}$$

is the first-step Bohr transition frequency. With these definitions the detunings read

$$\Delta_n = (n-1)[\Delta - (n-2)a], \tag{15.62}$$

where $\Delta = \Delta_2 = \omega_B - \omega$ is the detuning of the carrier frequency from the first-step Bohr frequency.

When the anharmonicity a is small, and the laser is resonant with the first excitation step, $\Delta = \omega_B - \omega = 0$, population will proceed along the chain until the cumulative anharmonicity becomes somewhat larger than the Rabi frequency for the next step of excitation. States for larger n will have little population. One can increase the excitation by increasing the laser intensity. Figure 15.9 illustrates this behavior [Sho90, §15.11]. For Rabi frequencies smaller than the anharmonicity parameter a little population reaches states beyond $n = 2$. As the Rabi frequency increases, excitation proceeds further along the chain.

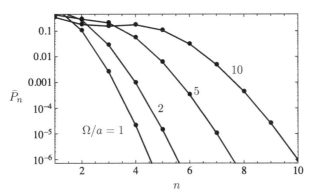

Figure 15.9 Time-averaged populations $\bar{P}_n$ vs. n for ten-state anharmonic oscillator for several values of the ratio of Rabi frequency Ω to anharmonicity a. As the Rabi frequency increases, excitation proceeds further along the excitation chain.

15.3.8 Multiphoton resonances

Rather than increase the laser intensity to access higher-excitation states, one can instead shift the laser frequency to allow a resonant transition between state 1 and state n. Between state 1 and state n a resonance occurs (meaning $\Delta_n = 0$) when the carrier frequency ω is equal to the Bohr frequency shifted by $n - 2$ anharmonicity units a,

$$\omega = \omega_B - (n-2)a, \quad \text{or} \quad \Delta = (n-2)a. \tag{15.63}$$

Figure 15.10 illustrates the multiphoton excitations of an anharmonic ladder.[2] Excitation is only appreciable when the resonance condition $\Delta_n = 0$ holds.

When the laser frequency matches the first-step Bohr frequency ω_B population flows from state 1 to state 2 and back. Over many Rabi cycles the two probabilities are approximately equal. (Some population resides, on average, in states of higher excitation.) If one shifts the laser frequency toward smaller values the resonance condition $\Delta_2 = 0$ no longer holds, and the population remains confined to state 1. But if the laser shift is equal to the anharmonicity parameter a, then population flows between state 1 and state 3. Because the dynamics is that of a two-photon transition, the effective Rabi frequency is smaller and the cycling is slower. The resonance width is consequently smaller than for the 1–2 transition.

If the laser frequency is moved toward still smaller values, a succession of resonances, separated by the anharmonicity a, are encountered [Sho90, §15.11]. These are successively narrower, in keeping with the diminishing Rabi frequency for higher-order multiphoton transitions.

Figure 15.11 illustrates examples of these multiphoton transitions for the ten-state ladder. The figure shows time-averaged populations for steady lossless coherent excitation, and exhibits how, as the multiphoton order increases from one-photon to n-photon, the resonance

---

[2] Were the final state in the continuum these would be examples of REMPI.

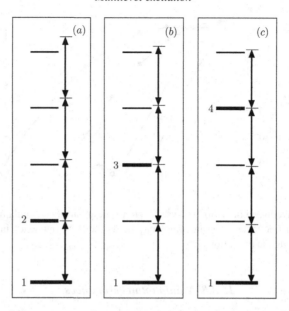

Figure 15.10 Single-frequency excitation of anharmonic ladders. The energy levels grow closer together with increasing excitation. To achieve resonance with state n it is necessary to adjust the frequency to achieve $(n-1)$-photon resonance. (*a*) One-photon resonance between states 1 and 2. (*b*) Two-photon resonance between states 1 and 3. (*c*) Three-photon resonance between states 1 and 4.

width diminishes. That is, the range of detunings that produce appreciable excitation to state n grows dramatically smaller as n grows larger. Although the calculations here are for coherent excitation, the conclusions have relevance to incoherent excitation. In those situations the requirement for resonant excitation is that the multiphoton line width (the inverse of the multiphoton Rabi frequency) must be less than the radiative lifetime of the excited state. With each additional excitation step this requirement becomes more restrictive. This property of multiphoton resonance allows highly selective excitation; for example, selectively detecting trace elements or trace isotopes [Sho03; Lu03; Lin04].

15.3.9 Two-state behavior in an N-state chain

Some N-state chain-linkage systems, describable by a constant tridiagonal RWA Hamiltonian, exhibit dynamics of a simple two-state system, characterized by Rabi oscillations between only two states, with negligible excitation present in other states [Sho79; Sho81a]. The two states may be the terminal states 1 and N; in general they may be any pair of states separated by one or more intermediate states. The characteristics of the Hamiltonian that lead to this simplification are recognizable from the pattern of detunings Δ_n and Rabi frequencies Ω_n.

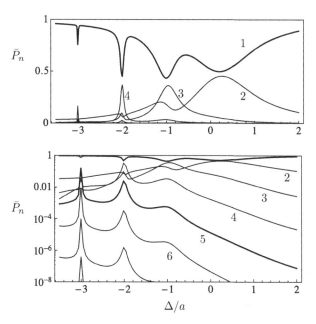

Figure 15.11 Time-averaged populations $\bar{P}_n$ vs. detuning Δ in units of anharmonicity parameter a. Upper frame shows linear scale, lower frame shows log scale. Resonance between state 1 and state n occurs for $\Delta/a = (n-2)a$; there is a succession of resonances, successively narrower.

Large intermediate detunings. One such case occurs when the intermediate cumulative detunings Δ_n, those for $1 < n < N$, are all much larger than any of the various Rabi frequencies or the final-step cumulative detuning δ [Sho90, §18.7]:

$$\Delta_n = \{0, \Delta_2, \Delta_3, \ldots, \Delta_{N-1}, \delta\}, \quad \Omega_n = \{\Omega_1, \Omega_2, \ldots, \Omega_{N-1}\}. \tag{15.64}$$

The system then behaves, for initial population in either state 1 or state N, as a two-state system, with effective detuning and effective Rabi frequency given as the constructions

$$\Delta^{\text{eff}} = \delta + \frac{|\Omega_{N-1}|^2}{4\Delta_{N-1}} - \frac{|\Omega_2|^2}{4\Delta_2}, \quad \Omega^{\text{eff}} = \frac{1}{2^{N-2}} \frac{\Omega_1 \Omega_2 \cdots \Omega_{N-1}}{\Delta_2 \Delta_3 \cdots \Delta_{N-1}}. \tag{15.65}$$

The two-state system undergoes an $(N-1)$-photon excitation, subject to dynamic Stark shifts expressed by the effective detuning Δ^{eff}.

Small terminal Rabi frequencies. A second example [Sho79; Sho81a] occurs when each step of the chain is resonant, except possibly the last, and the two chain-terminating Rabi frequencies Ω_1 and Ω_N are much smaller than all the others [Sho90, §18.7], as in the patterns

$$\Delta_n = \{0, 0, 0, \ldots, 0, \delta\}, \qquad \Omega_n = \{s_1, L_1, L_2, \ldots, L_{N-2}, s_{N-1}\}. \tag{15.66}$$

When the number of states N is an even integer and the large elements are much larger than the small elements, $|L_n| \gg |s_k|$, the dynamics is that of a two state system that has the original N-state detuning δ and has an effective Rabi frequency that is the ratio of products of odd-numbered Rabi frequencies to even-numbered frequencies,

$$\Delta^{\mathrm{eff}} = \delta, \qquad \Omega^{\mathrm{eff}} = \frac{s_1 L_3 L_5 \cdots L_{N-3} s_{N-1}}{L_2 L_4 \cdots L_{N-2}}. \tag{15.67}$$

Only for an even number of states will two-state behavior occur.

Choice of paired states. The restriction to the first and last states of a chain, as assumed above, is not essential. We can add additional links before and after these two states without altering the formula for the effective Rabi frequency. However, such bordering states will alter the effective detuning by contributing additional dynamic Stark shifts [Sho79; Sho81a].

15.3.10 Generalized pi pulses

When there occurs a resonant multiphoton transition between the initial state 1 and an excited state n of a chain, as illustrated in Fig. 15.11, the population will periodically be placed into, and return from, the excited state. The excited state need not be the terminus of the chain, nor does the RWA Hamiltonian need to be constant, so long as all elements have the same time dependence; see Sec. 15.2.2 above. In the absence of detuning (including dynamic Stark shifts) the excitation probability of the effective two-state system for a pulse of duration T is expressible as

$$\mathcal{P}_n = \sin^2\left[\tfrac{1}{2}\mathcal{A}(T)\right], \tag{15.68}$$

where the generalized area $\mathcal{A}(T)$ is the integral of the effective two-state Rabi frequency,

$$\mathcal{A}(T) = \int_0^T dt\, \Omega^{\mathrm{eff}}(t). \tag{15.69}$$

This integral can be expressed in terms of the adiabatic energies associated with the two states [Hol94]

$$\mathcal{A}(T) = \int_0^T dt\, [\varepsilon_1(t) - \varepsilon_n(t)]. \tag{15.70}$$

Complete population transfer occurs whenever $\mathcal{A}(T)$ is an odd-integer multiple of π, a *generalized pi pulse* that is independent of pulse shape [Hol94].

15.4 Branches

As the number of states increases, so too do the number of ways in which these can be linked. For four or more states the RWA Hamiltonian may describe not only simple chains but branched linkages, in which more than two states connect with a particular state[3] [Sho90, §21.4]. The properties of branched chains have consequences that seem quite surprising when first seen, e.g. a branch can completely block population flow along a chain [Ein79; Sho84]. The simplest system in which branches can occur is one with four states and three fields, termed a tripod.

15.4.1 The tripod linkage

The inclusion of one additional linkage from the excited state, using a third field, produces a *tripod* linkage pattern. Figure 15.12 illustrates equivalent examples of a four-state branched linkage, in which four quantum states are linked by three independent radiative transitions to a single excited state, but not amongst themselves.[4]

The linkage patterns of the three frames of Fig. 15.12 all lead to an interaction matrix W having the form of the first of the two matrices below; by reordering the states, or renumbering them, one obtains the second form, a bordered matrix:

$$
\begin{bmatrix}
. & X & . & . \\
X & . & X & X \\
. & X & . & . \\
. & X & . & .
\end{bmatrix}
\begin{matrix} 1 \\ 2 \\ 3 \\ 4 \end{matrix}
\quad , \quad \text{or} \quad
\begin{bmatrix}
. & X & X & X \\
X & . & . & . \\
X & . & . & . \\
X & . & . & .
\end{bmatrix}
\begin{matrix} 2 \\ 1 \\ 3 \\ 4 \end{matrix}
\quad . \tag{15.71}
$$

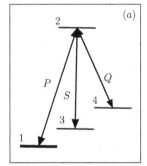 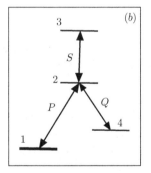 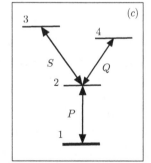

Figure 15.12 Examples of equivalent branched linkages for $N = 4$ states and three fields. The three states (1, 3, 4) each are linked to a single common state, 2, by the three independent interactions P, S, and Q. The linkage patterns are: (a) tripod, (b) inverted-Y, (c) letter-Y.

---

[3] The corresponding graphs are *trees*.
[4] As with other RWA linkages, the vertical placement of the several energies is irrelevant.

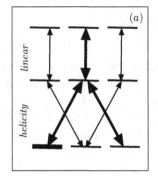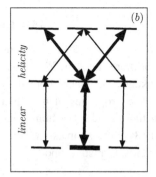

Figure 15.13 Examples of cases (*b*) and (*c*) of Fig. 15.12 found in a degenerate angular momentum sequence $J = 1 \leftrightarrow 1 \leftrightarrow 1$. In each example there occur two independent systems of states, one of which (heavy arrows) is a linkage of Fig. 15.12. (*a*) First step in helicity basis, second step linearly polarized. The linkage of light arrows is a generalized vee of five states. (*b*) Second step in helicity basis, first step linearly polarized. The linkage of light arrows is a generalized lambda of five states.

Angular momentum tripods. The linkages of the tripod pattern can occur with chains of angular momentum states, as illustrated in Fig. 15.13. Here are shown the linkage patterns obtained for excitation along the degenerate three-level chain $J = 1 \leftrightarrow 1 \leftrightarrow 1$, using a combination of helicity states and linear-polarization states of the field, as is appropriate when the two laser beams are perpendicular. The dark linkages are those of frames (*b*) and (*c*) of Fig. 15.12. Lighter lines show the additional linkages, an independent five-state chain, that will occur along with these. By imposing Zeeman (or Stark) shifts on the sublevels and adjusting the three laser frequencies appropriately the lighter-line linkages can be made nonresonant and hence of negligible effect (or they could be made the dominant linkage pattern).

Excitation with loss. A simple example of the dramatic effect branches can have is evident when we consider a lossy system. We first consider the three-state ladder linkage that occurs when the Q field of Fig. 15.12 is absent. Let loss (e.g. photoionization) occur from the uppermost state 3 at rate Γ. Let the two Rabi frequencies Ω_P and Ω_S be constant and equal. Then population flows along the ladder from initial state 1 to final state 3 with loss during each cycle. If the loss is comparable to the Rabi frequencies, then only a few cycles will be completed before all of the population is lost. Figure 15.14(*a*) shows an example of the population histories for such a situation.

The addition of a second link to the intermediate state, state 4 linked to state 2, makes the linkage pattern that of the tripod system. Now only a portion of the population is lost; the remainder is left untouched by the fields; it is in a dark state that cycles between states 1 and 4. Figure 15.14(*b*) illustrates this history.

As the next subsection shows, there are two degenerate dark states for this system; in this example they are superposed to give the oscillatory behavior.

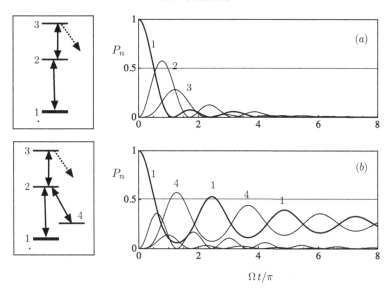

$$\Omega t/\pi$$

Figure 15.14 (*a*) Population histories for three-state system, unit Rabi frequencies, with loss $\Gamma = 0.5$ from uppermost state 3. (*b*) Population histories when there is an additional linkage, unit Rabi frequency, from state 2 to state 4. Population loss is now incomplete; the laser fields create a dark state that remains unaffected by the light. This behavior only occurs when there is a two-photon resonance between states 1 and 4; otherwise all population will eventually be lost. [Redrawn from Fig. 68 of B. W. Shore, Acta Physica Slovaka, **58**, 243–486 (2008) with permission.]

15.4.2 The tripod dark states

The tripod system provides a good example for the use of adiabatic (or dressed) states; in the scenario just discussed the RWA Hamiltonian is constant in time, and these eigenstates have fixed components.

Exact adiabatic eigenstates are known for the tripod system, and their properties have been discussed in several papers [Una98; Una99; Kis01a; Pas02; Kar04; Una04]. Such a pattern occurs with the $J = 1$ to $J = 0$ excitation transition with arbitrary polarization. To simplify some of the analysis it proves useful to assign the state 1 label to the excited state, so that states 2–4 are low-lying stable or metastable states. Figure 15.15 shows the labeling of states and fields used in the following.

To make the needed RWA we introduce a rotating and phased basis, taking the excited-state energy $E_1 = 0$ as the reference energy,

$$
\begin{aligned}
\boldsymbol{\psi}_1'(t) &= \mathrm{e}^{-\mathrm{i}\varphi_P}\boldsymbol{\psi}_1, \\
\boldsymbol{\psi}_2'(t) &= \mathrm{e}^{+\mathrm{i}\omega_P t}\boldsymbol{\psi}_2, \\
\boldsymbol{\psi}_3'(t) &= \mathrm{e}^{+\mathrm{i}\omega_S t}\,\mathrm{e}^{\mathrm{i}(\varphi_P-\varphi_S)}\boldsymbol{\psi}_3, \\
\boldsymbol{\psi}_4'(t) &= \mathrm{e}^{+\mathrm{i}\omega_Q t}\,\mathrm{e}^{\mathrm{i}(\varphi_P-\varphi_Q)}\boldsymbol{\psi}_4,
\end{aligned}
\tag{15.72}
$$

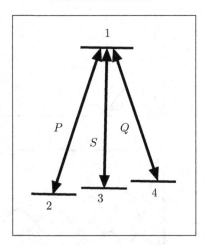

Figure 15.15 Linkage pattern for tripod system, showing labeling convention for states and fields. The states 2–4 need not be degenerate, but they are here assumed to be resonant with the linking field.

where φ_k is the phase of field envelope $\mathcal{E}_k(t)$. This construction makes any superposition of states 2, 3, 4 dependent only on phase differences $\varphi_P - \varphi_S$ and $\varphi_P - \varphi_Q$.[5] We here assume all fields are resonant, so that

$$E_2 + \hbar \omega_P = E_3 + \hbar \omega_S = E_4 + \hbar \omega_Q = E_1, \qquad (15.73)$$

and the RWA Hamiltonian matrix has the bordered form

$$\mathsf{W}(t) = \frac{1}{2} \begin{bmatrix} 0 & \Omega_P(t) & \Omega_S(t) & \Omega_Q(t) \\ \Omega_P(t) & 0 & 0 & 0 \\ \Omega_S(t) & 0 & 0 & 0 \\ \Omega_Q(t) & 0 & 0 & 0 \end{bmatrix}. \qquad (15.74)$$

This Hamiltonian has four eigenstates. Two of them are dark states (lacking a component in the excited state 1). These can be taken as [Una98]

$$\Phi_1^D(t) = \begin{bmatrix} 0 \\ \cos\Theta(t) \\ -\sin\Theta(t) \\ 0 \end{bmatrix}, \quad \Phi_2^D(t) = \begin{bmatrix} 0 \\ \sin\phi(t)\sin\Theta(t) \\ \sin\phi(t)\cos\Theta(t) \\ \cos\phi(t) \end{bmatrix}, \qquad (15.75)$$

---

[5] Unless the several fields derive, via coherent optical manipulations, from a single laser field, these phases are uncontrollable and can be taken to be zero, a choice associated with the arbitrary choice of the moment of zero crossing of each electric field.

where the two angles are

$$\tan\Theta(t) = \frac{\Omega_P(t)}{\Omega_S(t)}, \quad \tan\phi(t) = \frac{\Omega_Q(t)}{\sqrt{\Omega_P(t)^2 + \Omega_S(t)^2}}. \tag{15.76}$$

Consequent to our choice of energy zero point, each of these adiabatic states has a null eigenvalue. Because these states are degenerate one must consider the general dark state, a superposition [Una98]

$$\Phi(t) = D_1(t)\Phi_1^D(t) + D_2(t)\Phi_2^D(t). \tag{15.77}$$

When such a state describes the initial condition, and the Rabi frequencies are constant (or sufficiently slowly varying that the subsequent evolution is adiabatic), then only these two states contribute to $\Psi(t)$. However, they are resonantly coupled and so the coefficients $D_k(t)$ will generally vary with time.

To understand the effect of the degeneracy of dark states, let us assume the initial population resides in state 2, and that this initially coincides with state $\Phi_1(t)$. We further suppose that, after the pulse sequence, we can make the identifications $\Phi_1(t) = \psi_2'(t)$ and $\Phi_2(t) = \psi_3'(t)$. Then the result of adiabatic passage is the superposition

$$\Phi(t) = \cos\Theta_\infty \psi_2'(t) + \sin\Theta_\infty \psi_3'(t), \tag{15.78}$$

where the asymptotic mixing angle is

$$\Theta_\infty = \int_{-\infty}^{+\infty} dt \, \sin\phi(t) \frac{d}{dt}\Theta(t) = -\int_{-\infty}^{+\infty} dt \, \Theta(t) \frac{d}{dt}\sin\phi(t). \tag{15.79}$$

Note that, from the definition of $\psi_3(t)$, the superposition depends on the phase difference $\varphi_P - \varphi_S$ between the S and P fields. When $\omega_P = \omega_S \equiv \omega$ the superposition reads, in the nonrotating basis,

$$\Psi(t) = e^{-i\omega t}\left[\cos\Theta_\infty \psi_2 + \sin\Theta_\infty e^{i(\varphi_P - \varphi_S)}\psi_3\right]. \tag{15.80}$$

Thus the superposition is stationary. The composition of the superposition, parametrized by the mixing angle Θ_∞, does not depend on details of pulse history, only on integral properties of the pulses.

15.4.3 Longer branches

More generally, one may deal with systems in which a central chain is interrupted by one or more branches, each of which comprises one or more links. Interestingly, the effect of such a branch on the main chain depends dramatically on whether the branch has an even or an odd number of elements: an odd-element branch will act to sever a resonant chain [Ein79; Sho84]. Figure 15.16 shows two examples of four-state chains onto which there is, at state 3, a branch, either of one link, (*a*), or of two links, (*b*).

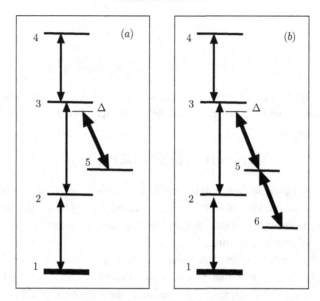

Figure 15.16 Examples of chains with single branches. (*a*) Single link in branch. (*a*) Two links in branch.

Figure 15.17 shows the population histories for the two linkage patterns of Fig. 15.16 for resonant excitation by constant Rabi frequencies when the branch Rabi frequencies are five times as strong as those of the main chain. When there is an odd number of branch states, as in frame (*a*), then the branch acts to sever the chain; population does not proceed further than the branching link, in this case $n = 2$. When there is an even number of links in the branch as in frame (*b*), then population flow proceeds along the original chain.

The explanation for this difference in behavior rests on an examination of the eigenvalues of the RWA Hamiltonian for the states of the branch [Sho84] [Sho90, §21.4]. The weak fields of the main chain act to probe the strongly coupled states of the branch, which include state 3 of the main chain. With an even number of branch links, as in frame (*b*), the branch has an odd number of states, and there will be a null eigenvalue, meaning there is a resonant connection from state 2 to state 3. With an odd number of branch links, as in frame (*a*), there is no such null eigenvalue and the probing of state 3 from state 2, when $\Delta = 0$, lies between peaks of Autler–Townes–type resonances – the link is effectively broken when $\Delta = 0$.

15.4.4 Fan linkages

When one, and only one, state has links to multiple levels the linkage pattern is a *fan*. Figure 15.12 showed examples in which state 2 had such links. Figure 15.18 shows examples of five-state fan systems.

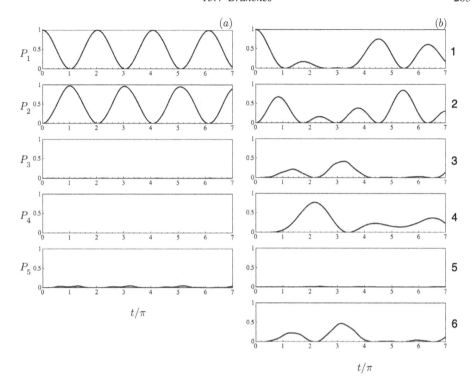

Figure 15.17 Population histories P_n vs. t/π for frames (a) and (b) of Fig. 15.16 with strong Rabi frequencies in the branch links and null detuning, $\Delta = 0$. Main chain Rabi frequencies are $\Omega_n = 1$, $n = 1, 3$. Branch Rabi frequencies are $\Omega_n = 5$, $n = 4, 5$.

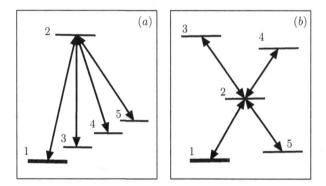

Figure 15.18 Examples of five-state fans in which state 2 is coupled to 4 states. The linkage structure of the Hamiltonian is the same for each of these examples. (a) State 2 has largest energy. (b) State 2 has an intermediate energy; letter-X linkage.

These fan structures, multistate generalizations of the tripod, appear in various ways on an energy diagram, just as in Fig. 15.12. Different choices for the numbering of the states lead to different patterns of the nonzero elements of the interaction matrix. With appropriate numbering of the states it appears as a bordered matrix, as was the case with the four-state tripod of Sec. 15.4.1, see eqn. (15.71).

Both of the patterns in Fig. 15.18 have equivalent structure of the RWA Hamiltonian and will undergo the same dynamics for a given set of Rabi frequencies and detunings. When each field is resonant the interaction matrices have zeros on the diagonal. The following sections discuss some examples.

15.4.5 Quasicontinuum

When treating excitation of a molecule one often encounters a situation in which the laser resonantly excites one particular state which, in turn, couples via nonradiative interactions to a set of excited states whose energies are closely spaced. To treat this situation it is desirable to carry out a preliminary diagonalization of the nonradiative interaction. The result is a set of closely spaced excited states, a quasicontinuum, each of which has radiative links to the ground state [Sho90, Chap. 16.1] [Bix68; Ste72; Ebe82; Sho83; Mil83; Pas00]. Figure 15.19 illustrates the change of linkages that accompany this basis revision. The figure shows only the linkages, not the detunings.

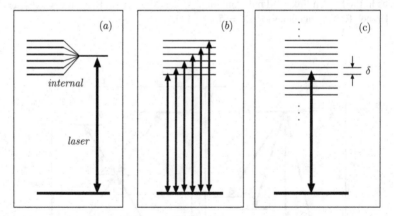

Figure 15.19 Alternative depictions of an excited state linked radiatively to the ground state and nonradiatively to a set of closely spaced levels. (*a*) A single radiative interaction links the ground state to one excited state, which couples nonradiatively to a set of excited states of similar energy. (*b*) Equivalent picture, using a basis of states that diagonalize the nonradiative interaction. The ground state now has links to each excited state; these are not all resonant. (*c*) The quasicontinuum fan linkage, a set of equidistantly spaced excited states each linked to the ground state. This diagram emphasizes the detunings but does not exhibit the individual linkages that couple the ground state to each excited state as in (*b*).

Phase relationships are important in the quasicontinuum model. The ground state excites into a coherent superposition of excited states,

$$\frac{d}{dt}C_1(t) = -\frac{i}{2}\sum_{n=2}^{N}\Omega_n(t)C_n(t), \tag{15.81a}$$

while the excited states each draw population from the same state,

$$\frac{d}{dt}C_n(t) = -i[\Delta_n + \tfrac{1}{2}\Omega_n(t)]C_1. \tag{15.81b}$$

Although practical examples deal with a finite number of excited states, only those closest to resonance are appreciably affected, and so often it is acceptably accurate to consider the limit of an infinite number of excited states, infinitesimally separated, as a model of a true continuum.

Typically the excited states, after diagonalization, are assumed to be evenly separated, as in Fig. 15.19(c) where the energy differences are multiples of $\hbar\delta$. The excited state whose energy is closest to resonance with the pulse carrier is taken as a reference to define the detuning Δ_0. In the RWA the other detunings differ by $m\delta$, where m is an integer, and the RWA Hamiltonian is a banded matrix of the form [Sho83]

$$\mathsf{W} = \frac{1}{2}\begin{bmatrix} 0 & \Omega_1 & \Omega_2 & \Omega_3 & \Omega_4 & \cdots \\ \Omega_1 & 2\Delta_0 & 0 & 0 & 0 & \cdots \\ \Omega_2 & 0 & 2(\Delta_0 + 1\delta) & 0 & 0 & \cdots \\ \Omega_3 & 0 & 0 & 2(\Delta_0 - 1\delta) & 0 & \cdots \\ \Omega_4 & 0 & 0 & 0 & 2(\Delta_0 + 2\delta) & \cdots \\ \vdots & \vdots & \vdots & \vdots & \vdots & \cdots \end{bmatrix}. \tag{15.82}$$

Although the nonradiative interaction amongst excited states may have complicated structure, the Rabi frequencies Ω_n can be taken as real, by suitable choice of basis-state phases [Sho83]. Often they are taken to be equal to a common value, thereby modeling a structureless quasicontinuum, or they are taken to mimic a Lorentzian distribution [Ste72; Mak78].

The use of a single discrete (*interloper state*) coupling to a set of evenly spaced excited states dates back to work of Rice [Ric29], who used it to model molecular predissociation, photodissociation, and unimolecular decomposition. An interloper embedded in a true continuum was treated by Fano in modeling autoionization [Fan61; Fan65]. Although the early uses of a quasicontinuum model primarily regarded this as a means of approximating a continuum, studies starting in the 1970s examined properties of analytic solutions. These found a variety of effects that are not observed with excitation into a true contin-

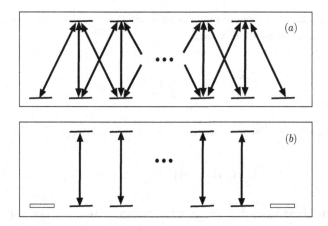

Figure 15.20 (*a*) Angular momentum linkage pattern for general elliptic polarization at arbitrary angle to quantization axis, $J \leftrightarrow J - 1$. (*b*) Equivalent linkage after Morris–Shore transformation. There are $J - 1$ independent pairs of coupled states and two dark states. [From Fig. 73 of B. W. Shore, Acta Physica Slovaka, **58**, 243–486 (2008) with permission.]

uum, such as segments of piecewise nonexponential probability loss from the ground state, recurrences, and quasiperiodicity [Sho90, §16.1]. Phase relationships between statevector components are important in such effects. It is only in an appropriate limit that one finds behavior associated with traditional "golden-rule" rate equations.

15.4.6 Multiple branches: Bright and dark states

As discussed in Sec. 12.4, a simple rotation of the physical coordinates used for labeling directions can produce different linkage patterns in the Hamiltonian matrix. When we deal with two degenerate levels in an angular momentum basis and the polarization is linear, the choice of quantization (z) axis along the polarization axis presents a simple picture of independent two-state linkages, rather than the more elaborate ones of, say, Fig. 14.21, that occur when the polarization is elliptical or is linear but not along the quantization axis. Similarly, right- or left-circular polarization appears, when expressed in a coordinate frame parallel to the propagation direction, as linkages between pairs of states; see Figs. 12.10 and 12.11. That is, for particular polarizations it is possible to choose a quantization axis such that the tripod linkage appears as a two-state linkage.

Such simplification is not possible, within an angular momentum basis, for general elliptical polarization, as represented by an arbitrary point on the Poincaré sphere (see Sec. 12.2.2) and parametrized by two angles, θ describing the relative magnitudes of the two spherical components and ϕ describing the phase between them; see [Few93]. Only specific points on the Poincaré sphere permit the linkage simplification.

Although such simplification does not occur through choice of quantization axis, it is possible to introduce a set of basis states that do produce such two-state simplifications, as shown in Fig. 15.20, for arbitrary polarization, and discussed in App. F.

The basis-state transformation is a generalization of the treatment of the lambda linkage by introducing a combination of states 1 and 3 such that one of them (the dark state) had no linkage with the excited state 2, while the other (the bright state) had all of the oscillator strength of the transition into state 2. The procedure, now known as the *Morris–Shore transformation* [Mor83; Vit00a; Vit03; Kis04; Iva06; Ran06], see App. F, replaces a complicated linkage pattern by independent two-state linkages, together with unlinked *spectator states*. It can be applied whenever the relevant Hamiltonian matrix has the following properties:

I. The complete set of states can be separated into two sets, say, N_A of set A (ground, including the initial state) and N_B of set B (excited). There are no couplings within A set or B set; only couplings between A and B.[6] These may be of any form, not necessarily linking only a single state with at most three other states, as occurs with electric-dipole radiation between angular momentum states.

II. The states within each set are degenerate. In the RWA this means that the states of each set share a single common detuning; overall there are only two detunings, Δ_A and Δ_B. Either or both of these may be zero.

When these conditions hold, it is possible to introduce a new set of basis states, separate superpositions within the A set and within the B set, such that the original Hamiltonian appears in block diagonal form: It consists of a set of two-state submatrices, plus additional unlinked portions that describe spectator states. Appendix F discusses the transformation.

When there are equal numbers of A and B states, $N_A = N_B$, then the resulting transformation produces N_A pairs of two-state interactions, each with a unique Rabi frequency but all sharing the same detunings. When there are more states in set A (the unexcited states), then there are N_B bright states and $N_A - N_B$ dark states.

Figure 15.21 illustrates this transformation for the four-state tripod linkage induced by three independent fields. After the transformation all the dynamics is concentrated in a two-state system, between the excited state e and a bright combination b of states 1, 2, and 3. There are two degenerate dark states d_1 and d_2 that are not directly affected by the field, although their composition does depend on the three fields.

15.5 Loops

When the number of fields equals the number of states, then closed loops may occur in the linkage pattern. Figure 13.2 showed the loop linkage of a three-state system. Figure 15.22 illustrates an example of a single loop in a four-state system. The two frames have identical linkages, a *box* or *diamond* pattern, representable by the same interaction matrix; they differ only in the ordering of the energies (irrelevant within the RWA).

---

[6] The graph corresponding to this linkage pattern is said to be *bipartite*.

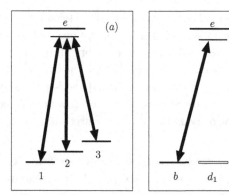

Figure 15.21 (*a*) The four-state tripod linkage. (*b*) Equivalent linkage after transformation into a single bright state *b* and two dark states d_1 and d_2 using the MS transformation. (Vertical locations of the energy levels are irrelevant.)

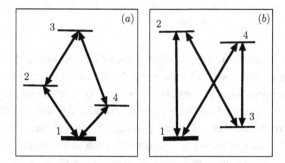

Figure 15.22 Examples of loop linkages for $N = 4$ states and four fields. The two diagrams have the same Hamiltonian linkage pattern when the states are appropriately labeled. (*a*) A diamond (or box) pattern, with $E_3 > E_4$. (*b*) A crossed box pattern, with $E_3 < E_4$.

With different numbering choices the matrix has the structure

$$\begin{bmatrix} . & X & . & X \\ X & . & X & . \\ . & X & . & X \\ X & . & X & . \end{bmatrix}\begin{matrix} 1 \\ 2 \\ 3 \\ 4 \end{matrix} , \quad \text{or} \quad \begin{bmatrix} . & . & X & X \\ . & . & X & X \\ X & X & . & . \\ X & X & . & . \end{bmatrix}\begin{matrix} 1 \\ 3 \\ 2 \\ 4 \end{matrix} . \tag{15.83}$$

Although the RWA imposes constraints on the frequencies of any looped linkage, allowable loops can occur in a number of simple situations [Una97; Fle99; Sho08].

15.5.1 *Loops from polarization links*

The following figures illustrate an example of a loop system obtained by treating linearly polarized light in a helicity basis.

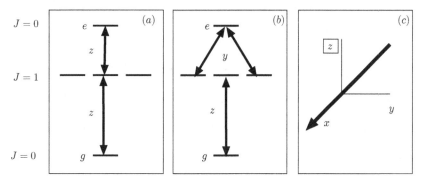

Figure 15.23 Examples of three-level linkage patterns for $J = 0 \leftrightarrow 1 \leftrightarrow 0$ sequence. (*a*) Parallel linear polarization, z, z. (*b*) Crossed linear polarization, z, y. No excitation can occur to state e. (*c*) The geometry: the propagation axis is y. [Redrawn from Fig. 69 of B. W. Shore, Acta Physica Slovaka, **58**, 243–486 (2008) with permission.]

$J = 0 \leftrightarrow 1 \leftrightarrow 0$. Figure 15.23 shows the linkages for a pair of resonant transitions involving linearly polarized light and the excitation sequence $J = 0 \leftrightarrow 1 \leftrightarrow 0$. The coordinates for quantization appear in frame (*c*). Frame (*a*) shows the linkages when both fields have the same polarization direction, and this is taken as the quantization axis z. There occurs a direct connection between the ground state g and the most excited state e, a simple three-state ladder. When the two Rabi frequencies are equal, population flows periodically between state g and state e, with regular complete population transfer into state e.

Frame (*b*) shows the linkage pattern when the first-step polarization is along the z axis, and the second-step polarization is x, i.e. the two fields are *cross polarized*. With the choice of quantization axis used here we must express the x polarization as a coherent superposition of right- and left-circular polarization, with the consequent linkage pattern shown. As can be seen, there is no connection between state g and the most excited state e; population undergoes two-state Rabi oscillations out of state g, but no population can reach excited state e.

Figure 15.24 presents these same two cases but with each polarization expressed in a helicity basis: each linearly polarized field is a phased superposition of right- and left-circularly polarized light. Now the links form a closed loop; they differ only in the relative phases of the paths. It is not immediately obvious from the linkage diagrams of this figure how the excitation will proceed. However, the possibility of excitation into state e cannot depend upon our choice of coordinates; the result is easy to recognize with the choice of basis fields used for Fig. 15.23.

The phases of frame (*a*), obtained when the two polarizations are parallel (e.g. both along x), interfere constructively to allow population transfer into state e. The phases of frame (*b*), obtained when the two linear polarizations are orthogonal (x and y), interfere destructively to prevent population from reaching state e.

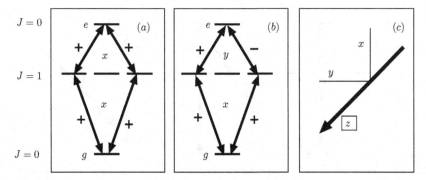

Figure 15.24 Examples of three-level linkage patterns for $J = 0 \leftrightarrow 1 \leftrightarrow 0$ sequence. (*a*) Parallel linear polarization, x, x, in helicity basis. (*b*) Crossed linear polarization, x, y, in helicity basis. (*c*) The geometry: the propagation axis is the quantization axis z. [Redrawn from Fig. 70 of B. W. Shore, *Acta Physica Slovaka*, **58**, 243–486 (2008) with permission.]

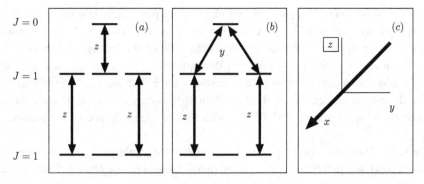

Figure 15.25 Examples of three-level linkage patterns for $J = 1 \leftrightarrow 1 \leftrightarrow 0$ sequence. (*a*) Parallel linear polarization, z, z. No excitation can occur to state e because a selection rule prevents the first-step transition $M = 0 \rightarrow M = 0$. (*b*) Crossed linear polarization, z, y. (*c*) The geometry: the propagation axis is y. [Redrawn from Fig. 71 of B. W. Shore, *Acta Physica Slovaka*, **58**, 243–486 (2008) with permission.]

$J = 1 \leftrightarrow 1 \leftrightarrow 0$. The association of destructive interference with crossed polarization is specific to the particular sequence of angular momentum states $J = 1 \leftrightarrow 0 \leftrightarrow 0$. Alternative sequences give different results. Figure 15.25 depicts the linkage pattern, as expressed in the Cartesian coordinates of Fig. 15.23, for the sequence $J = 1 \leftrightarrow 1 \leftrightarrow 0$. With this sequence of J values a selection rule prohibits the first-step $M = 0 \leftrightarrow 0$ transition and so the parallel polarization case of frame (*a*) does not allow excitation to reach state e. With crossed polarization, as seen in frame (*b*), there occurs a direct link between two of the initially populated sublevels and state e. In this situation the destructive interference of a loop portrayal of the interaction will occur with parallel polarizations.

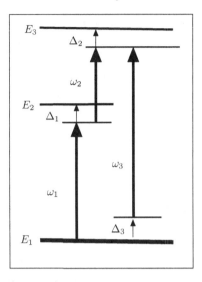

Figure 15.26 A three-state loop, involving frequencies ω_1, ω_2, and ω_3, showing the three detunings Δ_k.

15.5.2 Loops and the RWA; nonlinear optics

The introduction of the RWA accompanies the use of rotating coordinates in Hilbert space and the neglect of various exponentials, such as $\exp(\pm i\, 2\omega t)$. When linkage loops occur it is generally not possible to eliminate entirely the exponential time variations, except when the frequencies satisfy specific resonance conditions. To illustrate these requirements, consider a three-state loop, as suggested by Fig. 15.26. Such a linkage pattern will not occur when the interactions are all via electric dipole transitions and the system has a center of symmetry, because then each transition must occur between states of opposite parity. However, in the absence of a center of symmetry such a linkage loop becomes possible; such is the situation for atoms embedded in a matrix.

Let us write the statevector as

$$\Psi(t) = \exp(-i\, E_1 t/\hbar)\Big[C_1(t)\psi_1 + C_2(t)\,e^{-i\omega_1 t}\psi_2 + C_3(t)\,e^{-i\omega_1 t - i\omega_2 t}\psi_3\Big], \quad (15.84)$$

and take the Rabi frequencies asociated with carriers ω_k to be Ω_k, for $k = 1, 2, 3$, in keeping with the Hilbert-space rotations appropriate to the three-state ladder. Define the following detunings:

$$\hbar\Delta_1 = E_2 - E_1 - \hbar\omega_1,$$
$$\hbar\Delta_2 = E_3 - E_1 - \hbar\omega_1 - \hbar\omega_2, \quad (15.85)$$
$$\Delta_3 = \omega_1 + \omega_2 - \omega_3.$$

With the neglect of terms $\exp(\pm i2\omega_n t)$ compared with unity the Schrödinger equation translates into the following equations:

$$\frac{d}{dt}C_1 = -\frac{i}{2}\,\check{\Omega}_1^* C_2 - e^{-i\Delta_3 t}\,\check{\Omega}_3^* C_3,$$

$$\frac{d}{dt}C_2 = -i\Delta_1 C_2 - \frac{i}{2}\,\check{\Omega}_1 C_1 - \frac{i}{2}\,\check{\Omega}_2^* C_3, \qquad (15.86)$$

$$\frac{d}{dt}C_3 = -i\Delta_2 C_3 - \frac{i}{2}\,\check{\Omega}_2 C_2 - \frac{i}{2}e^{i\Delta_3 t}\,\check{\Omega}_3 C_1.$$

Unless the frequencies obey the resonance condition $\omega_3 = \omega_1 + \omega_2$ there will occur exponential variations at the detuning Δ_3, and the Hamiltonian, in the rotating basis, will not appear slowly varying. Under such situations it is not possible to assume slow variation, as is required for adiabatic passage.

In this system the coherence $C_1 C_3^*$ is associated with a dipole moment at frequency $\omega_1 + \omega_2$. As discussed in Sec. 21.4, this dipole moment serves as a source for growth of radiation at this frequency, thereby converting a photon of energy $\hbar\omega_1$ and another of energy $\hbar\omega_2$ into a single photon of energy $\hbar\omega_3 = \hbar\omega_1 + \hbar\omega_2$. By maximizing the relevant coherence one maximizes the frequency conversion.

A straightforward extension of this discussion to a chain of four states allows treatment of CARS and other four-wave mixing examples of nonlinear optics. For CARS the coherence $C_1 C_4^*$ is responsible for producing a field of frequency $\omega_4 = 2\omega_P - \omega_S$. For a three-state ladder linkage, with $E_1 < E_2 < E_3$, the coherence $C_1 C_4^*$ produces a field of frequency $\omega_4 = \omega_1 + \omega_2 + \omega_3$. In either case the frequency conversion can be optimized by maximizing the relevant coherence [Obe06].

15.6 Multilevel adiabatic time evolution

A variety of time-varying multilevel RWA Hamiltonians have useful applications, as generalizations of either the two-state adiabatic following that occurs with frequency-swept (chirped) detuning or with multiple overlapping resonant pulses, as with STIRAP. The following sections discuss some of those applications.

15.6.1 Chirped detuning in multistate systems

Just as a properly adjusted monotonic sweep of detuning (a chirp) can produce complete population transfer in a two-state system via adiabatic following, so too can an appropriate detuning sweep produce complete population transfer in a multistate system. The details of such adiabatic following depend upon the linkage pattern and upon the direction of the chirp.

Chirped fan. When a single state ψ_1 couples to a set of closely spaced energy levels that have no links amongst themselves the RWA Hamiltonian is a bordered matrix. Let us assume that only the single detuning $\Delta_1(t)$ varies with time. The RWA Hamiltonian matrix then has the structure

$$
\mathbf{W}(t) = \frac{1}{2}
\begin{bmatrix}
2\Delta_1(t) & \Omega_1(t) & \Omega_2(t) & \cdots & \Omega_{N-1}(t) \\
\Omega_1(t) & 2\Delta_2 & 0 & \cdots & 0 \\
\Omega_2(t) & 0 & 2\Delta_3 & \cdots & 0 \\
\vdots & \vdots & \vdots & \ddots & 0 \\
\Omega_{N-1}(t) & 0 & 0 & \cdots & 2\Delta_N
\end{bmatrix}.
\tag{15.87}
$$

There are $N-1$ constant diabatic energies, with values $\hbar\Delta_2 \dots \hbar\Delta_N$, and one varying diabatic energy $\Delta_1(t)$. When this is chirped its time-varying curve will cross each of the other values successively. If the detuning is an increasing function of time then population will adiabatically transfer from state 1 into state 2; if the detuning decreases with time the transfer is from state 1 into state N.

Figure 15.27 shows an example of the time-varying eigenvalues, frame (a), and populations, frame (b), for this system, with $N = 6$. For adiabatically increasing detuning complete population transfer occurs between states 1 and 2, as seen in frame (b).

Chirped bow tie. When the states are linked in a single chain of evenly spaced energies, excited by a single frequency, then the phase choice $\zeta_n = \zeta_{n-1} + \omega$ produces a RWA Hamiltonian matrix having the structure

$$
\mathbf{W}(t) = \frac{1}{2}
\begin{bmatrix}
2\Delta(t) & \Omega_1(t) & 0 & \cdots & 0 \\
\Omega_1(t) & 4\Delta(t) & \tfrac{1}{2}\Omega_2(t) & \cdots & 0 \\
0 & \Omega_2(t) & 6\Delta(t) & \cdots & 0 \\
\vdots & \vdots & \vdots & \ddots & \vdots \\
0 & 0 & 0 & \cdots & 2N\Delta(t)
\end{bmatrix}.
\tag{15.88}
$$

Assume that all the Rabi frequencies share a common pulse envelope $f(t)$, and that the detuning is linearly chirped. The pattern of adiabatic energies then is that of a star – a set of lines that intersect at $t = 0$. The presence of the pulsed interaction spreads these energies into a "bow tie" pattern. Figure 15.28(a) shows an example of this linkage pattern.

When the detuning is an increasing function of time then population will adiabatically transfer from state 1 into state N. For the example of Fig. 15.28 this adiabatic passage produces complete population transfer between states 1 and 6, as displayed in frame (b).

15.6.2 STIRAP in chains

The original STIRAP procedure aimed to produce complete population transfer through a chain of three states. Subsequently the underlying principle of adiabatic following has been extended to multistate chains (*chain-STIRAP*), as discussed in several reviews [Vit01b;

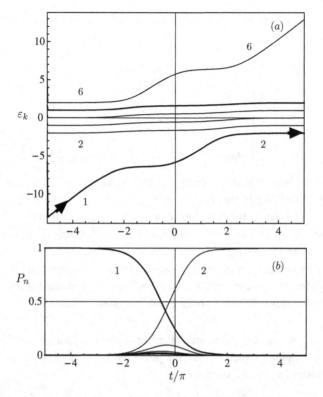

Figure 15.27 (a) Adiabatic eigenvalues ε_n and (b) population histories P_n for chirped detuning of a fan linkage. Population transfers adiabatically from state 1 to state 6.

Vit01c]. The discussion below notes the basic principles involved in extending the original three-state model to longer chains.

Specifically, we consider situations in which the RWA Hamiltonian matrix $W(t)$ has two distinct time dependences, originating in two pulse envelopes $f_P(t)$ and $f_S(t)$, that apply to alternate linkages,

$$\Omega_n(t) = \Omega_n(0) \times \begin{cases} f_P(t) & n \text{ odd}, \\ f_S(t) & n \text{ even}. \end{cases} \tag{15.89}$$

As with conventional STIRAP we regard the initial state as the first of the chain, state 1, and we adjust the phase of the associated unit vector so that the detuning is zero, $\Delta_1 = 0$. We require, as with STIRAP, that the last link of the chain also has zero detuning,

$$\Delta_1 = 0, \qquad \Delta_N = 0. \tag{15.90}$$

To implement STIRAP we require an adiabatic eigenvector that initially aligns with ψ_1 and later, upon the termination of both pulses, aligns with ψ_N. Such an adiabatic eigenvector only exists when the chain length N is an odd integer. It is also necessary that, in addition

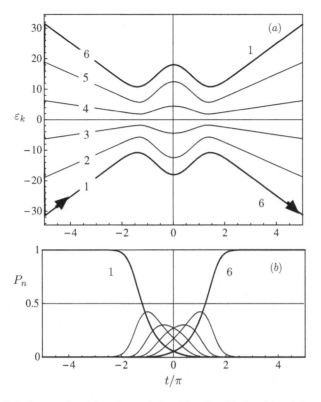

Figure 15.28 Adiabatic energies (*a*) and population histories (*b*) for chirped detuning of a chain linkage. Population flows adiabatically from state 1 to state 2.

to the N-state resonance condition of eqn. (15.90), all odd-n detunings Δ_n must vanish; nonzero detunings may occur only for even-n [Sho91a; Mar91; Smi92].

15.6.3 Chains of odd-integer length

The linkage pattern of the RWA Hamiltonian for a chain of an odd number of states, in which alternate detunings vanish, $\Delta_{2n+1} = 0$, can be viewed as a sequence of serially connected lambda systems, each of which maintains two-photon resonance. The even-numbered states are the apexes of the lambda links. Figure 15.29 illustrates the pattern, as it might be implemented in a $J = 2 \leftrightarrow 2$ angular momentum transition in which Zeeman shifts remove the degeneracy.

The RWA Hamiltonian has a zero eigenvalue and the corresponding dark adiabatic state $\Phi_0(t)$ is a time-dependent coherent superposition of the odd states in the chain,

$$\Phi_0(t) = A_1(t)\psi_1 + A_3(t)\psi_3 + \cdots + A_N(t)\psi_N, \qquad (15.91)$$

in which the coefficients $A_{2n+1}(t)$ are products of Rabi frequencies. For example, the RWA Hamiltonian of a five-state chain with constant detunings is

$$
W(t) = \frac{1}{2}
\begin{bmatrix}
0 & \Omega_1(t) & 0 & 0 & 0 \\
\Omega_1(t) & 2\Delta & \Omega_2(t) & 0 & 0 \\
0 & \Omega_2(t) & 0 & \Omega_3(t) & 0 \\
0 & 0 & \Omega_3(t) & 2\Delta & \Omega_4(t) \\
0 & 0 & 0 & \Omega_4(t) & 0
\end{bmatrix}.
\tag{15.92}
$$

The Rabi frequencies $\Omega_1(t)$ and $\Omega_3(t)$ will follow the time dependence $f_P(t)$ while the Rabi frequencies $\Omega_2(t)$ and $\Omega_4(t)$ have the time dependence $f_S(t)$. There is a two-photon resonance between states 1 and 3 and between states 3 and 5. The even-numbered states, 2 and 4, have arbitrary but equal detunings Δ.

The multilevel dark state of this system has the construction [Mor83; Hio85b; Sho91a; Mil96; Vit98b; Mil99]

$$
\Phi_0(t) = \frac{1}{\mathcal{N}(t)}\Big[\Omega_2(t)\Omega_4(t)\,\psi_1 - \Omega_1(t)\Omega_4(t)\,\psi_3 + \Omega_1(t)\Omega_3(t)\,\psi_5\Big],
\tag{15.93}
$$

where $\mathcal{N}(t)$ is a normalization factor. This state has null eigenvalue and has no component of the even-integer states 2, 4 that form the apexes of the linkage lambdas: it is "dark" for those states. The time dependence of this adiabatic state, expressed with the two pulses, is

$$
\Phi_0(t) = \frac{1}{\mathcal{N}(t)}\Big[\Omega_2(0)\Omega_4(0)\,f_S(t)^2\,\psi_1
$$
$$
- \Omega_1(0)\Omega_4(0)\,f_S(t)f_P(t)\,\psi_3 + \Omega_1(0)\Omega_3(0)\,f_P(t)^2\,\psi_5\Big].
\tag{15.94}
$$

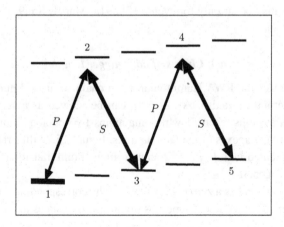

Figure 15.29 Chain-STIRAP for a five-state chain, as it might be implemented with angular momentum states subject to Zeeman shifts.

Hence, if the S pulse precedes the P pulse (counterintuitive pulse ordering) then $\Phi_0(t)$ will at early times be aligned with ψ_1 while at late times it will be aligned with ψ_5. Thus it serves as an adiabatic transfer state between states 1 and 5.

As in the original STIRAP, during adiabatic evolution the intermediate states (here states 2 and 4) remain unpopulated throughout the transfer. However, the intermediate odd states in the chain (here state 3) do acquire some transient populations.

Chain-STIRAP has proved to be a useful tool for a variety of applications, most notably for transferring coherent increments of transverse momentum to an atomic beam as it crosses a laser beam (see Sec. 17.2.2). It has been demonstrated experimentally by several groups [Pil93; Gol94; The98].

15.6.4 Chains of even-integer length

The failure of even-N chains to support a STIRAP-like population transfer can be seen most readily by considering the fully resonant chain, when all detunings vanish,

$$\Delta_1 = \Delta_2 = \Delta_3 = \ldots = \Delta_{N-1} = \Delta_N = 0. \qquad (15.95)$$

For even-N systems the determinant of the RWA Hamiltonian matrix is

$$\det \mathsf{W}(t) = (-)^{N/2} \Omega_1(t)^2 \, \Omega_3(t)^2 \ldots \Omega_{N-1}(t)^2, \qquad (15.96)$$

and this only vanishes when the pulse $f_P(t)$ vanishes. Therefore the RWA Hamiltonian for even-N does not maintain a null eigenvalue, unlike the odd-N Hamiltonian. More importantly, $\mathsf{W}(t)$ does not possess an adiabatic transfer state – an adiabatic eigenvector that changes alignment from ψ_1 to ψ_N with passing time. Consequently, even when such a system is driven adiabatically by counterintuitively ordered resonant pulses, there can be no STIRAP-like population transfer between the initial and final states of the chain. Instead, the final-state population exhibits Rabi-like oscillations as the pulse intensities increase [Ore92; Vit98a; Ban91b]. These oscillations occur because at early times as well as late times the state is a superposition of adiabatic states (rather than a single adiabatic state as in STIRAP); hence there occurs interference between the different paths from the initial state to the final state.

For example, in a resonantly driven four-state chain the four eigenvalues are $\pm\varepsilon_+$ and $\pm\varepsilon_-$, where

$$\varepsilon_\pm(t) = \sqrt{\frac{1}{2}\left(\Omega_T(t)^2 \pm \sqrt{\Omega_T(t)^4 - 4\Omega_1(t)^2\,\Omega_3(t)^2}\right)} \qquad (15.97)$$

and

$$\Omega_T(t)^2 = \Omega_1(t)^2 + \Omega_2(t)^2 + \Omega_3(t)^2. \qquad (15.98)$$

The initial state ψ_1 cannot be identified with a single adiabatic state at early times $t \to -\infty$, but is instead a superposition of the two adiabatic states corresponding to the smallest eigenenergies $-\varepsilon_-$ and ε_-. The same applies to the final state ψ_4 at $t \to +\infty$:

$$\psi_1 = \frac{1}{\sqrt{2}}[\Phi_{-\varepsilon_-}(-\infty) + \Phi_{\varepsilon_-}(-\infty)], \tag{15.99}$$

$$\psi_4 = \frac{1}{\sqrt{2}}[\Phi_{-\varepsilon_-}(+\infty) + \Phi_{\varepsilon_-}(+\infty)]. \tag{15.100}$$

If the system starts its evolution from state ψ_1, then the adiabatic solution for the final populations of the bare states is [Vit98a]

$$P_1(\infty) \approx \cos^2 \Theta \cos^2 \varphi,$$
$$P_2(\infty) \approx 0, \tag{15.101}$$
$$P_3(\infty) \approx \sin^2 \Theta \cos^2 \varphi,$$
$$P_4(\infty) \approx \sin^2 \varphi,$$

where

$$\tan \Theta = \lim_{t \to +\infty} [\Omega_1(t)/\Omega_2(t)] \quad \text{and} \quad \varphi = \int_{-\infty}^{\infty} dt \, \varepsilon_-(t). \tag{15.102}$$

Hence $P_4(t)$ oscillates as we approach the end of the pulse. There will be complete population removal, $P_1(\infty) = 0$, if the pulse $\Omega_2(t)$, responsible for coupling states 2 and 3, vanishes after the other pulses. Conversely, if this pulse vanishes first, there will occur complete population return, $P_3(\infty) = 0$,

$$\text{if } \Omega_1(t)/\Omega_2(t) \overset{t \to +\infty}{\longrightarrow} 0 \quad \text{then } P_1(\infty) \to 0,$$
$$\tag{15.103}$$
$$\text{if } \Omega_2(t)/\Omega_1(t) \overset{t \to +\infty}{\longrightarrow} 0 \quad \text{then } P_3(\infty) \to 0.$$

If the $\Omega_2(t)$ pulse that couples states 2 and 3 is always much stronger than the other couplings then $\Theta \approx 0$. Hence, although the process differs from STIRAP, it is possible to transfer population between states 1 and 4, bypassing the intermediate states 2 and 3 [Mal97].

16

Averages and the statistical matrix (density matrix)

16.1 Ensembles and expectation values

Equations of motion provide predictions of time-evolving variables of a dynamical system, and thereby enable us to predict the changing behavior. These equations are characteristic of a particular type of system, say a classical harmonic oscillator or a two-state quantum system. To complete the description of behavior one must specify initial conditions that select, from the many possible solutions to the equations of motion, a particular realization.

In the example of a classical point particle the equation is second order in time, and so we need two initial conditions, of position and velocity, to specify a particular solution. These two numbers, taken with the classical equation of motion, provide a complete description of the behavior of a specific harmonic oscillator. Similarly the two-state quantum system requires two initial values to complete its definition.

The initial conditions are never defined with infinite precision, even for classical mechanics. More generally we deal with an ensemble, that is, a collection of similar dynamical systems (e.g. oscillators) that differ in their preparation. The individual cases share some common attributes, but not necessarily all. Thus we have only incomplete information about any single system. We might, for example, consider planetary motion in which we know only the mean energy and the orbital plane. The loci of positions of the planets then form ellipses that make up the ensemble.

Although it is possible to prepare a single two-state quantum system in which the initial state is known (apart from an overall phase), often one deals with ensembles of such systems, each of which may differ in some uncontrollable property. For example, in an atomic beam the atoms originate from a thermal source in which all degenerate states have equal populations; one must consider each such degenerate state separately and then average the results. When the system is an impurity atom embedded in a solid matrix there generally are many such atoms, each with a slightly different environment and hence with slightly different energies and Bohr frequencies. Again one must average over particular cases that make up the ensemble.

It is not only the initial conditions that are subject to uncertainty, but also some characteristics of the environment may be incompletely controllable. What one observes are average properties of the system. The following subsections discuss examples of typical ensembles and the required averaging.

16.2 Statistical averages

This chapter begins discussion of ensemble averages by considering a quantum system subject to uncertainty of initial conditions. Following that will come a discussion of environmental uncertainties that affect the subsequent time evolution.

16.2.1 Initial conditions: Pure and mixed states

The probability amplitudes that serve as variables of the TDSE are the expansion coefficients in the expression of a time-dependent statevector $\Psi(t)$ within an abstract vector space. When the coordinate vectors are fixed, time-independent basis vectors ψ_n, the expansion reads

$$\Psi(t) = \sum_n c_n(t)\,\psi_n. \tag{16.1}$$

In particular, when the statevector is represented by a single basis state (say state ψ_1), so that only a single nonzero initial amplitude occurs, the initial condition takes the form

$$\Psi(0) = e^{i\zeta}\,\psi_1 \quad \text{or} \quad c_n(0) = e^{i\zeta}\,\delta_{n1}. \tag{16.2}$$

The phase ζ is usually not determined, and is taken as zero (but see Chap. 20). The initial state need not have the lower energy of the two states; the theory applies to deexcitation as well as excitation.

More generally, we allow the initial condition of the two-state system to be some coherent superposition of the two basis states, say as in the Bloch-vector representation of eqn. (5.31) in Sec. 5.6.1,

$$\Psi(\theta;0) = e^{-i\zeta}\left[\cos(\theta/2)\,\psi_1 + e^{-i\varphi}\sin(\theta/2)\,\psi_2\right], \tag{16.3}$$

a parametrization that clearly retains the normalization $|\Psi(\theta;0)|^2 = 1$ for any real θ. This initial state, as well as the state $\Psi(\theta;t)$ that evolves from it with passing time, depends upon our choice of parameter θ: the choice $\theta = 0$ makes ψ_1 the initial state, whereas the choice $\theta = 180°$ makes the initial state ψ_2. Different values of θ lead to different values of the expansion coefficients at any given time t, a fact that we can emphasize by introducing θ into the notation for the probability amplitude and rewriting eqn. (16.1) as

$$\Psi(\theta;t) = \sum_n c_n(\theta;t)\,\psi_n. \tag{16.4}$$

These probability amplitudes obey the TDSE, starting from the initial condition

$$c_1(\theta;0) = e^{-i\zeta}\cos(\theta/2), \quad c_2(\theta;0) = e^{-i\zeta}\sin(\theta/2). \tag{16.5}$$

For any fixed θ the probability $P_n(\theta; t)$ of finding state n at time t is just the absolute square of the probability amplitude $c_n(\theta; t)$:

$$P_n(\theta; t) = |\langle \psi_n | \Psi(\theta; t) \rangle|^2 = c_n(\theta; t) c_n(\theta; t)^*. \tag{16.6}$$

Thus, apart from the modified initial condition on our coupled equations, the procedure for obtaining probabilities is exactly that of previous discussions: we solve a single Schrödinger equation for the probability amplitudes, whose absolute squares yield the probabilities.

Mixed states. Equation (16.6) assumes we know that the system exists initially in a definite quantum state, namely the coherent superposition state $\Psi(\theta; 0)$ parametrized by the value θ. Suppose, however, that we lack such complete information, and that we know only that the system initially has probability $p(\theta)$ of being in the state $\Psi(\theta; 0)$. In this situation we must obtain $P_n(t)$ by summing all possible squares $|c_n(\theta; t)|^2$, each weighted by the appropriate probability $p(\theta)$. If the possible values of θ are discrete, then the prescription reads

$$\bar{P}_n(t) = \sum_k p(\theta_k) |c_n(\theta_k; t)|^2, \qquad \sum_k p(\theta_k) = 1. \tag{16.7a}$$

When θ can take a continuum of values the resulting expression for the averaged excitation probability is an integral,

$$\bar{P}_n(t) = \int_0^{2\pi} d\theta \, p(\theta) |c_n(\theta; t)|^2, \qquad \int_0^{2\pi} d\theta \, p(\theta) = 1. \tag{16.7b}$$

The summation (or integration) over possible initial states here involves only a single indexing parameter θ. More generally, there may be several parameters whose values we know imprecisely. Such a situation, when we lack information concerning the quantum state of the system, is termed a *mixed state*, as distinguished from a *pure* quantum state (a single state or a coherent superposition of basis states) that is representable by a single statevector $\Psi(t)$. A mixed state cannot be regarded, in any coordinate system of the abstract vector space, as a single statevector. It must be viewed as a distribution of statevectors.

The evaluation of mixed-state probabilities entails considerably more effort than is needed for evaluating probability amplitudes for a pure state: it requires that we obtain the amplitude $c_n(\theta; t)$ for each possible θ value and then compute a weighted average of the sum of the absolute squares to obtain $P_n(t)$. Section 16.6 presents an alternative approach based on statistical matrices, which greatly simplifies the task.

16.2.2 Thermalized start

The simplest occurrence of averaged initial conditions is that of thermodynamic equilibrium, in which there is no information about statevector phase, and states are populated according to energy E, in proportion to a Boltzmann factor $\exp(-E/k_B T)$ that depends on the ratio of energy to temperature. An example occurs when the systems are atoms in a collimated beam that intersects a laser beam. The atomic beam emerges from a thermal source, so that it includes, along with ground-state atoms, a small fraction of excited atoms. In any thermal environment there is a probability

$$p(n) = \frac{\exp(-E_n/k_B T)}{\mathcal{Z}(T)}, \qquad \mathcal{Z}(T) = \sum_n \exp(-E_n/k_B T), \tag{16.8}$$

that population will be found in state n with energy E_n. This thermal distribution of populations, normalized by the partition function $\mathcal{Z}(T)$, is a simple example of a mixed state. Within the two-state approximation both the excited state and the ground state are acted upon by the same Hamiltonian; only the initial condition differs in the two cases. The probability for finding population in state n at time t is a weighted sum of the squares of probability amplitudes $U_{nm}(t)$ for starting in state m and subsequently arriving in state n:

$$\bar{P}_n(t) = \sum_m p(m) |U_{nm}(t)|^2. \tag{16.9}$$

Here the probability amplitudes (elements of the time-evolution matrix or propagator of Sec. 7.5) satisfy the equation

$$\frac{d}{dt}U(t) = -iW(t)U(t), \qquad U_{nm}(0) = \delta_{mn}. \tag{16.10}$$

16.3 Environmental averages

Let us apply these notions to the description of atomic excitation in the RWA, recognizing that observed properties, such as excitation probabilities and dipole moments, originate in an ensemble of systems, each of which differs in some parameter that affects the excitation. Each individual system within the ensemble evolves subject to a distinct RWA Hamiltonian, say $W(e; t)$, that depends upon a set of environmental parameters, here denoted e. The systems evolve independently, and thus the observable probabilities are averages over probabilities obtained from considering separate time evolutions. The required algorithm, for a discrete set of environmental parameters e_j, is

$$\bar{P}_n(t) = \sum_j p(e_j) |C_n(e_j; t)|^2, \qquad \sum_j p(e_j) = 1, \tag{16.11}$$

where $C_n(e_j; t)$ is the solution to the TDSE with RWA Hamiltonian $W(e_j; t)$,

$$\frac{d}{dt}C(e_j; t) = -iW(e_j; t)C(e_j; t), \tag{16.12}$$

and $p(e_j)$ is the probability of finding the parameters e_j.

It is important to recognize that there usually is no single "average" RWA Hamiltonian $\bar{W}(t)$, i.e. no *average atom*, that will reproduce the averaging of eqn. (16.11); one cannot usually write

$$\bar{P}_n(t) = |\bar{C}_n(t)|^2, \quad \text{with} \quad \frac{d}{dt}\bar{C}(t) = -i\bar{W}(t)\bar{C}(t), \tag{16.13}$$

where $\bar{\mathsf{W}}(t)$ is an averaged RWA Hamiltonian matrix,

$$\bar{\mathsf{W}}(t) = \sum_j \mathrm{p}(e_j) \mathsf{W}(e_j; t). \tag{16.14}$$

16.3.1 Beam-particle velocities

Inevitably the atomic beam comprises atoms that have a range of velocities, usually as a result of thermal processes in the beam source. Each different velocity produces a slightly different radiation environment, with different Doppler-shifted frequencies and different durations of radiation. Each atom, identified by velocity $\mathbf{v}$, is an independent system, governed by a TDSE together with initial conditions. Because the Hamiltonian depends upon the radiation frequency in the rest frame of the atom, each different velocity corresponds to a different Hamiltonian. For each atom we obtain an excitation probability $P_n(\mathbf{v}; t)$ using the relevant velocity-dependent Hamiltonian. To obtain the ensemble average we must average these probabilities.

An atom moving nonrelativistically with velocity $v_\perp$ along the axis of a laser beam experiences the carrier frequency Doppler shifted. This appears in the RWA Hamiltonian $\mathsf{W}(v_\perp; t)$ as an altered detuning. Each different velocity provides a different exposure time (a different number of Rabi cycles) and each different path through the beam encounters a different peak intensity. Observed excitation is an average over various pulses.

16.3.2 Orientation averages

Any atom that has nonzero angular momentum J has $2J + 1$ possible discrete orientations with respect to any arbitrary direction (see Chap. 12), for example, the direction of the electric field of linearly polarized light. Each distinct orientation of the atom corresponds to a distinct projection of the dipole moment upon the electric field, i.e. a distinct value of $\mathbf{d} \cdot \mathbf{E}$. Each of these projections has a different energy, a different interaction Hamiltonian, and consequently a different Rabi frequency. Unless special preparation techniques are used, each of these orientations will be present in the initial ensemble of quantum states, and predictions of atomic behavior must include an average over all possible orientations, i.e. an average over a set of different Rabi frequencies.

16.3.3 Static environments

Atoms that are encapsulated as impurities within a crystalline environment retain many of the structural properties of free atoms. In particular it is often possible to observe, and induce, transitions between discrete states of electrons bound within the impurity. However, the neighboring crystalline structure may differ, albeit slightly, at different sites. Each impurity atom may have a different set of near neighbors, and hence will be subject to a different set of static interactions. The resulting absorption and emission spectra may, with high resolution, reveal this distribution of environments as distinct lines, one for each environment. Each atom undergoes excitation (perhaps coherently) independent of neighbors. The bulk properties are sums of the individual changes.

Each such environment produces a different shift of the basic energy levels E_n, and hence a different value of the detuning. The observed properties of an ensemble of impurity atoms must average over the detuning distribution, as in eqn. (16.11).

16.3.4 Homogeneous decoherence

The averages required for a mixed state typically sum contributions from different Rabi frequencies and detunings. The averaging tends to diminish the oscillations of the population that occur for each pure state, although they need not entirely eliminate them. This damping of oscillations is an example of *inhomogeneous* decoherence, so called because each atom of the ensemble has a different environment.

Rabi oscillations will also be damped when the phase of the laser field undergoes random fluctuations. Such variations have an effect that is identical for each atom. The loss of coherence is then known as *homogeneous* decoherence.

The two forms of decoherence have historically been studied for their effects on the broadening of spectral lines [Sob95; Pea06]. Inhomogeneous broadening occurs because the source atoms exist in a variety of environments, each of which produces a definite shift of the Bohr frequency. Homogeneous broadening occurs because each atom undergoes spontaneous emission at the same mean rate.

16.4 Expectation values

When we have available only an incomplete description of a system, because we deal with an ensemble, theory can only provide statistical information about the dynamics. The following paragraphs define the types of information available, first for a system whose motion follows classical dynamics and then for a system governed by quantum mechanics.

16.4.1 Classical expectation values; moments

Let us first consider a classical system, that is, a system in which Newton's or Hamilton's equations of motion provide a complete description of the dynamics for specified initial conditions. Let x denote a set of classical dynamical variables, say positions and momenta of the several particles that make up the system. Inevitably the finite size of our measuring instruments prevents us from determining initial conditions with infinite precision. We can only specify a probability distribution for possible initial values. The uncertainty of initial conditions leads, with passing time, to uncertainty in subsequent positions and momenta. That is, we cannot specify with infinite accuracy the positions or momenta, but must be content with probabilistic statements. This property makes the dynamical variables examples of *stochastic variables*, i.e. quantities whose values are specified by probabilities. Let us emphasize this stochastic nature of the variables by a tilde, $\tilde{x}$. Any particular observation of the set of variables $\tilde{x}$ will yield some definite set of numerical values. The totality of possible values comprises the *sample space* of these stochastic variables. For simplicity, let us assume that the observable values for $\tilde{x}$ form a discrete sample space $x_1, x_2, \ldots,$ as is the case for measurements with apparatus having finite resolution. A succession of measurements upon systems from an ensemble produces a sequence of values from this sample space. The values may all be exactly the same, or they may be distributed around some mean value, depending upon the preparation of the ensemble and the resolution of the measurement.

Now consider some function of these variables (e.g. the kinetic energy or angular momentum). This function depends upon stochastic variables, and so it too is a stochastic variable, say $\tilde{M}(\tilde{x})$. From each set of values x_n in the sample space of $\tilde{x}$ we can determine a value of the function. These values, $M(x_n)$, constitute the sample space of the variable $\tilde{M}$. From many successive observations, upon systems drawn from an ensemble, we can determine an average value of the stochastic variable $\tilde{M}$. This average, the classical *expectation value* $\langle \tilde{M} \rangle$, is expressible as the sum of all possible values $M(x_n)$, each weighted by the probability $p(x_n)$ that those values occur in the ensemble:

$$\langle \tilde{M} \rangle = \sum_n p(x_n) M(x_n). \tag{16.15}$$

As an example, the expectation values for various powers m of the stochastic variable provide the *moments* of x,

$$\langle \tilde{x}^m \rangle = \sum_n p(x_n) (x_n)^m. \tag{16.16}$$

These moments include the *mean* or *mean value*,

$$\langle \tilde{x} \rangle = \sum_n p(x_n) x_n, \tag{16.17}$$

and the *mean-square value*,

$$\langle \tilde{x}^2 \rangle = \sum_n p(x_n) (x_n)^2. \tag{16.18}$$

The root-mean-square (rms) is the square root of the mean-square value. The *variance* of the distribution is the expectation value of the squared difference between the variable and its mean value,[1]

$$\sigma(x) \equiv \langle (\tilde{x} - \langle \tilde{x} \rangle)^2 \rangle = \langle \tilde{x} \rangle^2 - \langle \tilde{x}^2 \rangle. \tag{16.19}$$

The *standard deviation* of a distribution is the square root of this variance.

16.4.2 Quantum expectation values

The Schrödinger equation, governing probability amplitudes, provides a means for evaluating the probability at time t that an atom will be found in an excited state. Yet the theory also permits sharply defined quantities, in the sense that probability may be unity for certain

---

[1] The symbol σ is traditional for variance, for cross section, and for Pauli matrices.

observations. For example, we might be certain that at time $t = 0$ the atom was in a single well-defined state of internal energy. This idealization may not apply in all situations; we may not be able to assign such certainty to the initial condition. Thus we may need to compound the probabilities inherent within the Schrödinger equation with other probabilities expressing the imprecision of our initial conditions.

16.4.3 Operator expectation values

Excitation probabilities are but one class of variables with which one can characterize atomic behavior. An equally valuable class of parameters are mean values – the ensemble averages – of various dynamical variables, such as total energy and total angular momentum.

Let us assume first that we deal with a system (say, the two-state atom) that is in a definite quantum state, as represented by the statevector $\Psi(t)$ of eqn. (16.1). It is a basic tenet of quantum theory that any repeated observation of some characteristic of the system (i.e. an *observable*) will produce a mean value (the *expectation value* of the observable) that is some weighted sum of products of probability amplitudes. Let us denote by M_{nm} the coefficient of the product $c_m^*(t)c_n(t)$ in such a sum. Then the expectation value $\langle \hat{M} \rangle$ of some observable $\hat{M}$ has the form of a summation

$$\langle \hat{M}(t) \rangle = \sum_{nm} c_m(t)^* c_n(t) M_{mn}. \qquad (16.20)$$

The amplitudes $c_n(t)$ incorporate the statistical and dynamical properties of the system (as derived from initial conditions and the Hamiltonian), independent of the particular observable; the numbers M_{nm} characterize the specific observation (e.g. excitation or dipole moment), independent of the dynamics. They can be regarded as the elements of a matrix M.

As an example, the matrix elements of a single-particle position-dependent observable $M(\mathbf{T})$ associated with the wavefunctions of two quantum states obtain from the spatial integral

$$M_{nm} = \int d\mathbf{r}\, \psi_n(\mathbf{r})^* M(\mathbf{r}) \psi_m(\mathbf{r}). \qquad (16.21)$$

For quantum states involving several particles the wavefunction description is less useful; we write

$$M_{nm} = \langle \psi_n | \hat{M} | \psi_m \rangle, \qquad (16.22)$$

implying a number associated with the two quantum states n and m and a particular observable. Because the basis states are stationary in Hilbert space, these numbers are independent of time. The expectation value of $\hat{M}$, when the system is known to be in the state described by the statevector $\Psi(t)$, is

$$\langle \hat{M}(t) \rangle = \langle \Psi(t) | \hat{M} | \Psi(t) \rangle. \qquad (16.23)$$

The connection between the expectation value of eqn. (16.23) and probability amplitudes occurs through expression of the statevector as a basis-state expansion, eqn. (16.1). Let the initial conditions of the statevector be parametrized by a set of environmental parameters e, as

$$\Psi(e;t) = \sum_n c_n(e;t)\,\psi_n. \tag{16.24}$$

The expectation value, when the system is known to be in the state described by statevector $\Psi(e;t)$, is

$$\langle \hat{M}(e;t) \rangle = \langle \Psi(e;t) | \hat{M} | \Psi(e;t) \rangle, \qquad \text{pure state.} \tag{16.25}$$

Expressed in terms of probability amplitudes this reads

$$\langle \hat{M}(e;t) \rangle = \sum_{nm} c_m(e;t)^* M_{mn} c_n(e;t), \qquad \text{pure state.} \tag{16.26}$$

For a mixed state the expectation value is

$$\langle \hat{M}(t) \rangle = \oint_e \mathrm{p}(e) \langle \hat{M}(e;t) \rangle, \qquad \text{mixed state,} \tag{16.27}$$

meaning a summation over discrete values and an integration over continuous values of the parameter set e. Expressed in terms of probability amplitudes the formula reads

$$\langle \hat{M}(t) \rangle = \sum_{nm} \oint_e \mathrm{p}(e) c_m(e;t)^* M_{mn} c_n(e;t), \qquad \text{mixed state.} \tag{16.28}$$

Implicit in all of these expressions is the assumption that the matrix elements M_{mn} are independent of time (i.e. the Hilbert-space coordinate system is stationary) and of the environmental variables e.

Section 16.6 below simplifies these expressions for expectation values by treating the numbers $\mathrm{p}(e)\,c_m(e;t)^*\,c_n(e;t)$ as elements of a matrix, the statistical or density matrix.

16.5 Uncertainty relations

The variance $(\Delta A)^2$ of an observable that is associated with an operator $\hat{A}$ is defined as

$$(\Delta A)^2 = \langle (\hat{A} - \langle \hat{A} \rangle)^2 \rangle = \langle \hat{A}^2 \rangle - \langle \hat{A} \rangle^2. \tag{16.29}$$

The *uncertainty* of the operator is ΔA, the square root of the variance. This uncertainty depends upon both the operator and the state or ensemble of states over which the expectation value $\langle \hat{A} \rangle$ is evaluated. The uncertainty ΔA is zero if and only if the states of the ensemble are all eigenstates of the operator $\hat{A}$ corresponding to the same eigenvalue.

Let two operators have the commutator

$$[\hat{A}, \hat{B}] = \mathrm{i}\hat{C}. \tag{16.30}$$

Then the product of the uncertainty of two operators satisfies the inequality[2]

$$\Delta A \times \Delta B \geq \tfrac{1}{2} |\langle \hat{C} \rangle|. \tag{16.31}$$

---

[2] As shown by Gottfried [Got66] this inequality represents a lower bound. In general the inequality includes the anticommutator $\hat{A}\hat{B} + \hat{B}\hat{A}$.

Applied to position and momentum this inequality produces the *Heisenberg uncertainty relations*

$$\Delta x \, \Delta p \geq \tfrac{1}{2} \hbar. \tag{16.32}$$

A quantum state for which the equality

$$\Delta A \times \Delta B = \tfrac{1}{2} |\langle \hat{C} \rangle| \tag{16.33}$$

holds for noncommuting operators is a *minimum-uncertainty state*.

A set of mutually commuting operators can all have zero uncertainty – there are then quantum states that are simultaneously eigenstates of all of these operators. The maximum information we can obtain about a quantum state is that which is associated with the largest set of commuting operators. The corresponding eigenvalues provide a complete characterization of a quantum state.

The eigenvalues of operators are numbers, often involving integers. These *quantum numbers* provide information about the state and are used to label the quantum states. The eigenvalues of a set of mutually commuting operators that commute with a Hamiltonian are termed "good" quantum numbers for that Hamiltonian.

16.6 The density matrix

The array of numbers M_{nm} appearing in eqn. (16.20) characterizes a particular observable, associated with operator $\hat{M}$, say excitation probability or dipole moment, and a pair of quantum states, n, m. The numbers M_{mn} are here assumed to be independent of time (because the Hilbert-space basis vector are fixed) and of the parameters e over which the statistical average is carried. They depend upon our choice of basis states, but they do not depend upon the initial conditions.

The environmental characteristics of the expectation value derive from the probability amplitudes, the numbers $c_n(e; t)$ that comprise the components of the statevector in the chosen basis. These numbers characterize the particular system – a particular set of initial conditions and subsequent excitation. They too depend upon our choice of basis states, but they are the same for all observables.

The statistical matrix. Rather than display explicitly the summation over environmental parameters, as in eqn. (16.28), we simplify the discussion by collecting the coefficients of the elements M_{mn} into a matrix $\rho(t)$ having as elements the averages over products of probability amplitudes,

$$\rho_{nm}(t) = \sum_{e} p(e) \, c_m(e; t)^* \, c_n(e; t) \equiv \{c_n(t) c_m(t)^*\}_{av}. \tag{16.34}$$

Here the symbol $\{\cdots\}_{av}$ denotes a statistical average. We can then express the expectation value $\langle \hat{M}(t) \rangle$ as a summed matrix product

$$\langle \hat{M}(t) \rangle = \sum_{n,m} \rho_{nm}(t) M_{mn}, \qquad (16.35)$$

in which M_{mn} are the elements of a matrix M. The sum is expressible as a trace,

$$\langle \hat{M}(t) \rangle = \text{Tr}\Big[\rho(t)\,\mathsf{M}\Big] = \text{Tr}\Big[\mathsf{M}\rho(t)\Big]. \qquad (16.36)$$

Although in general $\rho\mathsf{M}$ does not equal $\mathsf{M}\rho$, the trace of any matrix product does not depend upon the ordering of the factors.

The matrix $\rho(t)$, known as the *statistical matrix* or *density matrix* [Fan57; Blu96; Bar06], contains the information concerning the particular quantum system.[3] The diagonal elements of the matrix $\rho(t)$ are the populations,

$$P_n(t) = \rho_{nn}(t). \qquad (16.37)$$

The off-diagonal elements are termed *coherences*. The numerical values of the matrix elements (and whether the matrix is diagonal) will depend upon our choice of basis states $\boldsymbol{\psi}_n$.

Although the matrix elements $\rho_{nm}(t)$ incorporate all of the temporal behavior needed to determine population histories, their implicit dependence on the environmental parameters e occurs only in the summation that produces the environmental average.

If the quantum state that describes the system is one of the basis states, say $\boldsymbol{\Psi} = \boldsymbol{\psi}_n$, then the statistical matrix has only one element, unity in the (n, n) place. That is, the expectation value in this case is

$$\langle \hat{M}(t) \rangle = \langle \psi_n | \hat{M} | \psi_n \rangle = M_{nn}. \qquad (16.38)$$

More generally, the quantum state will be some coherent superposition of basis states, as discussed in the previous section, and the matrix ρ will not be diagonal.

It is important to note that in writing eqn. (16.36) we assume that the matrix elements of $\hat{M}$ in the basis $\boldsymbol{\psi}_n$ are independent of the environmental parameters e. This assumption is not always valid. For example, if e denotes the velocity of a moving atom and the operator $\hat{M}$ depends upon the frequency observed in the rest frame of the atom, then the simplification of eqn. (16.36) is not possible. We then must obtain the expectation value by averaging over density matrices,

$$\langle \hat{M}(t) \rangle = \oint_e \mathrm{p}(e) \sum_{n,m} \rho_{nm}(e; t) M_{mn}(e) = \oint_e \mathrm{p}(e)\,\text{Tr}[\rho(e; t)\mathsf{M}(e)], \qquad (16.39)$$

where the elements of the density matrix are unaveraged products,

$$\rho_{nm}(e; t) = c_n(e; t)\, c_m(e; t)^*. \qquad (16.40)$$

---

[3] Because it comprises products of probability amplitudes the density matrix contains no information about the overall phase of a statevector.

The statistical operator. The elements of the statistical matrix can be regarded as the matrix elements of a time-dependent *statistical operator* $\hat{\rho}(t)$ in a fixed basis,

$$\rho_{nm}(t) = \langle \psi_n | \hat{\rho}(t) | \psi_m \rangle. \tag{16.41}$$

When the statistical operator has the form

$$\hat{\rho}(t) = |\Psi(t)\rangle \langle \Psi(t)|, \quad \text{pure state}, \tag{16.42}$$

it describes a pure state, whereas when the statistical operator is the weighted sum of such pure-state operators,

$$\hat{\rho}(t) = \oint_e |\Psi(e;t)\rangle \, p(e) \, \langle \Psi(e;t)|, \quad \text{mixed state}, \tag{16.43}$$

it describes a mixed state. From expression (16.43) we evaluate any particular element as

$$\rho_{nm}(t) = \langle \psi_n | \hat{\rho}(t) | \psi_m \rangle = \oint_e p(e) \langle \psi_n | \Psi(e;t) \rangle \langle \Psi(e;t) | \psi_m \rangle, \tag{16.44}$$

in keeping with the definition in eqn. (16.34).

16.6.1 Example: Thermal excitation

An important example of a mixed state is the statistical matrix for a system in thermodynamic equilibrium (a *thermal state*) at temperature T. This statistical operator can be written as

$$\hat{\rho}^{therm} = \exp(-\hat{H}/k_B T)/\mathcal{Z}(T), \tag{16.45}$$

where $\hat{H}$ is the Hamiltonian energy operator and $\mathcal{Z}(T)$ is the partition function

$$\mathcal{Z}(T) = \text{Tr}\left[\exp(-\hat{H}/k_B T)\right]. \tag{16.46}$$

When the Hamiltonian is that of the bare atom and we employ energy states (eigenstates of the Hamiltonian) as our basis states, this statistical matrix produces the Boltzmann distribution of populations: It is a diagonal matrix whose elements are

$$\rho_{nm}^{therm} = \exp(-E_n/k_B T)/\mathcal{Z}(T)\delta_{nm}. \tag{16.47}$$

The expectation value of any observable $\hat{M}$ is then a weighted sum of diagonal elements,

$$\langle \hat{M} \rangle = \sum_n M_{nn} \exp(-E_n/k_B T)/\mathcal{Z}(T). \tag{16.48}$$

16.6.2 Example: Two states

To illustrate some of the properties of the density matrix consider the coherent superposition state of eqn. (5.31) in which θ and φ are Bloch angles of Sec. 5.6.1. The statistical matrix for this state,

$$\boldsymbol{\rho}(\theta,\varphi) = \begin{bmatrix} \cos^2(\theta/2) & \cos(\theta/2)\sin(\theta/2)\,e^{i\varphi} \\ \cos(\theta/2)\sin(\theta/2)\,e^{-i\varphi} & \sin^2(\theta/2) \end{bmatrix}, \quad (16.49)$$

provides information about the two parameters θ and φ, but not about the overall phase of the statevector ζ; it is an incomplete transcription of the statevector. This density matrix refers to a definite quantum state, whose statevector is $\boldsymbol{\Psi}(\theta,\varphi)$, not to a statistical mixture, and so it is appropriately labeled by θ and φ. If we have no knowledge of the initial values of these angles we must average the density matrix over a uniform distribution of values, each of which we must assume equally likely to occur. This procedure produces the incoherent statistical matrix

$$\boldsymbol{\rho}^{inc} = \frac{1}{2\pi^2} \int_0^\pi d\theta \int_0^{2\pi} d\varphi \, \boldsymbol{\rho}(\theta,\varphi) \quad (16.50)$$

that represents a *completely incoherent* superposition of states 1 and 2. This density matrix is a multiple of the two-dimensional unit matrix,

$$\boldsymbol{\rho}^{inc} = \frac{1}{2}\begin{bmatrix} 1 & 0 \\ 0 & 1 \end{bmatrix}, \quad (16.51)$$

and it yields for expectation values the result

$$\langle \hat{M} \rangle = \tfrac{1}{2}[M_{11}+M_{22}]. \quad (16.52)$$

This is the expression we expect for a classical system. It cannot be obtained from a single statevector.

The result of eqn. (16.52) contrasts with the pure-state statistical matrix that we obtain for the particular angles $\theta = 90°$ and $\varphi = 0$,

$$\boldsymbol{\rho}(90°) = \frac{1}{2}\begin{bmatrix} 1 & 1 \\ 1 & 1 \end{bmatrix}. \quad (16.53)$$

This matrix, representing a coherent superposition state, obtains from a statevector that has equal contributions from each component (the Bloch vector lies in the equatorial plane),

$$\boldsymbol{\Psi}(90°) = \frac{1}{\sqrt{2}}\big[\boldsymbol{\psi}_1 + \boldsymbol{\psi}_2\big]. \quad (16.54)$$

With this system the expectation value is

$$\langle \hat{M} \rangle = \tfrac{1}{2}[M_{11}+M_{22}+M_{12}+M_{21}]. \quad (16.55)$$

Thus the expectation value in a pure state includes off-diagonal matrix elements of M (i.e. coherences) that are absent when the state specification is incomplete.

Density matrix and Bloch vector. A Bloch vector can be defined from the elements of a density matrix, a generalization of eqn. (5.35),

$$r_1 = 2\text{Re}\{c_1 c_2^*\}_{av}, \quad r_2 = 2\text{Im}\{c_1 c_2^*\}_{av}, \quad r_3 = \{|c_2|^2 - |c_1|^2\}_{av}. \tag{16.56}$$

Thus for a lossless system the density matrix can be expressed as

$$\rho = \frac{1}{2}\begin{bmatrix} r_3 - 1 & r_1 + ir_2 \\ r_1 - ir_2 & r_3 + 1 \end{bmatrix}. \tag{16.57}$$

16.6.3 General properties of statistical matrices

These examples of density matrices exhibit several fundamental properties of all statistical matrices $\rho(t) \equiv \rho$ that are discussed in several reviews and texts [Fan57; Blu96; Bar06]. First, the diagonal elements of ρ are the excitation probabilities

$$\rho_{nn} = P_n. \tag{16.58}$$

This interpretation holds even when ρ has off-diagonal elements. It follows that the diagonal elements of any density matrix are non-negative numbers,

$$\rho_{nn} \geq 0, \tag{16.59}$$

and that, when there is no probability loss, the trace of the density matrix ρ (the sum of these probabilities) is unity:

$$\text{Tr}[\rho] = 1 \qquad \text{or} \qquad \sum_n P_n = 1. \tag{16.60}$$

Second, if the matrix ρ represents a *pure* quantum state, as does $\rho(\theta)$ of eqn. (16.49), then it is *idempotent*:

$$\rho\rho = \rho, \tag{16.61}$$

and hence for a pure state the trace of any power of ρ is, like the trace of ρ, equal to unity:

$$\text{Tr}[\rho^n] = 1 \qquad \text{for a pure state.} \tag{16.62a}$$

A mixed state has the property, for $n > 1$,

$$\text{Tr}[\rho^n] < 1 \qquad \text{for a mixed state.} \tag{16.62b}$$

For example, for $n = 2$ this inequality,

$$\text{Tr}[\rho^2] \leq \text{Tr}[\rho], \tag{16.63}$$

provides the constraint

$$\sum_{ij} \rho_{nm}\rho_{mn} \leq \sum_k \rho_{kk}. \tag{16.64}$$

This expression generalizes the observation that the sum of the squares of the eigenvalues must be less than unity, unless only a single eigenvalue is nonzero (as occurs with a pure state). Finally, any density matrix is Hermitian,

$$\boldsymbol{\rho}^{\dagger} = \boldsymbol{\rho} \qquad \text{meaning} \qquad \rho_{mn}^{*} = \rho_{nm}. \tag{16.65}$$

This condition not only requires that diagonal elements be real, but they must in fact be non-negative. The constraints of eqn. (16.65) diminish the number of quantities that are needed to specify a density matrix. The relationship

$$\sum_{n} \rho_{nn} \equiv \text{Tr}\,[\rho] = 1 \tag{16.66}$$

further reduces the number of real independent variables, from $2N^2$ to $N^2 - 1$.

Two dimensions. For a two-dimensional density matrix the most general form in which no off-diagonal elements appear is

$$\boldsymbol{\rho} = \frac{1}{a+b} \begin{bmatrix} a & 0 \\ 0 & b \end{bmatrix}. \tag{16.67}$$

This has unit trace, as required. However, the square of this matrix has trace

$$\text{Tr}\,[\boldsymbol{\rho}^2] = \frac{a^2 + b^2}{(a+b)^2}. \tag{16.68}$$

This can only be a pure state, describable by a single statevector, if one of the diagonal elements vanishes. That is, a pure state whose density matrix has more than one diagonal element must also have off-diagonal elements (coherences).

Need for density matrix. We must use a density matrix whenever our information about an ensemble is incomplete – when we only know the fraction of systems that have a particular property, i.e. the fraction that will be found in a particular quantum state. We can regard the mixed state as an ensemble of pure quantum states in which the state $\boldsymbol{\psi}_n$ occurs with probability p_n. The state is most highly mixed, i.e. is least like any single quantum state, when all these probabilities are equal. The density matrix is then a multiple of the unit matrix.

16.7 Density matrix equation of motion

The density matrix can be considered to be the average of bilinear products of probability amplitudes

$$\rho_{nm}(t) = \{c_n(t)c_m(t)^*\}_{av}, \tag{16.69}$$

each of which is governed by a TDSE. When the Hamiltonian is not affected by the averaging, then the time variation of the density matrix obeys the equation

$$\hbar \frac{d}{dt} \{c_n(t)c_m(t)^*\}_{av} = -\mathrm{i} \sum_k H_{nk}(t) \{c_n(t)c_m(t)^*\}_{av}$$

$$+ \mathrm{i} \sum_k H_{mk}(t)^* \{c_n(t)c_k(t)^*\}_{av}. \qquad (16.70)$$

With the assumption that the Hamiltonian is Hermitian this equation can be written in terms of the commutator of the Hamiltonian matrix with the density matrix

$$\hbar \frac{d}{dt}\boldsymbol{\rho}(t) = -\mathrm{i}\,\mathsf{H}(t)\boldsymbol{\rho}(t) + \mathrm{i}\,\boldsymbol{\rho}(t)\mathsf{H}(t) \equiv -\mathrm{i}\Big[\mathsf{H}(t),\,\boldsymbol{\rho}(t)\Big]. \qquad (16.71)$$

This matrix equation of motion, known as the *quantum Liouville equation*, is the quantum-mechanical counterpart of the classical Liouville equation that governs the time evolution of phase-space density.

The Liouville equation for an N-state system involves N^2 coupled variables, but it does not depend on the number of environmental parameters nor their probability distribution. These enter the theory only through the initial conditions on the elements of the density matrix.

The Liouville equation generalizes to allow treatment of inhomogeneous incoherent processes such as spontaneous emission. When these occur the Schrödinger equation no longer provides a description of the dynamics. The density matrix serves as a bridge between the completely coherent description of the Schrödinger equation and the completely incoherent rate equations[Fan57; Ack77; Blu96; Bar06] [Sho90, Chap. 6].

Initial conditions. The statistical matrix is particularly useful for describing the initial conditions, at $t = 0$, of a system in which we do not know the initial state precisely: we know the magnitudes of the probability amplitudes, i.e. the probabilities $P_n(0)$, but we cannot specify their relative phases. Thus we have insufficient information to construct a statevector: we cannot ensure from a statevector that

$$\{c_n(0)c_m(0)^*\}_{av} = 0 \qquad \text{if } n \neq m. \qquad (16.72)$$

Inhomogeneous relaxation. As presented here, the Liouville equation simply reexpresses the TDSE in a form better suited to incorporating statistical distributions of initial conditions. However, we can recognize from the derivation that it applies equally well to statistical averages of *any* parameter α that *does not appear explicitly in the Hamiltonian*.

Conversely, it does *not* apply to averages over parameters that do appear in the Hamiltonian. For such a variable α we require the equation

$$\hbar \frac{d}{dt}\boldsymbol{\rho}(\alpha; t) = -\mathrm{i}\,[\mathsf{H}(\alpha; t),\, \boldsymbol{\rho}(\alpha; t)], \qquad (16.73)$$

to be used with the formula

$$\langle \hat{M}(t) \rangle = \oint_{\alpha} p(\alpha) \operatorname{Tr} [\boldsymbol{\rho}(\alpha; t) M]. \tag{16.74}$$

This is the expression we require when evaluating averages over Doppler shifts, for example. In that application α denotes the velocity v and $p(v)$ is a Maxwellian velocity distribution; $\boldsymbol{\rho}(v; t)$ is the time-evolved density matrix for a particular velocity. Generally it is not possible to express the dynamics as that of an *average atom* in the form

$$\hbar \frac{d}{dt} \{\boldsymbol{\rho}(t)\}_{av} = -i \, [\{H(t)\}_{av}, \{\boldsymbol{\rho}(t)\}_{av}], \tag{16.75}$$

in which both the density matrix and the Hamiltonian are averaged separately,

$$\{\hat{H}(t)\}_{av} = \oint_{\alpha} p(\alpha) \hat{H}(\alpha; t), \qquad \{\boldsymbol{\rho}(t)\}_{av} = \oint_{\alpha} p(\alpha) \boldsymbol{\rho}(\alpha; t). \tag{16.76}$$

16.7.1 The Liouville operator

The time evolution of the density matrix expressed in eqn. (16.71) can be viewed as the action of a *Liouville operator* $\hat{L}(t)$ acting upon $\boldsymbol{\rho}(t)$,

$$\frac{d}{dt} \boldsymbol{\rho}(t) = -i \hat{L}(t) \boldsymbol{\rho}(t). \tag{16.77}$$

To interpret this equation let us organize the elements of $\boldsymbol{\rho}(t) \equiv \boldsymbol{\rho}$ into a column-vector array, say the transpose of the row vector

$$\boldsymbol{\rho}^T = [\rho_{11}, \rho_{22}, \ldots, \rho_{12}, \rho_{21}, \ldots]. \tag{16.78}$$

Then the Liouville equation appears as a set of linear equations for the N^2 components of a vector. The operator $\hat{L}(t) \equiv \hat{L}$ takes the form of a matrix, whose elements are combinations of elements of the Hamiltonian matrix. For example, the equations for the two-state system can be organized into the form (here I omit the time argument)

$$\frac{d}{dt} \begin{bmatrix} \rho_{11} \\ \rho_{22} \\ \rho_{12} \\ \rho_{21} \end{bmatrix} = -\frac{i}{\hbar} \begin{bmatrix} 0 & 0 & -H_{21} & H_{12} \\ 0 & 0 & H_{21} & -H_{12} \\ -H_{12} & H_{12} & (H_{11} - H_{22}) & 0 \\ H_{21} & -H_{21} & 0 & (H_{22} - H_{11}) \end{bmatrix} \begin{bmatrix} \rho_{11} \\ \rho_{22} \\ \rho_{12} \\ \rho_{21} \end{bmatrix}. \tag{16.79}$$

As with the Schrödinger equation, we deal with a set of coupled linear ODEs. Whereas the N-state Schrödinger equation leads to N equations for N complex-valued coefficients, the N-state Liouville equation provides N^2 equations. However, the number of independent real-valued variables is $N^2 - 1$.

Tetradics; superoperators. An alternative view of the Liouville equation, frequently adopted, retains the two indices upon the elements of ρ. The resulting form for the Liouville equation,

$$\frac{d}{dt}\rho_{nm}(t) = -i\sum_{jk} L_{nm,jk}(t)\rho_{jk}(t), \tag{16.80}$$

requires four indices for the Liouville operator. This viewpoint regards the array ρ as elements in a "doubled-atom" space (or *Liouville* space) and refers to $\mathcal{L}$ as a *tetradic* operator or "superoperator". The elements of $\hat{\mathcal{L}}$ are

$$\hbar L_{nm,jk}(t) = H_{nj}(t)\delta_{mk} - H_{km}(t)\delta_{jn}. \tag{16.81}$$

While such viewpoints can assist both formal analysis and computational evaluation, they should not be allowed to obscure the essential properties of the Liouville equation as a set of coupled linear first-order ODEs amenable to solution by techniques appropriate to any such sets – it is only necessary to arrange the elements of the density matrix in an ordered list to convert it into a vector.

For example, when the Hamiltonian is independent of time we can write the matrix solution to eqn. (16.80) as

$$\rho(t) = \exp\left[-i\,\hat{\mathcal{L}}t\right]\rho(0). \tag{16.82}$$

To this equation we can apply any of the techniques that are applicable to the equation

$$\Psi(t) = \exp\left[-i\,\hat{H}t/\hbar\right]\Psi(0). \tag{16.83}$$

The Lie derivative. The time derivative formed by taking the commutator with the Hamiltonian H, termed here the Liouville operator $\mathcal{L}$, is also known as the *Lie derivative*. The time-evolution operator for the Liouville operator is known as a *Lie transformation* [Ste86a].

16.7.2 Some constants of motion

It is often useful to write the density matrix in the form

$$\rho(t) = \mathsf{U}(t)\rho(0)\mathsf{U}^{-1}(t), \tag{16.84}$$

where the matrix $\mathsf{U}(t)$ is the time-evolution operator for the Schrödinger equation, the solution to

$$\hbar\frac{d}{dt}\mathsf{U}(t) = -i\,\hat{H}(t)\mathsf{U}(t), \qquad \mathsf{U}(0) = \mathbf{1}. \tag{16.85}$$

These equations presume that the Hamiltonian $\hat{H}(t)$ is Hermitian, so that the matrix $\mathsf{U}(t)$ is unitary. Expression (16.84) for $\rho(t)$ facilitates formal manipulation of operators. As an example, it provides the result

$$\rho(t)^2 \equiv \rho(t)\,\rho(t) = \mathsf{U}(t)\rho(0)\mathsf{U}^{-1}(t)\,\mathsf{U}(t)\rho(0)\mathsf{U}^{-1}(t)$$
$$= \mathsf{U}(t)\rho(0)^2\mathsf{U}^{-1}(t) \tag{16.86}$$

or, more generally, the result

$$\rho(t)^n = \mathsf{U}(t)\rho(0)^n\mathsf{U}^{-1}(t). \tag{16.87}$$

It follows from this expression that the trace of any power of the density matrix is independent of time:

$$\text{Tr}\left[\boldsymbol{\rho}(t)^n\right] = \text{Tr}\left[\boldsymbol{\rho}(0)^n\right] = c_n. \tag{16.88}$$

This property provides a set of constants of motion, c_n [Elg80; Hio81; Hio83; Hio84a]. The first of these constants, for $n = 1$, is just the total probability, unity. In a pure state all powers of $\boldsymbol{\rho}(t)$ are equal to $\boldsymbol{\rho}(t)$, and so all traces are unity, but in a mixed state this constraint no longer holds. For systems with more than two states, the higher powers provide independent constants of the motion for mixed states.

Another set of constants can be obtained when the Hamiltonian is independent of time, as it can be in the RWA. These constants, b_n, are expectation values of powers of $\hat{H}$, [Hio84a]:

$$\text{Tr}\left[\boldsymbol{\rho}(t)(\hat{H})^n\right] = b_n. \tag{16.89}$$

For a slowly varying (adiabatic) Hamiltonian these are slowly varying, i.e. adiabatic constants.

16.8 Incorporating incoherent processes

The notion of an isolated atom (one free from disturbing influences) is an idealization at best. In practice, the atom may collide with other atoms or ions, thereby exchanging energy and momentum. Collisions (or momentary fluctuations of the environment) also produce momentary energy shifts that alter the relative phases of probability amplitudes. Furthermore, the laser radiation that we conveniently idealize as monochromatic (or as a smooth pulse) inevitably has some finite bandwidth – the field typically has uncontrolled fluctuations of phase and amplitude characterized by a finite coherence time. The density matrix provides a simple means of incorporating into a single-atom equation of motion many of the effects of collisions with other particles, as well as fluctuations in the radiation field.

16.8.1 Bath states and reduced density matrices

One simple way to model the effects of surroundings upon an atom is to enlarge the system of interest to include not only the single atom (still idealized by a few discrete states) but also a complete set of states for additional particles and fields in the surroundings. These additional particles are not observed directly, nor are they controllable beyond general properties such as composition, density, and temperature. They will be termed a *bath*, and will be assumed to provide a reservoir that is unaffected by the single atom of interest.[4]

Let the quantum states of the atom be $|\psi_a\rangle \equiv |a\rangle$ and let the quantum states of the larger bath (the *bath states*) be $|\phi_b\rangle \equiv |b\rangle$. The label a on the atom states denotes a particular state from a discrete set (e.g. for a two-state system $a = 1, 2$) while the label b denotes

---

[4] The bath may also include the infinite set of initially unpopulated field modes that are responsible for spontaneous emission.

one of a very large number of states, typically including a continuum. We can describe the composite system, atom and bath, with the use of simple product states. Let us denote by $\hat{\rho}^{AB}$ the statistical operator of the composite atom–bath system. Pure states of this system have statistical operators of the form

$$\hat{\rho}^{AB}(a, b) = |a, b\rangle\langle a, b| \equiv |\psi_a\rangle|\phi_b\rangle\langle\phi_b|\langle\psi_a|. \tag{16.90}$$

The operators of interest (dipole moments, population projections, etc.) act within the space of atomic states and leave bath states unaffected. Therefore the expectation value of such an atomic operator $\hat{M}^A$ can be written as

$$\langle\hat{M}^A\rangle = \text{Tr}(\hat{M}^A\hat{\rho}^{AB}) = \sum_{aba'b'}\langle ab|\hat{M}^A|a'b'\rangle\langle a'b'|\hat{\rho}^{AB}|ab\rangle$$

$$= \sum_{aa'}\langle a|\hat{M}^A|a'\rangle\sum_b\langle a'b|\hat{\rho}^{AB}|ab\rangle. \tag{16.91}$$

Because the operator $\hat{\rho}^{AB}$ factors we can introduce a *reduced density matrix*, obtained from the statistical operator $\hat{\rho}^A$ by summing over the unobserved bath states,

$$\langle a'|\hat{\rho}^A|a\rangle = \sum_b\langle a'b|\hat{\rho}^{AB}|ab\rangle \quad \text{or} \quad \hat{\rho}^A = \text{Tr}_B(\hat{\rho}^{AB}). \tag{16.92}$$

By using this matrix, we can write the expectation value in the form previously employed, a trace over atomic states:

$$\langle\hat{M}^A\rangle = \sum_a\langle a|\hat{M}^A\hat{\rho}^A|a\rangle = \text{Tr}_A(\hat{M}^A\hat{\rho}^A). \tag{16.93}$$

The distinction between these two expressions (16.91) and (16.93) for $\langle\hat{M}^A\rangle$ is significant. In eqn. (16.91) the trace involves a statistical operator $\hat{\rho}^{AB}$ for the large but closed atom–bath system. In eqn. (16.93) the trace is over variables of the atom system alone.

The statistical operator for the full system, comprising atom, laser radiation, and bath, obeys the Liouville equation

$$\hbar\frac{d}{dt}\hat{\rho}^{AB}(t) = -\text{i}\left[\hat{H}^{AB}(t), \hat{\rho}^{AB}(t)\right], \tag{16.94}$$

where $\hat{H}^{AB}(t)$ is the total Hamiltonian of the atom (interacting with radiation), the bath, and the atom–bath interaction:[5]

$$\hat{H}^{AB}(t) = \hat{H}^A(t) + \hat{H}^B + \hat{V}^{AB}. \tag{16.95}$$

In eqn. (16.93) the evaluation of $\text{Tr}_A(\hat{M}^A\hat{\rho}^A)$ involves the operator of a reduced density matrix $\hat{\rho}^A$ for the atom alone. This operator does not satisfy a Liouville equation. Instead

---

[5] The operator $\hat{H} = \hat{H}^A$ describes the free atom and its interaction with laser radiation. It may also include all other radiation modes and so be the traditional radiation–atom Hamiltonian, but it is more convenient for the present discussion to include these weakly populated modes amongst the bath states. Then H^A is the coherent excitation Hamiltonian of previous sections.

it obeys the equation

$$\hbar \frac{d}{dt} \hat{\rho}^A(t) = -\mathrm{i} \left[\hat{H}^A(t), \hat{\rho}^A(t) \right] - \mathrm{i} \, \mathrm{Tr}_B \left[\hat{H}^B + \hat{V}^{AB}, \hat{\rho}^{AB}(t) \right], \qquad (16.96)$$

obtained by taking the trace over bath variables of the Liouville equation for $\hat{\rho}^{AB}$. The first term, $\left[\hat{H}^A(t), \hat{\rho}^A \right]$, is the Liouville operator of coherent atomic excitation, acting upon the reduced density matrix. This operator combines the free atom, the free radiation field (if appropriate), and the radiation–atom interaction responsible for coherent excitation. It is the operator that would apply to the atom in the absence of the bath.

The remaining term represents, exactly, the uncontrolled effects of the bath upon the atom. In particular, it must describe collisions between the atom and bath particles. If the Hamiltonian $\hat{H}^A(t)$ does not include weakly populated field modes, then this second term must describe spontaneous emission.

All of these processes have the effect of causing the elements of the reduced density matrix to relax towards steady equilibrium values. It is the final term of eqn. (16.96) that is responsible for this relaxation.

Other uses of reduced density matrices. By definition, a reduced density matrix is a statistical matrix of some subsystem of a larger system, obtained by taking a trace over a portion of the quantum labels of the full system. The concept can be applied to any system that comprises at least two parts. Chapter 17 discusses examples.

In a two-electron system one can define a single-electron reduced density matrix by taking the trace over a set of one-electron quantum numbers. One can also define two-electron reduced density matrices, needed in the evaluation of Coulomb repulsion between electrons, in a many-electron atom.

In the system of atom plus radiation one can define a reduced density matrix either for the atom or for the field. For all such systems one may start with a Liouville equation (equivalent to the Schrödinger equation) for the full system. By taking the trace over attributes of a portion of this system one obtains an equation for the subsystem of interest. This reduced equation is not equivalent to a Schrödinger equation for the subsystem: A portion of the equation describes the influence on the atom of the remaining portion of the system.

16.8.2 Relaxation as fluctuations

The previous discussion regarded the unobserved variables of the atomic surroundings as a bath of particles (and photons) whose properties were described by ensemble averages. Those averages appear as the mathematical operation Tr_B, in which we sum over bath variables weighted by probabilities. This ensemble average may also be regarded as the result of averaging over uncontrolled temporal fluctuations produced by moving perturbers. With this viewpoint we regard the full Hamiltonian $\hat{H}^{AB}(t)$ as comprising two parts:

$$\hat{H}^{AB}(t) = \hat{H}^A(t) + \tilde{V}(t). \qquad (16.97)$$

The first *mean-field* portion $\hat{H}^A(t)$ controls the behavior of the atom subject to an analytically described semi-classical radiation field (a *sure* field, as opposed to a *stochastic* field). The remaining portion of the Hamiltonian $\tilde{V}(t)$ expresses uncontrollable random fluctuations in the Hamiltonian (the *stochastic* part). This latter portion describes variations in atomic energies that occur as a result of collisions, i.e. transitions between atomic states produced by the fields during a collision, as well as fluctuations in the laser field that appear as a finite bandwidth. The reduced density matrix of the atom $\boldsymbol{\rho}^A(t) = \boldsymbol{\rho}(t)$ is obtained by averaging the full density matrix $\boldsymbol{\rho}^{AB}(t)$ over such fluctuations:

$$\boldsymbol{\rho}(t) = \{\boldsymbol{\rho}^{AB}(t)\}_{av}. \tag{16.98}$$

The average $\{\cdots\}_{av}$ appearing here may be regarded as either an ensemble average or a time average. With the former interpretation, this equation merely introduces a new notation for Tr_B. When we apply the averaging procedure to the Liouville equation the operator $\hat{H}^A(t)$ is not affected (by definition),

$$\{H_{nm}^A(t)\}_{av} = H_{nm}(t), \tag{16.99}$$

and we obtain the equation for the density matrix of the atom, $\hat{\rho}^A(t) \equiv \hat{\rho}(t)$,

$$\hbar \frac{d}{dt} \boldsymbol{\rho}(t) = -\mathrm{i}\,[H(t), \boldsymbol{\rho}(t)] - \mathrm{i}\{[\tilde{V}(t), \boldsymbol{\rho}^{AB}(t)]\}_{av}. \tag{16.100}$$

This is basically eqn. (16.96). It describes time evolution produced by a precisely defined time-varying Hamiltonian, the atom–radiation Hamiltonian $\hat{H}^A(t)$, together with relaxation. The interpretation of the two equations differs, however. The density matrix $\boldsymbol{\rho}^A(t)$ of eqn. (16.96) refers to the portion of a supposedly complete density matrix $\boldsymbol{\rho}^{AB}(t)$ that remains after we sum over whatever variables are needed to describe the bath. The density matrix $\boldsymbol{\rho}^A(t)$ of eqn. (16.100) refers to the matrix that remains after we average over all of the fluctuations of the surroundings. The assumption that the two definitions produce the same density matrix is essentially the assumption that the entire system, atom plus radiation plus bath, is *ergodic*.

Markov processes. Equation (16.100), when expressed in terms of matrix elements, has the structure

$$\hbar \frac{d}{dt} \rho_{nm}(t) = -\mathrm{i} \sum_k H_{mk}(t)\rho_{kn}(t) + \mathrm{i} \sum_k \rho_{mk}(t)H_{nk}(t) - \mathrm{i} V_{nm}(t). \tag{16.101}$$

The final term $V_{nm}(t)$ incorporates all of the properties of the bath or the fluctuations that produce changes to the matrix elements $\rho_{nm}(t)$, a discrete set of values, during the differential time increment dt. Because the underlying physical processes that produce $V_{nm}(t)$ are uncontrollable and are regarded as random, this function is a stochastic variable, characterizable only by means of statistical averages and the resulting parameters. In typical models of such dynamics it is assumed that the changes produced by $V_{nm}(t)$ depend only on the state of the system at time t, not upon any previous history of the system. This assumed property makes the stochastic variables examples of a *Markov process*, i.e. a stochastic process that has no memory and whose characterizing probabilities depend on the instantaneous state of the system. The bath, or the environment, is said to be *Markovian*. With this assumption, valid for spontaneous emission into vacuum modes or brief collisions (impacts), the final term takes the form of a set of time-independent linkages between matrix elements,

$$V_{nn'}(t) \to \sum_{kk'} V_{nn,kk'} \, \rho_{kk'}(t). \tag{16.102}$$

The resulting equation of motion for the atomic density matrix is sometimes called the generalized *master equation*.

Lindblad equations. When the randomizing environment is Markovian the equation of motion (the master equation) for the statistical operator of the atom $\hat{\rho} = \hat{\rho}^A$ is sometimes written as

$$\hbar \frac{d}{dt} \hat{\rho} = -\mathrm{i}[\hat{H}, \hat{\rho}] + \sum_{\mu} \hbar \left(\hat{L}_{\mu} \hat{\rho} \hat{L}_{\mu}^{\dagger} - \tfrac{1}{2} \hat{L}_{\mu}^{\dagger} \hat{L}_{\mu} \hat{\rho} - \tfrac{1}{2} \hat{\rho} \hat{L}_{\mu}^{\dagger} \hat{L}_{\mu} \hat{\rho} \right), \tag{16.103}$$

in which the $\hat{L}_{\mu}$ are *Lindblad operators* [Lin76] that act within the space of bath states. Each Lindblad operator represents a "quantum jump", detectable were we to monitor the environment fully. The master equation adjusts the density matrix for decoherence, as is required when A is part of an open system. The effect of the Lindblad operators is to diminish the phase coherences of the atomic states, i.e. to introduce *decoherence*.

16.9 Rotating coordinates

It often simplifies the analysis to introduce rotating Hilbert-space coordinates by writing the statevector as in eqn. (7.20),

$$\Psi(t) = \sum_{n} e^{-\mathrm{i}\zeta_n(t)} C_n(t) \psi_n = \sum_{n} C_n(t) \psi'_n(t). \tag{16.104}$$

The formula for an expectation value then displays the phases $\zeta_n(t)$ explicitly,

$$\langle \hat{M}(t) \rangle = \sum_{nm} \{C_m(t)^* M_{mn} C_n(t)\}_{av} \, e^{\mathrm{i}\zeta_m(t) - \mathrm{i}\zeta_n(t)}. \tag{16.105}$$

We define a statistical (density) matrix $\mathbf{s}(t)$ in rotating coordinates as having elements

$$s_{nm}(t) = \{C_n(t) C_m(t)^*\}_{av}, \tag{16.106}$$

so that the relationship with $\boldsymbol{\rho}(t)$ is

$$\rho_{nm}(t) = s_{nm}(t) \, e^{\mathrm{i}\zeta_m(t) - \mathrm{i}\zeta_n(t)}. \tag{16.107}$$

The expectation values will depend upon the choice of phases $\zeta_n(t)$,

$$\langle \hat{M}(t) \rangle = \sum_{n,m} s_{nm}(t) \, M_{mn} \, e^{\mathrm{i}\zeta_m(t) - \mathrm{i}\zeta_n(t)}. \tag{16.108}$$

In the RWA the probability amplitudes, and the consequent elements $s_{nm}(t)$, are usually slowly varying. However, carrier frequencies appear in the phases and so the expectation values typically vary with the carrier frequencies; e.g. eqns. (7.30) and (10.5).

16.9.1 Equations of motion

Equations of motion for the density matrix $s(t)$ involve the action of the coherent RWA Hamiltonian as well as parametrized actions of other interactions – radiative interactions that produce spontaneous emission and uncontrollable actions of the neighboring environment. Derivation of these equations, or of eqn. (16.116), may be found in numerous texts and articles ([Fai69; Bar06] [Sho90, §22]).

Coherent excitation. Changes to the density matrix produced by the coherent interaction derive from the time-dependent Schrödinger equation,

$$\frac{d}{dt}s_{mn}(t) = -i\sum_k W_{mk}(t)s_{kn}(t) + i\sum_k s_{mk}(t)W_{kn}(t)^*, \tag{16.109}$$

in which the RWA Hamiltonian matrix $\hbar W$ treats all laser-induced changes and detunings.

Population relaxation. Alterations of the density matrix from incoherent interactions are representable as Markov processes, i.e. the effects occur without memory of previous history. In the absence of coherent excitation the changes to the populations $P_n(t) = s_{nn}(t)$ obey the rate equations

$$\frac{d}{dt}s_{nn}(t) = \sum_{k \neq n} s_{kk}(t)\mathcal{R}_{k \to n} - \Gamma_n s_{nn}(t). \tag{16.110}$$

The first of these terms describes the flow of population into state m from all other states. The *rate coefficient* $\mathcal{R}_{n \to m}$ that parametrizes the population flow from state n to state m embodies all of the influences upon the atom that are not given by the coherent-excitation Hamiltonian $W(t)$. For example, it describes inelastic collisions that alter populations as well as spontaneous emission from higher-lying states.

The rate Γ_n appearing in the final term is the sum of all rate coefficients for processes that remove population from state n,

$$\Gamma_n = \sum_{k \neq n} \mathcal{R}_{n \to k}. \tag{16.111}$$

These include inelastic collisions and spontaneous emission to lower-lying states. They may also include photoionization into continuum states, and therefore be time dependent. None of these final states k need be part of the small set of essential states that are treated by the TDSE.

The values of $\mathcal{R}_{k \to n}$ are typically obtained by means of perturbation theory from Fermi's famous Golden Rule, see Sec. 9.6. For present purposes they can be regarded as empirical parameters.

Complex-valued energies. Loss of probability into states other than those of the three states used for constructing the statevector $\Psi(t)$ can be treated by taking the energies E_n to be complex, with concomitant changes to the detunings,

$$E_n \to E_n - i\hbar\gamma_n, \qquad \Delta_n \to \Delta_n - i\gamma_n, \qquad \gamma_n = \tfrac{1}{2}\Gamma_n. \tag{16.112}$$

The complex energies mean that, in the absence of coherent excitation fields, the probabilities diminish exponentially

$$P_n(t) = \exp[-2\gamma_n t]P_n(0). \tag{16.113}$$

Relaxation of coherences. Within the Markov approximation incoherent changes to the off-diagonal elements of the density matrix (the coherences) obey the equation

$$\frac{d}{dt}s_{nm}(t) = -\Gamma_{nm}s_{nm}(t), \qquad n \neq m, \tag{16.114}$$

where the rate Γ_{nm} expresses the losses from each of the paired states n and m as well as an additional contribution Γ_{nm}^{ad} that originates with elastic collisions or other adiabatic interactions with the environment,

$$\Gamma_{nm} = \tfrac{1}{2}[\Gamma_n + \Gamma_m] + \Gamma_{nm}^{ad}, \qquad n \neq m. \tag{16.115}$$

The coefficients Γ_{nm} combine photoionization rates $\mathcal{R}_n^{ion}$, population transfer rates $\mathcal{R}_{n\to k}$ (sums of radiative and collisional excitation or deexcitation rates), and collisional coherence relaxation rate γ^c. Thus they are generally larger than the relaxation rates that occur in the population equation (16.117).

The effect of incoherent processes on coherences is not fully treated by simply using complex energies. A complete treatment requires not only the modeling of probability loss but also the presence of incoherent population transfer, such as occurs with spontaneous emission or inelastic collisions.

Master equation. The coherent and incoherent changes described above combine to produce the generalized master equation

$$\frac{d}{dt}s_{m'm}(t) = -i\sum_n [W_{m'n}(t)s_{nm}(t) - s_{m'n}(t)W_{nm}(t)]$$

$$-\sum_{nn'}\Gamma_{m'mn'n}s_{n'n}(t). \tag{16.116}$$

Two states. As an example, the equation for the ground-state population change of a two-state atom reads

$$\frac{\partial}{\partial t}s_{11}(t) = -\Gamma_1 s_{11}(t) + s_{22}\mathcal{R}_{2\to 1} - i[W_{12}s_{21}(t) - W_{21}s_{12}(t)], \tag{16.117}$$

where the relaxation rate, including the photoionization rate $\mathcal{R}_1^{ion}$, is

$$\Gamma_1 = \mathcal{R}_{1\to2} + \mathcal{R}_1^{ion}. \tag{16.118}$$

A similar equation, with the replacement $1 \leftrightarrow 2$, applies to the excited state population $s_{22}(t)$. The equation for the coherence is

$$\frac{d}{dt}s_{12}(t) = i[W_{22}(t) - W_{11}(t)]s_{12}(t) - \Gamma_{12}s_{12}(t) - iW_{12}(t)(s_{22} - s_{11}), \tag{16.119}$$

with the relaxation rate

$$\Gamma_{12} = \tfrac{1}{2}(\mathcal{R}_{1\to2} + \mathcal{R}_{2\to1} + \mathcal{R}_1^{ion} + \mathcal{R}_2^{ion}) + \gamma^c. \tag{16.120}$$

When the only inelastic incoherent process is spontaneous emission at rate A the various parameters become

$$\Gamma_1 = 0, \qquad \Gamma_2 = A, \qquad \Gamma_{12} = \tfrac{1}{2}A + \gamma^c. \tag{16.121}$$

16.10 Multilevel generalizations

Several extensions and generalizations of the formalism of density matrices have application to treatments of coherent excitation, either to the description of the dynamics or to the analysis of the resulting excitation, as discussed in Chap. 19. In presenting the theory it is useful to recognize that, as with the vector model of Chap. 10 both the density matrix and the Hamiltonian matrix require parametrization. A useful approach is by means of elementary transition matrices, defined in Sec 16.10.1 below. These allow a generalization of the Bloch vector of Sec. 16.10.2 to a multidimensional coherence vector and a corresponding multidimensional torque equation. When placed within the formalism of angular momentum theory the transition matrices lead to the introduction of multipole moments, for both the density matrix and the Hamiltonian. The following sections present the needed theory.

16.10.1 Elementary transition matrices

The properties of a Hamiltonian or density matrix are often exhibited most simply when they are expressed as combinations of constant elementary basis matrices. The simplest such set is formed from the elementary transition matrices,

$$\boldsymbol{\pi}(n, m) \equiv |\psi_n\rangle\langle\psi_m| \equiv |n\rangle\langle m|, \tag{16.122}$$

and the associated idempotent projection operators,

$$\Pi(n) \equiv \boldsymbol{\pi}(n, n), \quad \Pi(n)\Pi(m) = \Pi(n)\delta_{nm}. \tag{16.123}$$

Using these matrices as a basis we obtain the presentation (decomposition) of the Hamiltonian as

$$\mathsf{H}(t) = \sum_{nn} H_{nm}(t)\boldsymbol{\pi}(n, m) = \sum_{m,n} |\psi_n\rangle H_{nm}(t)\langle\psi_m| \equiv \sum_{m,n} |n\rangle H_{nm}(t)\langle m|. \tag{16.124}$$

Similarly, we write the density matrix as

$$\rho(t) = \sum_{n,m} \rho_{nm}(t)\, \boldsymbol{\pi}(n,m). \tag{16.125}$$

The measurement of the probability of finding state n can be represented by a filter that selects only component n of a statevector. Thus the probability is the expectation value of a projection operator,

$$P_n(t) = \langle \Psi(t)|\Pi(n)|\Psi(t)\rangle, \quad \text{or} \quad P_n(t) = \mathrm{Tr}[\rho(t)\Pi(n)]. \tag{16.126}$$

16.10.2 The N-state coherence vector

The presentation of coherent excitation dynamics as motion of a vector in an abstract space offers opportunities for gaining insight into the response of the system to various pulses or pulse sequences. Such pictures need not be limited to the two-state vector model of motion of a point on the Bloch sphere. One higher-dimensional model that offers a simple pictorial description of excitation dynamics is the pseudospin model of Sec. 15.3.5. Other authors have generalized the notion of the Bloch vector, descriptive of two-state behavior, to obtain in N dimensions a vector (termed by Hioe and Eberly the *coherence vector* [Hio81]) that obeys a generalization of the torqued motion of the Bloch vector [Elg80; Hio81; Ore84a].

The starting point for this generalization is the set of elementary transition matrices $\boldsymbol{\pi}(n,m)$ and projection operators $\Pi(n)$ used to present the Hamiltonian as

$$\mathsf{H}(t) = \sum_{n} E_n\,\Pi(n) + \sum_{n \neq m} V_{nm}(t)\,\boldsymbol{\pi}(n,m). \tag{16.127}$$

We combine these to form new operators, for $n, m \leq N$ and $L < N$,

$$\begin{aligned}
\mathsf{u}(n,m) &\equiv [\boldsymbol{\pi}(n,m) + \boldsymbol{\pi}(m,n)], \\
\mathsf{v}(n,m) &\equiv -\mathrm{i}\,[\boldsymbol{\pi}(n,m) - \boldsymbol{\pi}(m,n)], \\
\end{aligned} \tag{16.128}$$

$$\mathsf{w}(n) \equiv \sqrt{\frac{2}{L(L+1)}}\,[\Pi(1) + \Pi(2) + \cdots + \Pi(L) - L\Pi(L+1)].$$

In the special case of two dimensions these are the three Pauli matrices, i.e. twice the spin-half matrices of Appendix A.1.3,

$$\mathsf{u}(1,2) = \begin{bmatrix} 0 & 1 \\ 1 & 0 \end{bmatrix}, \quad \mathsf{v}(1,2) = \begin{bmatrix} 0 & \mathrm{i} \\ -\mathrm{i} & 0 \end{bmatrix}, \quad \mathsf{w}(1) = \begin{bmatrix} 1 & 0 \\ 0 & -1 \end{bmatrix}. \tag{16.129}$$

Like the matrices $\boldsymbol{\pi}(n,m)$ these provide links between every pair of states. We then form an ordered array from these $N^2 - 1$ matrices, the column vector

$$\mathsf{s} = [\mathsf{u}(1,2),\ldots,\mathsf{v}(1,2),\ldots,\mathsf{w}(1),\ldots,\mathsf{w}(N-1)]^T. \tag{16.130}$$

The components of this operator have the commutation property

$$[\mathsf{s}_n, \mathsf{s}_m] = 2\mathrm{i} \sum_k \epsilon_{nmk} \mathsf{s}_k, \tag{16.131}$$

where ϵ_{nmk} is the completely antisymmetric structure constant of the $SU(N)$ group [Ore84a]. Using these matrices we have for the density matrix $\mathsf{s}(t)$ and the Hamiltonian $\mathsf{H}(t)$ the constructions

$$\mathsf{s}(t) = \frac{1}{N}\mathbf{1} + \frac{1}{2}\sum_{n=1}^{N^2-1} S_n(t)\mathsf{s}_n, \quad \text{where } S_n(t) = \mathrm{Tr}\,[\mathsf{s}(t)\mathsf{s}_n],$$

$$\tag{16.132}$$

$$\mathsf{H}(t) = \frac{1}{2}\hbar \sum_{n=1}^{N^2-1} \Upsilon_n(t)\mathsf{s}_n, \quad \text{where } \hbar\Upsilon_n(t) = \mathrm{Tr}\,[\mathsf{H}(t)\mathsf{s}_n].$$

The equation of motion for the functions $S_n(t)$ is

$$\frac{d}{dt}S_n(t) = \sum_{mk} \epsilon_{nmk}\, \Upsilon_m(t)\, S_k(t). \tag{16.133}$$

This is a generalization of the torque equation satisfied by the Bloch vector, to which it reduces when $N = 2$. The quantities $S_n(t)$ are elements of a coherence vector $\mathbf{S}(t)$ in $N^2 - 1$ dimensions, for which we can write the equation of motion as a torque equation,

$$\frac{d}{dt}\mathbf{S}(t) = \boldsymbol{\Upsilon}(t) \times \mathbf{S}(t). \tag{16.134}$$

The set of $N^2 - 1$ elements of the coherence vector are not all independent. As pointed out by Hioe and Eberly [Hio81], the motion of this vector is constrained by a set of conserved quantities

$$c(N,m) \equiv \mathrm{Tr}[\mathsf{s}(t)^m], \quad n \leq N, \tag{16.135}$$

where N is the dimensionality of the Hamiltonian. The first of these, $c(N, 1)$, simply states that probability is conserved.

Irreducible tensor form. An alternative form for the coherence vector was proposed by Oreg and Goshen [Ore84b]. It is based on using state labels that run, not from 1 to N, but from $-S$ to $+S$ for $N = 2S + 1$, as in the pseudospin model [Coo79; Hio87]. Using three-j symbols we construct, for $k = 0, \ldots, 2S$ and $q = -k, \ldots, +k$, the N^2 matrices

$$\mathsf{s}_q^{(k)} = \sum_{mm'} (-1)^{S-m}\sqrt{2k+1}\begin{pmatrix} S & k & S \\ m & q & m' \end{pmatrix}|m\rangle\langle m'|. \tag{16.136}$$

These are *irreducible tensors* in which k labels the order, and q labels the component. They are infinitesimal generators of the unitary group $U(N)$; those for $k \neq 0$ are infinitesimal generators of $SU(N)$. With this basis the RWA Hamiltonian has the construction

$$\mathsf{W}(t) = \sum_{kq} \Upsilon_q^{(k)}(t)\, \mathsf{s}_q^{(k)}, \qquad \Upsilon_q^{(k)}(t) = (-1)^q \operatorname{Tr} [\mathsf{W}(t)\mathsf{s}_{-q}^{(k)}], \tag{16.137}$$

and the coherence vector, based on the density matrix $s(t)$ in rotating coordinates becomes

$$S_q^{(k)}(t) = (-1)^q \operatorname{Tr} [s(t)\mathsf{s}_{-q}^{(k)}]. \tag{16.138}$$

The equation of motion for the coherence vector, in the RWA, has the form

$$\frac{d}{dt} S_q^{(k)}(t) = \sum_{k'q'k''q''} f_{q'q''q}^{k'k''k} \, \Upsilon_{q'}^{(k')}(t)\, S_{q''}^{(k'')}(t), \tag{16.139}$$

where the structure constants $f_{q'q''q}^{k'k''k}$ are expressible in terms of three-j and six-j symbols. When the Hamiltonian is restricted by suitable symmetry properties, the equation of motion breaks into independent blocks, one for each multipole order k.

Although irreducible tensors arise quite naturally in the theory of angular momentum (see App. A.2.1), they have use for any system, as the preceding equations demonstrate. The following paragraphs specialize the theory to angular momentum states.

16.10.3 State multipoles

In treating excitation of free atoms or molecules it is common to employ eigenstates of angular momentum operators, and to label basis states with angular momentum eigenvalues, as $|\alpha J M\rangle$, where α denotes additional quantum numbers needed to specify the state completely. In such a basis we express a statevector as

$$\Psi(t) = \sum_{\alpha J M} C_{\alpha J M}(t) \exp[-i\zeta_{\alpha J}(t)]\, \psi_{\alpha J M}, \tag{16.140}$$

where the phases $\zeta_{\alpha J}(t)$, independent of M, are chosen to allow the RWA. The density matrix for such a system bears similar labels. In the rotating basis it has the elements

$$s(\alpha' J' M', \alpha J M; t) = \{C_{\alpha J M}(t)^* C_{\alpha' J' M'}(t)\}_{av} \tag{16.141}$$

so that expectation values require evaluation of the expression

$$\langle \hat{M}(t)\rangle = \sum_{\alpha J M \alpha' J' M'} \langle \alpha J M | \hat{M} | \alpha' J' M'\rangle\, s(\alpha' J' M', \alpha J M; t)\, \exp(-i\zeta_{\alpha' J'} + i\zeta_{\alpha J}). \tag{16.142}$$

From the elements of such a density matrix we create coupled tensorial sets, or *state multipoles* [Omo77; Blu96; Bay06b; Auz10], by coupling the two angular momenta into possible resultants, using the three-j symbols (or the Clebsch–Gordan coefficients),

$$s(\alpha' J', \alpha J)_Q^{(K)} = \sum_{MM'} (-1)^{J'-M'} \sqrt{2K+1} \begin{pmatrix} J' & J & K \\ M' & -M & -Q \end{pmatrix} s(\alpha' J' M', \alpha J M). \tag{16.143}$$

Table 16.1. *Nomenclature for density-matrix multipoles*

	Two-state	General	$K = 0$	$K = 1$	$K = 2$	$\cdots$
Polarizations	$s(1,1)$	$s(\alpha J, \alpha J)_Q^{(K)}$	population	orientation	alignment	$\cdots$
Coherences	$s(1,2)$	$s(\alpha' J', \alpha J)_Q^{(K)}$	monopole	dipole	quadrupole	$\cdots$

The inverse relationship, obtaining density matrix elements from state multipoles, is[6]

$$s(\alpha' J' M', \alpha J M) = \sum_{KQ} (-1)^{J'-M'} \sqrt{2K+1} \begin{pmatrix} J' & J & K \\ M' & -M & -Q \end{pmatrix} s(\alpha' J', \alpha J)_Q^{(K)}.$$

(16.144)

State multipoles $s(\alpha' J', \alpha J)_Q^{(K)}$ are defined for all values of K in the range $|J' - J| \leq K \leq J' + J$.

The particular sets of elements for which $\alpha' J' = \alpha J$, i.e. multipoles derived from the diagonal elements of the density matrix, provide a description of the population distribution amongst the various magnetic sublevels of a given level. These state multipoles are known as *populations* (when $K = 0$) or as *polarizations* (when $K > 0$). They contrast with the *coherences*, state multipoles derived from off-diagonal elements of the density matrix and for which $\alpha' J' \neq \alpha J$.

The state multipole $s(\alpha J, \alpha J)_q^{(1)}$ has three components, and has the properties of a vector. It is known as the *orientation vector*: If at least one of the components of this vector is nonzero the system is said to be *oriented*.

The state multipole $s(\alpha J, \alpha J)_q^{(2)}$ has five components and is known as the *alignment tensor*. If it is nonzero, the system is said to be *aligned*. These are special cases of *polarized populations*, for which at least one polarization is nonzero. Table 16.1 summarizes this nomenclature.

These formulas are independent of any phase choice, i.e. they hold in any picture. Because the density matrix is Hermitian, the state multipoles have the following property:

$$s(\alpha' J', \alpha J)_Q^{(K)*} = (-1)^{J'-J-Q} s(\alpha J, \alpha' J')_{-Q}^{(K)}.$$
(16.145)

Populations. The state multipoles with $\alpha' J' = \alpha J$ and $Q = 0$ are linear combinations of the magnetic sublevel populations $P_{\alpha J M}$. Specifically, we have the formula

$$s(\alpha J, \alpha J)_0^{(K)} = \sum_M (-1)^{J-M} (J M, J - M | K 0) s(\alpha J M, \alpha J M) \equiv \sum_M c_M P_{\alpha J M}.$$

(16.146)

---

[6] The state multipoles defined here are in the rotating frame of the RWA.

For even K the coefficients c_M are symmetric about $M = 0$, whereas for odd K the coefficients are antisymmetric (and the coefficient vanishes for $M = 0$). Except for $K = 0$, all coefficients sum to zero. The monopole ($K = 0$) expectation value is proportional to the total population of the level (the sum of sublevel populations):

$$s(\alpha J, \alpha J)_0^{(0)} = \frac{1}{\sqrt{2J+1}} \sum_M P_{\alpha J M} = \frac{P_{\alpha J}}{\sqrt{2J+1}}. \tag{16.147}$$

For thermally equilibrated populations at temperature T the probability $P_{\alpha J}$ is proportional to a statistical weight $2J + 1$, an exponentiated energy $E(\alpha J)$, and the inverse of a partition function $\mathcal{Z}(t)$. The relevant formula is

$$s(\alpha J, \alpha J)_0^{(0)} = \sqrt{2J+1} \exp[-E(\alpha J)/k_B T]/\mathcal{Z}(T). \tag{16.148}$$

Representing operators in multipole form. Operator matrix elements in an angular momentum basis can be expressed in state multipole form,

$$M(\alpha' J', \alpha J)_Q^{(K)} = \sum_{MM'} (-1)^{J'-M'} \sqrt{2K+1} \begin{pmatrix} J' & K & J \\ -M' & Q & M \end{pmatrix} \langle \alpha' J' M' | \hat{M} | \alpha J M \rangle. \tag{16.149}$$

This expression is easily evaluated whenever the operator $\hat{M}$ has been expressed as an irreducible tensor operator. For example, let $\hat{M}$ be a superposition of tensor operators $M_q^{(k)}$, combined with numerical coefficients c_q^k in the form

$$\hat{M} = \sum_{kq} (-1)^q c_{-q}^k M_q^{(k)}. \tag{16.150}$$

By applying the Wigner–Eckart theorem we obtain the matrix elements of this operator as

$$\langle \alpha' J' M' | \hat{M} | \alpha J M \rangle = \sum_{kq} (-1)^q c_{-q}^k (-1)^{J'-M'} (\alpha' J' || M^{(k)} || \alpha J) \begin{pmatrix} J' & k & J \\ -M' & q & M \end{pmatrix}. \tag{16.151}$$

Using this result we find that eqn. (16.149) involves the product of two three-j symbols. We use the orthogonality property of these symbols to evaluate the summation over magnetic quantum numbers, with the result

$$M(\alpha' J', \alpha J)_Q^{(K)} = (-1)^Q c_{-Q}^K \frac{(\alpha' J' || M^{(K)} || \alpha J)}{\sqrt{2K+1}} = (-1)^{J'-J+Q} M(\alpha J, \alpha' J')_{-Q}^{(K)\dagger}. \tag{16.152}$$

Thus the required numbers are reduced matrix elements multiplied by numerical coefficients.

Operator expectation values. When we take expectation values the expansion coefficients $M(\alpha' J', \alpha J)_Q^{(K)}$ embody the properties of the particular operator $\hat{M}$, for any system, while the state multipoles express the properties of the particular system, for any operator. It follows that the time-dependent expectation value of the operator $\hat{M}(t)$ can be written

$$\langle \hat{M}(t) \rangle = \sum M(\alpha' J', \alpha J)_Q^{(K)^\dagger} s(\alpha' J', \alpha J; t)_Q^{(K)} \exp[i\zeta_{\alpha J} - i\zeta_{\alpha' J'}]. \qquad (16.153)$$

This is the extension to multipoles of the simple expression

$$\langle \hat{M}(t) \rangle = \sum_{mn} M_{mn} s_{nm}(t) \exp[i\zeta_m - i\zeta_n]. \qquad (16.154)$$

As an example, an operator expressible as a sum of tensor operators, as in eqn. (16.150), has an expectation value given by the formula

$$\langle \hat{M}(t) \rangle = \sum_{\alpha' J' \alpha J k} (-1)^{J'-J} \frac{(\alpha J \| M^{(k)} \| \alpha' J')}{\sqrt{2k+1}}$$

$$\times \sum_q c_q^k s(\alpha' J', \alpha J; t)_q^{(k)} \exp(i\zeta_{\alpha J} - i\zeta_{\alpha' J'}). \qquad (16.155)$$

Note that the dependence upon spatial projection quantum number q occurs in the numerical coefficients c_q^k and in the state multipoles but not in the (reduced) matrix elements of $\hat{M}$.

Equations of motion for state multipoles. Numerous articles present equations of motion for state multipoles [Sho90, §21.9] [Ham68; Row75; Coo80]. These generalize the more elementary Bloch equations for nondegenerate two-state systems to allow treatments of angular momentum degeneracy. As with equations of motion for nondegenerate density matrices, the state multipole equations relate changes of populations to coherences, and changes of coherences to populations.

17

Systems with parts

Quantum systems typically involve three-dimensional motion, often of several electrons or particles bound together, and thus the physics involves several degrees of freedom, meaning that the dynamics involves more than a single generalized coordinate. Each degree of freedom – each spatial coordinate – requires its own multidimensional Hilbert space. The several independent degrees of freedom $A, B, \ldots$, are associated with independent Hilbert spaces $\mathcal{H}^A, \mathcal{H}^B, \ldots$. When combined they require a product space (see Sec. 17.3.1 below) in which the factors describe the separate Hilbert spaces. The present section discusses some quantum properties associated with multiple degrees of freedom.

17.1 Separability and factorization

17.1.1 Separable coordinates

The Laplacian operator ∇^2 of the Helmholtz equation and kinetic energy operator is separable in various coordinate systems, and for each there exist well-studied analytic solutions to the scalar eigenvalue equation

$$\nabla^2 F(\mathbf{r}) = -k^2 F(\mathbf{r}) \tag{17.1}$$

in which the eigenfunctions $F(\mathbf{r})$ maintain constant values over surfaces defined by constant values of one coordinate [Mor53]. In Cartesian coordinates x, y, z the eigenfunction equation (17.1) reads

$$\left[\frac{\partial^2}{\partial x^2} + \frac{\partial^2}{\partial y^2} + \frac{\partial^2}{\partial z^2} - k^2 \right] F(x, y, z) = 0. \tag{17.2}$$

In cylindrical coordinates ρ, z, φ the equation reads

$$\left[\frac{\partial^2}{\partial \rho^2} + \frac{1}{\rho} \frac{\partial}{\partial \rho} + \frac{1}{\rho^2} \frac{\partial^2}{\partial \varphi^2} + \frac{\partial^2}{\partial z^2} - k^2 \right] F(\rho, z, \varphi) = 0. \tag{17.3}$$

In spherical coordinates r, θ, ϕ the equation is

$$\left[\frac{1}{r^2} \frac{\partial}{\partial r} r^2 \frac{\partial}{\partial r} + \frac{\hat{\mathbf{L}}^2}{r^2} - k^2 \right] F(r, \theta, \phi) = 0, \tag{17.4}$$

where $\hat{\mathbf{L}}^2$ is the squared angular momentum operator,

$$\hat{\mathbf{L}}^2 = \frac{1}{\sin\theta}\left[\frac{\partial}{\partial\theta}\sin\theta\frac{\partial}{\partial\theta} + \frac{\partial^2}{\partial\phi^2}\right]. \tag{17.5}$$

For each such choice of coordinates x_1, x_2, x_3 we can factor the eigenfunction $F(\mathbf{r})$ into three factors, one for each independent coordinate,

$$F(\mathbf{r}) = f_a^A(x_1) f_b^B(x_2) f_c^C(x_3). \tag{17.6}$$

Here the superscripts A, B, C identify families of solutions associated with a particular coordinate variable. When we impose constraining boundary conditions, as we do for a bound particle or a field cavity, each of the separate functions has a discrete set of solutions, identified here by labels a, b, c. The surfaces of constant function value may be either stationary (standing waves) or moving (traveling waves).

The functional factorization of eqn. (17.6) underlies the construction of single-electron wavefunctions in a central-field potential and the construction of mode fields. The labels that are here denoted a, b, c may take integer values, descriptive of the number of nodes in the associated coordinate, or they may be from a continuum for traveling waves.

17.1.2 Separable Hilbert spaces

Each degree of freedom has an associated Hilbert space whose basis vectors describe the possible states of that degree of freedom. To describe M degrees of freedom we require M independent Hilbert spaces. Any quantum state must specify attributes from each of these. A possible quantum state of the system might therefore have the form of a product, written variously as

$$\boldsymbol{\psi}_{abc...} = \boldsymbol{\psi}_a^A \boldsymbol{\psi}_b^B \boldsymbol{\psi}_c^C \cdots \equiv |a\rangle^A |b\rangle^B |c\rangle^C \cdots \equiv |a, b, c, \cdots\rangle^{ABC...}, \tag{17.7}$$

where the labels $A, B, C, \ldots$ identify the various degrees of freedom (or the various particles), i.e. the independent Hilbert subspaces $\mathcal{H}^A, \mathcal{H}^B, \mathcal{H}^C, \ldots$. The subscripts $a, b, c, \ldots$ specify particular basis states within the subspaces. More generally the statevector will be some superposition of such products, say[1]

$$\boldsymbol{\Psi} = \sum_{a,b,c,\cdots} C_{abc...}\hat{S}\boldsymbol{\psi}_{abc...} = \sum_{a,b,c,\cdots} C_{abc...}\hat{S}|a, b, c, \cdots\rangle^{ABC...}. \tag{17.8}$$

Here $\hat{S}$ denotes any necessary symmetrizing (or antisymmetrizing) operator, such as would be needed if the degrees of freedom are those associated with indistinguishable particles (e.g. electrons within an atom). For example, with two identical particles the needed symmetry is

$$\boldsymbol{\psi}_{abc} = \hat{S}|a, b\rangle^{AB} = \frac{1}{\sqrt{2}}[|a, b\rangle^{AB} \pm |b, a\rangle^{AB}], \tag{17.9}$$

---

[1] Here the labels on $C_{abc...}$ do not partake of the permutation.

where the plus sign occurs with bosons (integer spins) and the minus sign with fermions (half-integer spins). The system quantum state is *separable* if we can write the statevector as a simple product, as in eqn. (17.7).

17.2 Center of mass motion

In treating systems of bound particles, such as the electrons and nuclei of atoms and molecules, we treat the charges and currents as multipole moments, evaluated at the center of mass. The motion of that center is treated separately. The following paragraphs discuss the separation of particle motions into that of a collective center together with motion relative to that center.

17.2.1 Center of mass

Classical point-particles. In a classical system of N particles there will be N three-dimensional equations of motion. The force on particle i will arise both from interactions between other particles, $\mathbf{F}_{ij}$, and from fields external to the system, $\mathbf{F}_i^{ext}$,

$$m_i \frac{d^2}{dt^2} \mathbf{r}_i = \mathbf{F}_i^{ext} + \sum_j \mathbf{F}_{ij}. \tag{17.10}$$

We define the *center of mass* coordinate

$$\mathbf{R} = \frac{m_1 \mathbf{r}_1 + m_2 \mathbf{r}_2 + \cdots}{m_1 + m_2 + \cdots}, \tag{17.11}$$

for which the equation of motion reads

$$(m_1 + m_2 + \cdots) \frac{d^2}{dt^2} \mathbf{R} = \sum_i \mathbf{F}_i^{ext}. \tag{17.12}$$

In the absence of external forces the motion of the center of mass is that of a free particle whose mass equals the total mass of the system.

For the description of two particles we replace the coordinates $\mathbf{r}_1$ and $\mathbf{r}_2$ by the center of mass $\mathbf{R}$ and the relative separation $\mathbf{r}_{12} = \mathbf{r}_1 - \mathbf{r}_2$. The equation for this relative variable is

$$m \frac{d^2}{dt^2} \mathbf{r}_{12} = \mathbf{F}_{12}, \tag{17.13}$$

where m is the *reduced mass*,

$$m = \frac{m_1 m_2}{m_1 + m_2}, \quad \text{or} \quad \frac{1}{m} = \frac{1}{m_1} + \frac{1}{m_2}. \tag{17.14}$$

When the two masses are very different the reduced mass is very nearly the smaller mass. As an example, the reduced mass of the electron–nucleus system is approximately the electron mass, but with a small correction for the specific nuclear mass, observable as an isotope shift of bound-state energies.

Quantum particles; center-of-mass wavefunction. When constructing wavefunctions of two or more particles we consider motion described by vector $\mathbf{r}$ about a center of mass (cm) at $\mathbf{R}$. An overall composite wavefunction for a spinless particle can be written as a product

$$\psi_{cn\ell m}(\mathbf{R}, \mathbf{r}) = \psi_c(\mathbf{R}) \psi_{n\ell m}(\mathbf{r}), \qquad (17.15)$$

where c denotes appropriate labels describing the cm motion. When there are no confining forces on the particle the cm can move freely and the Hamiltonian for mass M is proportional to the Laplacian, here written as $\partial^2/\partial\mathbf{R}^2$,

$$\hat{H}^{cm} = -\frac{\hbar^2}{2\mathrm{M}} \frac{\partial^2}{\partial\mathbf{R}^2}. \qquad (17.16)$$

An appropriate wavefunction is one with well-defined linear momentum $\hbar\mathbf{k}$,

$$\psi_{\mathbf{k}}(\mathbf{R}) = \frac{1}{\sqrt{(2\pi)^3}} \exp(i\mathbf{k} \cdot \mathbf{R}). \qquad (17.17)$$

Alternatively, when the cm is constrained by confining forces (i.e. the particle is bound) then the relevant cm wavefunctions form a discrete set and the label c identifies one of the motional eigenstates.[2] Section 17.2 below discusses aspects of cm motion.

Motion of the cm has several important implications. It is with respect to the cm that the transition-inducing fields are to be evaluated. Thus motion of this reference frame appears as a Doppler shift.

Two other aspects of cm motion are relevant to coherent excitation. They differ in the nature of the wavefunction with which we describe the cm motion, a distinction analogous to that of traveling waves and standing waves for radiation.

The first occurs when we deal with excitation of a freely moving atom, such as occurs in an atomic beam moving across a laser beam. The cm coordinates of an atom then appear in a wavefunction appropriate to traveling waves of well-defined momentum. For such particles absorption of a traveling-wave photon adds both internal energy $\hbar\omega$ and cm momentum $\hbar\mathbf{k}$. Section 17.2.2 discusses this situation.

Alternatively, the cm may be constrained by static or nonresonant cw fields that trap the quantum system. Examples include ions held in a magneto-optical trap. The wavefunction describing cm motion then deals with bound motion. Section 17.2.5 summarizes this situation.

17.2.2 Atom optics

The center of mass motion of a nonrelativistic free particle, of mass M and velocity v, has wavelike properties characterized by the de Broglie wavelength

$$\lambda_d = 2\pi\hbar/\mathrm{M}v. \qquad (17.18)$$

---

[2] The choice of cm wavefunctions noted here is analogous to the traveling-wave and standing-wave mode fields discussed in App. D.2.5.

The possibility of producing particle beams whose velocity is so slow that the de Broglie wavelength is of macroscopic size makes possible the duplication of effects equivalent to those of classical optics but using matter waves, i.e. atom optics [Mey01; Hen06].

The absorption of a traveling-wave photon of frequency ω brings with it the addition of an increment in linear momentum $\hbar\omega/c$ in the direction of beam propagation. Resonant radiation therefore exerts a force on an atom [Kaz90; Coh98; Met99]. This force, induced by a single photon, deflects an atom of mass M by the angle

$$\theta = \arctan[\hbar\omega/\text{M}vc] = \arctan[\lambda_d/\lambda], \tag{17.19}$$

where $\lambda = 2\pi c/\omega$ is the optical wavelength. Subsequent deexcitation by *stimulated* emission will remove this momentum increment, thereby leaving the atom velocity as it was prior to excitation. However, *spontaneous* emission produces a radiation field that is distributed in direction (because it is a multipole field), and hence the associated momentum change is distributed over a range of angles. When spontaneous emission events are important, as is the case with excitation over a sufficiently long time, then the combination of laser excitation followed by spontaneous emission produces radiation pressure [Ash80]. The combination of absorption and spontaneous emission can serve as a cooling mechanism by diminishing cm motion [Ste86b; Met99; Jav06].

When two laser beams are present, propagating in opposite directions, then atoms can absorb radiation from one beam and return it to the other. A sequence of such events, appearing as Rabi oscillations of the internal excitation energy, will be accompanied by a coherent alteration of the cm motion of each atom. To describe fully the effect of radiation upon moving atoms it is therefore necessary to supplement the internal degrees of freedom (two or three excitation states, for example) with a continuum of linear momentum states for the cm coordinate.

Momentum states. Eigenstates $|\mathbf{p}\rangle$ of the cm linear momentum operator $\hat{\mathbf{p}}$ are defined, apart from a phase, by the properties

$$\hat{\mathbf{p}}\,|\mathbf{p}\rangle = \mathbf{p}\,|\mathbf{p}\rangle, \qquad \langle\mathbf{p}|\mathbf{p}'\rangle = \delta(\mathbf{p}, \mathbf{p}'). \tag{17.20}$$

Using Cartesian coordinates X, Y, Z for the cm, the wavefunction for cm coordinate $\mathbf{r}$ corresponding to a state of momentum $\mathbf{p}$, with components p_x, p_y, p_z, is factorable as

$$\langle\mathbf{r}|\mathbf{p}\rangle = \frac{1}{\sqrt{(2\pi)^2}} \exp(-\mathrm{i}\,p_x x/\hbar)\exp(-\mathrm{i}\,p_y y/\hbar)\exp(-\mathrm{i}\,p_z z/\hbar). \tag{17.21}$$

From this formula we evaluate matrix elements of the spatial part of the field. For example for the z component $p_z = p$

$$\langle p|\mathrm{e}^{\pm\mathrm{i}k\hat{z}}|p'\rangle = \int dz\,\langle p|z\rangle \exp(\pm\mathrm{i}kz)\,\langle z|p\rangle = \delta(p, p' \mp \hbar k). \tag{17.22}$$

That is, the field dependence $\exp(\mathrm{i}kz)$ acts to decrease the cm momentum in the z direction by $\hbar k$.

To account for such momentum changes we must include in the statevector expansion the product of states of internal excitation and of cm motion, in the form

$$\Psi(t) = \sum_n \int d\mathbf{p} \, C_{\mathbf{p},n}(t) \, |\mathbf{p}\rangle \, e^{-i\zeta_n} \psi_n. \tag{17.23}$$

Typically the field is idealized as a single traveling wave, as two counterpropagating waves, or as a standing wave. Under such circumstances it is only necessary to consider a discrete set of momentum states, differing by increments $\hbar\mathbf{k}$ where $\mathbf{k}$ is the wavevector appearing in the field construction. The statevector then has the form

$$\Psi(t) = \sum_{n,\nu} C_{\nu,n}(t) \, |\mathbf{p}_0 + \nu\hbar\mathbf{k}\rangle \, e^{-i\zeta_n} \psi_n, \tag{17.24}$$

where $\mathbf{p}_0$ is the initial momentum of the cm. From this expansion, given an expression for the field, we obtain a set of coupled equations for the amplitudes $C_{\nu,n}(t)$. These express both the probability of excitation (change of n) and of deflection, i.e. change of momentum by ν increments.

Photons. The discrete increments of momentum exchange do not require quantization of the electromagnetic field: Equation (17.22) provides the required discretization without the need for photons. However, the interpretation of the equations of motion and the observable phenomena is simpler with their use, as illustrated in the following paragraphs.

17.2.3 Deflection

A simple example will illustrate the basic principles. Because momentum changes occur in discrete increments it is useful to consider photons as the increments, and to deal with photon states. To treat the interaction of an atom with quantized radiation we require a Hamiltonian comprising three parts,

$$\mathsf{H} = \mathsf{H}^{rad} + \mathsf{H}^{at} + \mathsf{V}, \tag{17.25}$$

describing, respectively, the radiation field, the atom, and the interaction.

Two traveling waves. Consider a two-state atom of mass M traveling in the x, z plane and passing through monochromatic counterpropagating plane-wave laser beams, directed along the x axis. These have mode fields $\exp(\pm ikx)$. Quantization leads to two sets of photon states, $\phi_{n\pm}^{(\pm)}$, for these two fields. The field Hamiltonian for this system is the energy of the two fields,

$$\mathsf{H}^{rad} = \hbar\omega[\hat{a}_+^\dagger \hat{a}_+ + \hat{a}_-^\dagger \hat{a}_-]. \tag{17.26}$$

Let us assume that the atom is only slightly deflected from its initial direction, and so only the changing motion in the x direction need be evaluated; the original momentum maintains a constant value, say $\hbar K$. Changes of x momentum can only occur in discrete increments, of value $\hbar k$ added to an initial value $\hbar k_0$, and so the cm momentum state, for x motion, can be defined (apart from normalization) through the property

$$\hat{p}_x \boldsymbol{\psi}_\nu^{cm} = \hbar(k_0 + \nu k)\boldsymbol{\psi}_\nu^{cm}, \qquad (17.27)$$

where $\boldsymbol{\psi}_\nu^{cm} \equiv |\mathbf{p}_o + \nu\hbar\mathbf{k}\rangle$. To describe the combined state of the internal atomic structure, the cm motion, and the state of the two fields we express the statevector as

$$\Psi(t) = \sum_{a,\nu,n_+,n_-} \exp[-\mathrm{i}(n_+ + n_-)\omega t]\, C_{a,\nu,n_+,n_-}(t)\, \boldsymbol{\psi}_a'(t)\, \boldsymbol{\psi}_\nu^{cm}\, |n_+,n_-\rangle, \qquad (17.28)$$

where the internal-excitation states are taken as

$$\boldsymbol{\psi}_1'(t) = \exp(\mathrm{i}\omega t/2)\,\boldsymbol{\psi}_1, \qquad \boldsymbol{\psi}_2'(t) = \exp(-\mathrm{i}\omega t/2)\,\boldsymbol{\psi}_2, \qquad (17.29)$$

and the photon states are from a two-mode Fock space (App. D.3),

$$\hat{a}_+ |n_+,n_-\rangle = \sqrt{n_+}\,|n_+,n_-\rangle, \qquad \hat{a}_- |n_+,n_-\rangle = \sqrt{n_1}\,|n_+,n_- - 1\rangle. \qquad (17.30)$$

The Hamiltonian for the atom can, with suitable basis and in the RWA, be constructed from photon creation and annihilation operators along with the elementary transition operators $\pi(m,n)$,

$$\mathsf{H}^{at} = \frac{1}{2\mathrm{M}}\left[\hat{p}_x - \hbar k(\hat{a}_+^\dagger \hat{a}_+ - \hat{a}_-^\dagger \hat{a}_-)\right]^2 + E_1\,\pi(1,1) + E_2\,\pi(2,2). \qquad (17.31)$$

This accompanies an interaction

$$\mathsf{V} = \frac{\hbar\Omega_0}{2\sqrt{2}}\pi(2,1)[\hat{a}_+ \exp(\mathrm{i}k\hat{z}) + \hat{a}_- \exp(-\mathrm{i}k\hat{z})]$$

$$+ \frac{\hbar\Omega_0}{2\sqrt{2}}\pi(1,2)[\hat{a}_+^\dagger \exp(-\mathrm{i}k\hat{z}) + \hat{a}_-^\dagger \exp(\mathrm{i}k\hat{z})]. \qquad (17.32)$$

That is, each excitation of the atom requires loss of a photon from one of the beams, while each atomic deexcitation adds a photon into one of the beams. Each change of photon number changes the transverse momentum of the atom by $\hbar k/2\mathrm{M}$.

From this Hamiltonian, and the statevector expansion of eqn. (17.28), we obtain a set of coupled equations for the probability amplitudes. Their solution describes the progressive change of transverse momentum $\hat{p}_x$ with Rabi cycling of the field-induced excitation and deexcitation [Ber81].

Standing wave. The preceding discussion treated a model in which two freely traveling waves interact with the atom. If the radiation is, instead, a standing wave, then the two exponentials combine to produce the standing wave. Instead of two types of photons, one for each direction of propagation, we deal with a single type of photon, and a single pair of creation and annihilation operators $\hat{a}_s, \hat{a}_s^\dagger$. The radiation Hamiltonian becomes

$$\mathsf{H}^{rad} = \hbar\omega[\hat{a}_s^\dagger \hat{a}_s]. \qquad (17.33)$$

Instead of eqn. (17.32) we have

$$V = \frac{\hbar\Omega_0}{4}[\boldsymbol{\pi}(2,1)\hat{a}_s + \boldsymbol{\pi}(1,2)\hat{a}_s^\dagger][e^{ik\hat{z}} + e^{-ik\hat{z}}]. \qquad (17.34)$$

The absorption of a laser photon now brings a symmetric change of atom momentum: the action of the field cannot give a net momentum to the atom. However, the field can change the mean-square momentum. Over time, with the completion of Rabi cycles of excitation and deexcitation, the probability distribution of atom momentum alters and develops the appearance of a diffraction pattern [Ber81; Mey89; Sho91b][Sho90, §15.12].

17.2.4 Mirrors and beam splitters

Interesting effects occur when a three-state atom undergoes stimulated Raman transitions effected by collinear but counterpropagating pump and Stokes beams. Complete population transfer will then alter the cm motion, in the direction of the laser beams, by one increment from each beam. Typically the atoms move perpendicular to the laser beams, and so the result will be observable as an alteration of the atomic beam direction. This action, of complete beam deflection, is an atomic mirror, albeit for atoms at grazing incidence [Sho91b].

When the excitation produces a superposition of internal energy states, then it also will produce a superposition of cm linear momentum states. This action can serve as an atomic beam splitter, a matter-wave analog of an optical beam splitter [Mar91; Ash07]. For small deflection angles, when longitudinal momentum is much larger than photon recoil, the problem can be modeled as a one-dimensional Schrödinger equation for the transverse motion of the atoms.

17.2.5 Trapped ions

Ions held within a magneto-optic trap, or atoms held within an optical lattice, are describable by a separable Hilbert space that combines the discrete states of internal excitation for the trapped particle and cm motion within the trapping potential. To a first approximation this potential can be treated as quadratic in the displacement from equilibrium position, so the cm coordinates are effectively harmonic oscillator variables, and the statevector can be expressed as the sum of products of states appropriate to the internal excitation (e.g. the A space) and to the cm oscillators (the B space). The Hamiltonian of the trapped particle will include a coupling term that describes the effect of internal excitation upon cm motion [Lei03; For07; Dua10].

17.3 Two parts

The simplest illustration of a separable system is that of a quantum system that has two independent degrees of freedom (i.e. it is *bipartite*), say A and B. Examples include those of the following table:

A	B	A + B
electron orbit L, M_L	electron spin S, M_S	total ang. mom. J, M
electron J, M	nuclear spin I, M_I	hyperfine F, M_F
atom internal	cm motion	trapped atom
atom	quantized field	dressed states
atom	environment	

Each subsystem has a set of basis states, say ψ_a^A for set A and ψ_b^B for set B. The statevector can always be written as a sum of products, in the form

$$\Psi(t) = \sum_{ab} C_{ab}(t) \, \psi_a^A \psi_b^B. \tag{17.35}$$

An example occurs when we treat the electromagnetic field as quantized. Then for a single-mode field the full Hilbert space can be taken as the product of a space having atomic unit vectors ψ_a and a Fock space having photon-number states ϕ_n as a basis; we write

$$\Psi(t) = \sum_{an} C_{an}(t) \, \psi_a \phi_n. \tag{17.36}$$

For a bipartite system a separable statevector has the form

$$\Psi(t) = \Phi^A(t) \Phi^B(t), \tag{17.37}$$

where the individual factors are expressible as vectors within the respective subspaces $\mathcal{H}^A$ and $\mathcal{H}^B$,

$$\Phi^A(t) = \sum_a C_a^A(t) \, \psi_a^A, \qquad \Phi^B(t) = \sum_b C_b^B(t) \, \psi_b^B. \tag{17.38}$$

For any such state both A and B have definite properties; separate measurements give no additional information.

The quantum state is *not* separable if such a factoring is not possible; we cannot find separate vectors $\Phi^A(t)$ and $\Phi^B(t)$ such that eqn. (17.37) holds,

$$\Psi(t) \neq \Phi^A(t) \Phi^B(t). \tag{17.39}$$

Under such conditions a measurement of either A or B provides new information: we can deduce properties of B from measurements of A and so the subsystem parts are *correlated*.

The eigenstates of a Hamiltonian that links two subsystems can be classified by their correlation properties. Let the full Hamiltonian for a two-part system be the sum of the following parts,

$$\begin{aligned} \mathsf{H}(t) = {}& \mathsf{H}^A & & \textit{part A, eigenstates } \psi_n^A \\ & + \mathsf{H}^B & & \textit{part B, eigenstates } \psi_m^B \qquad (17.40) \\ & + \mathsf{V}^A(t) + \mathsf{V}^B(t) + \mathsf{V}^{AB} & & \textit{interaction.} \end{aligned}$$

For further simplification, suppose that each part – each subsystem – has just two quantum states, as occurs with a spin one-half particle or a two-state atom:

$$\psi_+^A, \psi_-^A \quad \text{and} \quad \psi_+^B, \psi_-^B.$$

When the interaction has the form $\mathbf{V}^A \cdot \mathbf{V}^B$, or

$$\mathsf{H}^{AB} = -V_-^A V_+^B + V_0^A V_0^B - V_+^A V_-^B \quad \text{where} \quad V_j \psi_k = v_{j,k} \psi_{j+k}, \tag{17.41}$$

then the eigenstates of H^{AB} are a triplet (threefold degenerate)

$$\Phi_{+1}^T = \psi_+^A \, \psi_+^B,$$
$$\Phi_0^T = \left(\psi_+^A \psi_-^B + \psi_-^A \psi_+^B\right)/\sqrt{2} \quad \text{correlated}, \tag{17.42}$$
$$\Phi_{-1}^T = \psi_-^A \psi_-^B,$$

and a singlet

$$\Phi_0^S = \left(\psi_+^A \psi_-^B - \psi_-^A \psi_+^B\right)/\sqrt{2} \quad \text{correlated}. \tag{17.43}$$

The terminology of singlet and triplet comes from the spectroscopic description of the collective spin states for two electrons.

Coupled angular momenta. The electrons of a multielectron atom offer an example of the interactions between separate degrees of freedom. The collective orbital angular momentum **L** describes a circulating electric current and therefore it is associated with a magnetic moment proportional to **L**. The collective spin similarly produces a magnetic moment proportional to **S**. The interaction energy of these two magnetic moments is proportional to the scalar product $\mathbf{L} \cdot \mathbf{S}$. By introducing the total angular momentum $\mathbf{J} = \mathbf{L} + \mathbf{S}$ and writing that interaction as

$$\mathbf{L} \cdot \mathbf{S} = \tfrac{1}{2}[\mathbf{J}^2 - \mathbf{L}^2 - \mathbf{S}^2], \tag{17.44}$$

we recognize that the interaction operator $\mathbf{L} \cdot \mathbf{S}$ commutes with the separate operators $\mathbf{J}^2, \mathbf{L}^2, \mathbf{S}^2$ and so it is diagonal in a coupled angular momentum basis $|LSJM\rangle$, see App. A.2.

17.3.1 Product space

The set of products exhibited in eqn. (17.35) is an example of a *direct product space* $\mathcal{H}^A \otimes \mathcal{H}^B$ of the separate Hilbert spaces $\mathcal{H}^A$ and $\mathcal{H}^B$ spanned by the basis vectors ψ_A^A and ψ_b^B. The operators acting on such a product space can be regarded as matrices obtained as a *Kronecker product* of matrices that act on individual parts. From matrices A and B we construct the Kronecker product $\mathsf{A} \otimes \mathsf{B}$ by replacing each element A_{nm} of A with the matrix $A_{nm}\mathsf{B}$. For example, let A and B be 2×2 matrices. The Kronecker product is a 4×4 matrix containing all possible combinations of individual elements of the two constituents,

$$\mathsf{A} \otimes \mathsf{B} = \begin{bmatrix} A_{11}\mathsf{B} & A_{12}\mathsf{B} \\ A_{21}\mathsf{B} & A_{22}\mathsf{B} \end{bmatrix} = \begin{bmatrix} A_{11}B_{11} & A_{11}B_{12} & A_{12}B_{11} & A_{12}B_{12} \\ A_{11}B_{21} & A_{11}B_{22} & A_{12}B_{21} & A_{12}B_{22} \\ A_{21}B_{11} & A_{21}B_{12} & A_{22}B_{11} & A_{22}B_{12} \\ A_{21}B_{21} & A_{21}B_{22} & A_{22}B_{21} & A_{22}B_{22} \end{bmatrix}. \qquad (17.45)$$

Note that the ordering is important, $\mathsf{A} \otimes \mathsf{B} \neq \mathsf{B} \otimes \mathsf{A}$: the separate constituents of the product act on specific systems.

Operators $\hat{A}$ or $\hat{B}$ that act only on the respective spaces A and B can be represented as Kronecker products in which one of the pairs is a unit matrix, $\mathbf{1}$. For two-dimensional matrices we have the construction

$$\mathsf{A} \otimes \mathbf{1}^B = \begin{bmatrix} A_{11} & 0 & A_{12} & 0 \\ 0 & A_{11} & 0 & A_{12} \\ A_{21} & 0 & A_{22} & 0 \\ 0 & A_{21} & 0 & A_{22} \end{bmatrix}, \quad \mathbf{1}^A \otimes \mathsf{B} = \begin{bmatrix} B_{11} & B_{12} & 0 & 0 \\ B_{21} & B_{22} & 0 & 0 \\ 0 & 0 & B_{11} & B_{12} \\ 0 & 0 & B_{21} & B_{22} \end{bmatrix}. \qquad (17.46)$$

The space of basis vectors is also representable as a Kronecker product. Continuing the two-dimensional example we have the construction

$$\Psi = \begin{bmatrix} a_1 \\ a_2 \end{bmatrix} \otimes \begin{bmatrix} b_1 \\ b_2 \end{bmatrix} = \begin{bmatrix} a_1 b_1 \\ a_1 b_2 \\ a_2 b_1 \\ a_2 b_2 \end{bmatrix}. \qquad (17.47)$$

The effect on such a column matrix produced by operators acting on the individual systems is

$$\mathsf{A} \otimes \mathbf{1}^B + \mathbf{1}^A \otimes \mathsf{B} = \begin{bmatrix} A_{11} + B_{11} & B_{12} & A_{12} & 0 \\ B_{21} & A_{11} + B_{22} & 0 & A_{12} \\ A_{21} & 0 & A_{22} + B_{11} & B_{12} \\ 0 & A_{21} & B_{21} & A_{22} + B_{22} \end{bmatrix}. \qquad (17.48)$$

17.3.2 Interactions between parts

When the system has many parts (i.e. subsystems), either from several degrees of freedom or from multiple-particle composition (e.g. atoms of a molecule or pairs of ions in a trap), the Hamiltonian for the full system has a block structure, in which matrices replace the elements that elsewhere appear as the numbers W_{nm}. The resulting matrix structure derives from two classes of interactions.

Within each subsystem there may occur interactions that link the states that form a basis for that subsystem and which can induce changes in this part alone. It is with such interactions that the preceding chapters dealt, particularly the construction of basis states (dressed and adiabatic) that eliminate such linkages.

But there may also occur interactions between the parts – between particles or between degrees of freedom.When the interactions are between nearest neighbors, then the block structure of the Hamiltonian matrix is a generalization of the tridiagonal matrices that occur in the chain linkage. When all the parts interact with a common part, as occurs when trapped particles interact with a common collective mode (a "bus"), then the block structure is a generalization of the bordered matrix. The general structures of these two classes of matrices are

$$
\mathsf{W}^{tri} =
\begin{bmatrix}
D_1 & V_1 & O & O & \cdots & O \\
V_1 & D_2 & V_2 & O & \cdots & O \\
O & V_2 & D_3 & V_3 & \cdots & O \\
\vdots & \vdots & \vdots & \vdots & \ddots & O \\
O & O & O & O & \cdots & D_N
\end{bmatrix},
$$

$$
\mathsf{W}^{bord} =
\begin{bmatrix}
D_1 & V_1 & V_2 & V_3 & \cdots & V_{N-1} \\
V_1 & D_2 & O & O & \cdots & O \\
V_2 & O & D_3 & O & \cdots & O \\
\vdots & \vdots & \vdots & \vdots & \ddots & O \\
V_{N-1} & O & O & O & \cdots & D_N
\end{bmatrix}.
$$

$$(17.49)$$

In all these cases the linkage patterns between parts generalize the linkage patterns between states, and similar techniques can be used to express the effect of a Hamiltonian, e.g. the use of spin matrices in App. A.1.3.

17.3.3 Accessing individual parts

Many of the quantum-state manipulations of interest treat composite systems in which the laser-excitation interaction alters only one of the parts. For example, rotational, vibrational, and electronic excitation of a molecule typically occur at very different wavelengths and so, to a first approximation, it is possible to alter these degrees of freedom independently. When the parts are two identical atoms it is necessary to focus the laser beam onto a single one of them; manipulations of the centers of mass can first separate and then recombine the atoms. In all such cases the part of the system that is unaffected provides a reference phase for comparison with the affected part; see Chap. 20.

The possibility of introducing product states, or separable wavefunctions, does not imply that we can treat the parts as entities that are separately accessible: An interaction may act simultaneously on two degrees of freedom. This happens, for example, in molecular excitation: A transition may simultaneously alter electronic, vibrational, and rotational quantum numbers. To treat such situations we reorganize the product states into new bases, with probability amplitudes that treat the specific transitions. The underlying product states no longer appear in the formalism; as in the preceding chapters, we deal with an ordered list of N essential states. The underlying electronic, vibrational, and rotational quantum numbers are abbreviated to simple integers $n = 1, 2, \ldots, N$ that label the N basis states of the Hilbert space of interest.

17.4 Correlation and entanglement

Under appropriate conditions the degrees of freedom can become *correlated*. That is, if by learning that the A degree of freedom is definitely in state n, then it may follow that the B portion of the system must be in state m.[3] When one of the following conditions holds such correlation is known as *entanglement* [Nie00; Rai01; Bla08; Bar09].

I. One degree of freedom represents internal structure while another represents center of mass motion (e.g. of a trapped particle).
II. The degrees of freedom are associated with distinct particles (or photons) which, though initially together, are observed physically separated.

17.4.1 Entanglement

A general statevector of a multipartite system can always be written as a superposition,[4]

$$\Psi = \sum_{abc\ldots} C_{abc\ldots} \, \psi_a^A \, \psi_b^B \, \psi_c^C \cdots . \qquad (17.50)$$

This expression is *separable* if, and only if, the coefficients $C_{abc\ldots}$ can be written as products

$$C_{abc\ldots} = C_a^A C_b^B C_c^C \cdots \qquad (17.51)$$

because then we can write Ψ as a product of independent factors, as in eqn. (17.7),

$$\Psi = \left[\sum_a c_a^A \psi_a^A \right] \left[\sum_b c_b^B \psi_b^B \right] \left[\sum_c c_c^C \psi_c^C \right] \cdots \equiv \Phi^A \, \Phi^B \, \Phi^C \cdots , \qquad (17.52)$$

one for each dimension of the system. When the quantum statevector is not a product of N subsystem statevectors then in general it is not possible to assign a single statevector to any of the subsystems.

---

[3] We do not regard subsystems to be correlated if it is known *a priori* that they must be, respectively, in states n and m; they must be in a state for which different possible measurement outcomes are possible.

[4] Equation (17.8) is a special case in which the subscripts $abc\ldots$ on $C_{abc\ldots}$ do not participate in the permutation of labels on the product state $\psi_{abc\ldots}$.

Several mathematical tools are available for quantifying the amount of entanglement present between a subsystem and the remainder of the system [Nie00; Rai01; Bla08; Bar09]. The following subsections discuss some of these.

17.4.2 Reduced density matrices

From a density matrix of a multipartite composite system, involving attributes of subsystems $A, B, C, \ldots$ we obtain a reduced density matrix for subsystem A by taking the trace over all the other subsystem variables,

$$\rho^A = \text{Tr}_{BC\cdots}[\rho]. \tag{17.53}$$

When the system is bipartite the required trace, over subsystem B, is

$$\rho^A = \text{Tr}_B[\rho] = \sum_b \langle \psi_b^B | \rho | \psi_b^B \rangle. \tag{17.54}$$

When the density matrix ρ^{AB} represents a pure state Ψ^{AB} then the density matrix can be written

$$\rho^{AB} = |\Psi^{AB}\rangle \langle \Psi^{AB}|. \tag{17.55}$$

When further the statevector is a single product, as in eqn. (17.7), then the reduced matrices are themselves those of pure states,

$$\rho^A = |\Phi^A\rangle \langle \Phi^A|, \qquad \rho^B = |\Phi^B\rangle \langle \Phi^B|. \tag{17.56}$$

When the reduced density matrix represents a mixed state (rather than a pure state) then the subsystem A is entangled with subsystem B.

A mixed state can be regarded as an ensemble of pure states characterized by the probabilities p_n that the system will be found in state n. We have the least information about the system, and the subsystem is *maximally entangled*, when these probabilities are equal, i.e. when the reduced density matrix is a multiple of the unit matrix.

17.4.3 Entropy

It is often useful to quantify the amount of entanglement by a single number. Several alternatives have been proposed. The *von Neumann entropy* associated with a density matrix ρ is

$$S(\rho) = -\text{Tr}\, \rho \log_2 \rho. \tag{17.57}$$

Let the eigenvalues of the density matrix be λ_n. When the density matrix is expressed in a basis of its eigenvectors then the preceding expression becomes the *Shannon entropy*

$$S(\rho) = -\sum_n \lambda_n \log_2 \lambda_n. \tag{17.58}$$

The von Neumann entropy is equal to the Shannon entropy when the basis states are orthogonal; it is otherwise less than the Shannon entropy.

The entropy of any pure state is zero. The entropy is maximal for a mixed state in which every constituent state is equally likely, so that the density matrix for an N-state system is a multiple of the unit matrix,

$$\rho^{max} = \frac{1}{N} \mathbf{1}_N. \tag{17.59}$$

A subsystem A is in a maximally entangled state if there exists some local basis such that the reduced density matrix ρ^A is proportional to the unit matrix $\mathbf{1}_A$.

If the overall system is in a pure state, the entropy of one subsystem can be used to measure its degree of entanglement with the other subsystems. The quantum *mutual information* for a bipartite system is defined as

$$I(\rho^{AB}) = S(\rho^A) + S(\rho^B) - S(\rho^{AB}). \tag{17.60}$$

17.4.4 Other entanglement measures

A number of other measures of entanglement have been useful. Let the density matrix of the full system be expressed as

$$\rho^{AB} = \sum_i |\psi_i\rangle\langle\psi_i|. \tag{17.61}$$

Let the reduced density matrix for subsystem A be

$$\rho^A = \mathrm{Tr}_B \rho^{AB} = \sum_i \rho_i^A, \qquad \rho_i^A = \mathrm{Tr}_B |\psi_i\rangle\langle\psi_i|. \tag{17.62}$$

Let $S(\rho^A)$ be the von Neumann entropy of the reduced density matrix ρ^A,

$$S(\rho^A) = -\rho^A \log_2 \rho^A. \tag{17.63}$$

Then the *creation entanglement* is

$$E^c(A; \rho) = \min \sum_i p_i S(\rho_i^A). \tag{17.64}$$

17.4.5 Quantum information; qubits

The indivisible unit of classical information is the *bit*, which takes one of two possible Boolean values, typically taken to be 0 and 1 (alternatively T and F).

The corresponding unit of quantum information is called the "quantum bit" or *qubit* [Nie00; Ved06; Lam06; Bar09]. It is the information required to describe a quantum state of a two-state system. The two states are typically labeled 0 and 1 (alternatively $+$ and $-$ or $\uparrow$ and $\downarrow$).

A qubit can be realized by any pair of normalized and mutually orthogonal quantum states, labeled for this purpose as $|0\rangle$ and $|1\rangle$. The two states form a "computational basis": any other (pure) state of the qubit can be written as a superposition

$$\alpha|0\rangle + \beta|1\rangle \tag{17.65}$$

for some complex-valued numbers α and β such that

$$|\alpha|^2 + |\beta|^2 = 1. \tag{17.66}$$

Mathematically, a qubit may be regarded as a vector in a two-dimensional Hilbert space – a complex vector space with inner product. A physical qubit is typically a microscopic system, such as an atom, a nuclear spin, or a polarized photon. A collection of N qubits is called a quantum register of size N.

17.4.6 Maximally entangled states

A subsystem is maximally entangled if the reduced density matrix is a multiple of the unit matrix. In two dimensions the entangled state

$$\psi_\pm = (|01\rangle \pm \epsilon|10\rangle)/\sqrt{1 + |\epsilon|^2} \tag{17.67}$$

is maximally entangled if and only if $|\epsilon| = 1$. An example of an N-dimensional maximally entangled state is

$$\Phi_+ = \frac{1}{\sqrt{N}} \sum_n |n\rangle. \tag{17.68}$$

For N dimensions the *Greenberger–Horne–Zeilinger (GHZ) state* [Pan98]

$$\Phi = \frac{1}{\sqrt{2}} [|000\cdots\rangle + |111\cdots\rangle] \tag{17.69}$$

is an example of a maximally entangled state.

For a bipartite system four maximally entangled states (*Bell states* or *Einstein–Podolsky–Rosen (EPR) states*) are

$$\phi_\pm = \frac{1}{\sqrt{2}}\left(|00\rangle^{AB} \pm |11\rangle^{AB}\right), \qquad \psi_\pm = \frac{1}{\sqrt{2}}\left(|01\rangle^{AB} \pm |10\rangle^{AB}\right). \tag{17.70}$$

The second pair of these states are the triplet and singlet states of eqns. (17.42) and (17.43),

$$\psi_+ = \Phi_0^T, \quad \psi_- = \Phi_0^S. \tag{17.71}$$

If we measure only the properties of one subsystem, say A, then we find states $|0\rangle$ and $|1\rangle$ with equal probability. However, simultaneous measurements for both subsystems are perfectly correlated.

18

Preparing superpositions

The first interests in coherent excitation by laser pulses concentrated on producing complete population transfer – typically inversion of a two-state system, starting from the ground state. In recent years attention has shifted towards more general quantum-state manipulation, such as the preparation of a specified coherent superposition of two or more quantum states.

Theorists have long been interested in coherent superpositions [Moy99; Gro98; Sil08], particularly of quantum states whose extent is macroscopic – the so-called "Schrödinger cat" states [Kni93; Ger97]. As with the procedures for producing population transfer by means of coherent excitation, the techniques for more general quantum-state manipulation are of two classes: the abrupt change, in which an abruptly applied resonant pulse induces a controlled partial Rabi oscillation, and adiabatic changes [Mar91; San04; Kis05]. I will concentrate on the latter technique, and will discuss only superpositions of atomic states, as contrasted with superpositions of photon states.

18.1 Superposition construction

In considering superpositions it is essential to distinguish between superpositions of *degenerate* states (those that share a common unperturbed energy $E_1 = E_2 = \cdots = E_N$) and superpositions of *nondegenerate* states. The distinction becomes clear when we write the statevector, in the absence of any laser interaction, as

$$\Psi(t) = \sum_n \exp(-i\, E_n t/\hbar)\, C_n\, \psi_n, \tag{18.1}$$

with constant C_n. This is the most general form for the statevector of the undisturbed atom: a fixed superposition of bare states ψ_n, each multiplied by a periodic phase factor $\exp(-i\, E_n t/\hbar)$. Although the probabilities determined from this superposition are constant,

$$P_n(t) = |C_n|^2, \tag{18.2}$$

other properties are not. In particular, the coherences undergo oscillations at Bohr frequencies. For example, the two-state coherence associated with the statevector of eqn. (18.1) is

$$\rho_{12}(t) = \mathrm{Re}\Big\{ C_1 C_2^* \exp[i\, (E_2 - E_1)t/\hbar] \Big\}. \tag{18.3}$$

Between nondegenerate pairs of states there generally exists some nonzero radiative transition probability and hence some nonzero probability of population loss via spontaneous emission. However, this need not be that of an electric-dipole interaction; the spontaneous emission rate may be that of a "forbidden" transition, e.g. electric quadrupole or octupole, or magnetic dipole. Thus it may well be that the excited state is sufficiently long lived (i.e. a *metastable state*) that laser-induced excitation (via electric-quadrupole or magnetic-dipole interaction) can be regarded as coherent excitation. Under such situations one may take interest in creating a two-state superposition

$$\Psi(t) = \exp[-i\,E_1 t/\hbar]\Big[C_1\psi_1 + C_2\psi_2'(t)\Big], \quad \psi_2'(t) \equiv \exp[-i\,(E_2-E_1)t/\hbar]\psi_2, \quad (18.4)$$

in which, apart from an overall phase, the structure appears constant in a reference frame that rotates at the Bohr frequency for the transition. However, during one Bohr period the actual superposition will, for example, cycle between $\psi_1 + \psi_2$ and $\psi_1 - \psi_2$.

The energies occurring in this equation are eigenvalues of the Hamiltonian in the absence of laser radiation. Therefore they include Zeeman shifts produced (deliberately or randomly) by magnetic fields, and Stark shifts produced by electric fields. Even though these may be small, they will in due time alter the phase of the superposition.

18.2 Nondegenerate states

The presence of nondegenerate states in a superposition produces time-varying probabilities even when there is no external interaction – no Rabi frequency. When the individual states are treated as wavefunctions the resulting time dependence appears as spatial distortion as the dominant probability shifts from one position to another. If only two states are involved the probability density cycles between two patterns. As more states are added to the superposition the wavefunctions form a wavepacket, whose changes may mimic that of a moving particle or a flexing structure.

18.2.1 Wavepackets

The construction of nondegenerate superpositions has been particularly useful in studies of vibrational excitation of molecules. For such situations one uses a single laser pulse to excite a single vibrational state into a set of closely spaced vibrational states of an excited electronic state. The relevant RWA Hamiltonian can be exhibited as a bordered matrix,

$$\mathsf{W} = \frac{1}{2}
\begin{bmatrix}
0 & \Omega_1 & \Omega_2 & \cdots & \Omega_{N-1} \\
\Omega_1 & 2\Delta_2 & 0 & \cdots & 0 \\
\vdots & \vdots & \vdots & \ddots & \vdots \\
\Omega_{N-1} & 0 & \cdots & 0 & 2\Delta_N
\end{bmatrix}, \quad (18.5)$$

where the magnitudes of the detunings Δ_n increase steadily with increasing n. After pulsed excitation with constant detunings,[1] the wavefunction for vibrational coordinate x then has the form

$$\Psi(t,x) = \sum_n \exp(-i\, E_n t/\hbar)\, C_n\, \psi_n(x) \qquad (18.6)$$

and the probability of observing the value x is

$$P(x) = \left| \sum_n \exp(-i\, E_n t/\hbar)\, C_n\, \psi_n(x) \right|^2. \qquad (18.7)$$

This distribution offers the opportunity to construct *wavepackets* – localized spatial distributions that move with time [Gar02; Wol05; Tan05].

The vibrational energies of a diatomic molecule are expressible as a power series in the index n,

$$E_n = E_0 + n\hbar\omega_a + n^2\hbar\omega_b + \cdots. \qquad (18.8)$$

A short pulse of duration $\Delta\tau$ will contain a range of frequencies $\Delta\omega \approx 2\pi/\Delta\tau$. Vibrational states that lie within the interval $\Delta\omega$ of resonance with the pulse carrier will undergo some excitation; the coefficient C_n depends upon the overlap of the initial wavefunction with the wavefunction $\psi_n(x)$ describing the vibrational coordinate x, weighted by the dipole transition moment (the Franck–Condon factor). The initial pulse-produced superposition represents a radial distribution of excitation localized at a position dictated by the initial wavefunction. As time passes the time-dependent phase factors $\exp(-i\, E_n t/\hbar)$ produce a moving wavepacket. This moves towards larger x, but slows and then returns as the centroid of the packet reaches the classical turning point on the vibrational energy curve.

In a typical application one creates an initial wavepacket constructed from electronically excited vibrations, allows this to evolve undisturbed, and then at an appropriate time induces a second set of transitions, this time back to low-lying electronic states. Such procedures allow, in principle, the coherent control of unimolecular chemical reactions, such as dissociation.

18.2.2 Creating superpositions with fractional STIRAP

The STIRAP dark state varies with the mixing angle; with this variation comes the changing dark-state composition that produces population transfer. It is possible to "freeze" this composition – and hence the statevector orientation in Hilbert space – by holding constant the ratio of pump and Stokes amplitudes. Obviously the fields must eventually vanish, but if they do so while maintaining constant ratio, then the dark state, and the attached statevector, will remain fixed. The result will be the superposition

$$\Psi(t) = \cos(\theta/2)\,\psi_1 - e^{-i\varphi}\sin(\theta/2)\,\psi_3'(t), \qquad (18.9)$$

---

[1] Excitation with swept detunings can produce complete population transfer from the initial state into a single final state, see Sec. 15.6.2.

where φ is the difference between Stokes- and pump-field phases and θ parametrizes the population inversion,

$$P_2(t) - P_1(t) = \cos\theta. \tag{18.10}$$

Such a procedure can be used to provide a coherent superposition of two degenerate Zeeman sublevels, starting from one of them, as discussed next. It contrasts with the technique of Sec. 14.7.6.

18.3 Degenerate discrete states

When the superposed quantum states are degenerate, then the time variation of the state-vector, in the absence of laser radiation, is a single overall oscillatory phase. Expressed in terms of a wavefunction, the construction reads

$$\Psi(t, x) = \exp(-i E_1 t/\hbar) \sum_n c_n \psi_n(x). \tag{18.11}$$

Not only the probabilities but also the coherences are static; one has a stationary superposition.

Such a stationary situation cannot occur in the simple two-state system excited via interaction by laser pulses, although such superpositions can be produced by a pulsed DC electric field (i.e. one with null carrier frequency). The simplest linkage that allows creation of a degenerate two-state superposition using laser fields is that of the lambda linkage, as exhibited with the stimulated Raman RWA Hamiltonian. With this system we can create superpositions of degenerate Zeeman sublevels of an atom characterized by angular momentum quantum numbers J and M.

There are several possibilities for constructing a superposition of Zeeman sublevels via two-stage lambda-like linkages and STIRAP-like pulses, see Sec. 14.7.6. To illustrate these, consider a system having $J = 1$ for the ground level (with states 1 and 3 having, respectively, $M = \mp 1$) and $J = 0$ for the excited state (state 2, with $M = 0$). We excite the system using a single carrier frequency but various choices of elliptical polarizations. Suppose we wish to construct a superposition of states 1 and 3, starting from state 1. We require two polarizations, σ_+ linking states 1 and 2 (the P field), and σ_- linking states 2 and 3 (the S field).

One possibility is to have large single-photon detuning Δ_P but resonant two-photon detuning, so that state 2 can be adiabatically eliminated. The result is a simple two-state system, with transitions between states 1 and 3 driven by the complex-valued two-photon Raman Rabi frequency[2]

$$\Omega_R(t) e^{i\varphi_R} = \frac{\check{\Omega}_P(t)\check{\Omega}_S(t)^*}{2\Delta_P}. \tag{18.12}$$

___
[2] The two-photon interaction generally has a phase factor $\exp(-i\varphi_P + i\varphi_S)$ multiplying the Rabi frequency.

We can apply the two pulses simultaneously, with individual Rabi angles adjusted such that the two-photon Raman Rabi angle

$$\mathcal{A}_R = \int_{-\infty}^{+\infty} dt \, \Omega_R(t) \tag{18.13}$$

produces the desired superposition upon completion of the pulse.

Another possibility is to employ a STIRAP-type pulse sequence, with the S pulse preceding the P pulse. However, unlike STIRAP, here we proceed with the adiabatic evolution only until the mixing angle has reached the value needed for the predetermined superposition. From that moment on, we force the two fields to maintain a constant ratio, so as to maintain this mixing angle, as they diminish. In this way we produce a *fractional STIRAP* (F-STIRAP) [Mar91; Vit99a; Kis05]. It is also possible to use SCRAP to produce superpositions (in a "half-SCRAP" [Yat02b]).

The full STIRAP procedure can be used to produce specified superpositions of degenerate Zeeman sublevels. Consider a linkage between sublevels of $J = 1$, through a single excited state having $J = 0$ via P field, and continuing via S field to metastable sublevels of $J = 1$, as shown in Fig. 14.19. By choosing polarizations of the two fields appropriately, the STIRAP procedure can produce a variety of sublevel superpositions.

18.4 Transferring superpositions

Once a two-state superposition has been created, various pulse sequences can alter its composition; the vector model of Chap. 10 offers a simple prescription for selecting an angular velocity vector, and pulse duration, that will produce any desired rotation of the Bloch vector, i.e. will convert any initial superposition state into a prescribed final superposition.

Often it proves useful to consider the set of states within which superpositions are to be prepared or transferred (the *target states* or *working states*) as a subspace of a larger Hilbert space, obtained by including a set of *ancillary* states which, though not populated initially or finally, participate in the statevector transformation.[3]

The simplest example is that of the lambda system: we can use a F-STIRAP technique to create a specified superposition of the two ground states with the aid of an ancillary excited state. More generally, we partition the full Hilbert space into two sets of states, with links only between them, not within them; see Sec. 17.3. Although we are interested only in manipulating states of the target set, we do so by means of unitary transformations that involve the ancillary states.

Such procedures allow us, for example, to transfer a superposition from one pair of states to a second pair of states, with the aid of a single ancillary state, as depicted in the five-state fan linkage of Fig. 18.1. This linkage pattern permits manipulation by means of a degenerate STIRAP-like adiabatic passage involving two simultaneous S-field components and two

---

[3] The ancillary states must be *essential* states – their linkages to the working states must be part of the full Hamiltonian – though initially and finally they hold no probability.

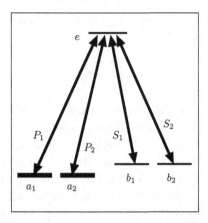

Figure 18.1 Adjustment of P and S pulses can transfer a superposition from states a_j to states b_j through ancillary excited state e. [Redrawn from Fig. 75 of B. W. Shore, Acta Physica Slovaka, **58**, 243–486 (2008) with permission.]

delayed P-field components. With suitable adjustment of the two pairs of amplitudes, transfer takes place between amplitudes of states a_j and b_j.

18.5 State manipulations using Householder reflections

Consider the transformation of a specified array of N probability amplitudes **C** by a succession of pulses to produce a specified array **C'**. This transformation of one coherent superposition into another can be expressed as the result of a unitary operator U,

$$\mathbf{C'} = \mathsf{U}\mathbf{C}. \tag{18.14}$$

We denote by u_{nm} the elements of the $N \times N$ matrix that produces this change,

$$\mathsf{U} = \begin{bmatrix} u_{11} & u_{12} & \cdots \\ u_{21} & u_{22} & \cdots \\ \vdots & \vdots & \ddots \end{bmatrix}. \tag{18.15}$$

These elements are obtained by solving the set of N linear equations (18.14). Given these elements, we wish to devise a pulse, or a succession of pulses, whose effect is described by U.

Because U is unitary, meaning that the inverse U^{-1} is equal to the Hermitian conjugate $\mathsf{U}^{\dagger}$, it can be regarded as rotation in N-dimensional Hilbert space, and can be constructed as the product of a succession of simple rotations affecting pairs of coordinates. Alternatively, it can be constructed from a succession of Hilbert-space reflections that involve coordinates in smaller-dimension subspaces. The following subsections examine the latter approach.

18.5.1 Diagonalization using Householder reflections

The key to constructing the desired transformation is the mathematical technique, termed *Householder reflections* [Hou58a; Hou58b], whereby we can successively eliminate columns and rows of U to obtain, in the end, a diagonal matrix. Each Householder reflection has the form [Iva06; Iva08; Kyo07; Iva07; Ran08]

$$M(v) = \mathbf{1} - 2|v\rangle\langle v|, \tag{18.16}$$

where $\mathbf{1}$ is the unit matrix and the vector $|v\rangle$ defines the normal to the reflection plane. A generalized Householder reflection (GHR) is defined as

$$M(v; \varphi) = \mathbf{1} + (e^{i\varphi} - 1)|v\rangle\langle v|. \tag{18.17}$$

This reduces to the simple Householder reflection when $\varphi = \pi$. The reflection has the property $M(v)^{-1} = M(v)$. The GHR has the property $M(v; \varphi)^{-1} = M(v; -\varphi)$.

The first step in this mathematical reduction considers the first column of U, expressed as the N-component column vector

$$|u\rangle_1 = [u_{11}, u_{21}, \ldots, u_{N1}]^T, \tag{18.18}$$

and the N-component unit vector

$$|e\rangle_1 = [1, 0, \ldots, 0]^T, \tag{18.19}$$

to construct the unit vector

$$|v_1\rangle = \frac{|u\rangle_1 - e^{i\phi_1}|e\rangle_1}{\sqrt{2[1 - Re(u_{11}e^{-i\phi_1})]}}, \tag{18.20}$$

where $\phi_1 = \arg[u_{11}]$. With this we construct the Householder reflection $M(v_1)$; acting on U it produces the result

$$M(v_1)U = \begin{bmatrix} e^{i\phi_1} & \mathbf{0} \\ \mathbf{0} & U^{(N-1)} \end{bmatrix}. \tag{18.21}$$

The square matrix $U^{(N-1)}$ appearing here has dimension $N - 1$. It has elements

$$U^{(N-1)} = \begin{bmatrix} u'_{22} & u'_{32} & \cdots \\ u'_{23} & u'_{33} & \cdots \\ \vdots & \vdots & \ddots \end{bmatrix}. \tag{18.22}$$

That is, this reflection has eliminated the first row and column of U, leaving only a diagonal element $e^{i\phi_1}$. To improve the construction, by replacing this exponential with unity, we use the GHR constructed with the vector

$$|v_1\rangle = \frac{|u\rangle_1 - |e\rangle_1}{\sqrt{2[1 - Re(u_{11})]}} \tag{18.23}$$

and the angle

$$\varphi_1 = 2 \arg(1 - u_{11}) - \pi \tag{18.24}$$

to produce the result

$$M(v_1; -\varphi_1)U = \begin{bmatrix} 1 & 0 \\ 0 & U^{(N-1)} \end{bmatrix}. \tag{18.25}$$

For the next step we use the first column of the matrix $U^{(N-1)}$

$$|u_2\rangle = [0, u'_{22}, u'_{23}, \cdots, u'_{N2}]^T \tag{18.26}$$

and the unit vector

$$|e_2\rangle = [0, 1, 0, \cdots, 0]^T \tag{18.27}$$

to construct the vector

$$|v_2\rangle = \frac{|u\rangle_2 - |e\rangle_2}{\sqrt{2[1 - \operatorname{Re}(u'_{22})]}}. \tag{18.28}$$

With these and the angle

$$\varphi_2 = 2 \arg(1 - u'_{22}) - \pi \tag{18.29}$$

we construct the GHR operator which reduces the matrix U further, producing the result

$$M(v_2; -\varphi_2)M(v_1; -\varphi_1)U = \begin{bmatrix} 1 & 0 & 0 \\ 0 & 1 & 0 \\ 0 & 0 & U^{(N-2)} \end{bmatrix}, \tag{18.30}$$

where $U^{(N-2)}$ has dimension $N-2$.

Proceeding in this way for N reflections we obtain the unit matrix

$$M(v_N; -\varphi_N) \cdots M(v_2; -\varphi_2)M(v_1; -\varphi_1)U = 1. \tag{18.31}$$

The final step involves only two states, and simply alters the phase of state N. From this we deduce that the original unitary matrix can be expressed as a succession of N GHRs,

$$U = M(v_1; \varphi_1)M(v_2; \varphi_2) \cdots M(v_N; \varphi_N). \tag{18.32}$$

The succession of reflections needed for the construction of eqn. (18.32) begins with a single ground state, continues with two ground states (a lambda linkage), then three (a tripod linkage), and continues with ever larger linkage patterns until, for the final step, all ground states are involved.

18.5.2 Implementation with a sequence of simultaneous pulses

A simple implementation of the Householder-reflection procedure is possible with the fan linkage pattern of Fig. 18.2, in which $N-1$ states couple to a single excited state.[4]

---

[4] For typographical simplicity I take N to be the total number of states; the working Hilbert subspace, in which quantum state manipulation occurs, has dimension $N-1$.

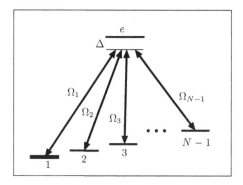

Figure 18.2 The "fan" linkage pattern for $N-1$ states linked to a single excited ancillary state.

This system has, as the RWA Hamiltonian, the bordered matrix

$$
W(t) = \frac{1}{2}
\begin{bmatrix}
0 & 0 & \cdots & \check{\Omega}_1(t)^* \\
0 & 0 & \cdots & \check{\Omega}_2(t)^* \\
\vdots & \vdots & \ddots & \vdots \\
\check{\Omega}_1(t) & \check{\Omega}_2(t) & \cdots & 2\Delta
\end{bmatrix}
\begin{matrix}
1 \\ 2 \\ \vdots \\ e
\end{matrix}.
\tag{18.33}
$$

Our objective is to transfer an initial superposition of the $N-1$ low-lying states into an alternative prescribed superposition; no population is to exist in the excited state.

The linkage pattern of this system has $N-2$ dark states together with a single bright state linked to the excited state. The time evolution of the bright state is governed by a RWA Hamiltonian (in the M-S basis) having the form

$$
w(t) = \frac{1}{2}
\begin{bmatrix}
0 & \check{\Omega}(t)^* \\
\check{\Omega}(t) & 2\Delta
\end{bmatrix}.
\tag{18.34}
$$

At the end of the pulse the time-evolution matrix for this two-state system can always be expressed with Cayley–Klein parameters as[5]

$$
U =
\begin{bmatrix}
a & b \\
-b^* e^{-i\chi} & a^* e^{-i\chi}
\end{bmatrix},
\tag{18.35}
$$

with $|b|^2 = 1 - |a|^2$ and

$$
\chi = \int_{t_1}^{t_2} dt' \, \Delta(t).
\tag{18.36}
$$

Of particular interest are pulses for which $b = 0$, so that all population has returned to the initial state (i.e. coherent population return, or CPR).

---

[5] The exponential factor $e^{-i\delta}$ occurs because the detuning has been placed asymmetrically with state 2, rather than as $\pm \Delta/2$ in states 1 and 2.

An important situation occurs when Δ is very large, so that $|b| \ll 1$ and little population transfer occurs [Kyo07]. Then the a parameter is

$$a = e^{\mathrm{i}\phi} \tag{18.37}$$

with phase

$$\phi = \int_{-\infty}^{+\infty} dt \, \frac{\Omega(t)^2}{\Delta}. \tag{18.38}$$

Thus it is always possible, with large detuning and suitable pulse envelope, to produce a predetermined phase.

We particularize this with the requirement that all of the Rabi frequencies have the same time dependence,

$$\Omega_n(t) = f(t) X_n. \tag{18.39}$$

For resonant interactions, $\Delta = 0$, we require that these pulses should be such that the Cayley–Klein parameter a is $a = -1$. This means that the pulse time dependence must satisfy the integral constraint

$$\int_{-\infty}^{+\infty} dt \, f(t) = (2k+1)2\pi/X, \tag{18.40}$$

where $X \equiv \sqrt{\sum_n |X_n|^2}$. Then the action of the Householder reflection $\mathsf{M}(v)$ is produced by taking the set of Rabi amplitudes X_n to be the elements of the vector $|v\rangle$.

To obtain a GHR $\mathsf{M}(v; \varphi)$ it is necessary that there be nonzero detuning, $\Delta \neq 0$, and that the Cayley–Klein parameter be $a = e^{\mathrm{i}\varphi}$. It is possible, for some analytic expressions for $f(t)$, to design the needed pulses, but in general it is necessary to resort to numerical simulation.

19

Measuring superpositions

The nature of measurements, and their place within quantum theory, has engaged physicists and philosophers for generations [Bra92; Sch03]. Much of that interest centered on variables such as position and momentum of free particles, whose values form a continuum. The present monograph deals with discrete quantum states; the measurements are those required to specify as completely as possible a particular discrete quantum state Ψ or, more generally, a density matrix ρ, defined within a finite-dimensional Hilbert space.

19.1 General remarks

General system. At the outset we assume that the possible quantum states are a small number – the N essential states used in formulating the time-dependent Schrödinger equation or specifying the dimensions of the density matrix. To completely characterize a density matrix for such a system we require the N^2 elements. Of these the N diagonal elements are real valued, while the off-diagonal elements of the upper right side are complex conjugates of those on the lower left. Thus with allowance for the requirement of unit trace a total of $N^2 - 1$ real numbers suffice to completely specify the density matrix. These values must be consistent with the constraints discussed in Sec. 16.6.3.

Pure state. If it is known that the system is in a pure state, we require the magnitude and phase of N probability amplitudes. These are constrained by normalization, and so only $2N - 1$ real numbers are needed. Out of these $2N - 1$ parameters the overall phase of the statevector is usually not of interest (but see Chap. 20). We then must devise measurement techniques that will provide satisfactory estimates of the remaining $2N - 2$ real numbers. For a two-state system the required $2N - 2 = 2$ quantities can be regarded as a pair of angles, as with the Poincaré description of polarization states (Sec. 3.6) or the Bloch-sphere presentation of probability amplitudes and density-matrix elements (Chap. 10). For larger systems we require additional quantities that are amenable to measurement.

Populations. It is relatively straightforward to obtain relative values for the N populations of the states (i.e. the absolute squares of probability amplitudes), either by observing the relative strength of a fluorescence signal or by measuring the relative photoionization signal,

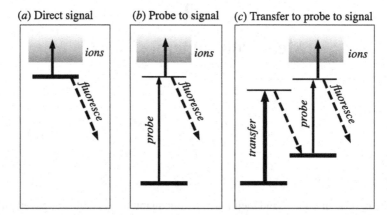

Figure 19.1 Various methods for observing populations. (a) Direct observation of fluorescence or photoionization. (b) Produce controlled excitation by a probe field, followed by observation of ionization or fluorescence. (c) Allow population transfer by spontaneous emission prior to probe step. [From Fig. 76 of B. W. Shore, Acta Physica Slovaka, **58**, 243–486 (2008) with permission.]

as discussed in Sec. 4.2 and shown schematically in Fig. 19.1. Such signals can come directly from the state of interest, by fluorescence or ionization, as indicated in Fig. 19.1(a), or they can be observed following a probe interaction (a *transfer field*) as in Fig. 19.1(b). They can also be evaluated after a sequence of events that include spontaneous emission, as in Fig 19.1(c). By suitably choosing the frequency and polarization of the probe field or the transfer field, it is possible to obtain signals directly proportional to the instantaneous population in a specified constituent of a superposition. However, such measurements give no information about the phase. To obtain that information it is necessary to map the phase onto a population which then produces a signal.

Measurement operators. When the system is known to be representable within a Hilbert space of dimension N then a series of discriminatory measurements will provide N values of populations – the diagonal elements of the density matrix, $\rho_{nn} = P_n$. Such measurements amount to obtaining expectation values of the projection operators of Sec. 16.10.1, $\Pi(n) \equiv |n\rangle\langle n|$ for $n = 1, N$,

$$P_n = \mathrm{Tr}[\rho\,\Pi(n)]. \tag{19.1}$$

The projection operators are Hermitian and orthogonal; they commute with each other and so they may be measured simultaneously. Such a measurement is known variously as a von Neumann projective measurement or as a *projection-valued measure* (PV) [Bra99].

To complete the specification of a density matrix or quantum state it is necessary to use additional operators that are not orthogonal. A common generalization of eqn. (19.1) introduces a set of non-negative Hermitian operators, represented by $N \times N$ matrices $\mathsf{A}(\mu)$ that sum to the identity operator (unit matrix)

$$\sum_{\mu} \mathsf{A}(\mu) = \mathbf{1}. \tag{19.2}$$

The mapping of a set onto the unit interval is an example of a (mathematical) *measure*. By extension, the mapping of a set of operators onto the unit matrix is termed a *positive operator-valued measure* or *probability-operator measure* (POVM) [Per95; Bra99; Bar09]. The operators are also referred to as Kraus operators [Kra83; Pec10]. Unlike the projection operators of eqn. (19.1) there is no limit on the number of independent measurements labeled by μ. From the measurement of the operator $\mathsf{A}(\mu)$ we obtain, instead of eqn. (19.1), the number

$$a_\mu = \text{Tr}[\rho\, \mathsf{A}(\mu)]. \tag{19.3}$$

This equation abbreviates a set of linear equations

$$a_\mu = \sum_{nm} \rho_{nm} A_{mn}(\mu) \tag{19.4}$$

in which the $A_{mn}(\mu)$ are known and the a_μ are measured. Given at least $N^2 - 1$ such independent expectation values a_μ it is possible, in principle, to solve the set of equations (19.4) for the unknowns ρ_{nm}; with more equations the solution is obtainable by least-squares procedures. In practice the usefulness of this formalism depends on the experimental availability of a suitable set of matrix operators. A minimal set of operators, sufficient for the determination of the density matrix, is termed a *quorum* [Bar09].

Quantum tomography. One approach to obtaining sufficient information to provide a full description of a quantum system from a succession of measurements of populations forms the discipline of *quantum tomography* [Her09]. In conventional tomography of a three-dimensional solid object the observer records a set of scans of intensity transmission along lines that slice the object. By inverting the arrays of data it is possible to infer the full three-dimensional density distribution of the object, thereby revealing the shapes and locations of embedded artifacts. Quantum tomography is similar: the observer measures populations for a set of parameters, often taken as generalized angles that define slices through a volume of Hilbert space, an example of eqn. (19.4). The resulting data array is inverted to obtain a density matrix.

Numerous papers have examined issues connected with measuring coherent super-positions – an aspect of the more encompassing concerns of quantum measurement theory [Bra92; Sch03]. Many of these articles suggested techniques based on adiabatic evolution [Una98; Vit99b; Vit00b; Kis01b; Vew03]. The sections below describe some of these.

19.2 Spin matrices and quantum tomography

The $N \times N$ basis matrices $\pi(n, m)$ of Sec. 16.10.1 suffice to access all the elements of any square matrix, and hence their expectation values, if measured, provide a complete characterization of the density matrix of an N-state quantum system. In characterizing the density matrix, and the underlying Hilbert space, two approaches to labeling the quantum states have use. In most of the present text I use integers $n = 1, N$ and the symbol ψ_n but several sections replace the integer n by the label $\mu = -s, s$ as is appropriate to a spin (or pseudospin) state (see Sec. 15.3.5), and use the symbol $\chi_\mu = |s, \mu\rangle$ for the associated quantum state. The connection is[1], as in eqns. (15.41)–(15.42),

$$\psi_1 = \chi_{-s}, \quad \dots, \quad \psi_n = \chi_\mu, \quad \dots, \quad \psi_N = \chi_s, \tag{19.5}$$

$$\mu = n - \tfrac{1}{2}(N+1), \qquad s = \tfrac{1}{2}(N-1). \tag{19.6}$$

With these labels we deal with spin-indexed elementary transition matrices $\pi(\mu, \nu)$. We can order the identifying subscripts $n, m = 1, N$, or $\mu, \nu = -s, s$ with $s = (N-1)/2$, into a single label $k = 1, N^2$ to obtain a set of independent basis matrices m_k, and with these write any density matrix as

$$\rho = \sum_k r_k \mathsf{m}_k, \qquad r_k = \mathrm{Tr}[\rho \mathsf{m}_k] = \langle \mathsf{m}_k \rangle. \tag{19.7}$$

Measurement of the complete set of expectation values $\langle \mathsf{m}_k \rangle$ provides the desired information. The elements of the coherence vector described in Sec. 16.10.2 offer one possibility. Various products of spin matrices offer another. The general measurement scheme discussed below was presented by Newton and Young [New68].

Two states. For $N = 2$ the spin matrices for $s = 1/2$ together with the unit matrix, denoted S_0, provide such a complete set of matrices. The four parameters r_k are then the components of the Bloch vector (relabeled) and unity, and the density matrix of eqn. (16.57) is parametrized as

$$\rho = \begin{bmatrix} \langle \mathsf{S}_0 \rangle - \langle \mathsf{S}_3 \rangle & \langle \mathsf{S}_1 \rangle + \mathrm{i}\langle \mathsf{S}_2 \rangle \\ \langle \mathsf{S}_1 \rangle - \mathrm{i}\langle \mathsf{S}_2 \rangle & \langle \mathsf{S}_0 \rangle + \langle \mathsf{S}_3 \rangle \end{bmatrix}. \tag{19.8}$$

Thus measurement of expectation values of the three spin matrices, equivalent to measurement of the three Bloch variables, $r_k = 2\mathsf{S}_k$, provides the desired characterization of the quantum state. The procedure is analogous to measurement of the Stokes parameters that characterize an electric field vector, see Secs. 3.6 and 19.3.

Larger N. When the Hilbert space is larger than $N = 2$ it is necessary to supplement the three spin matrices with products of spin matrices (or other matrices) to obtain a complete set of basis matrices; see Sec 16.10. By relating these matrix products to the elementary transition matrices $\pi(\mu, \nu)$ we obtain a connection between the elements of the density matrix and expectation values. It is then necessary to devise procedures that will measure the required expectation values.

For any spin s the diagonal matrix S_z is expressible as the sum of projection operators,

$$\mathsf{S}_z = \sum_\mu \mu \, \Pi(\mu), \qquad \Pi(\mu) = |s, \mu\rangle\langle s, \mu|, \tag{19.9}$$

---

[1] In the present context the quantum states χ_μ need have no connection with an intrinsic angular momentum, i.e. physical spin; in that sense they may be termed states of *pseudospin*. They are defined by the following equations.

and thus the expectation value of this matrix is the sum of N measurable probabilities[2]

$$\langle \mathsf{S}_z \rangle = \sum_\mu \mu\, P_\mu. \tag{19.10}$$

The measurement of this expectation value is analogous to the measure of spin orientation after an atomic beam has passed through a magnetic field that separates the trajectories of differing spin orientations – a *Stern–Gerlach measurement*.

Evaluation of expectation values for the other spin matrices requires a reorientation of the quantization axis, a rotation of the x or y axis into the vertical direction as parametrized by a rotation direction $\hat{\mathbf{n}}$.

Reoriented spin axes. The tomographic selection of discrete lines, slicing a multidimensional Hilbert-space volume, amounts to choosing a set of distinct quantization axes. The spin eigenstates $\boldsymbol{\chi}_\mu$ are defined with respect to a reference frame in which the quantization axis coincides with the z axis. Reorientation of the axis, along the direction $\hat{\mathbf{n}}$, introduces spin eigenstates whose polar and azimuthal angles are ϑ, φ, solutions to the eigenvalue equation

$$\mathsf{S} \cdot \hat{\mathbf{n}}\, \boldsymbol{\chi}_\mu(\hat{\mathbf{n}}) = \mu\, \boldsymbol{\chi}_\mu(\hat{\mathbf{n}}). \tag{19.11}$$

These states are constructed from the original three spin states with the use of elements of the reduced rotation matrix of order s,

$$\boldsymbol{\chi}_\mu(\hat{\mathbf{n}}) = \sum_\nu e^{i(\mu-\nu)\varphi}\, d^{(s)}_{\mu\nu}(\vartheta)\, \boldsymbol{\chi}_\mu. \tag{19.12}$$

Populations measured in the rotated reference frame are

$$P_\mu(\vartheta, \varphi) = \langle \Pi(\mu; \vartheta, \varphi) \rangle, \tag{19.13}$$

where the projection operators are

$$\Pi(\mu; \vartheta, \varphi) = |\chi_\mu(\hat{\mathbf{n}})\rangle \langle \chi_\mu(\hat{\mathbf{n}})|. \tag{19.14}$$

These measured quantities relate to the density matrix by the formula

$$P_\mu(\vartheta, \varphi) = \sum_{\mu\lambda} e^{i(\lambda-\nu)\varphi} d^{(s)}_{\nu,\lambda}(\vartheta)\, d^{(s)}_{\lambda,\mu}(\vartheta)\, \rho_{\lambda\nu}. \tag{19.15}$$

We carry out these measurements for a set of N^2 rotation angles $\hat{\mathbf{n}}_k$. This set of measured populations $P_\mu(k) \equiv P_\mu(\vartheta_k, \varphi_k)$ are related to the density matrix elements by a set of linear equations,

$$P_\mu(k) = \sum_{\mu\nu} c_{\mu k, \mu\nu}\, \rho_{\mu\nu}, \tag{19.16}$$

---

[2] The labeling here follows the convention of spin states, $\mu = -s, \ldots, s$ rather than that of the usual states, $n = 1, \ldots, N$.

in which the coefficients $c_{\mu k, \lambda \nu}$ are obtained from elements of the rotation matrix. We solve these equations (basically inverting the coefficient matrix) to obtain the elements of the density matrix $s_{\lambda \nu}$.

The probe Hamiltonian. This measurement procedure, the analog of a set of Stern–Gerlach measurements, as discussed by Newton and Young [New68], underlies the theory of quantum tomography. The filtering that occurs for physical spins must here be produced by a pulsed coherent-excitation Hamiltonian in which the adjustable parameters are Rabi frequencies and detunings, i.e. field polarizations, carrier frequencies and temporal pulse areas. Like the density matrix, the Hamiltonian is expressible as a set of unit matrices, and when these derive from products of spin matrices then the pulsed Hamiltonian is characterized by a set of angular momentum rotation angles. Starting from multiple copies of the unknown quantum state we measure the populations that result from the Hamiltonian parametrized by angles ϑ, φ. We make these measurements for a discrete set of angles, to obtain the probabilities of eqn. (19.15). The choice of rotation angles is somewhat arbitrary, although the polar angle θ should not be one of the zeros of the associated Legendre polynomial.

The excitation that is used to produce the measured probabilities need not be restricted to transitions amongst the N states being analyzed. We can tie these to other ancillary states during the excitation process, as long as these retain no permanent population.

19.3 Two-state superpositions

A two-state system provides an instructive example of possible techniques for obtaining the required information concerning the superposition amplitudes. The posed problem is analogous to the determination of field polarization characteristics, as embodied in the Stokes parameters, Sec. 3.6.2. Using a rotating reference frame in Hilbert space we parametrize the initial statevector at fixed time t with probability amplitudes or the two Bloch angles (cf. Sec. 5.6.1; the angle ϕ here need not be the field phase φ used there),

$$\Psi(t) = C_1 \psi'_1(t) + C_2 \psi'_2(t) = \cos(\theta/2)\, \psi'_1(t) + e^{-i\phi} \sin(\theta/2)\, \psi'_2(t). \qquad (19.17)$$

Alternatively, the information defining the quantum state can also be presented as a density matrix, thereby allowing more general mixed states and entangled systems:

$$\mathsf{s}(t) = \frac{1}{2} \begin{bmatrix} C_1 C_1^* & C_1 C_2^* \\ C_2 C_1^* & C_2 C_2^* \end{bmatrix} = \frac{1}{2} \begin{bmatrix} 1 + \cos\theta & e^{i\phi}\sin\theta \\ e^{-i\phi}\sin\theta & 1 - \cos\theta \end{bmatrix}. \qquad (19.18)$$

The density matrix in the rotating frame can also be described by components of the Bloch vector,

$$\mathsf{s}(t) = \frac{1}{2} \begin{bmatrix} r_3 - 1 & r_1 + ir_2 \\ r_1 1 i r_2 & r_3 + 1 \end{bmatrix}. \qquad (19.19)$$

Our objective is to devise a set of measurements from which to deduce the two angles or the four elements of the density matrix, thereby fully characterizing the quantum system (apart from an overall phase).

19.3.1 Direct excitation to signal

A straightforward procedure for a two-state system, suggested by Vitanov [Vit00b], is available when the two (working) states are degenerate and each has an excitation linkage to a single nondegenerate excited (ancillary) state, forming a lambda linkage pattern. The decay of the nondegenerate excited state – the decay rate and the angular distribution of radiation – does not depend on how it was excited, and hence any measurement of fluorescence intensity provides a measure of the population transferred and, in turn, the original population in the linked state. The links, labeled P and S, are uniquely fixed, either by polarization or, if the states are not degenerate, by frequency; we assume that these have the same time dependence, as will happen if they are elliptical polarization components.

We excite the system using either a single pulse or two simultaneous pulses, as described by the RWA Hamiltonian

$$\mathsf{W} = \frac{1}{2} \begin{bmatrix} 0 & 0 & \Omega_P e^{-i\varphi_P} \\ 0 & 0 & \Omega_S e^{+i\varphi_S} \\ \Omega_P e^{+i\varphi_P} & \Omega_S e^{-i\varphi_S} & 2\Delta \end{bmatrix} \begin{matrix} 1 \\ 2 \\ e \end{matrix} \qquad (19.20)$$

and parametrized by the two constant angles

$$\tan \Theta = \Omega_P(t)/\Omega_S(t), \qquad \beta = \varphi_P - \varphi_S. \qquad (19.21)$$

We measure the fluorescence signal $S(\Theta, \beta)$ produced from such excitation, for several choices of the controllable parameters Θ and β. One possible minimum sequence of measurements is shown in Fig. 19.2.

From the resulting signals we obtain the desired Bloch angles that define the superposition of eqn (19.17):

$$\tan(\theta/2) = \frac{\sqrt{S(0,0)}}{\sqrt{S(\pi/2,0)}},$$

$$\cos\phi = \frac{2S(\pi/4,0) - S(0,0) - S(\pi/2,0)}{2\sqrt{S(0,0)S(\pi/2,0)}}, \qquad (19.22)$$

$$\sin\phi = \frac{2S(\pi/4,\pi/2) - S(0,0) - S(\pi/2,0)}{2\sqrt{S(0,0)S(\pi/2,0)}}.$$

As with the determination of Stokes parameters, the procedure can also be applied to measurements that include an incoherent component and which therefore require a density matrix. When generalized, the method can be used to analyze any Zeeman coherence, for arbitrary numbers of sublevels, by means of excitation into an excited manifold of sublevels [Vit00a]. Again, control of polarization linkages produces different excitation

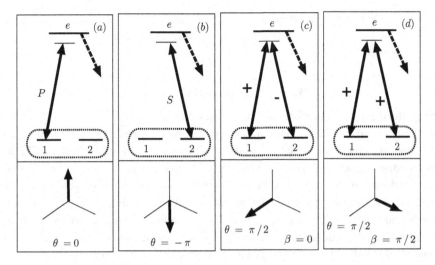

Figure 19.2 Set of pulsed interactions used to obtain fluorescence signal $\mathcal{S}(\Theta, \beta)$ for determination of superposition parameters. [Redrawn from Fig. 77 of B. W. Shore, Acta Physica Slovaka, **58**, 243–486 (2008) with permission.]

population. However, each excited sublevel radiates with a distinct, and different, dipole pattern. Thus the fluorescence intensity into a small solid angle depends upon both the total amount of excitation and the distribution of this excitation amongst sublevels. Extraction of the desired superposition parameters becomes more complicated, unless the experiment provides a signal that is independent of the sublevel distribution, e.g. photoionization.

19.3.2 Indirect excitation to signal

Rather than map the superposition directly into a signal-producing excited state we can instead use STIRAP to map it into a fixed lower-lying state whose population can subsequently be probed, either by spontaneous fluorescence, laser-induced fluorescence, or by photoionization. Figure 19.3 indicates the linkages involved: two simultaneous P fields, distinguished by polarization (magnitude and phase), and a single S field.

The analysis of a pair of states involves a sequence of STIRAP-like transitions that produce complete population transfers from initial states to a single analysis state. The concept relies on constructing a sequence of population transfers in which the P pulse has two components, thereby connecting two working states a_1 and a_2 with the excited state e. In turn a single S linkage connects this state to a single final-analysis state b. The overall linkage pattern is that of the tripod system.

19.4 Analyzing multistate superpositions

Kis and Stenholm [Kis01a] suggested a generalization of the procedure of Sec. 19.3.2 that provides information about the elements of a general N-state density matrix. Their idea is

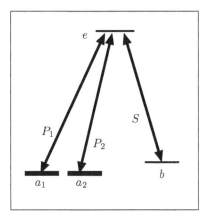

Figure 19.3 Linkage pattern for mapping a superposition of states a_1 and a_2 into population of a single state b by STIRAP with a two-component pump field $P_{\pm}$. Subsequent excitation and fluorescence from state a produces the signal. [Redrawn from Fig. 78 of B. W. Shore, Acta Physica Slovaka, **58**, 243–486 (2008) with permission.]

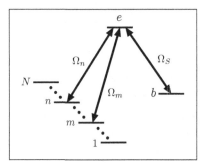

Figure 19.4 The Kis–Stenholm scheme [Kis01a] for analyzing the density matrix of N states. Pairs of pump pulses, Ω_n and Ω_m combine with a single Stokes pulse Ω_S to transfer population from states n and m into a single reference state b. Subsequent excitation and fluorescence from state a produces the signal.

to examine pairs of states from the superposition, each time with a set of P fields that will allow characterization of a 2×2 density matrix, as indicated in Fig. 19.4.

For each pair of sublevel states m, n a sequence of four measurements are needed, to obtain the values of the 2×2 density matrix. They propose four combinations of the two constituent P fields, parametrized as

$$\Omega_m(t) = \Omega_P(t)\cos\alpha, \qquad \Omega_n(t) = \Omega_P(t)\sin\alpha\, e^{i\beta}. \tag{19.23}$$

The result of such pulses is that the population in state b after the STIRAP sequence is related to the initial density matrix through the formula

$$\mathcal{P}_b = \cos^2\alpha\,\rho_{mm} + \sin^2\alpha\,\rho_{nn} + \mathrm{Re}\left\{\sin 2\alpha\,\mathrm{e}^{\mathrm{i}\beta}\,\rho_{mn}\right\}. \tag{19.24}$$

From measurements of population $\mathcal{P}_b$ transferred into state b for at least four choices of the parameters α, β there is sufficient information to evaluate the four elements of the density matrix involving states m, n.

19.5 Analyzing three-state superpositions

As the number of states increases, so too does the complexity of the measurement scheme needed to specify completely the characteristics of an unknown quantum state. With three states it is still possible to use polarization characteristics to provide the needed distinct probes. The simplest arrangement of linkages for this purpose is the tripod [Una98; Una99; Kis01a; Pas02; Kar04; Una04]. Figure 19.5 illustrates a tripod-linkage example with which to probe a degenerate three-state superposition through transitions to an excited state e followed by fluorescence. The superposed states here bear labels $-1, 0, +1$ appropriate to magnetic quantum number M of $J = 1$, rather than $1, 2, 3$, and in keeping with that interpretation we express the superposition using spin-state notation,

$$\Psi = c_-\chi_{-1} + c_0\chi_0 + c_+\chi_{+1}. \tag{19.25}$$

As with the two-state direct-excitation probe, the fluorescence from the nondegenerate excited state does not depend on how it was prepared. It therefore serves as a signal probe proportional to the population transferred.

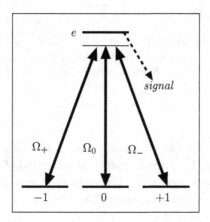

Figure 19.5 Mapping population in three degenerate Zeeman sublevels, $M = -1, 0, +1$, onto excitation followed by fluorescence signal from state e. [From Fig. 79 of B. W. Shore, Acta Physica Slovaka, **58**, 243–486 (2008) with permission.]

The RWA Hamiltonian for the set of linkages shown in Fig. 19.5 is a bordered matrix (the labels to the right show the ordering of states)

$$
\mathsf{W}(t) = \frac{1}{2}
\begin{bmatrix}
0 & 0 & 0 & \check{\Omega}_+(t)^* \\
0 & 0 & 0 & \check{\Omega}_0(t)^* \\
0 & 0 & 0 & \check{\Omega}_-(t)^* \\
\check{\Omega}_+(t) & \check{\Omega}_0(t) & \check{\Omega}_-(t) & 2\Delta
\end{bmatrix}
\begin{matrix}
-1 \\
0 \\
+1 \\
e
\end{matrix}
. \tag{19.26}
$$

The three Rabi frequencies appearing here are complex numbers, but we require that *they all have common time dependence*; we introduce amplitudes A_k and phases β_k, the control parameters, by writing

$$
\check{\Omega}_k(t) = A_k \exp(-\mathrm{i}\beta_k) f(t/T), \qquad k = +, 0, -. \tag{19.27}
$$

The procedure is to measure the fluorescence signal from state e for various choices of the control parameters A_k and β_k, and from these to deduce the unknown superposition parameters c_k.

The first step toward the theoretical description comes from recognizing that all of the transition strength of the three separate linkages can be combined into a transition into the excited state from a single "bright-state" superposition,

$$
\boldsymbol{\Phi}_b(t) = \frac{1}{2}\sqrt{(1-\eta)(1+\varepsilon)}\ \mathrm{e}^{\mathrm{i}\beta_+}\boldsymbol{\chi}_{-1} + \frac{1}{\sqrt{2}}\sqrt{1+\eta}\ \mathrm{e}^{\mathrm{i}\beta_0}\boldsymbol{\chi}_0
$$

$$
+ \frac{1}{2}\sqrt{(1-\eta)(1-\varepsilon)}\ \mathrm{e}^{\mathrm{i}\beta_-}\boldsymbol{\chi}_{+1}, \tag{19.28}
$$

where

$$
\varepsilon = \frac{A_+^2 - A_-^2}{A_+^2 + A_-^2}, \qquad \eta = \frac{A_0^2 - A_+^2 - A_-^2}{A_0^2 + A_+^2 + A_-^2}. \tag{19.29}
$$

This bright state and the excited state are linked, with detuning Δ, by the effective Rabi frequency

$$
\Omega(t) = \sqrt{A_0^2 + A_+^2 + A_-^2}\ f(t/T). \tag{19.30}
$$

The population in this bright state, expressed in terms of density matrices, is

$$
P_b(\varepsilon, \eta, \beta_{0+}, \beta_{0-}) = R_{-1-1}(\varepsilon, \eta)\rho_{-1-1} + R_{00}(\varepsilon, \eta)\rho_{00} + R_{11}(\varepsilon, \eta)\rho_{11}
$$

$$
+ R_{-10}(\varepsilon, \eta)\ |\rho_{-10}|\ \cos(\varphi_{-10} + \beta_{0+})
$$

$$
+ R_{01}(\varepsilon, \eta)\ |\rho_{01}|\ \cos(\varphi_{10} + \beta_{0-})
$$

$$
+ R_{-11}(\varepsilon, \eta)\ |\rho_{-11}|\ \cos(\varphi_{-11} + \beta_{-+}). \tag{19.31}
$$

Here the $R_{ij}(\varepsilon, \eta)$ are known functions.

The three probe fields act on this population to produce transitions into the excited state; the resulting population is proportional to the initial bright-state population and to a transition rate,

$$
P_e(t) = p_{b \to e}(t, t_i) P_b. \tag{19.32}
$$

The fluorescence signal S is proportional to this population (and to the probability η_b that this excited state will produce a fluorescence signal). Thus the signal can be related to the initial bright-state population as

$$S = \eta_b P_b \times \mathcal{P}_{b \to e}, \tag{19.33}$$

where $\mathcal{P}_{b \to e}$ is an appropriate time-integrated transition rate and η_b is the probability that the population in state b will produce a signal. For a short pulse the signal is

$$S = \eta_b P_e(t_f) = \eta_b P_b \times p_{b \to e}(t, t_i). \tag{19.34}$$

For a long pulse, producing optical pumping, the signal is

$$S = \eta_b \Gamma_e \int_0^\infty dt \, P_e(t) = \eta_b P_b \times \Gamma_e \int_0^\infty dt \, p_{b \to e}(t, t_i). \tag{19.35}$$

We conduct a series of measurements, each with a different setting of the excitation parameters. There results a set of linear equations relating the observed signal to combinations of the original density matrix elements,

$$\begin{aligned}
S(\varepsilon, \eta, \beta_{0+}, \beta_{0-}) = \eta \mathcal{P}_{b \to e} \Big[& R_{-1-1}\rho_{-1-1} + R_{00}\rho_{00} + R_{11}\rho_{11} \\
& + R_{-10} \, |\rho_{-10}| \, \cos(\varphi_{-10} + \beta_{0+}) \\
& + R_{01} \, |\rho_{01}| \, \cos(\varphi_{10} + \beta_{0-}) \\
& + R_{-11} \, |\rho_{-11}| \, \cos(\varphi_{-11} + \beta_{-+}) \Big].
\end{aligned} \tag{19.36}$$

Here the quantities $R_{ij}(\varepsilon, \eta)$ are known; the unknowns to be determined are ρ_{ij}, ϕ_{ij}, and η_b. In principle, one can evaluate the unknown parameters from a set of nine independent measurements of S [Vit00a; Vit00b].

19.6 Alternative procedures

Filtration. A somewhat different approach has been used in studies of superposition preparations in beams of metastable neon [Vew09]. Rather than vary the parameters of detection, as is done with conventional tomography, the system is first filtered. It is the filter whose properties are varied, thereby producing a signal whose value relates to the properties of the unknown quantum state. The filter field F selectively removes sublevel population by optical pumping. The particular removed sublevels depend on their definition in a coordinate system referenced to the F-field polarization rather than to the original coordinates used during preparation. Following this filtration step, the fixed field D transfers all remaining population into an excited state from which a cascade of spontaneous emissions produce the observed signals, proportional to the total population of all the post-filtration sublevels. By making measurements with a set of orientation angles for the F field the experimenter obtains sufficient information to specify the amplitudes and phases of the unknown quantum state.

Time reversal. Conceptually the simplest procedure for measuring a superposition that has been constructed with the aid of crafted laser pulses, starting from a single state, is to attempt to reverse the process using a second set of adjustable laser pulses. When these precisely mimic the preparation-laser set in reversed time ordering, then the system will return to the initial state. By noting the conditions of the reversing-laser pulses that maximize the population in the single state, one can deduce the nature of the superposition [Vew09].

When the superposition is produced in an atomic beam by a STIRAP transition into a set of magnetic sublevels, then in principle it is possible to reverse the procedure by passing the beam through a second pair of S and P beams but with the P pulse preceding the S pulse. If the polarizations of the second set of fields are identical to those of the first set, then all of the population will be returned to the single original state. When any other polarizations occur in the second set, then the population transfer will be incomplete. Thus a measurement of the population that returns to the original state, for various choices of polarizations in the second pair, will reveal the polarizations of the first pair of beams, and hence will reveal the parameters of the superposition [Una98].

Conditional measurements. Rather than employ a fixed set of measurements, as described here, one can instead proceed by a sequence of measurements that depend, at each step, upon the information gained from previous measurements [Buz97; Buz98]. Such a procedure allows progressive improvement in estimates and reduces the number of required measurements.

20

Overall phase; interferometry and cyclic dynamics

The basis vectors of the finite-dimensional Hilbert space used for describing quantum states typically take their identifying labels, in part, from eigenvalues of sets of commuting operators. Foremost of these operators is the atomic Hamiltonian $\hat{H}^{at}$, i.e. the full Hamiltonian without the interaction with laser fields and the environment. The discrete values of these energy eigenvalues provide the only needed label when the eigenvectors are not degenerate,

$$\hat{H}^{at}\boldsymbol{\psi}_n = E_n\boldsymbol{\psi}_n. \tag{20.1}$$

Alternatively, one can deal with instantaneous energy states (adiabatic states) defined as eigenvectors of the full Hamiltonian at time t,

$$\hat{H}(t)\boldsymbol{\Phi}_\nu(t) = \hbar\varepsilon_\nu(t)\boldsymbol{\Phi}_\nu(t). \tag{20.2}$$

When degeneracy occurs it is necessary to obtain other operators, commuting with $\hat{H}^{at}$ or $\hat{H}(t)$, whose eigenvalues supplement the energy as identifying labels. Zeeman sublevels, labeled by magnetic quantum number, are an example; see Sec 12.1. Thus the information that defines a particular quantum system derives from eigenvalues and, by extension, expectation values of suitable operators.

Using any of these complete sets of orthonormal vectors one can construct a statevector $\boldsymbol{\Psi}(t)$ by solving the time-dependent Schrödinger equation,

$$\frac{d}{dt}\boldsymbol{\Psi}(t) = -\frac{i}{\hbar}\hat{H}(t)\boldsymbol{\Psi}(t), \tag{20.3}$$

subject to initial conditions that specify $\boldsymbol{\Psi}(0)$ at $t = 0$. The initial construction may be expressed as a superposition of either the fixed basis vectors $\boldsymbol{\psi}_n$ or the energy eigenvectors $\boldsymbol{\Phi}_\nu(0)$. To fully characterize the subsequent system by its statevector we must measure not only the magnitudes of these constituents and their relative phases but also the overall phase of the statevector, relative to its initial value. This chapter discusses some of the issues associated with measuring and controlling that phase. The mathematics generalizes the discussion of electric field phase in Chap. 3.

20.1 Hilbert-space rays

A measurement that reveals a nondegenerate quantum system to have energy E_n implies that the quantum state at that moment is represented by a unit-norm eigenvector of the energy operator $\hat{H}^{at}$. Similarly, a measurement that reveals the system to have energy $\hbar\varepsilon_v(t)$ implies a statevector aligned with the unit vector $\mathbf{\Phi}_v(t)$. However, such an energy measurement does not (and cannot) define the phase of this unit vector: An eigenvalue equation defines a vector only within a constant scalar multiplier. For basis vectors $\boldsymbol{\psi}_n$ and statevectors $\mathbf{\Psi}(t)$ as well as instantaneous energy eigenvectors $\mathbf{\Phi}_v(t)$ the magnitude of this scalar is set by the requirement that the vectors have unit length,

$$\langle\psi_n|\psi_n\rangle = 1, \qquad \langle\Phi_v(t)|\Phi_v(t)\rangle = 1, \qquad \langle\Psi(t)|\Psi(t)\rangle = 1. \tag{20.4}$$

Although every statevector has unit length, the information coded into Hilbert space includes a phase. From measurements of expectation value E_n we can only determine that the quantum state is associated with some Hilbert-space unit vector $e^{i\phi}\boldsymbol{\psi}_n$, with phase ϕ undetermined. This same ambiguity, of incomplete information, applies to descriptions based upon a density matrix: The elements of a density matrix, like the elements of the Stokes vector, do not have information about overall phase.

Hilbert-space vectors having the same direction but different complex magnitudes are termed *rays*. Experiments that rely on measurements of expectation values do not give information about the phase, and so for such situations all elements of a ray are indistinguishable. Thus a physical state, as defined by its observables, can be regarded as associated with a ray of vectors, differing by a proportionality constant $e^{i\phi}$ of unit magnitude.

The phases of the fixed basis vectors $\boldsymbol{\psi}_n$ are arbitrary, and their components may be taken to be real. The overall phase of a statevector or an adiabatic state may be chosen initially to be zero. However, at later times or after a succession of operations on the quantum system, the phase may differ from this initial zero. Although the eigenvalue and expectation-value observables are unchanged, such a phase change is predictable and measurable by comparison with a reference state. For such a measurement it is necessary to establish a rule (a *connection*) for comparing phases. The mathematical tools are those used to treat the abstract structures of *projective spaces* and *fiber bundles*. The measurement techniques are generalizations of interference observations of superposed beams of coherent radiation.

Projective space. The normalized vectors of a given ray differ only by a phase. A change of vector phase is an example of a gauge transformation, a representation of the unitary group $U(1)$. Mathematically, a ray is an equivalence class of vectors differing only by amplitude or by phase. The space of rays is termed a *projective space* of the space of unit-norm vectors [Ana97].

Fiber bundles. The rays corresponding to physical states are complex lines (fibers) in Hilbert space. The set of all vectors, meaning all points of all lines, is an example of a mathematical *fiber bundle* or *line bundle*. Each ray is a fiber of the bundle, and the set of all rays is the *base* of the bundle. The eigenvectors associated with a given eigenvalue span a *sub-bundle* of all bundles [Zum87].

Let $\mathcal{N}$ be the space of normalizable vectors. Let $\mathcal{R}$ be the space of rays, in which elements only differ by a phase. The *natural projection* π maps each vector onto the ray on which it lies. The triplet of vectors, rays, and projections $(\mathcal{N}, \mathcal{R}, \pi)$ forms a *principal fiber bundle* over the base space of rays $\mathcal{R}$. Parallel transport defines a *natural connection* on this fiber bundle [Sam88].

20.2 Parallel transport

Objects that are constrained to move on surfaces, such as a ball rolling without slipping on a plane or over a sphere, have fewer degrees of freedom than would the unconstrained object, and their system space is correspondingly reduced. After such an object moves around a closed path on the constraining surface its system point need not return to the initial value; its difference is termed a *holonomy*.[1] To evaluate the effect it is necessary to consider motion of a full reference frame as it moves over its constrained surface. A simple example will illustrate the concept [Ber89].

Consider a position $\mathbf{r}$ on a unit sphere in three-dimensional Euclidean space, and place there a unit vector $\mathbf{e}$ perpendicular to $\mathbf{r}$. These two unit vectors, along with the unit vector

$$\mathbf{h} = \mathbf{e} \times \mathbf{r}, \tag{20.5}$$

provide a complete three-dimensional Cartesian reference frame of orthogonal unit vectors, in which $\mathbf{e}$ and $\mathbf{h}$ define a two-dimensional *tangent plane* to the sphere. (The set of all tangent planes form a *tangent bundle*.) Motion of this frame, when restricted to the surface of the sphere, is describable by an (instantaneous) angular velocity vector

$$\Upsilon = \mathbf{r} \times \frac{d}{dt}\mathbf{r}. \tag{20.6}$$

The motion of the frame is said to undergo *parallel transport* if the vector $\mathbf{e}$ remains tangent to the surface and the frame does not twist about the radius vector, meaning

$$\mathbf{e} \cdot \mathbf{r} = 0, \qquad \Upsilon \cdot \mathbf{r} = 0. \tag{20.7}$$

The resulting equation of motion for parallel transport is a torque equation:

$$\frac{d}{dt}\mathbf{e} = \Upsilon \times \mathbf{e}. \tag{20.8}$$

Upon defining the complex tangent-plane unit vector

$$\boldsymbol{\psi} = (\mathbf{e} + \mathrm{i}\mathbf{h})/\sqrt{2}, \qquad \boldsymbol{\psi} \cdot \mathbf{r} = 0, \tag{20.9}$$

---

[1] The term holonomy is established in the mathematical literature although Berry has proposed and advocated the term *anholonomy* for this difference [Ber90].

the parallel transport law reads [Ber89]

$$\text{Im}\langle\psi|\frac{d}{dt}\psi\rangle = 0. \tag{20.10}$$

That is, changes of the unit vector $\boldsymbol{\psi}$ are always orthogonal to $\boldsymbol{\psi}$ when the motion is parallel transport. The motion is *nonintegrable*: After the vector $\mathbf{r}$ completes a closed path on the sphere, $\mathbf{e}$ does not return to its initial direction; it has turned through an angle, the holonomy, equal to the solid angle subtended at the origin by the surface path. When expressed as a phase angle, the holonomy is termed a *geometric phase* [Ber84b; Sha89; Hol89; Gar10].

Examples. A holonomy angle is observable in the steady precession of the plane in which a Foucault pendulum swings. After 24 hours the pendulum has returned to its earlier position relative to the Sun, but the plane of swinging as viewed by an earthly observer has rotated by an angle (the Hannay angle), the holonomy of the motion.

As another example, the triad of vectors $\mathbf{E}, \mathbf{B}, \mathbf{k}$ with which one describes the vector properties of a laser beam will undergo parallel transport inside an unkinked but bent optical fiber as the propagation vector $\mathbf{k}$ changes direction while maintaining constant intensity [Chi86]. This holonomy is another example of the Hannay angle [Han85; Aga90].

Changes of electric field polarization for fixed propagation direction, may trace a closed path on the Poincaré sphere, as mentioned in Sec. 3.6.4. The holonomy then is the Pancharatnam angle [Pan56; Ber87; Sam88; Sha89].

The property of parallel transport presented here for three-dimensional Euclidean space readily generalizes to changes of Hilbert-space unit vector (i.e. basis vectors in N-dimensional Hilbert space). There parallel transport becomes adiabatic passage, and the geometric phase is often termed the Berry phase [Zwa90; Ana97] in recognition of influential papers by Berry.[2]

20.3 Phase definition

Measurements of phases require a comparison of two complex-valued functions or, for systems described by Hilbert-space rays, a superposition of two vectors brought into coincidence by suitable connection. The measurement then is one of vector length (interpreted as intensity or probability).

The Pancharatnam connection. Let $|a\rangle$ and $|b\rangle$ be two Hilbert-space vectors that are not orthogonal. The length of their sum (the interference of the two states or beams) $|c\rangle = |a\rangle + |b\rangle$ is

$$|\langle c|c\rangle|^2 = \langle a|a\rangle + \langle b|b\rangle + 2\text{Re}\langle a|b\rangle. \tag{20.11}$$

This length is maximum when the overlap $\langle a|b\rangle$ is real and positive. Under this condition the vectors (or states) are said to be "in phase", a relationship devised by Pancharatnam [Pan56;

---

[2] The work by Berry treated adiabatic motion, and some authors therefore regard the Berry phase as a particular case of the more general geometric phase.

Ber87; Sam88; Sha89]: A succession of vectors can be categorized by whether they are, with this definition, in phase.

Phase differences. We define the phase difference $\chi(a, b)$ between two states through the formula

$$\exp[\mathrm{i}\,\chi(a, b)] = \langle a|b\rangle / |\langle a|b\rangle|. \tag{20.12}$$

Using this phase we rewrite the interference expression as

$$|\langle c|c\rangle|^2 = \langle a|a\rangle + \langle b|b\rangle + 2|\langle a|b\rangle|\cos\chi(a, b). \tag{20.13}$$

This formula exhibits the traditional cosinusoidal variation of interference intensity as alterations of the constituent beam phases vary. Applied to the superposition of two statevectors differing only in phase,

$$\Psi' = (\Psi + e^{\mathrm{i}\chi}\Psi)/\sqrt{2}, \tag{20.14}$$

it produces the interferometry formula

$$|\langle\Psi'|\Psi'\rangle|^2 = \tfrac{1}{2}[1 + \cos\chi]. \tag{20.15}$$

Thus a measurement of probability (statevector length) can provide a phase.

Sequential operations. The phase difference defined above allows us to measure a phase change accompanying a succession of changes that return the state to its original ray in Hilbert space,

$$|a\rangle \to |b\rangle \to |c\rangle \to |a'\rangle. \tag{20.16}$$

This procedure allows measurement of phase changes produced by parallel transport, such as can be implemented by motion over the Poincaré sphere (where the phase change is known as the Pancharatnam phase) or the Bloch sphere (where it is termed a geometric phase). It is this procedure, using time evolution as the generator of operations, that allows us to define the overall phase of a statevector, relative to an initial value defined as zero.

20.4 Michelson interferometry

Interferometry techniques allow comparison of the phase of some system with that of a reference system by means of intensity measurements, as in eqn. (20.15). The prototype is a two-arm Michelson interferometer, schematically shown in Fig. 20.1 [Mic81; Har03; Cro09]. Its operation is as follows. An incoming wavepacket first encounters a beam splitter B_1 which divides the beam into two parts, one of which will serve as the reference beam. These beams travel through two separate arms of the apparatus, wherein they are reflected by mirrors M. The beams are then recombined by a second beam splitter B_2. Along one route the phase is modified, shown as ϕ, beyond whatever phase changes occurs to the reference beam. The intensity of the output varies according to the phase difference of the two paths – the apparatus maps a phase difference onto an intensity. In traditional Michelson interferometers the paths follow straight lines, and so the closed circuit involves mirrors, as

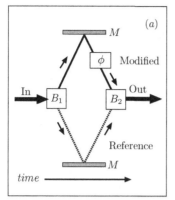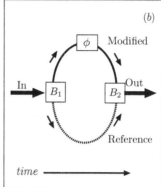

Figure 20.1 Schematic diagram of a Michelson interferometer: the input passes through a beam splitter B_1 where it divides into two separate paths, in one of which there is a phase alteration ϕ. The beams are recombined with a second beam splitter B_2, and the output intensity is measured. (a) The paths are straight lines, redirected by mirrors M. (b) The paths are through optical fibers.

shown in frame (a). When the optical paths are through fibers, as in frame (b), no mirrors are needed. With any such interferometer the phase difference ϕ includes any phase from differences in path length.

The essence of interferometry, as applied to quantum-state phase determination, generalizes the Michelson interferometer with the following steps [Cro09]:

Prepare: A specific state is prepared or selected.

Split: Coherent action (beam splitting) creates a phase superposition. In optical applications these are two momentum states. For atoms they are a superposition of two or more quantum states.

Alter: Modify one component of the superposition relative to the other.

Recombine: Coherently recombine the two components, thereby mapping phase information onto state populations.

Detect: Detect a specific population.

By varying the alteration stage the experimenter changes the phase of the recombining superposition components, thereby mapping phase onto populations and producing constructive (bright) or destructive (dark) interference. From the pattern of bright and dark interference fringes the experimenter determines the relative phase of the superposition components.

20.4.1 Matrix formalism

The operation of an interferometer readily fits a matrix description [Sjo00]. The effect of an idealized beam splitter that divides a beam into two equal parts is described by a matrix

U_B while a lossless mirror is described by U_M,

$$\mathsf{U}_B = \frac{1}{\sqrt{2}}\begin{bmatrix} 1 & 1 \\ 1 & -1 \end{bmatrix}, \qquad \mathsf{U}_M = \begin{bmatrix} 0 & 1 \\ 1 & 0 \end{bmatrix}. \tag{20.17}$$

The interferometer combines these actions as a succession of operations involving a beam splitter, a mirror, and a recombining beam splitter, $\mathsf{U}_B\mathsf{U}_M\mathsf{U}_B$. To account for differences in beam paths we alter the phase of one component, using a phase shifter

$$\mathsf{U}(\phi) = \begin{bmatrix} e^{i\phi} & 0 \\ 0 & 1 \end{bmatrix}. \tag{20.18}$$

The effect upon a statevector is the alteration

$$\mathbf{\Psi}' = \mathsf{U}_B\mathsf{U}_M\mathsf{U}(\phi)\mathsf{U}_B\mathbf{\Psi}, \tag{20.19}$$

from which we find the effect upon a density matrix as

$$\rho^{out} = \mathsf{U}_B\mathsf{U}_M\mathsf{U}(\phi)\mathsf{U}_B\,\rho^{in}\,\mathsf{U}_B\mathsf{U}(-\phi)\mathsf{U}_M\mathsf{U}_B. \tag{20.20}$$

The effect of this sequence upon a pure state 1 is

$$\rho^{out} = \frac{1}{2}\begin{bmatrix} 1+\cos\phi & i\sin\phi \\ -i\sin\phi & 1-\cos\phi \end{bmatrix}. \tag{20.21}$$

In particular, the element $\rho_{11}^{out} = \frac{1}{2}(1+\cos\phi)$ exhibits cosinusoidal dependence on the phase difference ϕ between the modified path and the reference path. When the phase difference is an integer multiple of 2π this element is maximum, and the two beams are said to be *in phase*.

20.4.2 Internal structure

Now consider a more general situation in which there are two independent degrees of freedom, say center of mass (cm) motion and internal changes (int). Let the cm motion be described by wavepackets traveling through an interferometer, as described above, and let the internal degrees of freedom be described by a discrete set of N energy states. We take the statevector to be the direct product of vectors from two independent Hilbert spaces, and correspondingly we take the evolution matrices to be direct products of matrices for each of these spaces (see Sec. 17.3.1):

$$\mathsf{U} = \mathsf{U}^{cm} \otimes \mathsf{U}^{int}. \tag{20.22}$$

By definition the mirrors and beam splitters do not affect the internal degrees of freedom, and so their matrices are

$$\mathsf{U}_M = \begin{bmatrix} 0 & 1 \\ 1 & 0 \end{bmatrix} \otimes \mathbf{1}^{int}, \qquad \mathsf{U}_B = \frac{1}{\sqrt{2}}\begin{bmatrix} 1 & 1 \\ 1 & -1 \end{bmatrix} \otimes \mathbf{1}^{int}, \tag{20.23}$$

where $\mathbf{1}^{int}$ denotes the unit matrix for internal degrees of freedom. Between the beam splitter and mirror we allow internal evolution to change one of the two arms via time evolution

$$\mathsf{U}_{\phi,T} = \begin{bmatrix} 0 & 0 \\ 0 & 1 \end{bmatrix} \otimes \mathsf{U}^{int}(T) + \begin{bmatrix} e^{i\phi} & 0 \\ 0 & 0 \end{bmatrix} \otimes \mathbf{1}^{int}. \tag{20.24}$$

The resulting effect on a density matrix is [Sjo00]

$$\rho^{out} = \frac{1}{4} \left\{ \begin{bmatrix} 1 & 1 \\ 1 & 1 \end{bmatrix} \otimes \mathsf{U}^{int}(T)\rho^{in}\mathsf{U}^{int}(-T) + \begin{bmatrix} 1 & -1 \\ -1 & 1 \end{bmatrix} \otimes \rho^{in} \right.$$
$$\left. + e^{i\chi} \begin{bmatrix} 1 & 1 \\ -1 & -1 \end{bmatrix} \otimes \rho^{in}\mathsf{U}^{int}(-T) + e^{-i\chi} \begin{bmatrix} 1 & -1 \\ 1 & -1 \end{bmatrix} \otimes \mathsf{U}^{int}(T)\rho^{in} \right\}. \tag{20.25}$$

The initial state we take to be

$$\rho^{in} = \begin{bmatrix} 1 & 0 \\ 0 & 0 \end{bmatrix} \otimes \rho^{int}(0). \tag{20.26}$$

The corresponding elements of the final state have variation proportional to the interference pattern

$$1 + |a|\cos[\phi - \chi],$$

where the amplitude and phase are defined by the equation

$$\text{Tr}[\mathsf{U}^{int}(T)\rho^{int}(0)] = a\,e^{i\chi}. \tag{20.27}$$

The interference oscillations of the reference phase ϕ are shifted by an amount χ.

20.5 Alternative interferometry

Rather than do interferometry with one state and two Hamiltonians, it is often preferable to use two states with one Hamiltonian, and to observe their superposition

$$|c\rangle = |a\rangle + |b\rangle. \tag{20.28}$$

After action of the Hamiltonian that alters the phase of the individual states their superposition is

$$|c'\rangle = |a'\rangle + |b'\rangle = e^{i\gamma_a}|a\rangle + e^{i\gamma_b}|b\rangle. \tag{20.29}$$

Then the length of this sum vector is

$$|\langle c'|c'\rangle|^2 = \langle a|a\rangle + \langle b|b\rangle + 2|\langle a|b\rangle|\cos(\chi + \gamma_a - \gamma_b), \tag{20.30}$$

where χ is the relative phase of the two states prior to the interaction.

20.6 Ramsey interferometry

The traditional optical interferometer relies on changes in propagation direction to separate the two beams – the two quantum states. Alternatively, one can employ a pair of pulses instead of the beam splitters, to create a Ramsey interferometer [Ram49]. The procedure, discussed in Sec. 10.2.5 and indicated schematically in Fig. 20.2, is as follows.

Prepare: Atoms are prepared, or selected, to be in state a.

Split: A $\pi/2$ pulse creates a superposition of the two states,

$$\Psi(0) = (\boldsymbol{\psi}_a + \boldsymbol{\psi}_b)/\sqrt{2}. \tag{20.31}$$

This is the counterpart of the beam splitter B_1 of Fig. 20.1.

Alter: The two states evolve freely for a time τ. During this interval they acquire different phases, ϕ_a and ϕ_b, proportional to their respective energies E_a and E_b and the time interval τ. The resulting statevector is

$$\Psi(\tau) = (e^{-i\phi_a}\boldsymbol{\psi}_a + e^{-i\phi_a}\boldsymbol{\psi}_b)/\sqrt{2}. \tag{20.32}$$

Recombine: A second $\pi/2$ pulse acts to produce the superposition

$$\Psi' = \tfrac{1}{2}(e^{-i\phi_a} - e^{-i\phi_b})\boldsymbol{\psi}_a + \tfrac{1}{2}(e^{-i\phi_a} + e^{-i\phi_b})\boldsymbol{\psi}_b. \tag{20.33}$$

The structure of this superposition depends on the relative phase of the two components at time τ. This, in turn, will depend on the relative energies of the two states in the regions between the pulses. Taken with a suitable beam filter, this is the counterpart of the beam splitter B_2 of Fig. 20.1.

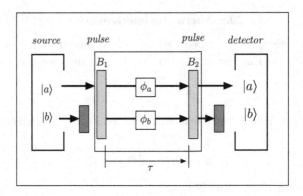

Figure 20.2 Ramsey pulses for a two-state system. The initial state a is subject to a first $\pi/2$ pulse that creates a superposition. The states evolve independently during time τ, during which they acquire different phases. A second $\pi/2$ pulse alters the superposition, restoring state a if the phase difference is π.

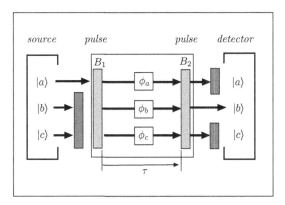

Figure 20.3 Ramsey pulses for a three-state system. The initial state a, subjected to the first pulse, becomes a superposition of the three states. Each evolves independently, acquiring a phase, until a second pulse restructures the superposition. The restructuring is phase sensitive. Detection here is of state b.

Detect: A detector measures the probability of state a,

$$P_a = |\langle \psi_a | \Psi' \rangle|^2 = \tfrac{1}{2}[1 + \cos(\phi_a - \phi_b)]. \tag{20.34}$$

This probability oscillates with phase difference, proportional to the interval τ, thereby exhibiting an interference pattern of *Ramsey fringes*.

Multiple states. The two-state Ramsey interferometer readily generalizes to N states, as suggested in Fig. 20.3 for $N = 3$. A first pulse, acting on initial state a at time $t = 0$, creates a superposition of the three states $\Psi(0)$, after which they evolve independently, each acquiring a phase. At time τ a second pulse, not necessarily repeating the first one, induces interactions that alter the superposition $\Psi(\tau)$, producing a state Ψ' whose components depend on the relative phases at time τ. Detection may involve any of the three states, here taken as state b.

A version of this general procedure has been discussed by Vewinger *et al.* [Vew07b; Vew09]. In that implementation the states are Zeeman sublevels and the first coherent superposition is produced by a STIRAP process. The second field is a probe field, adjusted to select particular superpositions. Alteration of the excitation parameters produces interference fringes whose analysis provides information about the superposition phases.

20.7 Cyclic systems

The coherent time evolution of a statevector that obeys the TDSE

$$\frac{d}{dt}\Psi(t) = -\frac{i}{\hbar}H(t)\Psi(t) \tag{20.35}$$

is expressible as the result of a unitary time-evolution operator $\mathsf{U}(t)$

$$\Psi(t) = \mathsf{U}(t)\Psi(0), \tag{20.36}$$

which obeys the equations

$$\frac{d}{dt}\mathsf{U}(t) = -\frac{i}{\hbar}\mathsf{H}(t)\mathsf{U}(t), \qquad \mathsf{U}(0) = \mathbf{1}. \tag{20.37}$$

A Hamiltonian is said to be *cyclic* if, after time T, it is indistinguishable from its initial value,

$$\mathsf{H}(T) = \mathsf{H}(0). \tag{20.38}$$

Its eigenvalues will have this same cyclic property. However, its eigenvectors may differ by a phase; Floquet theory is applicable, see App. G.

The statevector evolution is said to be cyclic, in the interval $[0, T]$, if and only if the matrix $\mathsf{U}(T)$, evaluated at the conclusion of the interval, $t = T$, is just a phase factor,[3] independent of the numerical value of T [Ana97],

$$\mathsf{U}(T) = \exp(i\phi). \tag{20.39}$$

This means that the statevector $\Psi(T)$ is an eigenstate of $\mathsf{U}(T)$, and that it has only changed from its initial position in Hilbert space by a phase factor involving the *cycle phase* ϕ,

$$\Psi(T) = e^{i\phi}\Psi(0). \tag{20.40}$$

As noted below, a nondegenerate adiabatic state is an example of a cyclic state.

Projective space. The Hilbert space $\mathcal{N}$ of unit-norm statevector solutions to the time-dependent Schrödinger equation has a subspace, a projective space $\mathcal{R}$, comprising the rays of $\mathcal{N}$. The time evolution of a statevector follows a curve in $\mathcal{N}$. If the statevector is cyclic it follows a closed curve $\mathcal{C}$ in the ray space (a projective mapping from $\mathcal{N}$).

To follow the development of the phase from its arbitrary initial value we define a new statevector that carries a phase $f(t)$ [Aha87; Eke00],

$$\Psi(t) = \exp[i\, f(t)]\,\widetilde{\Psi}(t). \tag{20.41}$$

Then the instantaneous phase $f(t)$ satisfies the equation

$$\frac{d}{dt}f(t) = -\frac{1}{\hbar}\langle\Psi(t)|\hat{H}(t)\Psi(t)\rangle + i\langle\widetilde{\Psi}(t)|\frac{d}{dt}\widetilde{\Psi}(t)\rangle. \tag{20.42}$$

We require that the change in $f(t)$ be the cycle phase, $f(T) - f(0) = \phi$, so that the phased statevector returns to its initial value, $\widetilde{\Psi}(T) = \widetilde{\Psi}(0)$. Then the cycle phase is the sum of two integrals,

$$\phi = a + b, \tag{20.43}$$

---

[3] As noted below with eqn. (20.56), when degeneracies are present the exponential becomes a more general unitary matrix.

that comprise distinct and very different contributions. The first integral, termed the *dynamical phase*, is minus the time integral of the expectation value of the Hamiltonian:

$$a = -\frac{1}{\hbar} \int_0^T dt \langle \Psi(t) | \hat{H}(t) \Psi(t) \rangle. \tag{20.44}$$

This portion of the cycle phase depends upon the details of the changing Hamiltonian and the speed with which changes occur. The second integral,

$$b = i \int_0^T dt \langle \tilde{\Psi}(t) | \frac{d}{dt} \tilde{\Psi}(t) \rangle, \tag{20.45}$$

discussed below, is independent of the rate at which the Hamiltonian changes: it is a *geometric phase* that depends only on the Hilbert-space path traced by the phased statevector, not the rate at which it travels.

It is instructive to apply the preceding theory to situations in which the statevector is initially aligned with an eigenstate of the Hamiltonian, i.e. a dressed or adiabatic state, and the time variations of the Hamiltonian are attributed to experimentally controlled changes of parameters. The following section discusses such dynamics.

20.7.1 Cyclic adiabatic evolution

Control space. The Hamiltonian that controls the changes to a statevector, and which force it to follow a particular trajectory in Hilbert space, is defined by a suitable set of parameters (e.g. angles and frequencies) over which the experimenter holds control, say $X_1, X_2, \ldots, X_M$. These can be regarded as components of a vector $\mathbf{X}$ in an M-dimensional *parameter space*. The motion of a point in this parameter space of the Hamiltonian (the *control space*) is linked, via the time-dependent Schrödinger equation, to the motion of the statevector in its Hilbert space.

As an experimenter adjusts the parameters $\mathbf{X}$, the Hamiltonian will vary, and with this change will come a change in the composition of the energy eigenvectors (the dressed states), as expressed in fixed Hilbert-space coordinates. That is, experimental controls establish a link between parameter space $\mathcal{P}$ and the Hilbert space of normalized statevectors $\mathcal{N}$.

Cyclic dynamics. Particular interest attaches to scenarios in which the system path in parameter space is *closed*, meaning that after some time T all parameters return to the values they had at time $t = 0$; the parameters are cyclic, $\mathbf{X}(T) = \mathbf{X}(0)$. This *closed cycle* in parameter space need not be repetitive (i.e. periodic), although that is a common example. The Hamiltonian is therefore cyclic, $\hat{H}(T) = \hat{H}(0)$. However, although the parameters and the Hamiltonian return to their initial values, the statevector need not; it can acquire a phase. When the control parameters complete a cycle the observables (as embodied by the statevector) may or may not return to their initial values; any difference is known as a *holonomy*. This is a geometric (topologic) effect, independent of any details of the parameter-space trajectory or the time required to follow it.

Nondegenerate adiabatic evolution. To understand the origin of the geometric phase we consider a cyclic Hamiltonian $\hat{H}(\mathbf{X}; t)$ whose time variation originates entirely with time-dependent parameters $\mathbf{X}(t)$ [Aha87; Ana97; Gar10]. The statevector associated with the Hamiltonian satisfies the equation

$$\hbar \frac{d}{dt} \boldsymbol{\Psi}(\mathbf{X}; t) = -\mathrm{i}\, \hat{H}(\mathbf{X}; t) \boldsymbol{\Psi}(\mathbf{X}; t). \tag{20.46}$$

Let us take the initial condition, at $t = 0$, as alignment between the statevector and an instantaneous eigenstate of the Hamiltonian (an instantaneous dressed state or adiabatic state),

$$\boldsymbol{\Psi}(\mathbf{X}; 0) = \boldsymbol{\Phi}_\nu(\mathbf{X}; 0), \quad \text{where} \quad \hat{H}(\mathbf{X}; t)\boldsymbol{\Phi}_\nu(\mathbf{X}; t) = \hbar\varepsilon_\nu(\mathbf{X}; t)\boldsymbol{\Phi}_\nu(\mathbf{X}; t). \tag{20.47}$$

That is, we know that initially the system is in a particular energy state which, for the moment, we assume to be nondegenerate.

The eigenvalue equation defines the adiabatic state $\boldsymbol{\Phi}_\nu(\mathbf{X}; t)$ only within a constant scalar multiplier. The magnitude of this scalar is set by the requirement that each eigenvector have unit length,

$$\langle \boldsymbol{\Phi}_\nu(\mathbf{X}; t) | \boldsymbol{\Phi}_\nu(\mathbf{X}; t) \rangle = 1. \tag{20.48}$$

We are at liberty to take the initial phase as zero. However, the phase at later times is not arbitrary, and remains to be determined.

Geometric phase. For a nondegenerate state the subsequent construction of the statevector after a cycle acquires a phase factor that can be partitioned into two parts, as in eqn. (20.43),

$$\boldsymbol{\Psi}(\mathbf{X}; T) = \exp\left[\mathrm{i}a_\nu + \mathrm{i}b_\nu\right] \boldsymbol{\Phi}_\nu(\mathbf{X}; t). \tag{20.49}$$

The first contribution is a dynamical phase equal to the time-integrated adiabatic eigenvalue,

$$a_\nu = -\int_0^T dt\, \varepsilon_\nu(\mathbf{X}; t). \tag{20.50}$$

The second contribution is the integral of eqn. (20.45),

$$b_\nu = \mathrm{i} \int_0^T dt\, \langle \boldsymbol{\Phi}_\nu(\mathbf{X}; t) | \frac{d}{dt} \boldsymbol{\Phi}_\nu(\mathbf{X}; t) \rangle. \tag{20.51}$$

The derivative appearing in the integrand is expressible as variation of the adiabatic state with parameter changes,

$$\frac{d}{dt} \boldsymbol{\Phi}_\nu(\mathbf{X}; t) = \frac{\partial}{\partial X_1} \boldsymbol{\Phi}_\nu(\mathbf{X}; t) \frac{dX_1}{dt} + \cdots + \frac{\partial}{\partial X_M} \boldsymbol{\Phi}_\nu(\mathbf{X}; t) \frac{dX_M}{dt}$$

$$\equiv \frac{d\mathbf{X}}{dt} \cdot \frac{\partial}{\partial \mathbf{X}} \boldsymbol{\Phi}_\nu(\mathbf{X}; t). \tag{20.52}$$

Thus the geometric phase can be rewritten as an integral over parameter values,

$$b_\nu = i \int_{\mathbf{X}(0)}^{\mathbf{X}(T)} d\mathbf{X} \cdot \langle \Phi_\nu(\mathbf{X}) | \frac{\partial}{\partial \mathbf{X}} \Phi_\nu(\mathbf{X}) \rangle. \tag{20.53}$$

Because the parameter change is cyclic the integral is over a closed loop $\mathcal{C}$ in parameter space, written as

$$b_\nu(\mathcal{C}) = i \oint_{\mathcal{C}} \langle \Phi_\nu(\mathbf{X}) | d\Phi_\nu(\mathbf{X}) \rangle, \tag{20.54}$$

where $d\Phi_\nu(\mathbf{X})$ denotes the differential of the adiabatic state,

$$d\Phi_\nu(\mathbf{X}) = d\mathbf{X} \cdot \frac{\partial}{\partial \mathbf{X}} \Phi_\nu(\mathbf{X}). \tag{20.55}$$

This integral is independent of the choice of parametrization and is independent of the rate at which the system point moves. It is a geometric phase, dependent only on the trajectory $\mathcal{C}$ followed by the system point in parameter space – the area enclosed by the trajectory or solid angle subtended by the area. When the changes of the Hamiltonian are adiabatic the geometric contribution to the cycle phase is termed the *Berry phase* [Ber84a; Ber90; Zwa90; Ana97].

Degenerate adiabatic evolution. When the states are degenerate the cyclic phase factor $e^{i\phi}$ of eqn. (20.39) generalizes to a unitary matrix, called the *Wilczek–Zee phase* [Wil84; Ana88; Una98; Fle99]. The form is

$$\Psi(T) = \exp\left[-i \int_0^T dt\, \varepsilon_\nu(\mathbf{X};t) \right] \sum_{\nu'} D_{\nu\nu'}(\mathbf{X}) \Phi_{\nu'}(\mathbf{X};T), \tag{20.56}$$

where the summation goes over all degenerate adiabatic states and the $D_{\nu\nu'}(\mathbf{X})$ are elements of a unitary matrix that replaces the simple exponential phase of eqn. (20.39). This situation applies to adiabatic evolution of the four-state tripod-linkage system, where there occur two degenerate dark states [Una98; Fle99].

20.7.2 Bloch vector

The motion of a Bloch vector and its Hamiltonian provide a simple example of geometric phase. Consider a two-state Hamiltonian parametrized by two angles, θ and φ,

$$\hat{H}(\theta, \varphi) = \frac{H_0}{2} \begin{bmatrix} \cos\theta & e^{-i\varphi}\sin\theta \\ e^{i\varphi}\sin\theta & -\cos\theta \end{bmatrix}. \tag{20.57}$$

This Hamiltonian is cyclic in both parameters, each with period 2π,

$$\hat{H}(\theta + 2\pi, \varphi) = \hat{H}(\theta, \varphi), \qquad \hat{H}(\theta, \varphi + 2\pi) = \hat{H}(\theta, \varphi). \tag{20.58}$$

Eigenvectors. To obtain eigenvectors of this Hamiltonian (i.e. dressed or adiabatic states) we rewrite this Hamiltonian, using the symbols $c \equiv \cos(\theta/2)$, $s \equiv \sin(\theta/2)$, as

$$\hat{H}(\theta, \varphi) = \frac{H_0}{2} \begin{bmatrix} (c^2 - s^2) & 2sc \; e^{-i\varphi} \\ 2sc \; e^{i\varphi} & -(c^2 - s^2) \end{bmatrix}. \tag{20.59}$$

We immediately see that two orthonormal eigenvectors $\mathbf{\Phi}_{\pm}(\theta, \varphi)$ can be written as

$$\mathbf{\Phi}_{+}(\theta, \varphi) = \begin{bmatrix} c \; e^{i\varphi} \\ s \end{bmatrix}, \qquad \mathbf{\Phi}_{-}(\theta, \varphi) = \begin{bmatrix} -s \; e^{i\varphi} \\ c \end{bmatrix}. \tag{20.60}$$

When $\theta = 0$ these are aligned with the unit vectors,

$$\mathbf{\Phi}_{+}(0, \varphi) = \begin{bmatrix} 1 \; e^{i\varphi} \\ 0 \end{bmatrix}, \qquad \mathbf{\Phi}_{-}(0, \varphi) = \begin{bmatrix} 0 \\ 1 \end{bmatrix}. \tag{20.61}$$

However, upon changing θ by 2π, the period of the Hamiltonian, these eigenvectors change sign,

$$\mathbf{\Phi}_{\pm}(2\pi, \varphi) = -\mathbf{\Phi}_{\pm}(0, \varphi). \tag{20.62}$$

To obtain single-valued eigenvectors we must include a phase factor,

$$\mathbf{\Phi}'_{\pm}(\theta, \varphi) = e^{i\theta} \, \mathbf{\Phi}_{\pm}(\theta, \varphi). \tag{20.63}$$

The statevector. The statevector for a two-state system is expressible in terms of angles θ and φ on the Bloch sphere. To make this construction single valued we include a phase factor $e^{i\theta}$ and write

$$\mathbf{\Psi} = e^{i\theta} \Big[\sin(\theta/2) \, \boldsymbol{\psi}_1 + \cos(\theta/2) \, e^{i\varphi} \, \boldsymbol{\psi}_2 \Big] = \mathbf{\Phi}'_{+}(\theta, \varphi). \tag{20.64}$$

For this statevector the geometric phase is

$$\gamma = i \oint_C \langle \Psi | d\Psi \rangle. \tag{20.65}$$

Suppose we fix θ and allow φ to vary. Then the statevector differential is

$$d\mathbf{\Psi} = i \, d\varphi \cos(\theta/2) \, e^{i\varphi} \, \boldsymbol{\psi}_2 \tag{20.66}$$

and so the integrand of the geometric phase is

$$\langle \Psi | d\Psi \rangle = i \, d\varphi \cos^2(\theta/2) = \tfrac{1}{2} i \, d\varphi \, (1 + \cos\theta). \tag{20.67}$$

The closed curve follows a circle at latitude θ for which the integral evaluates as

$$\gamma = -\frac{1}{2}(1 + \cos\theta) \int_0^{2\pi} d\varphi = -\pi \, (1 + \cos\theta). \tag{20.68}$$

Alternatively, we can fix φ and vary θ, for which the integrand is[4]

$$\langle \Psi | d \Psi \rangle = i \, d\theta. \tag{20.69}$$

Taken around a great circle of longitude this change produces the phase

$$\gamma = - \int_0^{2\pi} d\theta = -2\pi. \tag{20.70}$$

For these, and more general variations of θ and φ for the single-valued statevector, the geometric phase is minus half the solid angle enclosed by $\mathcal{C}$ on the Bloch sphere.

20.7.3 Spin system

A system in which the N-dimensional Hamiltonian is expressible as the sum of three spin matrices S_i (as is always the case for $N = 2$ and for pseudospin) offers a simple illustration of the notions of parallel transport and geometric phase. Following Berry, consider a system whose interaction Hamiltonian can be written as proportional to $\mathbf{S} \cdot \mathbf{B}$,

$$\hat{H} = -\kappa \, \mathbf{S} \cdot \mathbf{B}, \tag{20.71}$$

where $\mathbf{S}$ is a spin matrix in $N = 2s + 1$ dimensions, $\mathbf{B}$ is a unit vector in three dimensions, and κ is a proportionality constant. (The model suits a magnetic moment in a magnetic field, but holds quite generally for any Hamiltonian expressible as the sum of three N-dimensional spin matrices.) The eigenstates of the spin Hamiltonian are the angular momentum states $\chi_m = |s, m\rangle$ and the energy eigenvalues are

$$E_m = -\kappa \, m \, |\mathbf{B}|, \qquad m = -s, \ldots, +s. \tag{20.72}$$

When κ is constant (as is $|\mathbf{B}|$) then this statevector evolution is an example of parallel transport of a tangent plane over the surface of a parameter-space sphere (of radius $\kappa |\mathbf{B}|$) [Ana87]. More generally, let the system start in spin state χ_m and allow κ and $\mathbf{B}$ to be slowly varying parameters. Then, after adiabatic cycling of the magnitude $\kappa(t)$ and the unit vector $\mathbf{B}(t)$ from $\kappa(0), \mathbf{B}(0)$ to $\kappa(T), \mathbf{B}(T) = \kappa(0), \mathbf{B}(0)$, the statevector will acquire a geometric phase [Chi86]

$$\gamma_m(\mathcal{C}) = -m \, \Omega(\mathcal{C}), \tag{20.73}$$

where $\Omega(\mathcal{C})$ is the solid angle subtended at the coordinate origin by the loop. It is not necessary that the evolution be adiabatic, only that the parameters return to their original

---

[4] Had we omitted the phase factor from $\psi'(t)$ this integrand would be zero. The sign change associated with the cycle of θ then is found in the construction of the adiabatic state.

values. In particular, when the path is a circle around a sphere, at latitude θ, the angle is

$$\gamma_m(\mathcal{C}) = -m\,2\pi(1-\cos\theta). \tag{20.74}$$

When $N = 2$, so that there are two spin states (and $m = \pm\frac{1}{2}$) the phase is minus half the solid angle, giving a phase factor

$$\langle \Psi(T)|\Psi(0)\rangle = \exp[-\tfrac{1}{2}\Omega(\mathcal{C})]. \tag{20.75}$$

21

Atoms affecting fields

The major portion of this monograph discusses how given pulses of laser radiation affect individual atoms. This chapter inverts that relationship, describing how the matter alters the fields. Those field changes provide quantitative measures of the quantum-state changes produced in the atoms. Their description therefore is an adjunct to Chap. 19.

Basically the incident radiation produces excitation which, in turn, alters the various multipole moments of the atoms. When viewed as a macroscopic sample of matter, such changes alter the electric polarization field $\mathbf{P}$ and the magnetization field $\mathbf{M}$ of the matter through which the radiation must pass. The Maxwell equations, see App. C.1, provide the needed description of how the $\mathbf{P}$ and $\mathbf{M}$ fields alter the electric and magnetic fields $\mathbf{E}$ and $\mathbf{B}$. The combination of the Maxwell equations for the fields and the Schrödinger, Bloch, or Liouville equations for the atoms provide the tools needed to construct a self-consistent description of radiation passing through matter – atoms responding to a pulsed field and traveling waves being modified by the resulting atomic changes [All87; Ale92; Muk99; Die06; Vit01a]. The present chapter, drawing on [Sho90, Chap. 12] and App. C, discusses this theory.

Incoherent radiation passing through matter typically undergoes exponential attenuation in accord with eqn. (6.9). A measure of the incremental change of intensity in distance L by absorption coefficient κ is the optical depth κL. This parameter appears in rate-equation treatments of incoherent light attenuation. It also provides a convenient measure of propagation effects of coherent excitation, although the effects on pulses are often quite different from simple attenuation. Indeed, coherent effects can make even dense material transparent, as in the two-state *self-induced transparency* (SIT) (see Sec. 21.2.7) or in the *electromagnetically induced transparency* (EIT) (see Sec. 21.4.3) obtainable with three-state stimulated Raman processes.

21.1 Induced dipole moments; propagation

As radiation travels through matter it exerts an oscillating force on bound electrons. The effect is measurable as an oscillating induced electric dipole moment. In turn, this oscillating dipole moment serves as a source of radiation; in both classical and quantum treatments the radiated electric field is proportional to the acceleration of the dipole moment [Sho90,

§11.1]. A portion of the new radiation acts to alter the original field (slowing the propagation velocity and attenuating the amplitude), while other portions appear as scattered light.

The starting point for treating such radiation is the inhomogeneous wave equation of App. C for the electric field in the presence of polarizable material,

$$\left[\nabla^2 - \frac{1}{c^2}\frac{\partial^2}{\partial t^2}\right]\mathbf{E}(\mathbf{r},t) = \frac{1}{c^2\epsilon_o}\frac{\partial^2}{\partial t^2}\mathbf{P}(\mathbf{r},t),\tag{21.1a}$$

or for the magnetic field in the presence of magnetic material (but no electric moments),

$$\left[\nabla^2 - \frac{1}{c^2}\frac{\partial^2}{\partial t^2}\right]\mathbf{B}(\mathbf{r},t) = \mu_0\nabla^2\mathbf{M}(\mathbf{r},t),\tag{21.1b}$$

in which $c = 1/\sqrt{\epsilon_0\mu_0}$ is the speed of light in vacuum and the electric polarization field $\mathbf{P}(\mathbf{r},t)$ or the magnetization field $\mathbf{M}(\mathbf{r},t)$ expresses the effect of the matter [DeG69; Lan84].

In traditional application of these equations it is customary to regard matter as being distributed continuously, with number density $\mathcal{N}(\mathbf{r})$, and to write the electric polarization and the magnetization as a distribution of electric or magnetic dipole moments,

$$\mathbf{P}(\mathbf{r},t) = \mathcal{N}(\mathbf{r})\langle\mathbf{d}(\mathbf{r},t)\rangle, \qquad \mathbf{M}(\mathbf{r},t) = \mathcal{N}(\mathbf{r})\langle\mathbf{m}(\mathbf{r},t)\rangle.\tag{21.2}$$

To simplify the discussion I will consider just the fields $\mathbf{P}$ and $\mathbf{E}$.

To proceed we require a prescription for the time-varying induced dipole moment, from which to evaluate the electric polarization field. When the atomic response is steady and weak the induced time-varying dipole moment is proportional to the incident electric field. The effects are observable through attenuation and can be incorporated into a complex-valued index of refraction. Larger responses can occur when the field is resonant with a transition linked to the populated ground state of the atoms. Stronger fields, inducing significant coherent excitation, produce a variety of nonlinear effects in the propagation equation. In particular, if the atoms are prepared in a coherent-superposition state then a number of unusual effects can occur [Scu92; Pas02; Har97; Fle05].

The modeling of atomic response, and the complexity of the field changes, depends on our model of the atom – how many states N participate in the excitation dynamics and their Rabi frequencies and detunings. The simplest cases are two-level atoms acted upon by a single field, considered in the next section. Subsequent sections discuss first the general multistate system, and then specializations to examples of three and four states.

All cases introduce two common approximations for the basic variables: the slowly varying envelope approximation (SVEA) for the field variables and the rotating-wave approximation (RWA) for the atomic description. Each of these approximations neglects rapid variations: the RWA neglects terms whose temporal variation is comparable to the carrier frequency and the SVEA neglects also terms whose spatial variation is comparable to the wavelength.

21.2 Single field, $N = 2$

Consider a plane-wave field that travels along the z axis, as described by a fixed transverse unit vector $\mathbf{e}$, a carrier wave of frequency ω and wavevector magnitude k, and a complex-valued envelope $\check{\mathcal{E}}(z,t)$,

$$\mathbf{E}(z,t) = \mathrm{Re}\left\{ \mathbf{e}\,\check{\mathcal{E}}(z,t)\,\mathrm{e}^{\mathrm{i}(kz-\omega t)}\right\} = \mathbf{E}^{(+)}(z,t) + \mathbf{E}^{(-)}(z,t). \tag{21.3a}$$

The electric polarization field we write similarly as

$$\mathbf{P}(z,t) = \mathrm{Re}\left\{ \mathbf{e}\,\check{\mathcal{P}}(z,t)\,\mathrm{e}^{\mathrm{i}(kz-\omega t)}\right\} = \mathbf{P}^{(+)}(z,t) + \mathbf{P}^{(-)}(z,t). \tag{21.3b}$$

In each of these expressions the positive-frequency part is associated with the time dependence $\mathrm{e}^{-\mathrm{i}\omega t}$ and the relationship between ω and k is taken to be via a complex-valued refractive index $\check{\eta}$,

$$ck = \check{\eta}\omega, \tag{21.4}$$

where c is the speed of light in vacuum. The envelopes $\check{\mathcal{E}}(z,t)$ and $\check{\mathcal{P}}(z,t)$ are taken to be complex-valued functions of their arguments, varying more slowly than does the exponential. Implementing that assumption we make the SVEA (see App. C.5.1) meaning that we assume the spatial variations of the envelopes occur over distances that are much larger than a wavelength, and the time variations occur over many optical cycles, as expressed by the inequalities

$$\left| \frac{\partial^2 \mathcal{F}}{\partial z^2} \right| \ll k \left| \frac{\partial \mathcal{F}}{\partial z} \right|, \qquad \left| \frac{\partial^2 \mathcal{F}}{\partial t^2} \right| \ll \omega \left| \frac{\partial \mathcal{F}}{\partial t} \right|, \tag{21.5}$$

where $\mathcal{F}$ is either $\check{\mathcal{E}}$ or $\check{\mathcal{P}}$. We can also separate a steady linear response from a transient nonlinear response by introducing a constant complex-valued susceptibility $\check{\chi}$ and writing the polarization envelope as

$$\check{\mathcal{P}}(z,t) = \epsilon_0 \check{\chi}\check{\mathcal{E}}(z,t) + {}^{nl}\check{\mathcal{P}}(z,t). \tag{21.6}$$

The linear response, parametrized by the susceptibility, expressed the effect of those atoms that are only weakly and steadily affected by the radiation – the host atoms of a solid or the buffer gas of a vapor. The nonlinear polarization, denoted here as ${}^{nl}\check{\mathcal{P}}(z,t)$, expressed the response of atoms whose excitation we follow explicitly with the time-dependent Schrödinger equation. From the second-order wave equation we thereby obtain a first-order equation (the *reduced-wave equation*)

$$\left[\frac{\partial}{\partial z} + \frac{\check{\eta}}{c}\frac{\partial}{\partial t} \right] \check{\mathcal{E}}(z,t) = \mathrm{i}\frac{\omega}{2\epsilon_0 c\check{\eta}}{}^{nl}\check{\mathcal{P}}(z,t), \tag{21.7}$$

where the complex-valued refractive index $\check{\eta}$ is obtained from the susceptibility as

$$\check{\eta}^2 = 1 + \check{\chi}. \tag{21.8}$$

Missing from the first-order equation is any description of reflection: only the forward traveling wave is present. When the matter has discontinuities at surfaces that field must be included.

The separation of the polarization field into linear and nonlinear parts is, in the present context, somewhat arbitrary. When we treat the effects of near-resonant light on a simple two-state system it is the envelope ${}^{nl}\check{\mathcal{P}}(z,t)$ that holds interest. The linear response then expresses the effects of all other atoms that are present in the material – they provide a background susceptibility and a background refractive index. In much of the following I will largely neglect this, setting

$$\check{\chi} = 0, \qquad \check{\eta} = 1, \qquad ck = \omega. \tag{21.9}$$

21.2.1 Maxwell–Schrödinger equations

To evaluate the single-atom dipole moment, and from it the electric polarization envelope $\check{\mathcal{P}}(z,t)$, let us consider a two-state atom, whose center of mass is at position z, in a pure state described by the statevector[1]

$$\Psi(z,t) = e^{-i\zeta_1}\left[C_1(z,t)\boldsymbol{\psi}_1 + e^{i(kz-\omega t)}e^{-i\varphi}C_2(z,t)\boldsymbol{\psi}_2 \right], \tag{21.10}$$

with φ to be chosen below. The expectation value of the single-atom dipole moment is

$$\langle \mathbf{d}(z,t)\rangle = \langle \Psi(z,t)|\mathbf{d}|\Psi(z,t)\rangle$$
$$= \langle 1|\mathbf{d}|2\rangle\, e^{-i\varphi}\, C_1^* C_2\, e^{i(kz-\omega t)} + \langle 2|\mathbf{d}|1\rangle\, e^{-i\varphi}\, C_2^* C_1\, e^{-i(kz-\omega t)}. \tag{21.11}$$

From the coefficient of $e^{i(kz-\omega t)}$ in this expression we identify the envelope of the electric polarization field to be

$$\check{\mathcal{P}}(z,t) = 2\mathcal{N}\, d_{12}\, C_2(z,t)C_1(z,t)^*, \qquad d_{12} = \mathbf{e}^* \cdot \langle 1|\mathbf{d}|2\rangle\, e^{-i\varphi}. \tag{21.12}$$

We choose the phase φ to make the parameter d_{12} real. It follows that the complex-valued electric field envelope satisfies, in the laboratory (nonmoving) reference frame, the propagation equation

$$\left[\frac{\partial}{\partial z} + \frac{\check{\eta}}{c}\frac{\partial}{\partial t}\right]\check{\mathcal{E}}(z,t) = i\frac{\mathcal{N} d_{12}\omega}{\check{\eta}c\epsilon_0}\, C_2(z,t)C_1(z,t)^*. \tag{21.13a}$$

This equation accompanies the RWA equations for the probability amplitudes:

$$\frac{\partial}{\partial t}C_1(z,t) = -\frac{i}{2}\check{\Omega}(z,t)^* C_2(z,t), \tag{21.13b}$$

$$\frac{\partial}{\partial t}C_2(z,t) = -i\Delta C_2(z,t) - \frac{i}{2}\check{\Omega}(z,t)C_1(z,t). \tag{21.13c}$$

---

[1] The argument z here serves as a parameter, not as the coordinate of an electron, as would be the meaning for an electron wavefunction.

where the complex-valued Rabi frequency is,[2]

$$\check{\Omega}(z,t) = -d_{21}\,\check{\mathcal{E}}(z,t)/\hbar, \qquad d_{21} = \mathbf{e}\cdot\langle 2|\mathbf{d}|1\rangle\,\mathrm{e}^{\mathrm{i}\varphi} = d_{12}^{*}. \tag{21.14}$$

This set of three coupled partial differential equations constitute the RWA *Maxwell–Schrödinger equations* for a uniform distribution of two-state systems. Written in terms of the Rabi frequency the propagation equation reads, when $\check{\eta} = 1$,

$$\left[\frac{\partial}{\partial z} + \frac{1}{c}\frac{\partial}{\partial t}\right]\check{\Omega}(z,t) = -\mathrm{i}\,\varsigma\,C_2(z,t)C_1(z,t)^{*}, \tag{21.15}$$

where the scaling parameter for Rabi frequency propagation, with units of inverse (length × time), is[3]

$$\varsigma = \frac{\mathcal{N}|d_{12}|^2\omega}{\hbar c\epsilon_0}. \tag{21.16}$$

21.2.2 Analytic solutions

Exact solutions for the Maxwell–Schrödinger equations exist for a variety of situations including infinitely periodic trains of pulses. Solutions are simplest when the atoms are all resonant with the carrier frequency, $\Delta = 0$, so that the equations read

$$\left[\frac{\partial}{\partial z} + \frac{1}{c}\frac{\partial}{\partial t}\right]\Omega = -\varsigma\,\mathrm{i}\,C_2\,C_1,$$

$$\frac{\partial}{\partial t}C_1 = -\tfrac{1}{2}\Omega\,\mathrm{i}\,C_2, \tag{21.17}$$

$$\frac{\partial}{\partial t}\,\mathrm{i}\,C_2 = \tfrac{1}{2}\Omega C_1,$$

so that C_1 and $\mathrm{i}\,C_2$ as well as Ω are real-valued functions of z and t. The single-pulse solutions first described by McCall and Hahn [McC69; All87] are

$$\Omega(z,t) = \frac{2}{\tau}\,\mathrm{sech}X(z,t),$$

$$C_1(z,t) = \tanh X(z,t), \tag{21.18}$$

$$C_2(z,t) = \mathrm{i}\,\mathrm{sech}X(z,t),$$

where the dimensionless argument of the hyperbolic functions is

$$X(z,t) = \frac{t}{\tau} - \frac{z}{v\tau}. \tag{21.19}$$

The pulse envelope $\Omega(z,t)$ moves without changing form (it is *form stable*),

$$\left[\frac{\partial}{\partial z} + \frac{\check{\eta}}{c}\frac{\partial}{\partial t}\right]\Omega(z,t) = 0, \tag{21.20}$$

---

[2] Alternative definitions of the Rabi frequency may differ in sign and by a factor of two.
[3] As noted below, ς is expressible as the product $\varsigma = \kappa_{12}T^{*}$ of an absorption coefficient (the inverse Beer's length κ_{12}), and the effective relaxation time T^{*} used in evaluating that length.

at the group velocity $v = c/\check{\eta}$, definable in terms of a refractive index $\check{\eta}$ or an inverse time $\mathcal{K}$,

$$v = \frac{c}{\check{\eta}}, \qquad \check{\eta} = 1 + \mathcal{K}\tau, \qquad \mathcal{K} = \tfrac{1}{2}\varsigma c\tau. \tag{21.21}$$

The temporal pulse area, for any z, is

$$\mathcal{A}(z) = \int_{-\infty}^{\infty} dt\, \Omega(z,t) = 2\pi, \tag{21.22}$$

and so the pulse restores the atom to its initial state, whatever that may be. The hyperbolic secant pulse not only restores resonant population but, because it varies slowly, also will restore probability amplitudes for detuned atoms, $\Delta \neq 0$, as discussed in Sec. 8.4.7.

21.2.3 Maxwell–Bloch equations

When the system is described by Bloch variables or a density matrix we evaluate the electric polarization field from the atomic coherence, i.e. the off-diagonal elements of the density matrix. It is this coherence, rather than the population inversion, that alters the pulse envelope. Let us write the polarization envelope as

$$i\check{\mathcal{P}}(z,t) = \mathcal{N} d_{12}\, \check{Q}(z,t), \tag{21.23}$$

where the dimensionless complex-valued function $\check{Q}(z,t)$ is the two-state coherence, expressible in terms of Bloch variables $r_k(z,t)$, or probability amplitudes $C_n(z,t)$, as

$$\check{Q}(z,t) = r_2(z,t) + i r_1(z,t) = i\, 2C_2(z,t)C_1(z,t)^*. \tag{21.24}$$

The function $\check{Q}(z,t)$ defined by this expression has, as real and imaginary parts, the components $r_2 \equiv v$ and $r_1 \equiv u$ of the traditional Bloch vector (component $r_3 \equiv w$ is the atomic inversion) and so the magnitude of $\check{Q}$ cannot exceed unity. Using this definition we write the propagation equation for the complex-valued Rabi frequency as

$$\left[\frac{\partial}{\partial z} + \frac{1}{c}\frac{\partial}{\partial t}\right]\check{\Omega}(z,t) = -\tfrac{1}{2}\varsigma\, \check{Q}(z,t), \tag{21.25a}$$

where the scaling parameter ς is defined in eqn. (21.16). The equations for the atom are the optical Bloch equations. In the absence of relaxation these are, rewritten for complex variable $\check{Q}$:

$$\frac{\partial}{\partial t}\check{Q}(z,t) = -i\Delta\check{Q}(z,t) - \check{\Omega}(z,t)r_3(z,t), \tag{21.25b}$$

$$\frac{\partial}{\partial t}r_3(z,t) = \text{Re}\left\{\check{\Omega}(z,t)^*\, \check{Q}(z,t)\right\}. \tag{21.25c}$$

These three combined field–atom equations are known as the coupled *Maxwell–Bloch equations*. When the excitation is resonant ($\Delta = 0$) and the initial field envelope is real, then both $\check{Q}(z,t)$ and $\check{\Omega}(z,t)$ will remain real functions of their arguments; the variable $\check{Q}$ is then equal to the Bloch variable r_2.

21.2.4 Moving reference frame

A useful step in the presentation of propagation equations for pulses is the introduction of a reference frame that moves with the group velocity $v = c/\breve{\eta}$. In this frame the space and time coordinates are

$$Z = z, \qquad T = t - z/v, \qquad v = c/\breve{\eta}. \tag{21.26}$$

The derivatives of any function $F(z, t)$ of z and t evaluate as

$$\left[\frac{\partial}{\partial z} + \frac{1}{v} \frac{\partial}{\partial t} \right] F(z, t) = \left[\frac{\partial Z}{\partial z} \frac{\partial}{\partial Z} + \frac{\partial T}{\partial z} \frac{\partial}{\partial T} + \frac{1}{v} \frac{\partial T}{\partial t} \frac{\partial}{\partial T} \right] F(Z, T) = \frac{\partial}{\partial Z} F(Z, T). \tag{21.27}$$

Using this moving reference frame we write the propagation equation for the electric field amplitude as

$$\frac{\partial}{\partial Z} \breve{\mathcal{E}}(Z, T) = i \frac{\omega}{2c\epsilon_0 \breve{\eta}} {}^{nl}\breve{\mathcal{P}}(Z, T), \tag{21.28}$$

where the argument T refers to the time in the moving coordinate frame. The TDSE, when referred to the moving reference frame, reads

$$\frac{\partial}{\partial T} C_n(Z, T) = -i \sum_m W_{nm}(Z, T) C_m(Z, T). \tag{21.29}$$

That is, for any given position $Z = z$ the equation for probability amplitudes uses a time variable located with reference to a moving pulse. Typically $T = 0$ is taken to mark the peak value of the pulse. To simplify notation in the following presentation I will usually denote by z, t the space-time coordinates in the moving reference frame.

21.2.5 Phase and intensity equations; absorption (attenuation)

To make contact with treatments of incoherent propagation, based on intensity rather than field envelope, we write the complex-valued envelope as an amplitude (expressed as intensity) and a phase

$$\breve{\mathcal{E}}(z, t) = \sqrt{2I(z, t)/\epsilon_0 c} \; e^{i\varphi(z,t)}. \tag{21.30}$$

Rewriting the moving-frame propagation equation (21.28) as

$$\frac{\partial}{\partial z} \breve{\mathcal{E}}(z, t) = i \frac{\omega}{2c} \breve{\chi}(z, t) \breve{\mathcal{E}}(z, t), \tag{21.31}$$

where the instantaneous susceptibility is defined as

$$\breve{\chi}(z, t) \equiv \frac{\breve{\mathcal{P}}(z, t)}{\epsilon_0 \breve{\mathcal{E}}(z, t)} = i \frac{\mathcal{N} |d_{12}|^2}{\hbar \epsilon_0} \frac{\breve{Q}(z, t)}{\breve{\Omega}(z, t)}, \tag{21.32}$$

we obtain the result

$$i \frac{\partial}{\partial z} \varphi(z, t) + \frac{1}{2I} \frac{\partial}{\partial z} I(z, t) = \frac{\omega}{2c} \left[i \operatorname{Re} \breve{\chi}(z, t) - \operatorname{Im} \breve{\chi}(z, t) \right]. \tag{21.33}$$

From the real part of this equation we obtain an intensity propagation equation

$$\frac{\partial}{\partial z} I(z, t) = -I(z, t) \frac{\omega}{c} \text{Im} \, \check{\chi}(z, t), \qquad (21.34a)$$

while the imaginary part provides an equation for the phase,

$$\frac{\partial}{\partial z} \varphi(z, t) = \frac{\omega}{2c} \text{Re} \, \check{\chi}(z, t). \qquad (21.34b)$$

These two equations describe coherent plane-wave pulse propagation, through a uniform medium of single-species two-state atoms, in the SVEA and RWA.

The instantaneous absorption coefficient. By rewriting the intensity equation as

$$\frac{\partial}{\partial z} I(z, t) = -\kappa(z, t) I(z, t) \qquad (21.35)$$

we define an effective instantaneous absorption coefficient $\kappa(z, t)$.[4]   It has units of inverse length and is proportional to the ratio of atomic coherence $\check{Q}$ to complex-valued Rabi frequency $\check{\Omega}$, both of which vary with space and time,

$$\kappa(z, t) = -\frac{\omega}{c\epsilon_0} \text{Im} \, \frac{\check{P}(z, t)}{\check{\mathcal{E}}(z, t)} = \frac{\omega}{c} \text{Im} \, \check{\chi}(z, t) = \varsigma \, \text{Re} \, \frac{\check{Q}(z, t)}{\check{\Omega}(z, t)}. \qquad (21.36)$$

Here ς is the scaling parameter of eqn. (21.16). When $\kappa(z, t)$ is positive the intensity diminishes with distance; energy is removed from the beam of radiation. When $\kappa(z, t)$ is negative, the atoms add energy to the beam, i.e. the radiation undergoes *gain*. In the absence of any relaxation process the solutions to the Schrödinger equation, and the coherence $\check{Q}$, oscillate indefinitely, as long as there is any field. The result of one such Rabi cycle is the removal of energy from the field followed by replacement of this energy, as expressed by an absorption coefficient $\kappa(z, t)$ that varies between positive and negative.

Steady state. Relaxation processes, needed to produce steady-state solutions to the Bloch equation, are most simply treated by including in the coherence equation a term that diminishes the instantaneous value, as parametrized by a coherence relaxation rate $\gamma = 1/T_2$,

$$\frac{\partial}{\partial t} \check{Q}(z, t) = -(\gamma + i\Delta) \check{Q}(z, t) - \check{\Omega}(z, t) r_3(z, t). \qquad (21.37)$$

This is an example of homogeneous relaxation – it affects all atoms equally. Typically the resulting relaxation changes to the coherence are more rapid than changes to populations, and we therefore obtain the desired steady state by setting to zero its time derivative in this equation – an example of adiabatic elimination. We further regard the population inversion

---

[4] More aptly called the attenuation coefficient or extinction coefficient, this is often denoted by α.

r_3 as unaffected by the pulsed interaction; it may be a function of position z but not time. The result is that the coherence is proportional to the field, as expressed by the ratio

$$\frac{\check{Q}(\Delta; z, t)}{\check{\Omega}(z, t)} = -\frac{r_3(z)}{\gamma + i\Delta}. \tag{21.38}$$

The complex susceptibility obtained from this steady state is

$$\check{\chi}(\Delta; z) = -r_3(z)\frac{\mathcal{N}|d_{12}|^2}{\hbar\epsilon_0}\frac{\Delta - i\gamma}{\Delta^2 + \gamma^2}. \tag{21.39}$$

The steady-state absorption coefficient obtained from the imaginary part of this formula is a Lorentzian function of detuning Δ, concentrated within the width γ near resonance, $\Delta = 0$,

$$\kappa(\Delta; z) = -[P_2(z) - P_1(z)]\varsigma\frac{\gamma}{\gamma^2 + \Delta^2}. \tag{21.40}$$

This expression is positive when most population is unexcited, $P_1 > P_2$. The atoms then attenuate the radiation. It is negative when the population becomes inverted – the field then undergoes gain. The range of frequencies for which attenuation occurs is set by the homogeneous relaxation rate γ, centered on resonance with the Bohr frequency.

Inhomogeneous relaxation. The atomic dipole moments that serve as the source of the polarization field **P** for propagating fields originate with atoms in a variety of environments. These must be averaged, as discussed in Sec. 16.2,

$$i\check{P}(z, t) = \mathcal{N}d_{12}\{\check{Q}(z, t)\}_{av}. \tag{21.41}$$

An important class of environmental differences appear as variations of the two-state detuning, either because thermal motions cause Doppler shifts of the apparent carrier frequency ω or because different static environmental fields alter the Bohr transition frequency ω_B. The required average is then obtained by integrating over all possible detunings $\Delta \equiv \omega_B - \omega$,

$$\{\check{Q}(z, t)\}_{av} = \int_{-\infty}^{\infty} d\Delta\, g(\Delta)\, \check{Q}(\Delta; z, t). \tag{21.42}$$

The integrand weight, the probability $g(\Delta)$ of observing detuning Δ, is normalized to unit integral; its peak value, at $\Delta = 0$, is often taken to define an inhomogeneous relaxation time T^*:

$$\int d\Delta\, g(\Delta) = 1, \qquad \pi g(0) = T^*, \tag{21.43}$$

and the integrand $\check{Q}(\Delta; z, t)$ satisfies eqn. (21.37),

$$\frac{\partial}{\partial t}\check{Q}(\Delta; z, t) = -(\gamma + i\Delta)\check{Q}(\Delta; z, t) - \check{\Omega}(z, t)r_3(\Delta; z, t). \tag{21.44}$$

With this averaging the propagation equation for the Rabi frequency, in the moving reference frame, becomes

$$\frac{\partial}{\partial z}\check{\Omega}(z,t) = -\frac{1}{2}\varsigma\{\check{Q}(z,t))\}_{av}.$$ (21.45)

Similar averaging applies to the intensity equation and to the absorption coefficient.

Steady inhomogeneous relaxation. Expressions (21.39) and (21.40) incorporate only homogeneous relaxation effects, as parametrized by the coherence decay rate γ. To include inhomogeneous relaxation, either from Doppler shifts of a vapor atom or of environmental energy shifts in a solid, we average over detunings Δ. The required average is the integral of $\kappa(\Delta, z)$ weighted by $g(\Delta)$,

$$\bar{\kappa}(z) = \{\kappa(z)\}_{av} = \int d\Delta\, g(\Delta)\kappa(\Delta; z).$$ (21.46)

The averaging integral over $\gamma/(\Delta^2 + \gamma^2)$ can be regarded as an effective relaxation time $1/\gamma_e$ that combines homogeneous and inhomogeneous widths:

$$\frac{1}{\gamma_e} \equiv \left\{\frac{\gamma}{\Delta^2 + \gamma^2}\right\}_{av} = \int d\Delta\, g(\Delta)\frac{\gamma}{\Delta^2 + \gamma^2} \rightarrow \begin{cases} 1/\gamma = T_2 & \text{large } \gamma, \\ \pi g(0) = T^* & \text{small } \gamma. \end{cases}$$ (21.47)

Beer's length. A convenient measure of absorption (attenuation), the inverse *Beer's length*, is defined for effective relaxation rate γ_e as

$$\kappa_{12} = \frac{\varsigma}{\gamma_e} = \frac{\mathcal{N}|d_{12}|^2\omega}{\hbar c\epsilon_0\gamma_e} = 4\pi\alpha\frac{\mathcal{N}|r_{12}|^2\omega}{\gamma_e}.$$ (21.48)

The last expression, involving the dimensionless Sommerfeld fine-structure constant $\alpha \approx 1/137$, makes clear the dimension of κ_{12}. Using this parameter we express the general instantaneous absorption coefficient as

$$\kappa(z,t) = \kappa_{12}\gamma_e\,\text{Re}\,\frac{\check{Q}(z,t)}{\check{\Omega}(z,t)},$$ (21.49)

and the averaged steady-state value as

$$\bar{\kappa}(z) = -\kappa_{12}\, r_3(z).$$ (21.50)

Using the parameter κ_{12} we rewrite the propagation equation for the Rabi frequency, eqn. (21.25a), as

$$\frac{\partial}{\partial z}\check{\Omega}(z,t) = -\frac{1}{2}\kappa_{12}\gamma_e\,\check{Q}(z,t).$$ (21.51)

As is evident from the definition of eqn. (21.48), the product $\varsigma = \kappa_{12}\gamma_e$ that appears in the propagation equation is independent of the width γ_e.

21.2.6 Pulse reshaping

When the incident field is constant, it will produce oscillations of the Bloch variables. These oscillations will, in turn, modify the field that passes the atoms. The effect is most clearly seen when we consider a rectangular pulse whose carrier frequency is resonant with stationary two-state atoms. The atoms in the first small slice of matter are influenced by a suddenly imposed Rabi frequency, and consequently they begin Rabi oscillations. The population oscillations accompany an oscillating dipole moment, proportional to the Bloch variable r_2, that is phased to absorb energy from the field. Later in the Rabi cycle the dipole moment is phased to return energy to the field. Figure 21.1 illustrates the relationships between the atomic variables and the field variable.

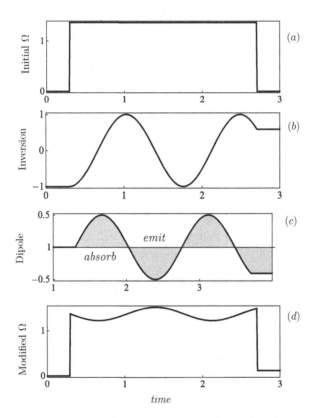

Figure 21.1 Schematic effect of atomic coherence on the propagation of a rectangular pulse, within the first spatial slice of a uniform material. (*a*) The incident pulse. (*b*) The population inversion r_3. (*c*) The coherence r_2, indicating alternating absorption and emission. (*d*) The modified pulse. The steady Rabi oscillations of population accompany an oscillation of the dipole moment that alternately removes and adds energy to the field. [Redrawn from Fig. 12.7-1 of B. W. Shore, *The Theory of Coherent Atomic Excitation*. Copyright Wiley-VCH Verlag GmbH & Co. 1990. Reproduced with permission.]

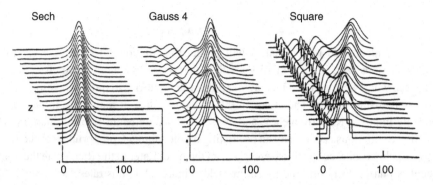

Figure 21.2 Examples of pulse propagation, for pulses whose envelopes are initially hyperbolic secant, Gauss-4 and rectangular. [From Fig. 12.7-2 of B. W. Shore, *The Theory of Coherent Atomic Excitation*. Copyright Wiley-VCH Verlag GmbH & Co. 1990. Reproduced with permission.]

When the pulse is intense but short, so that relaxation has no effect, pulse reshaping can be severe. Figure 21.2 illustrates the reshaping that occurs with different initial pulse shapes as they travel through a medium of stationary relaxation-free two-state atoms. In these examples each pulse has an initial area of 2π, so the atoms complete one full Rabi cycle during the pulse.

21.2.7 Self-induced transparency (SIT)

A resonant pulse whose temporal area is 2π produces one complete Rabi cycle of population inversion. The pulse returns all atoms to their initial state. Therefore the pulse produces no permanent excitation and leaves no energy in the medium. This behavior is known as *self-induced transparency* (SIT). Although the pulse will maintain constant area (2π) as it propagates, it will typically undergo reshaping until it eventually stabilizes in the shape of a hyperbolic secant pulse [McC69; All87]. Although the pulses must be shorter than any homogeneous relaxation time, the reshaping can take place when inhomogeneous relaxation is present, as noted below.

The area theorem. Although a pulse typically undergoes a variety of changes as it propagates, several general properties of any pulse are predictable. One of the significant quantifiers of a pulse is the total pulse area $\mathcal{A}(z)$, defined as the time integral of the real-valued Rabi frequency $\Omega(z,t)$:

$$\mathcal{A}(z) = \int_{-\infty}^{\infty} dt\, \Omega(z,t). \tag{21.52}$$

This quantity obeys the equation

$$\frac{\partial}{\partial z}\mathcal{A}(z) = -\frac{1}{2}\left[\frac{\mathcal{N}|d_{12}|^2\omega}{\hbar c\epsilon_0}\right]\int_{-\infty}^{\infty} dt\,\{\check{Q}(z,t)\}_{av}. \tag{21.53}$$

After the passage of the pulse the dipole moment oscillates at the detuning frequency Δ. In the infinite time limit these oscillations produce a null integral, except for the resonant contribution $\Delta = 0$ [McC69; All87]. The result is that, for pulses whose duration τ is much shorter than homogeneous relaxation times but longer than the inhomogeneous relaxation time T^*,

$$\tau \ll T_2, T_1 \quad \text{but} \quad \tau \gg T^*, \tag{21.54}$$

the change in total pulse area depends on the final value (after the pulse has passed) of the resonant component of the coherence,

$$\frac{\partial}{\partial z} \mathcal{A}(z) = -\frac{1}{2}\kappa_{12}\, Q(0; z, \infty), \tag{21.55}$$

parametrized by the absorption coefficient

$$\kappa_{12} = \frac{\mathcal{N}|d_{12}|^2 \omega}{\hbar c \epsilon_0} \pi g(0). \tag{21.56}$$

The measure of detuning distributions – the width of inhomogeneous broadening – is set by the parameter $\pi g(0) = T^*$. The resonantly excited Bloch vector, described by the coupled equations

$$\frac{\partial}{\partial t} r_2(0; z, t) = -\check{\Omega}(z, t)\, r_3(0; z, t),$$

$$\frac{\partial}{\partial t} r_3(0; z, t) = +\check{\Omega}(z, t)\, r_2(0; z, t), \tag{21.57}$$

traces out a portion of a great circle arc on the Bloch sphere; the components r_2 and r_3 undergo Rabi oscillations with starting value $r_2 = 0$. At the conclusion of the pulse the value of r_2 is set by the sine of the temporal pulse area $\mathcal{A}(z)$. Using this value we obtain the *McCall–Hahn area theorem* [McC69; All87],

$$\frac{\partial}{\partial z} \mathcal{A}(z) = \frac{1}{2} w_0 \kappa_{12} \sin \mathcal{A}(z), \tag{21.58}$$

where the initial resonant inversion $w_0 = r_3(0; z, -\infty)$ at $t \to -\infty$ is $+1$ for an initially inverted medium (an amplifier) and is -1 for an initially cold medium (an attenuator).

Equation (21.58) is the basic equation describing propagation of short pulses through a uniform medium of moving two-state atoms that individually undergo completely coherent excitation, but with a distribution of detunings. The properties of this equation depend significantly on whether the pulse is weak (small $\mathcal{A}$) or strong (large $\mathcal{A}$).

Weak pulses. For weak pulses we can replace the sine by its argument. When the medium is initially cold ($r_3(0) = -1$) this equation becomes an exponential attenuation law for pulse area

$$\frac{\partial}{\partial z} \mathcal{A}(z) = \frac{1}{2}\kappa_{12}\, \mathcal{A}(z). \tag{21.59}$$

This is the counterpart, for short pulses through a coherently excited medium, to the exponential attenuation of steady intensity (the Beer–Lambert law). To emphasize the similarity of the present result with conventional results for continuous-wave attenuation we can express this equation as

$$\frac{\partial}{\partial z}|\mathcal{A}(z)|^2 = -\kappa_{12}|\mathcal{A}(z)|^2. \tag{21.60}$$

The square of the pulse area is, apart from a numerical factor, the pulse fluence. Unlike the traditional continuous-wave attenuation, which occurs because the atoms remove energy from the beam and thereby increase their internal excitation energy, the attenuation of this short pulse does not accompany atomic excitation. It occurs because of destructive interference between dipole moments with different detunings.

Strong pulses. For strong pulses eqn. (21.59) predicts behavior that differs profoundly from propagation governed by traditional exponential attenuation. The area theorem predicts that as pulses propagate the total pulse area will eventually, as $z \to \infty$, approach a limiting value equal to an integral multiple of π. Specifically, the solutions have the property

$$\tan[\tfrac{1}{2}\mathcal{A}(z)] = \tan[\tfrac{1}{2}\mathcal{A}(0)]\exp[-\tfrac{1}{2}\kappa_{12}z]. \tag{21.61}$$

For the usual case of an initially unexcited medium (an attenuator) the asymptotic value is the *even* multiple of π that is closest to the initial pulse area. For the case of an amplifying medium, in which population is initially inverted, the pulse area tends to the *odd* multiple of π that is closest to the initial value. It follows from this property that pulses which initially differ only very slightly in area may, after propagation, differ in area by a multiple of 2π. In particular, in an attenuating medium a pulse with area slightly less than π will diminish in area to zero, while a pulse with area slightly greater than π will become a 2π pulse [McC69; All87]. Not only is the area fixed, so is the final pulse shape: it tends toward a hyperbolic secant pulse. Once formed, such a pulse travels as a solitary wave (soliton) without further change of shape, even in the presence of inhomogeneous broadening. It induces a brief single cycle of atomic excitation, but eventually leaves the atoms unexcited, just as they were initially.

21.2.8 Fluence and energy conservation

Propagating short pulses – those with duration shorter than relaxation times – undergo a variety of changes to their shape, not just simple attenuation as occurs for longer pulses for which atomic populations achieve steady state. It is useful to find parameters that characterize the pulses even as they become distorted. One of these parameters is the pulse fluence, or time-integrated intensity.

Fluence conservation. We start with the equations for propagating the intensity,

$$\frac{\partial}{\partial z}I(z,t) = -\tfrac{1}{2}\mathcal{N}\hbar\omega\,\mathrm{Re}\left\{\check{\Omega}^*(z,t)Q(z,t)\right\}. \tag{21.62}$$

We evaluate the right-hand side by using the equation for population changes in the presence of spontaneous emission, occurring between state 2 and state 1 at the rate A_{21},

$$\frac{\partial}{\partial t} P_2(z, t) = -A_{21} P_2(z, t) - \frac{1}{2} \mathrm{Re}\left\{ \check{\Omega}(z, t)^* Q(z, t) \right\}.$$ (21.63)

Combining these equations we obtain the result

$$\frac{\partial}{\partial z} I(z, t) = -\mathcal{N} \hbar \omega \left[\frac{\partial}{\partial t} P_2(z, t) + A_{21} P_2(z, t) \right].$$ (21.64)

This equation shows that it is the *change* in population inversion, not the population inversion, that alters the beam intensity (or photon flux). The time-integrated intensity within a pulse is the *pulse fluence*,

$$\mathcal{F}(z) = \int_{-\infty}^{\infty} dt \, I(z, t).$$ (21.65)

By integrating the intensity propagation equation we obtain the equation describing the change of fluence during propagation:

$$\frac{\partial}{\partial z} \mathcal{F}(z) = \mathcal{N} \hbar \omega \left[-[P_2(\infty) - P_2(-\infty)] + A_{21} \int_{-\infty}^{\infty} dt \, P_2(z, t) \right].$$ (21.66)

This equation states that the photons lost from the pulse appear as excitation or as scattered light (from spontaneous emission). For pulses that are sufficiently short that spontaneous emission is negligible, and that return population to the initial condition via completed Rabi oscillation ($2\pi n$ pulses), fluence is conserved as the pulse propagates [All87]. The fluence equation does not assume weak excitation. When the pulse produces excitation, then the final inversion $r_3(\infty)$ is not equal to the initial inversion $r_3(-\infty)$, and pulse energy decreases during propagation. To evaluate the change in inversion we must solve the population equations.

Energy conservation. An important example of a conserved quantity is the combined energy of atom and field. From the propagation equation

$$\frac{\partial}{\partial z} \check{\mathcal{E}}(z, t) = \frac{\mathcal{N} d_{12} \omega}{2c\epsilon_0} \int_{-\infty}^{\infty} d\Delta \, g(\Delta) \, Q(\Delta; z, t)$$ (21.67)

and the equation of motion for the population inversion

$$\frac{\partial}{\partial t} r_3(\Delta; z, t) = -\frac{1}{2\hbar} \left[d_{12} \mathcal{E}^* Q(\Delta; z, t) + d_{21} \mathcal{E} Q(\Delta; z, t)^* \right],$$ (21.68)

we obtain a conservation equation

$$c \frac{\partial}{\partial z} u^{em}(z, t) + \frac{\partial}{\partial t} u^{em}(z, t) + \frac{\partial}{\partial t} u^{at}(z, t) = 0,$$ (21.69)

relating the divergence of the electromagnetic energy density

$$u^{em}(z, t) = \frac{1}{2} \epsilon_0 |\check{\mathcal{E}}(z, t)|^2$$ (21.70)

and the temporal change of energy, both electromagnetic and that of atomic excitation, as expressed by the excitation energy density

$$u^{at}(z,t) = \frac{1}{2} \mathcal{N} \hbar \omega \int_{-\infty}^{\infty} d\Delta \, \mathsf{g}(\Delta) \, r_3(\Delta; z, t). \tag{21.71}$$

Equation (21.69) is a form of the Poynting theorem, expressed in terms of envelope functions and allowing energy exchange with matter.

21.3 Multiple fields

When multiple laser frequencies are present various multistate coherent excitations become possible, and with them dramatically altered time-varying induced dipole moments. The resulting nonlinear response of the atoms can produce a variety of novel effects. Treatment of concurrent multiple fields of different frequencies follows a straightforward extension of the single- and two-field theory. We express the field as a sum of separate terms, typically a plane-wave approximation. For waves traveling along the z axis we introduce the expansion

$$\mathbf{E}^{(+)}(z,t) = \sum_\lambda \mathbf{E}_\lambda^{(+)}(z,t) = \frac{1}{2} \sum_\lambda \mathbf{e}_\lambda \breve{\mathcal{E}}_\lambda(z,t) \exp[i\xi_\lambda(z,t)], \tag{21.72a}$$

with a similar expansion for the polarization field,

$$\mathbf{P}^{(+)}(z,t) = \sum_\lambda \mathbf{P}_\lambda^{(+)}(z,t) = \frac{1}{2} \sum_\lambda \mathbf{e}_\lambda \breve{\mathcal{P}}_\lambda(z,t) \exp[i\xi_\lambda(z,t)]. \tag{21.72b}$$

Here the envelopes $\breve{\mathcal{E}}_\lambda$ and $\breve{\mathcal{P}}_\lambda$ are taken as complex valued, and the exponential phases are

$$\xi_\lambda(z,t) = k_\lambda z - \omega_\lambda t, \tag{21.73}$$

with the wavevector k_λ and carrier frequency ω_λ related by the refractive index $\breve{\eta}_\lambda$ or the (linear) susceptibility χ_λ,

$$k_\lambda = \breve{\eta}_\lambda \omega_\lambda / c, \qquad (\breve{\eta}_\lambda)^2 = 1 + \chi_\lambda. \tag{21.74}$$

The individual unit vectors $\mathbf{e}_\lambda$ must each lie in the x, y plane in order that the field be transverse. The intensity associated with a particular pulse is

$$I_\lambda(z,t) = \frac{c\epsilon_0}{2} |\breve{\mathcal{E}}_\lambda(z,t)|^2. \tag{21.75}$$

We make the SVEA by neglecting second derivatives and we separate the linear response by introducing the susceptibility,[5]

$$\breve{\mathcal{P}}_\lambda(z,t) = \epsilon_0 \chi_\lambda \breve{\mathcal{E}}_\lambda(z,t) + {}^{nl}\breve{\mathcal{P}}_\lambda(z,t). \tag{21.76}$$

---

[5] In general the susceptibility, like the polarizability, is a tensor. Here it is simplified as a scalar.

The resulting equation for an envelope traveling in the positive direction, the reduced Maxwell's equation, reads

$$\left[\frac{\partial}{\partial z} + \frac{\check{\eta}_\lambda}{c}\frac{\partial}{\partial t}\right]\check{\mathcal{E}}_\lambda(z,t) = +i\frac{\omega_\lambda}{2c\check{\eta}_\lambda\epsilon_0}{}^{nl}\check{\mathcal{P}}_\lambda(z,t). \tag{21.77}$$

When the several refractive indices differ appreciably it is not possible to introduce a common set of moving windows that follow all the pulses.

The nonlinear polarization envelope exhibited here derives from the response of dipole-moment expectation values to the fields, as determined by the TDSE or a density-matrix equation. In posing those equations we make the RWA for the Hamiltonian of the single-atom statevector expressed as a superposition of quantum states in a rotating coordinate system,

$$\Psi(z,t) = \sum_n e^{-i\zeta_n(z,t)}C_n(z,t)\psi_n. \tag{21.78}$$

The polarization envelope function that results from this expression is

$${}^{nl}\check{\mathcal{P}}_\lambda(z,t) = 2\mathcal{N}(z)\sum_{n,m}\mathbf{e}_\lambda^*\cdot\mathbf{d}_{mn}\, s_{nm}(z,t)\exp[i(\xi_\lambda + \zeta_m - \zeta_n)], \tag{21.79}$$

where $s_{nm}(z,t)$ are elements of the density matrix in a rotating reference frame,

$$s_{nm}(z,t) = \{C_m^*(z,t)C_n(z,t)\}_{av}. \tag{21.80}$$

The RWA limits the summation of eqn. (21.79) to those terms for which the exponential vanishes, as occurs when the carrier frequency matches a Bohr transition frequency.

Each electric field envelope $\check{\mathcal{E}}_\lambda$ appears in eqn. (21.77) with its unique source of polarization field. However, each of those is linked to other electric fields through the TDSE that couples the elements of s_{nm}. To complete the self-consistent description of single atoms acted upon and influencing fields we must supplement Maxwell's equations, simplified to envelope-propagation equations, with equations that treat the atomic dynamics. This can be the TDSE or, to incorporate homogeneous relaxation such as spontaneous emission, the master equation for the density matrix. The full definition of the problem requires initial conditions: the values of the field envelopes $\mathcal{E}_\lambda(z,t)$ for all times t at some exterior position, say $z = 0$, and the value of all the probability amplitudes at $t = 0$ for all values of the positions $z \geq 0$. Fields that are initially absent may grow as a result of field-generated coherences.

21.4 Two or three fields, $N = 3$

The general properties of the coupled atom–field system appear when one considers the three-state atom. The unique couplings of a triangle loop allow three fields, which I label P

(for pump), linking states 1,2, S (for Stokes), linking states 2,3 and R (for Raman), linking states 3,1,

$$\mathbf{E}(z,t) = \mathrm{Re}\left\{\mathbf{e}_P\,\check{\mathcal{E}}_P(z,t)\,\mathrm{e}^{\mathrm{i}\,(k_P z - \omega_P t)}\right\} + \mathrm{Re}\left\{\mathbf{e}_S\,\check{\mathcal{E}}_S(z,t)\,\mathrm{e}^{\mathrm{i}\,(k_S z - \omega_S t)}\right\}$$
$$+\mathrm{Re}\left\{\mathbf{e}_R\,\check{\mathcal{E}}_R(z,t)\,\mathrm{e}^{\mathrm{i}\,(k_R z - \omega_R t)}\right\}. \tag{21.81}$$

The companion polarization field for near-resonant atoms (previously denoted as $^{nl}\mathbf{P}$), is

$$\mathbf{P}(z,t) = \mathrm{Re}\left\{\mathbf{e}_P\,\check{\mathcal{P}}_P(z,t)\,\mathrm{e}^{\mathrm{i}\,(k_P z - \omega_P t)}\right\} + \mathrm{Re}\left\{\mathbf{e}_S\,\check{\mathcal{P}}_S(z,t)\,\mathrm{e}^{\mathrm{i}\,(k_S z - \omega_S t)}\right\}$$
$$+\mathrm{Re}\left\{\mathbf{e}_R\,\check{\mathcal{P}}_R(z,t)\,\mathrm{e}^{\mathrm{i}\,(k_R z - \omega_R t)}\right\}. \tag{21.82}$$

In the SVEA the propagation equation (the reduced Maxwell's equation) is eqn. (21.77) with $\lambda = P, S, R$. To simplify the formulas I neglect the background atoms, setting $\check{\eta}_\lambda = 1$, and use a coordinate system that moves with the speed of light. The equation then reads

$$\frac{\partial}{\partial z}\check{\mathcal{E}}_\lambda(z,t) = \mathrm{i}\,\frac{\omega_\lambda}{2c\epsilon_0}\check{\mathcal{P}}_\lambda(z,t), \qquad \lambda = P, S, R. \tag{21.83}$$

Intensity propagation. From the propagation equation (21.83) there follows an equation for the intensity,

$$\frac{\partial}{\partial z}I_\lambda(z,t) = -\kappa_\lambda(z,t)I(z,t). \tag{21.84}$$

The instantaneous absorption coefficient appearing here is defined as

$$\kappa_\lambda(z,t) = -\frac{\omega_\lambda}{c\epsilon_0}\,\mathrm{Im}\,\frac{\check{\mathcal{P}}_\lambda(z,t)}{\check{\mathcal{E}}_\lambda(z,t)} = -\frac{\omega_\lambda}{c}\,\mathrm{Im}\,\chi_\lambda(z,t), \tag{21.85}$$

where the instantaneous susceptibility $\chi_\lambda(z,t)$ is defined through the relationship

$$\check{\mathcal{P}}_\lambda(z,t) = \epsilon_0\chi_\lambda(z,t)\check{\mathcal{E}}_\lambda(z,t). \tag{21.86}$$

To evaluate any of these expressions it is necessary to have a constitutive relationship for the single-atom dipole moment, from which to deduce the polarization field. This comes from the TDSE.

The dipole moment. With the introduction of rotating coordinates for the probability amplitudes $C_n \equiv C_n(z,t)$ the statevector for three states has the Hilbert-space structure

$$\mathbf{\Psi}(z,t) = \mathrm{e}^{-\mathrm{i}\zeta_1}\left[C_1\boldsymbol{\psi}_1 + \mathrm{e}^{\mathrm{i}(\zeta_1-\zeta_2)}C_2\boldsymbol{\psi}_2 + \mathrm{e}^{\mathrm{i}(\zeta_1-\zeta_3)}C_3\boldsymbol{\psi}_3\right], \tag{21.87}$$

and so the single-atom dipole moment has the construction

$$\langle\mathbf{d}(z,t)\rangle = 2\mathrm{Re}\left\{\mathbf{d}_{21}\,s_{12}\,\mathrm{e}^{\mathrm{i}(\zeta_2-\zeta_1)} + \mathbf{d}_{32}\,s_{23}\,\mathrm{e}^{\mathrm{i}(\zeta_3-\zeta_2)} + \mathbf{d}_{31}\,s_{13}\,\mathrm{e}^{\mathrm{i}(\zeta_3-\zeta_1)}\right\}, \tag{21.88}$$

where

$$s_{mn} = C_m C_n^* = s_{nm}^*. \tag{21.89}$$

are elements of the density matrix in rotating coordinates. For an atom in free space the parity selection rule prohibits the presence of three dipole moments linked in a loop, but such transitions occur in solids.

For the lambda linkage,

$$E_2 > E_1, \qquad E_2 > E_3, \qquad E_3 > E_1, \tag{21.90}$$

we make the phase choice $\hbar \dot\zeta_1 = E_1$ and

$$\begin{aligned}
\zeta_2 - \zeta_1 &= -(k_P z - \omega_P t), \\
\zeta_2 - \zeta_3 &= -(k_S z - \omega_S t), \\
\zeta_3 - \zeta_1 &= -(k_P z - \omega_P t) + (k_S z - \omega_S t + \varphi_S).
\end{aligned} \tag{21.91}$$

In typical applications, at most two of the fields are controlled directly by the experimenter. I take these to be the P and S fields. The third field, R, then arises from nonlinear processes. The following constraints will be used for this field:

$$k_R = k_P - k_S, \qquad \omega_R = \omega_P - \omega_S. \tag{21.92}$$

With these definitions we obtain the expression

$$\langle \mathbf{d}(z,t) \rangle = 2 \mathrm{Re} \left\{ \mathbf{d}_{12} s_{21} \, e^{i(k_P z - \omega_P t)} + \mathbf{d}_{32} s_{23} \, e^{i(k_S z - \omega_S t)} + \mathbf{d}_{13} s_{31} \, e^{i(k_R z - \omega_R t)} \right\}. \tag{21.93}$$

Propagation equations. By matching terms of expansion (21.93) with those of eqn. (21.82) that have the same exponentials we identify the polarization envelopes as

$$\check{\mathcal{P}}_P(z,t) = 2 \mathcal{N} d_{12} s_{21}, \qquad \check{\mathcal{P}}_S(z,t) = 2 \mathcal{N} d_{32} s_{23}, \qquad \check{\mathcal{P}}_R(z,t) = 2 \mathcal{N} d_{13} s_{31}, \tag{21.94}$$

where the dipole-moment components are

$$d_{12} = \mathbf{e}_P^* \cdot \mathbf{d}_{12}, \qquad d_{32} = \mathbf{e}_S^* \cdot \mathbf{d}_{32}, \qquad d_{13} = \mathbf{e}_R^* \cdot \mathbf{d}_{13}. \tag{21.95}$$

Associated with these transitions are the complex-valued Rabi frequencies,[6]

$$\begin{aligned}
\hbar \check\Omega_P(z,t) &= -d_{21} \check{\mathcal{E}}_P(z,t), \qquad \hbar \check\Omega_S(z,t) = -d_{23} \check{\mathcal{E}}_S(z,t), \\
\hbar \check\Omega_R(z,t) &= -d_{31} \check{\mathcal{E}}_R(z,t).
\end{aligned} \tag{21.96}$$

To parametrize the strengths of the interactions we use the inverse Beer's lengths

$$\kappa_P = \frac{\omega_P \mathcal{N} |d_{12}|^2}{\hbar c \epsilon_0 \check\eta_P \Gamma_P}, \qquad \kappa_S = \frac{\omega_S \mathcal{N} |d_{32}|^2}{\hbar c \epsilon_0 \check\eta_S \Gamma_S}, \qquad \kappa_R = \frac{\omega_R \mathcal{N} |d_{31}|^2}{\hbar c \epsilon_0 \check\eta_R \Gamma_R}, \tag{21.97}$$

where Γ is the width of the absorption profile. We express the needed atomic properties as the coherences

$$\check{Q}_P = 2 s_{21} = 2 C_2 C_1^*, \qquad \check{Q}_S = 2 s_{23} = 2 C_2 C_3^*, \qquad \check{Q}_R = 2 s_{31} = 2 C_3 C_1^*, \tag{21.98}$$

---

[6] Other authors define Rabi frequencies differing in sign and by a factor of two.

and write the propagation equations as

$$\left[\frac{\partial}{\partial z} + \frac{\breve{\eta}_P}{c}\frac{\partial}{\partial t}\right]\breve{\Omega}_P = -\frac{1}{2}\kappa_P\Gamma_P\,\breve{Q}_P,$$

$$\left[\frac{\partial}{\partial z} + \frac{\breve{\eta}_S}{c}\frac{\partial}{\partial t}\right]\breve{\Omega}_S = -\frac{1}{2}\kappa_S\Gamma_S\,\breve{Q}_S, \tag{21.99}$$

$$\left[\frac{\partial}{\partial z} + \frac{\breve{\eta}_R}{c}\frac{\partial}{\partial t}\right]\breve{\Omega}_R = -\frac{1}{2}\kappa_R\Gamma_R\,\breve{Q}_R.$$

Initial conditions; new fields. These equations require, for completion, a definition of the initial field envelopes $\breve{\mathcal{E}}_\lambda(z,t)$ for all t at some entrance to the material, say $z = 0$. If only one field is initially present, the atomic dynamics may produce nonzero elements of the density matrix that serves as a source of the other fields: they appear as the initial pulse propagates. Thus the excitation dynamics of the atoms generates new frequency components of the field, an example of transient nonlinearities. The polarization directions of the new fields are fixed by the properties of the two states between which the transition occurs. Thus the unit vectors for the P, S, and R fields are

$$\mathbf{e}_P = \frac{\mathbf{d}_{21}}{d_{21}}, \qquad \mathbf{e}_S = \frac{\mathbf{d}_{23}}{d_{23}}, \qquad \mathbf{e}_R = \frac{\mathbf{d}_{31}}{d_{31}}. \tag{21.100}$$

The Schrödinger equation. To complete the description of the system we use the TDSE for the probability amplitudes whose products comprise the coherences. With the RWA these read

$$\frac{\partial}{\partial t}C_1 = -\frac{i}{2}\breve{\Omega}_P^*C_2 - \frac{i}{2}\breve{\Omega}_R^*C_3,$$

$$\frac{\partial}{\partial t}C_2 = -i\Delta_P C_2 - \frac{i}{2}\breve{\Omega}_P C_1 - \frac{i}{2}\breve{\Omega}_S C_3, \tag{21.101}$$

$$\frac{\partial}{\partial t}C_3 = -i\Delta_R C_3 - \frac{i}{2}\breve{\Omega}_R C_1 - \frac{i}{2}\breve{\Omega}_S^*C_2,$$

subject to appropriate initial conditions on $C_n(z,t)$ for $t = 0$ and all z. Here, by constraint, $\Delta_R = \Delta_P - \Delta_S$ and the single-field detunings are

$$\hbar\Delta_P = E_2 - E_1 - \hbar\omega_P, \qquad \hbar\Delta_S = E_2 - E_3 - \hbar\omega_S. \tag{21.102}$$

Equations (21.99) and (21.101) constitute the Maxwell–Schrödinger equations for the Raman system. To complete the definition of the problem we must specify the probability amplitudes at the initial time $t = 0$ for all z and the values of the Rabi frequencies at the entry face $z = 0$ for all time t.

21.4.1 The stimulated Raman system

The traditional stimulated Raman system involves only the P and S fields. The relevant equations for the three probability amplitudes,

$$\frac{\partial}{\partial t} C_1 = -\frac{i}{2} \check{\Omega}_P^* C_2,$$

$$\frac{\partial}{\partial t} C_2 = -i \Delta_P C_2 - \frac{i}{2} \check{\Omega}_P C_1 - \frac{i}{2} \check{\Omega}_S C_3,$$

$$\frac{\partial}{\partial t} C_3 = -i (\Delta_P - \Delta_S) C_3 - \frac{i}{2} \check{\Omega}_R C_1 - \frac{i}{2} \check{\Omega}_S^* C_2, \tag{21.103}$$

derive from the RWA Hamiltonian

$$\mathsf{W} = \frac{1}{2} \begin{bmatrix} 0 & \check{\Omega}_P^* & 0 \\ \check{\Omega}_P & 2\Delta_P & \check{\Omega}_S \\ 0 & \check{\Omega}_S^* & 2(\Delta_P - \Delta_S) \end{bmatrix}. \tag{21.104}$$

The propagation equations are

$$\left[\frac{\partial}{\partial z} + \frac{\check{\eta}_P}{c} \frac{\partial}{\partial t} \right] \check{\Omega}_P(z,t) = i \frac{\mathcal{N} |d_{12}|^2 \omega_P}{\hbar c \epsilon_0 \check{\eta}_P} C_2(z,t) C_1(z,t)^*,$$

$$\left[\frac{\partial}{\partial z} + \frac{\check{\eta}_S}{c} \frac{\partial}{\partial t} \right] \check{\Omega}_S(z,t) = i \frac{\mathcal{N} |d_{32}|^2 \omega_S}{\hbar c \epsilon_0 \check{\eta}_S} C_2(z,t) C_3(z,t)^*. \tag{21.105}$$

Customarily the two pulses, P and S, enter unexcited material, so that the initial condition is $C_1(z,t) = 1$ for $t = 0$ for all z. However, it is possible to prepare the atoms in the excited state or in a coherent superposition of states 1 and 2, a situation that Scully has called *phaseonium* [Scu92].

21.4.2 Bright–dark basis

As with the earlier discussion of the three-state system excited by given fields, the Maxwell–Schrödinger equations simplify when we treat excitation that maintains two-photon resonance. The RWA Hamiltonian matrix is then

$$\mathsf{W}(z,t) = \frac{1}{2} \begin{bmatrix} 0 & \Omega_P\, e^{-i\varphi_P} & 0 \\ \Omega_P\, e^{i\varphi_P} & 2\Delta & \Omega_S\, e^{i\varphi_S} \\ 0 & \Omega_S\, e^{-i\varphi_S} & 0 \end{bmatrix}, \tag{21.106}$$

where the phases φ are chosen to make the Rabi frequencies real-valued functions of time and position,

$$\check{\mathcal{E}}_P = \mathcal{E}_P\, e^{i\varphi_P}, \qquad \check{\mathcal{E}}_S = \mathcal{E}_S\, e^{i\varphi_S}. \tag{21.107}$$

We define the mixing angle

$$\sin \Theta = \Omega_S / \Omega_T, \qquad \cos \Theta = \Omega_P / \Omega_T, \qquad \tan \Theta = \Omega_S / \Omega_P, \tag{21.108}$$

where Ω_T is the rms value of the two Rabi frequencies,

$$\Omega_T = \sqrt{|\Omega_P|^2 + |\Omega_S|^2},$$ (21.109)

and define bright- and dark-state amplitudes of Sec. 13.3.2,

$$C_B = e^{i\varphi_P}\cos\Theta\, C_1 + e^{i\varphi_S}\sin\Theta\, C_3,$$
$$C_D = -e^{i\varphi_P}\sin\Theta\, C_1 + e^{i\varphi_S}\cos\Theta\, C_3.$$ (21.110)

These amplitudes, together with the amplitude for state 2, satisfy the equations

$$\frac{\partial}{\partial t}C_2 = -i\,\Delta C_2 - \frac{i}{2}\Omega_T C_B,$$

$$\frac{\partial}{\partial t}C_B = -\frac{i}{2}\Omega_T C_2 - \frac{i}{2}\Omega_D C_D,$$ (21.111)

$$\frac{\partial}{\partial t}C_D = -\frac{i}{2}\Omega_D^* C_B,$$

in which the coupling between the dark and bright states is

$$\Omega_D = -2i\,\partial_t\Theta = 2i\,\frac{\Omega_S\,\partial_t\Omega_P - \Omega_P\,\partial_t\Omega_S}{|\Omega_P|^2 + |\Omega_S|^2}.$$ (21.112)

When the mixing angle Θ varies sufficiently slowly, compared with the total Rabi frequency Ω_T, then this coupling is negligible. The dark-state component remains unaltered by the radiation and the changing dynamics is that of a two-state atom.

When, in addition to this adiabatic approximation of $\partial_t\Theta = 0$, the system is initially in the dark state, then we obtain the approximations

$$C_D(z,t) \approx 1, \qquad C_B(z,t) \approx 0, \qquad C_2(z,t) \approx i\,\frac{2\partial_t\Theta}{\Omega_T}.$$ (21.113)

This leads to the approximations

$$C_1(z,t) \approx \cos\Theta, \qquad C_2(z,t) \approx i\,\frac{2\partial_t\Theta}{\Omega_T}, \qquad C_3(z,t) \approx -\sin\Theta.$$ (21.114)

These approximations are the starting point for treating an aspect of pulse propagation that has no classical counterpart, in which the presence of one field (a strong one, S) makes the material transparent to a second field (a weak probe field P). The phenomenon, discussed next, relies on Autler–Townes splitting of the resonance transition that is probed by the P field.

21.4.3 Electromagnetically induced transparency (EIT)

Let us consider the effect on a weak (probe) P field of a collinearly propagating S field. When the S field is strong and approximately constant we have the perturbative approximations

$$C_1(z,t) \approx 1, \qquad C_2(z,t) \approx i\,\frac{2\partial_t\Theta}{\Omega_T} = -i\,\frac{2\partial_t\Omega_P}{\Omega_T^2}.$$ (21.115)

This amplitude produces a homogeneous *P*-field propagation equation,

$$\left[\frac{\partial}{\partial z} + \frac{1}{c}\frac{\partial}{\partial t}\right]\check{\Omega}_P(z,t) = -\frac{\kappa_P\Gamma}{\Omega_T(z,t)^2}\frac{\partial}{\partial t}\check{\Omega}_P(z,t). \tag{21.116}$$

We define an index of refraction $\check{n}_P$ and a *group velocity* v_P for the *P* field

$$v_P = \frac{c}{1+\check{n}_P}, \qquad n_P = \frac{c\kappa_P\Gamma}{\Omega_T^2}, \tag{21.117}$$

and rewrite the *P*-field propagation equation as

$$\left[\frac{\partial}{\partial z} + \frac{1}{v_P}\frac{\partial}{\partial t}\right]\check{\Omega}_P(z,t) = 0. \tag{21.118}$$

That is, the *P* pulse is form stable at a (reduced) group velocity v_P. Within the lowest-order perturbative approximation the strong *S* field is not affected by the atoms, and satisfies the free-space equation

$$\left[\frac{\partial}{\partial z} + \frac{1}{c}\frac{\partial}{\partial t}\right]\check{\Omega}_S(z,t) = 0. \tag{21.119}$$

The only effect of the medium is a temporal delay of the pump pulse. No energy is lost by absorption or spontaneous emission: the medium is transparent. This is an example of *electromagnetically induced transparency* or EIT [Har97; Fle05]. At the leading edge of the *P* pulse the atoms absorb photons from this pulse and transfer them into the *S* field (which is much stronger and therefore effectively unchanged) through the STIRAP mechanism. The process reverses at the tail of the *P* pulse, i.e. all energy is returned to the *P* pulse.

21.4.4 Coherence transfer: Dark-state polaritons and stopped light

In fact it is possible to control the propagation velocity of the *P* pulse, to bring it to a full stop, and to reaccelerate it on demand. This behavior is associated with the existence of a quasiparticle called a *dark-state polariton* [Fle02], which is a coherent superposition of atomic and field components

$$F(z,t) = \cos\theta(z,t)\check{\Omega}_P(z,t) - \sin\theta(z,t)\sqrt{\kappa_P c\Gamma}\,C_3(z,t)C_1^*(z,t). \tag{21.120}$$

The angle θ (not to be confused with the mixing angle Θ used earlier) is defined by

$$\tan\theta(z,t) = \frac{\sqrt{c\kappa_P\Gamma}}{\check{\Omega}_S(z,t)}. \tag{21.121}$$

The dark-state polariton obeys the simple propagation equation

$$\left[\frac{\partial}{\partial t} + c\cos^2\theta(z,t)\frac{\partial}{\partial z}\right]F(z,t) = 0. \tag{21.122}$$

If $\check{\Omega}_S$ (and hence θ) is approximately uniform in *z*, then eqn. (21.122) describes a form-invariant propagation of the quasiparticle with propagation velocity

$$v(t) = c\cos^2\theta(t). \tag{21.123}$$

When $\theta(t)$ is adiabatically rotated from 0 to $\pi/2$ by externally controlling the amplitude of the S field, $\check{\Omega}_S$, an initially pure electromagnetic polariton ($F = \check{\Omega}_P$) is transformed into a pure atomic polarization ($F = \sqrt{\alpha c \Gamma} C_3 C_1^*$). At the same time the propagation velocity is changed from the vacuum speed of light to zero. The pump pulse is thus "stopped", meaning that its coherent information is transferred to collective atomic states [Phi01; Fle02; Liu01; Luk03]. The atomic polarization can be extracted by reversing the transfer process, i.e. rotating θ back from $\pi/2$ to 0 and recreating the pump pulse.

21.5 Four fields, $N = 4$; four-wave mixing

Amongst the useful nonlinear actions of coherently excited atoms is the generation of new frequencies. From three frequencies ω_1, ω_2, and ω_3 it is possible to generate a fourth [Kor03; Obe06; Vew07a],

$$\omega_4 = \omega_1 + \omega_2 + \omega_3. \tag{21.124}$$

An important implementation of such four-wave mixing allows the generation of short-wave radiation from several longer-wave pulses. Figure 21.3 diagrams a possible linkage pattern, in which frequencies ω_1 and ω_2 provide a resonant three-state ladder linkage. Acting on the uppermost state of this ladder the third frequency will generate a fourth field, at a frequency set by eqn. (21.124).

To treat such processes we assume that the atoms are uniformly distributed, with number density $\mathcal{N}$, throughout a medium in which three collinear plane waves pass. We express the plane-wave nature of the fields, and their distinguishability by frequency, through expansions (21.72a) and (21.72b), and obtain the propagation equations (21.77).

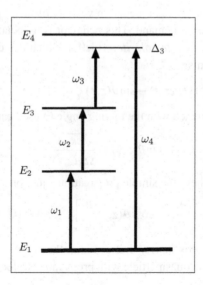

Figure 21.3 Excitation linkages for producing ω_4 from $\omega_1 + \omega_2 + \omega_3$.

To proceed we require a prescription for the time-varying induced dipole moment, from which to evaluate the polarization envelope (21.77). This we obtain from the TDSE, writing the statevector (at fixed position) as

$$
\begin{aligned}
\Psi(z,t) = \ & C_1(z,t)\psi_1 \\
& + C_2(z,t)e^{-i\omega_1 t + i\varphi_1}\psi_2 \\
& + C_3(z,t)e^{-i\omega_1 t - i\omega_2 t + i\varphi_1 + i\varphi_2}\psi_3 \\
& + C_4(z,t)e^{-i\omega_1 t - i\omega_2 t - i\omega_3 t + i\varphi_1 + \varphi_2 + \varphi_3}\psi_4.
\end{aligned} \tag{21.125}
$$

The dipole moment calculated from the probability amplitudes appearing here is

$$
\begin{aligned}
\langle d(z,t)\rangle = 2\mathrm{Re}\Big\{ & C_1(z,t)^* C_2(z,t) d_{12} e^{+i\omega_1 t} + C_2(z,t)^* C_3(z,t) d_{23} e^{+i\omega_2 t} \\
& + C_3(z,t)^* C_4(z,t) d_{34} e^{+i\omega_3 t} + C_4(z,t)^* C_1(z,t) d_{41} e^{-i\omega_4 t}\Big\}.
\end{aligned} \tag{21.126}
$$

Each of the four terms has a well-defined frequency, from which we can identify the polarization field associated with each of the incident electric fields and with the originally absent field of frequency ω_4. This latter source is maximum when $|C_4(t)| = |C_1(t)|$, i.e. when there is a coherent superposition of states 1 and 4.

The polarization fields deduced from the statevector, for fixed position, are

$$
\begin{aligned}
\check{P}_1(z,t) &= 2\mathcal{N}\mathrm{Re}\Big\{C_1(z,t)^* C_2(z,t) d_{12}\Big\}, \\
\check{P}_2(z,t) &= 2\mathcal{N}\mathrm{Re}\Big\{C_2(z,t)^* C_3(z,t) d_{23}\Big\}, \\
\check{P}_3(z,t) &= 2\mathcal{N}\mathrm{Re}\Big\{C_3(z,t)^* C_4(z,t) d_{34}\Big\}, \\
\check{P}_4(z,t) &= 2\mathcal{N}\mathrm{Re}\Big\{C_4(z,t)^* C_1(z,t) d_{41}\Big\}.
\end{aligned} \tag{21.127}
$$

We obtain the $C_n(z,t)$ from RWA Schrödinger equation

$$
\frac{\partial}{\partial t}C_n(z,t) = -i\sum_m W_{nm}(z,t) C_m(z,t). \tag{21.128}
$$

The RWA Hamiltonian needed here is

$$
\mathsf{W}(z,t) = \frac{1}{2}
\begin{bmatrix}
0 & \check{\Omega}_1(z,t)^* & 0 & \check{\Omega}_4(z,t) \\
\check{\Omega}_1(t) & 2\Delta_1 & \check{\Omega}_2(z,t)^* & 0 \\
0 & \check{\Omega}_2(z,t) & 2\Delta_2 & \check{\Omega}_3(z,t)^* \\
\check{\Omega}_4(z,t) & 0 & \check{\Omega}_3(z,t) & 2\Delta_2 + 2\Delta_3
\end{bmatrix}, \tag{21.129}
$$

where the Rabi frequencies are

$$
\check{\Omega}_n(z,t) = -d_{n,n+1}\check{\mathcal{E}}_n(z,t)/\hbar. \tag{21.130}
$$

We now assume that there is no resonance available, starting from state 4 for $\check{\Omega}_3$. With the assumption of large detuning Δ we can adiabatically eliminate state 4; its probability amplitude is

$$C_4(t) \approx \frac{\check{\Omega}_3(t)}{2\Delta_2 + 2\Delta_3} C_3(t) + \frac{\check{\Omega}_4(t)}{2\Delta_3} C_1(t). \qquad (21.131)$$

When $\check{\Omega}_4$ is very small, as it is initially, then

$$
\begin{aligned}
\langle d(t) \rangle \approx 2\mathrm{Re} \Big\{ & C_1(t)^* C_2(t) d_{12} \exp[+\mathrm{i}\omega_1 t] \\
& + C_2(t)^* C_3(t) d_{23} \exp[+\mathrm{i}\omega_2 t] \\
& + |C_3(t)|^2 \left[\frac{\check{\Omega}_3(t) d_{34}}{2\Delta_3} \right] \exp[+\mathrm{i}\omega_3 t] \\
& + C_3(t)^* C_1(t) \left[\frac{\check{\Omega}_3(t) d_{41}}{2\Delta_3} \right] \exp[-\mathrm{i}\omega_4 t] \Big\}. \qquad (21.132)
\end{aligned}
$$

The maximum growth at ω_4 occurs when $|C_1| = |C_3|$, i.e. when the coherence ρ_{13} is largest. The propagation equation for field 3 is

$$
\begin{aligned}
\left[\frac{\partial}{\partial z} + \frac{\check{\eta}_3}{c} \frac{\partial}{\partial t} \right] \mathcal{E}_3 &= \mathrm{i} \frac{\omega_3}{2c\check{\eta}_3\epsilon_0} \check{P}_3, \\
\check{P}_3 = 2\mathcal{N} C_3^* C_4 d_{34} &= -\mathcal{N} \left[|C_3|^2 \frac{(d_{34})^2}{\Delta_3} \right] \mathcal{E}_3,
\end{aligned}
\qquad (21.133)
$$

while that for field 4 expresses nonlinear growth, with field 3 contributing to the source

$$
\begin{aligned}
\left[\frac{\partial}{\partial z} + \frac{\check{\eta}_4}{c} \frac{\partial}{\partial t} \right] \mathcal{E}_4 &= \mathrm{i} \frac{\omega_4}{2c\check{\eta}_4\epsilon_0} \check{P}_4, \\
\check{P}_4 = 2\mathcal{N} C_4^* C_1 d_{41} &= -\mathcal{N} \left[C_3^* C_1 \frac{d_{34} d_{41}}{\Delta_3} \right] \mathcal{E}_3.
\end{aligned}
\qquad (21.134)
$$

Maximizing field production. From these equations we can devise the following procedure to maximize the growth of field 4. First we apply fields 1 and 2 to create a coherence adiabatically. The relevant RWA Hamiltonian is

$$
\mathsf{W} = \frac{1}{2} \begin{bmatrix} 0 & \check{\Omega}_1^* & 0 & 0 \\ \check{\Omega}_1 & 2\Delta_1 & \check{\Omega}_2^* & 0 \\ 0 & \check{\Omega}_2 & 2\Delta_2 & 0 \\ 0 & 0 & 0 & 2\Delta_3 \end{bmatrix}. \qquad (21.135a)
$$

Once this coherence forms, we apply field 3. The RWA Hamiltonian is now

$$
\mathsf{W} = \frac{1}{2}
\begin{bmatrix}
0 & 0 & 0 & \check{\Omega}_4^* \\
0 & 2\Delta_1 & 0 & 0 \\
0 & 0 & 2\Delta_2 & \check{\Omega}_3^* \\
\check{\Omega}_4 & 0 & \check{\Omega}_3 & 2\Delta_3
\end{bmatrix}.
\tag{21.135b}
$$

The effect of this Hamiltonian is to create the coherence that will generate the fourth field, via a two-photon transition.

21.6 Steady state; susceptibility

Completely coherent excitation, as described by the TDSE for nearly monochromatic radiation, exhibits ongoing Rabi oscillations that persist indefinitely. In reality, the atomic response eventually reaches a steady-state equilibrium with the applied fields, a consequence of various relaxation mechanisms. These include spontaneous emission as well as interactions with an environment that serves as a thermal bath. Over time intervals that are longer than a relaxation time these mechanisms bring the atom into a steady state, equilibrated with both the background and with averaged properties of the coherent fields. It is this steady state that produces the linear dependence of dipole moment upon electric field, as expressed by the traditional susceptibility.

Treatments that cover both the short-time coherent excitation of the TDSE and the incoherent regime of longer times make use of the density matrix, as evaluated in the rotating coordinate system that defines probability amplitudes C_n,

$$
s_{nm}(z,t) = \{C_n(z,t)C_m(z,t)^*\}_{av}.
\tag{21.136}
$$

I consider here the Raman system of Chap. 13.

21.6.1 Equations of motion for density matrix

Equations of motion for this density matrix involve the action of the coherent RWA Hamiltonian as well as parametrized actions of other interactions – radiative interactions that produce spontaneous emission and uncontrollable actions of the neighboring environment.

Coherent excitation. Changes to the density matrix produced by the coherent interaction derive from the TDSE,

$$
\frac{\partial}{\partial t} s_{mn} = -i \sum_k W_{mk} s_{kn} + i \sum_k s_{mk} W_{kn}^*.
\tag{21.137}
$$

In the RWA (without loss) the three-state matrix W has the structure given by the equation

$$
\mathsf{W} =
\begin{bmatrix}
0 & W_P & W_R^* \\
W_P & \Delta_P & W_S \\
W_R & W_S & \Delta_P - \Delta_S
\end{bmatrix}
\tag{21.138}
$$

in which the off-diagonal matrix elements $W_{12} = W_P$ and $W_{23} = W_S$ are taken to be real. The matrix elements are expressible as Rabi frequencies[7]

$$W_P = \tfrac{1}{2}\Omega_P, \qquad W_S = \tfrac{1}{2}\Omega_S, \qquad W_R^* = \tfrac{1}{2}\check{\Omega}_R^*, \tag{21.139}$$

where, with $\varphi = \varphi_P - \varphi_S - \varphi_R$, the Rabi frequencies are

$$\check{\Omega}_R^* = -d_{13}\mathcal{E}_R\, e^{i\varphi}/\hbar, \qquad \Omega_S = -d_{23}\mathcal{E}_S/\hbar, \qquad \Omega_P = -d_{12}\mathcal{E}_P/\hbar. \tag{21.140}$$

Relaxation. Incorporation of incoherent processes, for relaxation of time variations, takes place by adding inelastic and elastic rate coefficients to produce the generalized master equation discussed in Sec. 16.9.1,

$$\frac{\partial}{\partial t} s_{m'm}(t) = -i \sum_n [W_{m'n}(t)s_{nm}(t) - s_{m'n}(t)W_{nm}(t)]$$

$$- \hbar \sum_{nn'} \Gamma_{m'mn'n} s_{n'n}(t). \tag{21.141}$$

With the combination of coherent and incoherent changes the equations for the populations read

$$i(\partial_t + \gamma_{11})P_1 = +W_P(s_{21} - s_{12}) + W_R^* s_{31} - W_R s_{13} + P_2 \mathcal{R}_{2\to1} + P_3 \mathcal{R}_{3\to1},$$
$$i(\partial_t + \gamma_{22})P_2 = +W_P(s_{12} - s_{21}) + W_S(s_{32} - s_{23}) + P_3 \mathcal{R}_{3\to2}, \tag{21.142}$$
$$i(\partial_t + \gamma_{33})P_3 = +W_R s_{13} - W_R^* s_{31} + W_S[s_{23} - s_{32}] + P_2 \mathcal{R}_{2\to3}.$$

The equations for the coherences read

$$i[\partial_t + i\gamma_{12} - i\Delta_P]s_{12} = -W_S s_{13} - W_P(P_1 - P_2),$$
$$i[\partial_t + \gamma_{23} + i\Delta_S]s_{23} = +W_P s_{13} - W_S(P_2 - P_3), \tag{21.143}$$
$$i[\partial_t + i\gamma_{13} - i(\Delta_P - \Delta_S)]s_{13} = +W_P s_{23} - W_S s_{12}.$$

Here the relaxation rates γ_{nm} are the sums of the loss rates from states n and m,

$$\gamma_{nm} = \gamma_n + \gamma_n = \gamma_{mn}. \tag{21.144}$$

Steady state. Steady-state solutions for the coherences are obtained by setting the time derivatives to zero, thereby obtaining the following relationships:

$$(i\gamma_{12} + \Delta_P)s_{12} + W_S s_{13} = -W_P(P_1 - P_2), \tag{21.145a}$$
$$(i\gamma_{23} + \Delta_S)s_{23} - W_P s_{13} = -W_S(P_2 - P_3), \tag{21.145b}$$
$$(i\gamma_{13} + \Delta_P - \Delta_S)s_{13} - W_P s_{23} + W_S s_{12} = 0. \tag{21.145c}$$

---

[7] Other authors use the symbol $\check{\Omega}$, and the term "Rabi frequency" with definitions that differ by a factor of 2 and/or a minus sign.

Let us consider specifically the coherence s_{12} associated with the P-field transition from the ground state 1 (initially populated) to the excited state 2, as linked by the dipole moment d_{12}. We evaluate this for given steady-state values of the populations P_n.

21.6.2 Single steady P field

The simplest situation occurs when the S-field interaction is absent or very weak. With the identification

$$W_P = \tfrac{1}{2}\Omega_P = -\tfrac{1}{2}d_{12}\mathcal{E}_P/\hbar \qquad (21.146)$$

eqn. (21.145a) then gives the result

$$s_{12} = (s_{21})^* = -\frac{W_P(P_1 - P_2)}{(i\gamma_{12} + \Delta_P)} = \frac{d_{12}}{2\hbar}\frac{(P_1 - P_2)}{(i\gamma_{12} + \Delta_P)}\mathcal{E}_P. \qquad (21.147)$$

This coherence exhibits a Lorentzian variation with detuning Δ_P, over a width fixed by the damping constant γ_{12}. The complex susceptibility for the P field is

$$\chi_P(\Delta_P) = \frac{2\mathcal{N}d_{12}}{\epsilon_0\hbar}\frac{s_{21}}{\mathcal{E}_P} = \frac{\mathcal{N}|d_{12}|^2}{\epsilon_0\hbar}\frac{(P_1 - P_2)}{(-i\gamma_{12} + \Delta_P)}$$

$$= (P_1 - P_2)\frac{\mathcal{N}|d_{12}|^2}{\hbar\epsilon_0}\frac{(\Delta_P + i\gamma_{12})}{\Delta_P^2 + (\gamma_{12})^2}. \qquad (21.148)$$

The resonant value is purely imaginary

$$\chi_P(0) = i(P_1 - P_2)\frac{\mathcal{N}|d_{12}|^2}{\hbar\epsilon_0\gamma_{12}}. \qquad (21.149)$$

Figure 21.4 shows an example of susceptibility χ, normalized to unit peak value, for a two-state transition. The upper frame shows the imaginary part, which is proportional to the absorption coefficient. The profile is Lorentzian, centered about the Bohr frequency, and with a width set by the relaxation rate. The lower frame shows the real part of χ, from which one obtains the refractive index. This curve passes through zero on resonance.

This figure describes the susceptibility of a system in which there is a weak probe of a transition between states 1 and 2.

21.6.3 Strong steady S field

We take account of the S field, as it becomes stronger, by including the other three steady-state equations. By eliminating the coherences s_{23} and s_{13} we obtain the equation

$$\left[(i\gamma_{13} + \Delta_P - \Delta_S)(i\gamma_{12} + \Delta_P) - W_S^2 - \frac{(i\gamma_{12} + \Delta_P)}{(i\gamma_{23} + \Delta_S)}W_P^2\right]s_{12} \qquad (21.150)$$

$$= +W_P\left[\frac{W_P^2(P_1 - P_2) + W_S^2(P_2 - P_3)}{(i\gamma_{23} + \Delta_S)} - (i\gamma_{13} + \Delta_P - \Delta_S)(P_1 - P_2)\right].$$

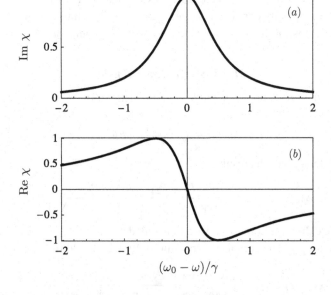

Figure 21.4 Example of relative χ for simple two-state system.

Typically we regard the P-field coupling as weak, thereby neglecting the terms proportional to W_P^2. We also regard the excited-state populations as negligible, thereby neglecting the terms P_2 and P_3 on the right-hand side. The result is the expression

$$s_{12} = -W_P \frac{(i\gamma_{13} + \Delta_P - \Delta_S)}{(i\gamma_{13} + \Delta_P - \Delta_S)(i\gamma_{12} + \Delta_P) - W_S^2}$$

$$= -W_P \frac{1}{i\gamma_{12} + \Delta_P - \dfrac{W_S^2}{(i\gamma_{13} + \Delta_P - \Delta_S)}}. \qquad (21.151)$$

We substitute this result into the formula for χ_P, using the definition of W_P, and obtain the result

$$\chi_P(\Delta_P, \Delta_S) = \frac{2\mathcal{N}d_{12}}{\epsilon_0} \frac{s_{21}}{\mathcal{E}_P} = i \frac{\mathcal{N}|d_{12}|^2}{\epsilon_0} \frac{1}{\gamma_{12} - i\Delta_P + \dfrac{(\check{\Omega}_S/2)^2}{\gamma_{13} - i(\Delta_P - \Delta_S)}}. \qquad (21.152)$$

When the excited state 2 has an additional, strong, coupling to state 3 by field S, then the resulting χ exhibits a doublet structure attributable to an Autler–Townes doublet. Figure 21.5 shows the relative χ for the probe field P, when the S field is resonant with the 2–3 transition.

The symmetry of these profiles occurs only when the S field is resonant. Figure 21.6 shows an example when it is detuned.

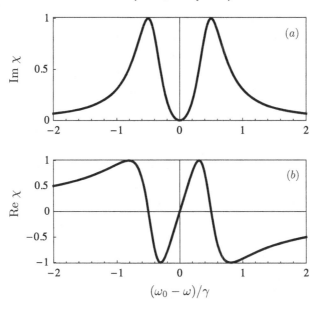

Figure 21.5 Example of χ for the 1–2 transition of a three-state system in which there is a strong 2–3 coupling by an S field that is resonant with the 2–3 transition.

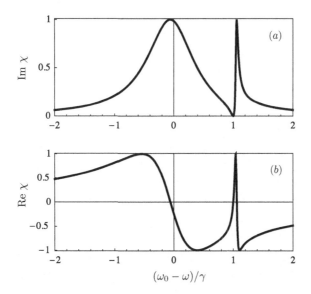

Figure 21.6 Example of χ as in the previous figure, for detuned S field.

21.6.4 Inhomogeneous broadening

The formulas appearing above deal with a collection of identical stationary atoms, each of which has the same energy levels, and each of which undergoes probability losses at the same rate. To treat a collection of atoms that have different energy levels, as occurs when the atoms are in various host environments of a solid matrix, we must average over the appropriate detunings. The simplest case occurs when the middle state of the chain, state 2, is shifted relative to states 1 and 3. Then both detunings Δ_P and Δ_S will be shifted by the same amount. We require the average

$$\bar{\chi} = \int dx\, g(x)\, \chi(\Delta_P + x, \Delta_S + x), \tag{21.153}$$

where $g(x)$ is the probability of detuning shift x. The result, plotted as a function of Δ_P, will exhibit *inhomogeneous* broadening of the function. When γ_{12} is very small, then the profile will have a width taken from the function $g(x)$.

22

Atoms in cavities

The fields within cavities differ in an important respect from those of a laser beam in free space: The bounding surfaces that define the cavity enclosure impose constraints that limit the fields to discrete modes, characterized in part by discrete frequencies. The surfaces enclosing a cavity are boundaries where the dielectric properties change abruptly; they are idealized as discontinuities of the susceptibility ϵ and permeability μ. Across any such surface the normal component of the **B** field is continuous, as is the transverse component of the **E** field. These conditions imply, for example, that along a perfectly conducting surface (an idealized mirror) the electric field has a node. The allowed fields (the discrete mode fields) are then particular solutions to the Maxwell equations, or their conversion into Helmholtz equations, that vanish along bounding surfaces.[1]

Figure 4.2 of Sec. 4.1.3 depicts two classes of cavities. Frame (*a*) shows a cylindrical cavity used for microwave radiation. The cavity completely encloses the field, apart from a small aperture through which the atoms pass. Frame (*b*) shows a prototype optical cavity, in which the cavity field is that of a beam confined along one axis. Idealized as perfect conductors, the confining mirrors permit only integral half waves between them, and the frequencies of such plane waves are correspondingly discrete. Because the enclosing end-mirrors are not perfectly reflecting there will occur some loss through them, along the cavity axis, and the field is not strictly monochromatic. The emerging field is a source of photons; see Sec. 22.3.5.

In each of these cases the radiation within the enclosure is most appropriately described as a superposition of standing waves rather than traveling waves. The cavity geometry may either enhance or suppress particular standing-wave modes. In turn, this alteration of possible field modes will affect the rate of spontaneous emission into what, in free space, is an unstructured continuum of radiation modes. The altered field structure may either increase or decrease the spontaneous emission rate, see Sec. 22.1.1.

The transfer of energy from an excited atom into energy of a confined field will alter in measurable ways the cavity field. The alteration can be detected either by sending a probe atom through the cavity or by observing a (small) portion of the field that leaks out. As

---

[1] As with discrete spectral lines, there occur finite widths of the frequency possibilities, because the cavity walls are not perfect conductors and the idealization of a discontinuity is only approximate.

will be evident, the coupling between field and atom is not well described by perturbation theory. However, a variety of alternative approaches, taken from the mathematics of coherent excitation by laser pulses, are applicable.

Although I present the image of a single cavity, such as occurs with microwaves, a variety of other constructions have similar properties. Photonic crystals [Joa95; Ang01] are an example. These extended quasiperiodic structures mimic many of the properties of crystals, but fields within them are inhibited as they would be in a network of waveguides.

It is the fields within cavities that exhibit most clearly the consequences of quantization, i.e. of treating radiation by means of dynamical variables represented by noncommuting operators. In turn, the response of atoms placed within a cavity to those fields provides a variety of novel opportunities for quantum-state manipulation. This chapter discusses some simple examples.

22.1 The cavity

For present purposes a cavity is an enclosure in which the electromagnetic field is constrained. Two viewpoints are useful, one based on classical (unquantized) fields, the other on fields that are quantized. The following sections discuss these two views.

22.1.1 Classical cavity

A one-dimensional cavity of length L with conducting mirrors at the ends allows only discrete longitudinal modes, characterized by fitting an integer number of half wavelengths into the cavity. The allowed wavelengths for propagation along that axis must satisfy the constraint $m\lambda = 2L$ for integer m, and so the allowed frequencies are

$$\omega_m = m\frac{2\pi c}{2L}. \qquad (22.1)$$

The separation between discrete frequencies is the *free spectral range*

$$\Delta\omega_{FSR} = \frac{\pi c}{L}. \qquad (22.2)$$

The mirrors are not completely reflective, but allow a small leakage of energy, quantified by the reflectivity $\mathcal{R}$. As a consequence of this loss, as well as losses by absorption to the cavity walls, the cavity will not retain radiation indefinitely. The loss of cavity field energy is often quantified by the cavity quality factor or *Q factor*, defined, for frequency ω, by the ratio

$$\frac{1}{Q} = \frac{\text{energy lost}}{\omega \times \text{energy stored}} = \frac{2\kappa}{\omega}. \qquad (22.3)$$

A consequence of the power loss is that the power spectrum of a cavity is not simply a discrete set of frequencies, but each of these has a spectral width equal to the cavity decay rate of 2κ. The separation of cavity frequencies, in units of this width, is the *finesse* of the cavity, $\mathcal{F}$:

$$\mathcal{F} = \Delta\omega_{FSR}/2\kappa. \qquad (22.4)$$

The cavity finesse is the average number of round trips in the cavity before a photon is lost through the end mirrors; it can be evaluated as

$$\mathcal{F} = \pi \sqrt{\mathcal{R}}/(1 - \mathcal{R}).\qquad(22.5)$$

Atoms undergoing spontaneous emission in a cavity (or in any confined space) must produce a field suited to the constrained surroundings – as exhibited by the mode structure of the field. As a consequence, spontaneous emission is altered from its free-space rate by the presence of any boundaries. The *Purcell factor* [Pur46; Goy83; Mil91]

$$f_P = \frac{3Q\lambda^3}{4\pi V}\qquad(22.6)$$

quantifies this change. Here Q is the quality factor of the cavity. The spontaneous emission is enhanced if $f_P > 1$, and is inhibited if $f_P < 1$.

22.1.2 Quantum cavity

As discussed in App. D the quantum field in a cavity is expressible by means of single-photon creation and annihilation operators together with mode functions that express the spatial and vectorial properties of the field. The positive-frequency part of a single-mode field at position $\mathbf{r}$ in the cavity is

$$\hat{\mathbf{E}}^{(+)}(\mathbf{r}) = \sqrt{2\pi\hbar\omega}\,\mathbf{e}\,U(\mathbf{r})\,\hat{a},\qquad(22.7)$$

where $\hat{a}$ is a photon annihilation operator, $\mathbf{e}$ is a unit vector, and the mode field $U(\mathbf{r})$ is a scalar solution to the Helmholtz equation, subject to suitable boundary conditions, and normalized as

$$\int d\mathbf{r}|U(\mathbf{r})|^2 = 1.\qquad(22.8)$$

The Hamiltonian operator for the single-mode field in the bare cavity, idealized as lossless, is taken to be (see App. D.4)

$$\hat{H}^R = \hbar\omega_c\,\hat{a}^\dagger\hat{a}.\qquad(22.9)$$

The field will not remain indefinitely within the cavity, but will eventually diminish as a consequence of cavity losses – absorption by the cavity walls or transmission through bounding mirrors. As with probability loss of atomic populations we model this by adding a supplementary non-Hermitian operator to the free-field. We take this operator to be proportional to the photon number operator $\hat{a}^\dagger\hat{a}$, as discussed below.

22.1.3 The atom

The basic Hilbert-space description of an atom in a cavity begins with the definition of free-space quantum states $\boldsymbol{\psi}_n$, eigenstates of an atomic Hamiltonian H^{at},

$$\hat{H}^{at}\boldsymbol{\psi}_n = E_n\boldsymbol{\psi}_n.\qquad(22.10)$$

When treating quantized fields by means of creation and annihilation operators it is often convenient to use operators to express changes of atomic states. The simplest of those, used in the following, are the elementary transition operators (see Sec. 16.10.1)

$$\boldsymbol{\pi}(n, m) = |\psi_n\rangle\langle\psi_m| \tag{22.11}$$

that replace atomic unit vector $\boldsymbol{\psi}_m$ by the unit vector $\boldsymbol{\psi}_n$. The projection operator $\boldsymbol{\Pi}(n) = \boldsymbol{\pi}(n, n)$ is a special case, in which $m = n$. Using these operators we express the atom Hamiltonian as

$$\hat{H}^{at} = \sum_n E_n \boldsymbol{\Pi}(n). \tag{22.12}$$

When the atom is in an excited state it may spontaneously emit radiation, either into unconfined field modes or into the cavity. The radiation into unconstrained modes can be modeled as a probability loss from each excited state, by introducing complex-valued energies and a concomitant non-Hermitian Hamiltonian expressed with the population operators $\boldsymbol{\Pi}(n)$, as discussed below.

22.1.4 The interaction

The various multipole interactions all involve the action of a single photon. They produce interaction Hamiltonians of the form

$$V = \tfrac{1}{2}\hbar\check{\Omega}_1\,\boldsymbol{\pi}(2, 1)\,[\hat{a}^\dagger + \hat{a}] + \tfrac{1}{2}\hbar\check{\Omega}_1^*\,[\hat{a}^\dagger + \hat{a}]\,\boldsymbol{\pi}(1, 2), \tag{22.13}$$

in which the complex-valued Rabi frequency $\check{\Omega}_1$ quantifies the strength of the single-photon interaction – the product of an atomic transition moment and a field strength. The interaction $\check{\Omega}^*$ is associated with a transition in which the atom loses energy, while Ω accompanies a gain of atomic energy.[2]

For an electric-dipole interaction the interaction strength evaluates as the product

$$\hbar\check{\Omega}_1 = -d_{12}\check{\mathcal{E}}_1, \qquad \hbar\check{\Omega}_1^* = -d_{21}\check{\mathcal{E}}_1^*, \tag{22.14}$$

where d_{12} is the projection of the dipole transition moment onto the direction of the cavity electric field, and $\mathcal{E}_1$ is the envelope of the single-photon electric field, as evaluated at the atom center of mass $\mathbf{R}$,

$$d_{12} = \langle 1|\mathbf{d}\cdot\mathbf{e}|2\rangle, \qquad \mathcal{E}_1 = \sqrt{2\pi\hbar\omega_c}\,U(\mathbf{R}). \tag{22.15}$$

This field varies with the position of the atom: It is largest when the atom is at an antinode of the mode function $U(\mathbf{r})$, and it vanishes when the atom is at a node. If the atom is moving, say entering or leaving the cavity, then this field varies with time. As an example,

---

[2] This Hamiltonian has not made the RWA, and so it includes terms that do not conserve overall energy – in which gain of atom energy accompanies gain of field energy. Section 22.2.2 introduces the RWA and neglects interactions in which there is a large deviation from resonance.

the maximum cavity field of a single-mode photon in a cube of side L is

$$\mathcal{E}_1 = \sqrt{\frac{8\pi\hbar\omega_c}{L^3}}. \tag{22.16}$$

22.2 Two-state atoms in a cavity

Consider a two-state atom in a cavity supporting a single mode, of frequency ω_c. For subsequent extension to three states, I will here use labels $g = 1$ and $x = 2$ for the two atomic states. We take as basis states for the two-state atom the eigenstates of the free-atom Hamiltonian H^{at}, having eigenvalues $E_g < E_x$. The difference of these energies defines the Bohr frequency

$$\hbar\omega_B = E_x - E_g. \tag{22.17}$$

Let Δ_c denote the detuning of the cavity from the Bohr frequency,

$$\hbar\Delta_c = E_x - E_g - \hbar\omega_c. \tag{22.18}$$

The regime of interest here is when the cavity frequency ω_c matches, or nearly matches, the Bohr frequency.

The field we describe by single-mode photon-number states $|n\rangle$ (see App. D.1.1). To treat the coupled system we use product states, $|\text{atom, field}\rangle$, and express the statevector as the superposition of all possible photon-number states, each paired with an atomic state,

$$\Psi(t) = \frac{1}{\sqrt{\mathcal{N}}} \sum_n e^{-i\zeta_n(t)} \left[C_{gn}(t)|g,n\rangle + C_{xn}(t)|x,n-1\rangle \right], \tag{22.19}$$

where the normalizing factor is obtained from the sum at the initial time $t = 0$,

$$\mathcal{N} = \sum_n \left[|C_{gn}(0)|^2 + |C_{xn}(0)|^2 \right], \tag{22.20}$$

and $\zeta_n(t)$ is a phase chosen for mathematical convenience. Below, these functions are taken to involve a constant frequency plus a static phase φ,

$$\hbar\zeta_n(t) = \left[E_1 + n\hbar\omega_c \right]t + \hbar\varphi. \tag{22.21}$$

This choice of phase of the statevector is tantamount to taking the zero of energy as that of atomic state E_1. In the sum of eqn (22.19) the two contributing product states in brackets have the energies

$$E_{gn} = \left[E_g + \hbar\omega \right] + (n-1)\hbar\omega_c, \qquad E_{xn} = \left[E_x \right] + (n-1)\hbar\omega_c. \tag{22.22}$$

The energies of these two atom–field states of eqn. (22.19) are equal when the cavity frequency ω_c matches the Bohr frequency, i.e. $\Delta_c = 0$. Our interest lies in this resonant case, or in the near-resonant regime when the cavity detuning is either zero or much smaller than the cavity frequency,

$$\Delta_c \ll \omega_c. \tag{22.23}$$

The sum then proceeds over pairs of states that have similar (or identical) energies, separated by ω_c.

22.2.1 The two-state Schrödinger equation

From these definitions and the operator properties

$$\hat{a}^\dagger|n\rangle = \sqrt{n+1}|n+1\rangle, \qquad \hat{a}^\dagger|n\rangle = \sqrt{n}|n-1\rangle,$$
$$\pi(x,g)|g\rangle = |x\rangle, \qquad \pi(g,x)|x\rangle = |g\rangle, \qquad (22.24)$$
$$\pi(x,g)|x\rangle = 0, \qquad \pi(g,x)|g\rangle = 0,$$

we obtain a set of coupled equations,

$$\frac{d}{dt}C_{gn}(t) = -i\frac{1}{2}\check{\Omega}_n^* C_{xn}(t) - i\left[\frac{1}{2}\check{\Omega}_n^* C_{x,n+1}(t)\right],$$
$$\frac{d}{dt}C_{xn}(t) = -i\Delta_c C_{xn}(t) - i\frac{1}{2}\check{\Omega}_n C_{gn}(t) - i\left[\frac{1}{2}\check{\Omega}_n C_{g,n-1}(t)\right], \qquad (22.25)$$

where

$$\hbar\check{\Omega}_n = -d_{12}\mathcal{E}_1\, e^{i\varphi}\sqrt{n} \qquad (22.26)$$

is the n-photon Rabi frequency. We will choose the arbitrary phase φ to make this Rabi frequency real.

22.2.2 The two-state RWA

In the preceding equations the bracketed term is associated with an atom–field state whose energy differs by $\hbar\omega_c$ from the energies of the unbracketed terms. These are the quantized-field versions of the counter-rotating terms of the classical interaction with a monochromatic field. They will be neglected in what follows. The result is the *Jaynes–Cummings model* (JCM),

$$\frac{d}{dt}C_{gn}(t) = -\frac{i}{2}\Omega_n C_{xn}(t), \qquad (22.27a)$$

$$\frac{d}{dt}C_{xn}(t) = -i\Delta_c C_{xn}(t) - \frac{i}{2}\,\Omega_n C_{gn}(t), \qquad (22.27b)$$

in which, as a consequence of our phase choice, eqn. (22.21), the cavity coupling Ω_n is real and constant. We write this in matrix form, using $\mathbf{C}^T = [C_{gn}, C_{xn}]$, as

$$\frac{d}{dt}\mathbf{C}(t) = -i\mathbf{W}\mathbf{C}(t), \qquad (22.28)$$

where the RWA Hamiltonian matrix appropriate to photon number n, is

$$\mathbf{W} = \begin{bmatrix} 0 & \frac{1}{2}\Omega_n \\ \frac{1}{2}\Omega_n & \Delta_c \end{bmatrix} \begin{matrix} g \\ x \end{matrix}. \qquad (22.29)$$

This pair of coupled equations are identical to those of a traditional two-state atom; the Rabi frequency Ω_n is proportional to the square root of the photon number, i.e. to the square root of the mean intensity, as with the classical field.

These are the equations appropriate to a single stationary two-state atom, immune to spontaneous emission, interacting with the single lossless mode of an ideal cavity from which no radiation escapes or is otherwise lost. Several modifications are needed to make this model a more realistic description of atom–cavity interactions.

Alternative phases. The expansion of eqn. (22.19) taken with phase (22.21), leads to the traditional mathematics of quantized radiation, in which the coupling is constant, parametrized by a Rabi frequency Ω_n that is real and constant. In this approach there occurs a diagonal element of the Hamiltonian, the detuning Δ_c. Alternatively, one can choose a phase for state $|x, n-1\rangle$ that differs from the phase of state $|g, n\rangle$ by the detuning Δ_c. The resulting Hamiltonian matrix will have null diagonal elements, but will have off-diagonal elements that have time variation $\exp(i\,\Delta_c t)$.

22.2.3 The Jaynes–Cummings model (JCM)

The pair of equations (22.27a) and (22.27b) are two of an infinite set of uncoupled equations that constitute the Jaynes–Cummings model [Jay63; Mil91; Sho93] of a quantum-mechanical atom interacting with a quantized field. The two equations are those obtained for the simple two-state atom in a constant-intensity field that has constant detuning, and hence they have analytic solutions involving sine and cosine of the effective Rabi frequency.

For an atom that is motionless within the cavity, the Rabi frequency is constant. If the atom is brought into or out of the cavity, the Rabi frequency changes with time, just as with the classical field model. In particular, if the atom is motionless and begins in the excited state $|x, N-1\rangle$ at $t=0$ the solution to the Schrödinger equation is one of Rabi oscillations,

$$C_{xn}(t) = i\,(\Omega_n/\Upsilon_n)\,\sin(\Upsilon_n t/2), \qquad (22.30)$$

where the n-photon flopping frequency is

$$\Upsilon_n = \sqrt{|\Omega_n|^2 + \Delta_c^2}. \qquad (22.31)$$

As can be recognized, there exist exact analytic solutions to the model of a stationary two-state system interacting, in the RWA, with a single quantized field mode. This idealization, the JCM, has been extended to treat a variety of refinements [Sho93].

Within the JCM only two photon numbers are involved in the Rabi oscillations: the photon number n associated with the ground state of the atom and the number $n-1$ associated with the excited state to which the atom–field state $|g, n\rangle$ links. However, when excited atoms enter the cavity they bring energy that may subsequently remain within the cavity, if the dynamics during the travel of the atom is such that an odd-integer number of Rabi cycles occur. Similarly, an atom can remove a quantum of energy if it is in an excited state when it leaves the cavity.

The probabilities of finding the system in atomic state g or x are obtained by squaring the projection of the statevector onto the appropriate atomic unit vector

$$P_g(t) = |\langle g|\Psi(t)\rangle|^2, \qquad P_x(t) = |\langle x|\Psi(t)\rangle|^2. \tag{22.32}$$

If the initial state is one in which the photon number is precisely known, say $|x, n-1\rangle$, then the probabilities are

$$P_g(t) = |C_{gn}(t)|^2, \qquad P_x(t) = |C_{xn}(t)|^2. \tag{22.33}$$

More generally the radiation field, though single mode, is not that of a well-defined photon number but is a superposition of photon numbers. If the initial condition is a mixed state, characterized by a probability $p(n)$ of n photons, then the excitation probability is

$$\bar{P}_x(t) = \sum_n p(n)|C_{xn}(t)|^2. \tag{22.34}$$

For example, when the atom is initially unexcited (and motionless) and the cavity is resonant, then the excitation probability is

$$\bar{P}_x(t) = \frac{1}{2} \sum_{n=0}^{\infty} p(n)\left[1 - \cos(\sqrt{n+1}\,\check{\Omega}_1 t)\right]. \tag{22.35}$$

The averaging superposes solutions having different Rabi frequencies. The result is an effective damping of the oscillations. However, under suitable conditions the "collapse" of the oscillations is followed by a "revival" [Nar81].

When the superposition is coherent the associated probability amplitudes are sums over all possible photon numbers, and so the probabilities are squares of sums,

$$\bar{P}_g(t) = \frac{1}{\mathcal{N}}\left|\sum_n e^{-i\zeta_n(t)} C_{gn}(t)\right|^2, \qquad \bar{P}_x(t) = \frac{1}{\mathcal{N}}\left|\sum_n e^{-i\zeta_n(t)} C_{xn}(t)\right|^2. \tag{22.36}$$

These sums generally produce interference effects.

The Jaynes–Cummings Hamiltonian. The equations of the JCM described above derive from an interaction Hamiltonian

$$V = +\tfrac{1}{2}\hbar\check{\Omega}_1\,\boldsymbol{\pi}(2,1)\hat{a} + \tfrac{1}{2}\hbar\check{\Omega}_1^*\,\hat{a}^\dagger\boldsymbol{\pi}(1,2). \tag{22.37}$$

In this approximation each emission of a photon accompanies a deexcitation of the atom, while each atomic excitation accompanies the loss of a photon. (Missing are transitions in which, e.g., atomic excitation accompanies loss of a photon.) The strength of this interaction is parametrized by the single-photon (or vacuum) Rabi frequency $\check{\Omega}_1$ for an atom whose center of mass is $\mathbf{R}$:

$$\hbar\check{\Omega}_1 = -2\sqrt{2\pi\hbar\omega_c}\,\langle 2|\mathbf{d}\cdot\mathbf{e}|1\rangle\,U(\mathbf{R}). \tag{22.38}$$

If the atom remains motionless this quantity is constant; if the atom moves through the cavity then this becomes time dependent. The full Hamiltonian, for a single two-state atom interacting with a single-mode quantized field, is the JCM

$$
\begin{aligned}
\hat{H} &= \hat{H}^R + \hat{H}^{at} + \mathsf{V} \\
&= \bar{E}\Big[\boldsymbol{\Pi}(1) + \boldsymbol{\Pi}(2)\Big] + \tfrac{1}{2}\hbar\omega_c + \tfrac{1}{2}\hbar\omega_B\Big[\boldsymbol{\Pi}(2) - \boldsymbol{\Pi}(1)\Big] \\
&\quad + \hbar\omega_c\hat{a}^\dagger\hat{a} + \tfrac{1}{2}\Big[\hbar\check{\Omega}_0^* \,\hat{a}^\dagger\boldsymbol{\pi}(1,2) + \hbar\check{\Omega}_0\,\boldsymbol{\pi}(2,1)\hat{a}\Big].
\end{aligned}
\tag{22.39}
$$

This Hamiltonian has two constants of motion, obtained as expectation values of the excitation-number operator

$$
\hat{N} = \boldsymbol{\Pi}(2) + \hat{a}^\dagger\hat{a},
\tag{22.40}
$$

and the exchange-number operator

$$
\hat{C} = \tfrac{1}{2}\Delta_c\,\boldsymbol{\Pi}(2) + \mathsf{V}.
\tag{22.41}
$$

22.2.4 Losses

If the cavity were completely surrounded by perfectly conducting walls, then the energy within the cavity would remain constant, and no energy would be detectable outside the cavity, nor would atoms be able to enter and leave the cavity. Microwave cavities can be reasonable approximations to this ideal. In practice some cavity losses occur, as quantified by the quality factor (or Q factor). Imperfect conduction allows the field energy to convert, irreversibly, into heat within the walls. Radiation can also escape, irreversibly, through the cavity aperture through which the atoms enter. This loss of energy we typically quantify by means of an energy decay rate, 2κ.

Cavities used for optical frequencies typically are formed by two highly reflecting mirrors that confine fields along only one axis. An atom in such a cavity can spontaneously emit radiation that emerges around the sides of the cavity, although radiation into a single longitudinal mode will fit the JCM. To model spontaneous emission into modes other than the confined cavity mode we can introduce a non-Hermitian Hamiltonian, as with traditional two-state descriptions, letting 2γ be the loss rate for probability in the excited state x.

In a microwave cavity the observations of the cavity field occur by examining atoms that pass through the cavity. In an optical cavity the loss of field energy occurs through one of the confining end mirrors: this emerging field provides a signal that probes the cavity field.

We take into account these two loss mechanisms by means of a non-Hermitian Hamiltonian of loss

$$\hat{H}^{loss} = -i\,\hbar\gamma\,\boldsymbol{\pi}\,(x,x) - i\,\hbar\kappa\,\hat{a}\hat{a}^{\dagger}, \tag{22.42}$$

parametrized by two real-valued positive constants:

γ quantifies the loss of atom population from excited state x by spontaneous emission into noncavity modes

κ quantifies loss of electric field from the cavity through the end mirrors.

With the addition of these loss mechanisms the two-state Schrödinger equation reads

$$\frac{d}{dt}C_{gn}(t) = -\kappa n\,C_{gn}(t) - \frac{i}{2}\check{\Omega}_n C_{xn}(t),$$
$$\frac{d}{dt}C_{xn}(t) = -[\kappa(n-1)+\gamma]\,C_{xn}(t) - i\,\Delta_c C_{xn}(t) - \frac{i}{2}\check{\Omega}_n C_{gn}(t). \tag{22.43}$$

The corresponding RWA Hamiltonian matrix, the Jaynes–Cummings Hamiltonian, is

$$\mathsf{W} = \begin{bmatrix} -i\kappa n & \tfrac{1}{2}\check{\Omega}_n \\ \tfrac{1}{2}\check{\Omega}_n & \Delta_c - i\gamma - i\kappa(n-1) \end{bmatrix} \begin{matrix} g,n \\ x,n-1 \end{matrix}. \tag{22.44}$$

This matrix serves as the starting point for a number of applications.

Rapid cavity loss. A limiting case, that of the "*bad cavity*", occurs when the cavity loss rate dominates the dynamics, as happens when $\kappa \gg g_0^2/\kappa \gg \gamma$. Under these conditions we can adiabatically eliminate state g by setting its time derivative to zero,

$$\frac{d}{dt}C_{gn}(t) = 0, \tag{22.45}$$

thereby obtaining the approximation

$$C_{gn} = -i\,\Omega_1\kappa\sqrt{n}\,C_{xn} \tag{22.46}$$

and, in turn, the equation of motion

$$\frac{d}{dt}C_{xn} = -[\gamma' + i\,\Delta_c]C_{xn}. \tag{22.47}$$

The effective damping rate occurring here, independent of n, is

$$\gamma' = \gamma + \frac{\Omega_1^2}{4\kappa}. \tag{22.48}$$

Here γ is the rate of emitting photons into noncavity modes, while $\Omega_1^2/4\kappa$ is the rate of stimulated emission into the cavity mode. In this regime the field is lost too quickly to support any Rabi oscillations: it emerges directly as it is produced.

22.3 Three-state atoms in a cavity

An important extension of the JCM involves three atomic states. Two of these states, g and x, of energies $E_g < E_x$, are coupled near-resonantly via the cavity field. We quantify that interaction by the vacuum Rabi frequency, here denoted $\check{\Omega}_1 = 2g$. The third state e, of energy $E_e < E_x$, couples to state x by a classical laser field, at carrier frequency ω_L. This field, evaluated at the atom center of mass $\mathbf{R}$, is

$$\mathbf{E}_L(\mathbf{R}, t) = \text{Re}\left\{\mathbf{e}_L \, \mathcal{E}_L(t) \exp(-i\omega_L t)\right\}. \tag{22.49}$$

We parametrize this interaction with the Rabi frequency $\check{\Omega}(t)$,

$$\check{\Omega}(t) = -\langle g|\mathbf{e}_L \cdot \mathbf{d}|x\rangle \mathcal{E}_L(t), \tag{22.50}$$

possibly complex, and the detuning Δ_L,

$$\hbar\Delta_L = E_x - E_e - \hbar\omega_L, \tag{22.51}$$

real but possibly time varying. The cavity field we describe by single-mode photon-number states $|n\rangle$ and a Rabi frequency $2g$.

We take as basis states for the atom the eigenstates of the free atom, H^{at},

$$\mathsf{H}^{at} = E_e \, \boldsymbol{\pi}(e, e) + E_x \, \boldsymbol{\pi}(x, x) + E_g \, \boldsymbol{\pi}(g, g). \tag{22.52}$$

The interaction Hamiltonian incorporates the laser field linking states e and x, and the cavity field linking states x and g. We assume that the conditions for the RWA hold for the two fields, $\Delta_c \ll \omega_c$, $\Delta_L \ll \omega_l$. In the RWA the interaction Hamiltonian is

$$\mathsf{V} = \hbar g \, \boldsymbol{\pi}(x, e)\, \hat{a} + \hbar g \, \hat{a}^\dagger \, \boldsymbol{\pi}(e, x) + \tfrac{1}{2}\hbar\check{\Omega}(t) \, \boldsymbol{\pi}(g, x) + \tfrac{1}{2}\hbar\check{\Omega}(t)^* \, \boldsymbol{\pi}(x, g). \tag{22.53}$$

The losses we model by the constant Hamiltonian $\hat{H}^{loss}$ of eqn. (22.42).

To treat the coupled system we use product states, $|\text{atom, field}\rangle$, and express the statevector of the atom–cavity system as

$$\Psi(t) = \frac{1}{\sqrt{\mathcal{N}}}\left[\sum_n e^{-i\zeta_n(t)} \left[C_{gn}(t)|g, n\rangle + C_{xn}(t)|x, n-1\rangle\right]\right.$$
$$\left. + \sum_n e^{-i\zeta'_n(t)} C_{en}(t)|e, n-1\rangle\right], \tag{22.54}$$

where the normalizing factor is obtained from the sum at time $t = 0$,

$$\mathcal{N} = \sum_n \left[|C_{gn}(0)|^2 + |C_{xn}(0)|^2 + |C_{en}(0)|^2\right]. \tag{22.55}$$

The phases ζ_n and ζ'_n are arbitrary; for subsequent mathematical convenience we take them to be

$$\hbar\zeta_n(t) = \left[E_g + n\hbar\omega_c\right]t + \hbar\varphi, \qquad \hbar\zeta'_n(t) = \hbar\,\zeta_n(t) - \hbar\omega_L t. \tag{22.56}$$

We choose the static phase φ to make the cavity–field coupling constant g be real but we do not here introduce any additional phase for the state $|e, n\rangle$, and so the laser coupling (the Rabi frequency $\check{\Omega}$) is not taken to be real.

With allowance for loss the three-state RWA Schrödinger equation then reads

$$\frac{d}{dt}C_{gn} = -\kappa n C_{gn} - ig\sqrt{n} C_{xn},$$

$$\frac{d}{dt}C_{xn} = -i[\Delta_c - i\gamma - i\kappa(n-1)]C_{xn} - ig\sqrt{n} C_{gn} - \frac{i}{2}\check{\Omega}C_{en}, \qquad (22.57)$$

$$\frac{d}{dt}C_{en} = -i[\Delta_c - \Delta_L - i\kappa(n-1)]C_{en} - \frac{i}{2}\check{\Omega}C_{xn}.$$

where g is the single-photon coupling constant for the cavity, denoted $\frac{1}{2}\check{\Omega}_1$ in the previous section, and $\check{\Omega}(t)$ is the Rabi frequency for the laser field. Written in matrix form,

$$\frac{d}{dt}\mathbf{C}(t) = -i\mathbf{W}(t)\mathbf{C}(t), \qquad (22.58)$$

with the definition $\mathbf{C}^T = [C_{gn}, C_{xn}, C_{en}]$, these equations have the RWA Hamiltonian (the labels at the right show the states on which the matrix acts)

$$\mathbf{W}(t) = \begin{bmatrix} -i\kappa n & g\sqrt{n} & 0 \\ g\sqrt{n} & \Delta_c - i\gamma - i\kappa(n-1) & \frac{1}{2}\check{\Omega}(t) \\ 0 & \frac{1}{2}\check{\Omega}(t)^* & \Delta_c - \Delta_L - i\kappa(n-1) \end{bmatrix} \begin{array}{l} g, n \\ x, n-1 \\ e, n-1 \end{array} \qquad (22.59)$$

Alternative phases. The preceding equations are those of the traditional RWA, in which the couplings are made as slowly varying as possible by placing detunings in the diagonal elements of the Hamiltonian. One can, alternatively, make all real-valued diagonal elements vanish (as in the Dirac picture), by allowing more rapid variation of the Rabi frequencies.

22.3.1 Adiabatic elimination

When the cavity detuning (or the spontaneous emission rate) is large we can adiabatically eliminate state x with the approximation

$$C_x(t) = -\frac{g}{\Delta_c - i\gamma}C_g(t) - \frac{1}{2}\frac{\check{\Omega}(t)}{(\Delta_c - i\gamma)}C_e(t), \qquad (22.60)$$

and obtain the two-state equation (here $\Delta_L = \Delta_c$)

$$\frac{d}{dt}C_g(t) = -i\left[\frac{g^2}{\Delta_c - i\gamma} - i\kappa\right]C_g(t) + \frac{i}{2}\frac{g\check{\Omega}(t)}{(\Delta_c - i\gamma)}C_e(t),$$

$$\frac{d}{dt}C_e(t) = +\frac{i}{2}\frac{g\check{\Omega}(t)^*}{(\Delta_c - i\gamma)}C_g(t) + \frac{i}{4}\frac{|\check{\Omega}(t)|^2}{(\Delta_c - i\gamma)}C_e(t). \qquad (22.61)$$

We have here a description of two-state behavior, including Rabi oscillations at the effective two-photon Rabi frequency

$$\check{\Omega}'(t) = \frac{1}{2} \frac{|\check{\Omega}(t)|^2}{(\Delta_c - i\gamma)}, \tag{22.62}$$

subject to an effective time-varying detuning

$$\Delta'_c(t) = \frac{g^2 + (1/4)|\check{\Omega}(t)|^2}{\Delta_c - i\gamma}, \tag{22.63}$$

as well as to loss, at rate κ, from state g.

22.3.2 Three-state dynamics

Under the influence of the constant cavity field and the applied classical field the atomic dynamics evolves as does a conventional three-state system. In particular there are two conventional regimes of excitation that follow initial preparation of the atom in state e in a cavity vacuum.

Abrupt initiation. When we apply, suddenly, a classical field which thereafter maintains constant Rabi frequency, we have the situation of three states linked by two constant couplings. There occur Rabi oscillations between the initial state e, through the intermediate state x, and into the final state g. The presence of atomic states g and x accompany the presence of a cavity photon. During the subsequent continued Rabi cycling this photon is removed as the system returns to atom state x and then state e. Thus there is a periodic appearance and disappearance of a cavity photon.

The population oscillations are damped by the spontaneous-emission loss from state x, and by cavity losses from state g. For an optical cavity the cavity loss describes the emergence of a traveling wave through the bounding mirrors, i.e. the production of a photon.

Adiabatic passage. The Hamiltonian matrix of eqn. (22.59) has the structure used for treating Raman processes: the presence of two fields makes possible the mechanism of adiabatic passage, such as occurs with STIRAP. For that application the cavity coupling g is present initially, and so it serves as the S field of STIRAP. We assume that initially the atom is in state e. To achieve adiabatic population transfer we slowly increase the classical laser field (the P field of STIRAP) and thereby shift the statevector from alignment with the e state to alignment with the g state, bypassing the lossy x state. The result is placement of a photon into the cavity. The following section describes this procedure.

22.3.3 Cavity-induced adiabatic passage

The traditional dark adiabatic state for the n-photon Hamiltonian of eqn. (22.59), in the absence of losses, has the structure

$$\Phi(n; t) = \cos \Theta_n |e, n-1\rangle - \sin \Theta_n |g, n\rangle, \tag{22.64}$$

where the mixing angle is defined by the relationship

$$\tan \Theta_n = \check{\Omega}/2g\sqrt{n}. \tag{22.65}$$

When the system is in this dark state the ratio of probabilities is

$$\frac{|\langle e, n-1|\Psi(t)\rangle|^2}{|\langle g, n|\Psi(t)\rangle|^2} = \frac{4\pi n g^2}{\check{\Omega}^2}. \tag{22.66}$$

The STIRAP process takes place as follows:

- The atom, initially in state e, enters the cavity.
- The laser field is off as the atom moves into the cavity; the S field (of the cavity) is present.
- The laser field (P) is turned on, becoming much larger than the S field.
- The time evolution of the statevector adiabatically follows the traditional dark state $\Phi(n; t)$.
- The statevector moves from alignment with state $|e, n-1\rangle$ to alignment with state $|g, n\rangle$.
- The statevector motion adds one photon to the cavity.

The cavity field will eventually be lost and a traveling-wave photon will emerge from the cavity, at a rate fixed by the cavity loss rate κ. This procedure can be used to produce single photons on demand.

22.3.4 *Excitation of vacuum*

An interesting example of this dark state and its application occurs when $n = 0$, i.e. the atom state e is paired with the cavity vacuum. Adiabatic passage then alters the adiabatic state into alignment with the state $|g, 1\rangle$, thereby placing one photon into the cavity, a process known as *vacuum STIRAP* or V-STIRAP.

Consider a single three-state atom in a cavity in which there can be at most one cavity photon. We express the statevector as a superposition of three atomic states together with either one or no cavity photons,

$$\Psi(t) = e^{-i\zeta_1(t)}\left[C_g(t)|g, 1\rangle + C_x(t)|x, 0\rangle + e^{-i\omega_L t}C_e(t)|e, 0\rangle\right]. \tag{22.67}$$

That is, amplitude C_g is associated with atomic state g and a single cavity photon, while atomic states e and x are associated with the photon vacuum. Here $\zeta_1(t)$ is a single-photon phase chosen for mathematical convenience; below these functions are taken to involve a constant frequency plus a static phase φ,

$$\hbar\zeta_1(t) = \left[E_g + \hbar\omega_c\right]t + \hbar\varphi. \tag{22.68}$$

We choose φ to make the cavity coupling g be real.

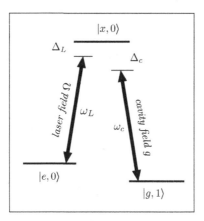

Figure 22.1 Linkages of three states for cavity system showing labels of atom–field states. The cavity field coupling g is always present, the laser field is a pulsed classical field.

The Schrödinger equation. When $n = 1$ the TDSE, in the RWA, reads

$$\frac{d}{dt}C_g(t) = -\kappa C_g(t) - \mathrm{i}gC_x(t),$$

$$\frac{d}{dt}C_x(t) = -\mathrm{i}[\Delta_c - \mathrm{i}\gamma]C_x(t) - \mathrm{i}gC_g(t) - \frac{\mathrm{i}}{2}\check{\Omega}(t)C_e(t), \qquad (22.69)$$

$$\frac{d}{dt}C_e(t) = -\mathrm{i}[\Delta_c - \Delta_L]C_e(t) - \frac{\mathrm{i}}{2}\check{\Omega}(t)^*C_x(t).$$

These equations have the RWA Hamiltonian

$$\mathsf{W}(t) = \begin{bmatrix} -\mathrm{i}\kappa & g & 0 \\ g & \Delta_c - \mathrm{i}\gamma & \frac{1}{2}\check{\Omega}(t) \\ 0 & \frac{1}{2}\check{\Omega}(t)^* & \Delta_c - \Delta_L \end{bmatrix} \begin{matrix} g, 1 \\ x, 0 \\ e, 0 \end{matrix} \cdot \qquad (22.70)$$

Figure 22.1 shows the linkage pattern of the three states and the two interactions.

22.3.5 Controlled single-photon emission

Once a photon has been placed into a cavity it will eventually be lost through the confining surface. When this loss occurs through a cavity end mirror the interior standing-wave field links to a traveling-wave field, appearing as a traveling photon. The emerging photon amplitude is therefore directly proportional to the time-varying cavity field, and its envelope is therefore controllable through control of the procedure that creates the cavity photon [Kuh01; Kuh10].

Traditionally one proposes a particular laser pulse and evaluates the resulting time dependence of the probability amplitudes. For the vacuum cavity the probability amplitude $C_g(t)$ is associated with the presence of a cavity photon and, as this field emerges through the cavity mirror, with a traveling-wave field – an emitted photon.

Atoms in cavities

A procedure described by Vasiliv *et al.* [Vas10] allows one to design a laser pulse that, acting on one leg of a lambda-type transition of a three-state system, will produce a cavity photon, and subsequently an emitted photon, of a desired temporal shape (within limits). For single photons $n = 1$, and resonance, the Hamiltonian is

$$W(t) = \begin{bmatrix} -i\kappa & g & 0 \\ g & -i\gamma & \frac{1}{2}\check{\Omega}(t) \\ 0 & \frac{1}{2}\check{\Omega}(t)^* & 0 \end{bmatrix} \begin{matrix} g \\ x \\ e \end{matrix} . \tag{22.71}$$

The emitted photon field has the time dependence

$$\mathcal{F}(t) = \sqrt{2\kappa}\, C_g(t), \tag{22.72}$$

and therefore we require that the Hamiltonian should produce the amplitude

$$C_g(t) = \frac{\mathcal{F}(t)}{\sqrt{2\kappa}}. \tag{22.73}$$

From the Schrödinger equation we obtain the result

$$C_x(t) = \frac{i}{g}[\partial_t C_g(t) + \kappa C_g]. \tag{22.74}$$

We also have the connection

$$C_e(t) = \sqrt{P_e(t)}, \tag{22.75}$$

with

$$P_e(t) = 1 - P_x(t) - P_g(t) - \int_0^t dt'\, [2\gamma P_x(t') + 2\kappa P_g(t')]. \tag{22.76}$$

These equations provide the connection between the three probability amplitudes and the photon wavefunction $\mathcal{F}(t)$. We then solve for the Rabi frequency, obtaining

$$\check{\Omega}(t) = i2\frac{\partial_t C_e(t)}{C_x(t)}. \tag{22.77}$$

From this time dependence we obtain the prescription for the laser pulse needed to produce a desired single-photon envelope $\mathcal{F}(t)$ from the cavity.

23

Control and optimization

The examples of quantum-state manipulation and coherent excitation discussed in this monograph present idealizations of actual quantum systems, simplifications that allow straightforward theoretical description. As one moves beyond the models of isolated atoms, few essential states, and transform-limited pulses to deal with more realistic models that can describe experimental reality, the basic tools described hitherto require elaboration and extension. Theoretical treatments of large molecules and chemical reactions involving laser-induced changes rely upon numerical simulation more than on analytic solutions. This final chapter discusses two of the themes applicable to that work: control theory and optimization.

23.1 Control theory

Classical control theory, as followed by mathematicians and engineers, deals with procedures for manipulating the input (the "controls") of a dynamically changing system to obtain a desired output of the system. In a *closed-loop control system* some device measures the output and, using a feedback loop, alters the input (via a *control element*) to bring the output closer to conformity with a goal. These techniques typically find application in control of experiments but they also work for theoretical modeling. An *open-loop control system* is one without such feedback; the controls are adjusted in accord with some established plan. Design of a suitable control mechanism (a *control function* ratioing output to input) must ensure that the system is *stable* (i.e. a finite input signal produces a finite output signal) and *controllable* (i.e. it is possible to obtain the desired output). It is also desirable that the control be *robust*: the control should not require much change if the actual system differs slightly from the mathematical model used in its design.

The most common form of classical controller, a proportional-integral-derivative (PID) controller, takes a time-dependent signal from a weighted sum of three signals: the output, its derivative, and its time integral. The controller may also take account of various state variables – variables that encode sufficient information to uniquely identify the state of the system. From this information the controller adjusts the input to the system, using some protocol.

The literature on control theory is voluminous, and replete with terminology. Adaptive control methods revise the control mechanism or its parameters during operation, thereby gaining robustness [Pea01; Bri10]. Intelligent control uses artificial intelligence approaches (e.g. fuzzy logic and evolutionary algorithms [Dav91; Mac03; Coe07]). Optimal control methods aim to achieve a desired output while minimizing chosen cost functions or maximizing value functions that quantify various desired constraints and benefits. Optimal control theory, an extension of the calculus of variations, deals with dynamical systems defined by sets of differential equations (equations of motion) that incorporate sets of parameters – the controls. Their time dependence is obtained as solutions to partial differential equations subject to constraints. The theory provides mathematical procedures for deriving control policies that will produce the desired optimized outputs.

Optimization. For all but the simplest idealizations, the parameter space (control space) is multidimensional, and can only be pictured via two-dimensional slices. Nevertheless, if we have but a single variable to optimize (say final population of a given state), then the values of this variable, as a function of the parameters, form a (control) landscape in which mountains, valleys, and cliffs are discernible [Cha07]. When presented as a maximization problem, the control problem seeks the highest peak on this terrain.

A conceptually simple procedure is to start from an initial set of parameters, a location on this landscape, pursue a path that leads always uphill, stopping only when any future step would take us downhill. In this journey it is necessary to evaluate, for each step, a generalized derivative – the change in height for incremental change of each control variable. To ensure that we find a global maximum, rather than a local one, we can make many initial starts, from points throughout the landscape. The various paths can, of course, be pursued simultaneously with a computer.

Alternatively we can consider a set of starting points and iteratively revise them to steadily improve their height. In this case an individual path may be a *Markov chain*, meaning the successive steps take no account of the past values – there is no memory of path history included in the choice of each step. Such algorithms include simulated annealing and various evolutionary algorithms.

23.2 Quantum control

The goal of many experimental manipulations of quantum states is to produce a predetermined quantum state from a given initial state – to produce a specified motion of the statevector. In chemical applications the objective may be a specific molecular reaction, characterized by a particular reaction product. More generally we wish to produce a specified superposition of quantum states from some given initial superposition. In achieving this goal, of maximizing some particular state, experimenters typically wish to minimize the laser fluence required and to minimize the contamination from undesired states. Such objectives fit the formalism of quantum optimal control theory, as explained in many articles by Rabitz and co-workers, e.g. [Hus91; Rab04; Bri10], and in other works discussing control of atomic and molecular dynamics [Gor97; Ass98; Ric00a; Bru03; Bri04; Wol05; Wer07; Win08].

Observables. An experimenter may state a goal to be the preparation of a particular quantum state, a specific statevector, or a set of elements of a density matrix, either during a time interval or at the end of that interval, subject to various constraints. Such statements must be translated into observational language in order to design an optimization procedure: There must be well-defined signals that signify success. In studies of molecular reactions the signal may involve filtering of molecules through a mass spectrometer, or selective photoionization. When details of a quantum superposition are to be controlled, it is necessary to devise a suitable set of observables that will identify successful preparation of the superposition. These may parametrize populations and, for more detail, coherences.

The optimization goal can be posed as the production of a predetermined set of expectation values $\bar{v}_j$ for system variables, perhaps at discrete times. The list of desired values, or *objectives*, form a vector **v**. To fit within a theoretical framework the observables are best defined in terms of expectation values of operators, such as the Kraus operators, eqn. (19.2) of Chap. 19. The observables may, for example, be populations expectation values of the projectors $\Pi(n)$ or coherences expectation values of the elementary transition operators $\pi(n, m)$ in various combinations.

Control parameters. The expectation values derive from a statevector (or density matrix), whose dynamics is governed by a Hamiltonian through the TDSE.

The tools of the experimenter are related to possible Hamiltonians. The parametric variables that define the Hamiltonian comprise the controls. These may include, for a given set of carrier frequencies and pulse shapes, the peak Rabi frequency and temporal pulse area, or they may specify pulse intensity at a discrete set of times or frequency components at a mesh of frequencies. However chosen, a set of adjustable parameters p_k, arranged as a vector **p**, fully define the Hamiltonian. The quantum control problem is to adjust the vector **p** to produce the output vector **v**, through the intermediary of the TDSE.

Connection. Conceptually the simplest examples of quantum control are those in which the objectives are to be evaluated at the conclusion of various pulses, at a final time T. The final statevector depends upon the set of parameters p_i that define the Hamiltonian, and so we can write the expectation value produced by a given pulsed Hamiltonian as

$$\langle v(\mathbf{p}) \rangle = \langle \Psi(\mathbf{p}; T) | \hat{v}_j | \Psi(\mathbf{p}; T) \rangle. \tag{23.1}$$

The goal of the experimenter is to adjust the parameter values that form the parameter vector **p**, thereby constructing an interaction $V(t)$ that brings these observed values as close as possible to the desired values $\bar{v}_j$.

In practice the experimenter imposes a variety of constraints. It is not possible to employ fields of arbitrarily high intensity; typically it is desirable to accomplish the excitation with as little fluence or intensity as possible, and within a time interval shorter than the relaxation time of the system. The task for the theorist modeling the activity is therefore a problem of constrained optimization: to find a statevector that satisfies the TDSE for a parametrized

Hamiltonian and minimizes the errors

$$\epsilon_j = |\langle v_j(\mathbf{p})\rangle - \bar{v}_j|^2, \tag{23.2}$$

subject to any constraints placed on the parameters.

For two- and three-state systems there are well-established procedures for accomplishing these objectives, using crafted laser pulses. As the quantum states become more numerous and the systems become more complicated, it becomes increasingly difficult to find procedures that will accomplish the desired population transfer, and to construct a multidimensional Hamiltonian matrix that will accomplish this task.

Controllability. From the TDSE it follows that, for amplitude $C_n(t)$ to change there must be Hamiltonian link $W_{nm}(t)$ from a nonzero amplitude $C_m(t)$, and that this, in turn, must link ultimately to a state $C_1(t)$ that is initially nonzero. When it is possible to obtain any desired probability amplitude $C_n(T)$ from any given initial amplitude $C_m(0)$ the system (Hilbert space together with pulsed Hamiltonian) is said to be *completely controllable* [Ram95; Tur01; Wer07].

In practice we implement the TDSE by introducing a finite set of basis states, defining thereby an N-dimensional Hilbert space spanned by unit vectors ψ_n. The Hamiltonian, represented by a time-varying $N \times N$ square matrix, acts to move the statevector along a trajectory on the surface of an N-dimensional unit sphere, from an initial point $\Psi(0)$ to a final location $\Psi(T)$. The task of quantum control theory is to devise a mapping, via the time-varying interaction matrix $V(t)$, of the initial statevector into the final statevector, as defined by various expectation values.

In presenting the statevector in an N-dimensional Hilbert space all information resides in probability amplitudes, the components of the statevector along the basis vectors ψ_n. The Hamiltonian, in turn, is expressible as various basis matrices m_μ, see Secs. 16.10.1 and 19.2,

$$H(t) = \sum_\mu h_\mu(t) m_\mu. \tag{23.3}$$

From this Hamiltonian we obtain the time-evolution matrix $U(t)$ as the solution to the TDSE,

$$\frac{d}{dt} U(t) = -i H(t) U(t), \qquad U(0) = 1. \tag{23.4}$$

If the construction of the Hamiltonian, i.e. the various linkages provided by the basis matrices, allows any given initial state n to connect, via $U_{mn}(t)$, with any specified final state m, then the system (i.e. the Hamiltonian within a given Hilbert space) is completely controllable.

It is not difficult to imagine simple systems that are not controllable in this sense. If the interaction is that of electric dipole transitions, then in a hydrogenic system starting from the 1s orbital the 2s orbital is not reachable. For complete controllability of the hydrogenic system it is necessary to allow additional interactions, as represented by other multipoles (or collisions) and by their relevant basis matrices.

Alternatively, one can delete this unreachable state from the finite set of states used to define the Hilbert space. Then the system, meaning the combination of Hilbert space and Hamiltonian, becomes controllable.

Controllability and graphs. The linkage structure of the Hamiltonian matrix – the rules that govern which quantum states are directly affected by other states – derives from both the totality of basis matrices m_ν used in its construction and the various time dependences $h_\nu(t)$ that characterize a specific Hamiltonian. The totality of linkages of the basis matrices define a *graph* [Die05] in which each of the N states is a *vertex* and the links between pairs of states are *edges*. For the system to be fully controllable, in the technical sense used above, the graph must be fully connected: there can be no vertex that is not tied to an edge, and there can be no unconnected subgraphs – there must exist at least one path from each vertex to every other vertex.

Graph theory can provide only part of the needed analysis of a particular Hamiltonian. It is also necessary to know whether the values of the various pulses $h_\nu(t)$ produce destructive interference between two (or more) paths that prevents transfer of probability [Tur01; Bro02].

Controllability and Lie algebras. Sets of square matrices provide an example of a Lie algebra, in which matrices (including the null matrix) are the elements and the defining operation between pairs of elements is the commutator. To form a Lie algebra it is necessary that every commutator be an element of the set. The matrices used to construct the Hamiltonian can be taken from a Lie algebra, although additional matrices may be required to complete the algebra.

The unitary matrices that provide the time-evolution operators for N probability amplitudes form elements of the unitary group $U(N)$. The equations these satisfy, the TDSE, can be written as

$$\frac{d}{dt}\mathsf{U}(t) = \mathsf{B}_0\mathsf{U}(t) + \sum_{\mu=1}^{K} f_\mu(t)\mathsf{B}_\mu\mathsf{U}(t), \qquad (23.5)$$

where the matrices B_0 and B_μ are skew-Hermitian matrices that can be considered elements of a Lie algebra. The matrix $\mathsf{U}(t)$, being unitary, belongs to the Lie algebra $u(N)$. All coherent superpositions of states can be achieved (i.e. the system is controllable) if the Lie algebra of the $K+1$ matrices $\mathsf{B}_0, \mathsf{B}_1, \ldots \mathsf{B}_m$ is $su(N)$ [Ram95].

23.3 Optimization

Given a Hamiltonian that allows complete control of an N-state quantum system, a common objective is to determine a pulse sequence for which a given process not only occurs but, in some sense, is produced optimally. More specifically one has a set of N coupled ODEs, the TDSE,

$$\frac{d}{dt}C_n(t) = -\mathrm{i} \sum_m W_{nm}(t)C_m(t), \tag{23.6}$$

involving, through the Hamiltonian matrix elements $W_{nm}(t)$, a set of controllable functions (time-dependent detunings and Rabi frequencies). One defines various measures of cost, such as the fluences or peak intensities of various laser pulses, and one defines measures of success, such as excitation probability or, more generally, statevector fidelity. One wishes to solve the multidimensional TDSE subject to the requirement of maximizing success and minimizing cost. In application to quantum-state manipulation this is termed a problem of *optimal control theory*.

This is an example of a problem of variational calculus (a *minimax* problem), in which the differential equation is supplemented by Lagrange multipliers, one for each variable to be maximized or minimized.

Numerical optimization. Although in simple cases it may be possible to find functional forms for detunings and Rabi frequencies that will produce some desired optimum, generally it is necessary to resort to numerical procedures. In essence these make a succession of trial solutions to the TDSE, each with different values of the control functions, and to seek results that are successively closer to meeting the desired criteria (and so are closer to being "optimal").

Typically the task of finding a suitable combination of basis states and Hamiltonian matrix requires numerical techniques, although these may be based upon analytic descriptions of the time-evolution matrix. As an example, in examining a two- or three-state system we may choose the pulses to have Gaussian time dependence, for a Hamiltonian that is parametrized by pulse areas and fixed detunings. More generally we might discretize the pulses, either in frequency or time, and use the values of discrete spectral components or time values as the parameters. We wish to find the values of these real numbers that will best satisfy some criteria, say maximizing the population transfer or, equivalently, minimizing the errors defined in eqn. (23.2).

Learning algorithms. The task can be regarded as development of a *learning algorithm*, in which a computer program, through trial and error, produces an optimized Hamiltonian. Each trial consists of a single integration of the TDSE for a single set of parameter values **p**, producing a single set of expectation values $\langle v_j(\mathbf{p}) \rangle$ and a consequent list of error values ϵ_j. Possibly taking into account the past history of error values (have these been increasing or decreasing), the computer program chooses a new list of parameter values, and redoes the integration.

In this iteration procedure it is desirable to have rules for systematically improving the set of parameters. These form a multidimensional vector space for expressing **p**: each type of parameter (e.g. detuning or pulse area) defines a coordinate axis, while the value of the parameter p_j is a component (a projection) of the vector **p**. The algorithm must find a path through this landscape to the point, or points, that provide the best fit to the objectives (e.g. a peak of population transfer or a valley of error) [Jud92; Rab00a; Rab00b; Mac03].

Genetic algorithms. Amongst the most popular of the numerical optimization procedures are *evolutionary algorithms*, known also as *genetic algorithms* (GA) [Dav91; Mac03; Coe07]. These are computational optimization schemes based on a metaphor of biological evolution in which genes become trial Hamiltonians, and the *fitness* for survival is inversely related to the error produced by the Hamiltonian. Each new biological generation is taken to be a revision of the trial Hamiltonian, using changes known variously as inheritance, *crossover* and *mutation*.Over many generations of evolution the population of individuals, i.e. the set of trial Hamiltonians, becomes more fit, i.e. the observables become closer to the objectives.

To carry out GA we begin by expressing the various parameters that define a particular pulsed Hamiltonian, say the carrier frequencies and a set of amplitudes at discrete times, as an ordered list of values, termed a *gene*. We create a set of *individuals*, each represented by a gene. For each of these "parent" genes we solve the TDSE, evaluate the expectation values, and record the errors (the fitness). Those genes whose errors are smallest (the fittest individuals) we take directly into the next generation. For genes that are less fit we perform a crossover, in which pairs of genes exchange groups of genetic information bits. During the crossover procedure a random mutation is occasionally introduced, corrupting one of the genetic bits at random. These paired offspring then fill the remainder of the next generation of individuals, ready for the next step of evolution. This random mutation is needed to prevent premature convergence to a single fit individual. The least fit individuals may be discarded.

For this new generation of individuals we determine the fitness, i.e. we solve the TDSE and evaluate the errors. We modify the gene pool and proceed, successively determining fitness and altering the genes. Eventually the set of genes produces the minimal errors possible within the model.

Appendix A

Angular momentum

Quantum theory, in common with several branches of applied mathematics, makes extensive use of operators. These may be represented by differentials, such as $\partial/\partial x$ that act on functions $f(x)$, they may be matrices that multiply vectors or other matrices, or they may be some combination of the two. The Hamiltonian is an important example of an operator, either in differential form acting on a wavefunction or in matrix form acting on a state-vector. A particularly important class of operators are the pairs of operators $\hat{a}$ and $\hat{a}^\dagger$ that serve as annihilation and creation operators for harmonic oscillator excitation increments as discussed in App. D.1. The most important set of three operators are those that represent the three Cartesian components of angular momentum. Typically we wish to determine properties of the eigenvectors and eigenvalues of operators, starting from known characteristics of the operators, i.e. their commutation properties. This appendix, taken from [Sho90, Chap. 18], summarizes the theory of quantum-mechanical angular momentum, on which there is an extensive literature [Zar88; Lou06; Bie09; Auz10].

Definition of angular momentum. Generalizing the canonical commutation relations for orbital angular momentum (eqn. (A.9) below) we take the cyclic commutation relations,

$$\left[\hat{J}_1, \hat{J}_2\right] = i\,\hat{J}_3, \qquad \left[\hat{J}_2, \hat{J}_3\right] = i\,\hat{J}_1, \qquad \left[\hat{J}_3, \hat{J}_1\right] = i\,\hat{J}_2, \qquad \text{(A.1)}$$

to define the three components of a general angular momentum operator. From these commutators there follow the values of angular momentum eigenvalues and the properties of angular momentum eigenstates.

A.1 Angular momentum states

Whenever three operators satisfy the commutation relations of eqn. (A.1) then these serve as components of a three-dimensional angular momentum vector.[1] It is customary to regard these operators as components of a Cartesian vector and to use the axis labels x, y, z in

---

[1] In atomic physics the total angular momentum $\hat{\mathbf{J}}$ is the sum of a part $\hat{\mathbf{L}}$ that acts to generate displacements of function values (i.e. *orbital angular momentum* represented by partial derivatives) and a part $\mathbf{S}$ that alters the components of vectors (i.e. *intrinsic spin*). The essential properties of any angular momentum operator are those of the commutation relations presented here.

place of the labels 1, 2, 3. However, the properties of angular momentum do not require that the operators refer to variables in ordinary (Euclidean) space. The operator

$$\hat{\mathbf{J}}^2 \equiv (\hat{J}_1)^2 + (\hat{J}_2)^2 + (\hat{J}_3)^2 = (\hat{J}_x)^2 + (\hat{J}_y)^2 + (\hat{J}_z)^2 \tag{A.2}$$

commutes with each of the three angular momentum operators $\hat{J}_k$, and so it is possible to obtain eigenstates of $\hat{\mathbf{J}}^2$ and any single $\hat{J}_k$. Conventionally this is chosen to be $\hat{J}_3 = \hat{J}_z$. The operator $\hat{\mathbf{J}}^2$ has as eigenvalues the numbers $J(J+1)$ where $2J$ is a non-negative integer. For given J the eigenvalues M of $\hat{J}_z$ differ by unity, ranging from $-J$ to $+J$. Let the angular momentum states be denoted by $|J, M\rangle$. Then the operators have the following effect:

$$\hat{\mathbf{J}}^2 |J, M\rangle = J(J+1)|J, M\rangle, \qquad \hat{J}_z |J, M\rangle = M|J, M\rangle. \tag{A.3}$$

The eigenstates of $\hat{J}_z$ are taken to be orthonormal,

$$\langle J, M | J, M' \rangle = \delta_{MM'}. \tag{A.4}$$

From the remaining two angular momentum operators we can construct the combinations

$$\hat{J}_{\pm 1} = \mp \frac{1}{\sqrt{2}} (\hat{J}_x \pm i \hat{J}_y), \qquad \hat{J}_0 = \hat{J}_z. \tag{A.5}$$

The operator $\hat{J}_{+1}$ is a *raising operator*, acting to increase the eigenvalue M by unity, while $\hat{J}_{-1}$ is a *lowering operator*, acting to decrease M by one. Specifically they have the effect

$$\sqrt{2}\hat{J}_{\pm 1} |J, M\rangle = \mp \sqrt{J(J+1) - M(M \pm 1)} \, |J, M \pm 1\rangle. \tag{A.6}$$

Thus by starting with the "stretched" state $|J, J\rangle$ we can construct any desired state by successive actions of a lowering operator:

$$|J, M\rangle = \sqrt{\frac{(J+M)!}{(2J)!(J-M)!}} \, (\hat{J}_x - i\hat{J}_y)^{J-M} \, |J, J\rangle. \tag{A.7}$$

This procedure provides a consistent phase to a set of angular momentum states.

A.1.1 Differential representations: Spherical harmonics

The operators of orbital angular momentum provide a simple example of operators that obey the commutation rules of an angular momentum. When these operators act upon functions of position, say $F(x, y, z)$, they appear as partial differential operators. These take simplest form in Cartesian coordinates, when they read

$$\hat{p}_x = -i\hbar \frac{\partial}{\partial x}, \qquad \hat{p}_y = -i\hbar \frac{\partial}{\partial y}, \qquad \hat{p}_z = -i\hbar \frac{\partial}{\partial z}, \tag{A.8}$$

so that the (orbital) angular momentum operators are

$$\hat{L}_z = \frac{1}{\hbar}(\hat{x}\hat{p}_y - \hat{y}\hat{p}_x) = -i\left(x\frac{\partial}{\partial y} - y\frac{\partial}{\partial x}\right) \quad et\ seq. \tag{A.9}$$

In spherical coordinates, where

$$x = r\cos\phi\sin\theta, \qquad y = r\sin\phi\sin\theta, \qquad z = r\cos\theta, \qquad (A.10)$$

the angular momentum operators are (see [Edm57; Bie81a]).

$$\hat{L}_x = i\sin\phi\frac{\partial}{\partial\theta} + i\cot\theta\cos\phi\frac{\partial}{\partial\phi},$$

$$\hat{L}_y = -i\cos\phi\frac{\partial}{\partial\theta} + i\cot\theta\sin\phi\frac{\partial}{\partial\phi}, \qquad (A.11)$$

$$L_z = -i\frac{\partial}{\partial\phi}.$$

The eigenstates (eigenfunctions) of orbital angular momentum are functions of the spherical angles θ and ϕ and, like any angular momentum states, they carry two angular momentum indices. The functions are the *spherical harmonics* $Y_{\ell m}(\theta,\phi)$, defined, apart from normalization, by the requirement that they be eigenfunctions of $\hat{\mathbf{L}}^2$ and $\hat{L}_z$:

$$\hat{\mathbf{L}}^2 Y_{\ell m} = \ell(\ell+1)Y_{\ell m}, \qquad \hat{L}_z Y_{\ell m} = m Y_{\ell m}. \qquad (A.12)$$

Orbital angular momentum eigenfunctions exist for all nonzero *integer* values of the orbital angular momentum quantum number ℓ, but not for half-integer values of ℓ. This restriction is seen most easily from the fact that the orbital angular momentum $\mathbf{L}$ has the property

$$\mathbf{r}\cdot\mathbf{L} = 0, \qquad (A.13)$$

from which we deduce that $m = 0$ is always a possible eigenvalue. It follows that orbital angular momentum states require integer values of the angular momentum quantum number ℓ.

The arguments θ and ϕ of the spherical harmonics are often expressed as either the unit vector $\hat{\mathbf{r}}$, as in $Y_{\ell m}(\hat{\mathbf{r}})$, or as the solid angle Ω, as in $Y_{\ell m}(\Omega)$. The spherical harmonics are normalized by the integral over all solid angles:

$$\int_0^\pi d\theta\sin\theta\int_0^{2\pi} d\phi\, Y_{\ell'm'}(\theta,\phi)^* Y_{\ell m}(\theta,\phi) = \delta_{\ell\ell'}\delta_{mm'}. \qquad (A.14)$$

Explicit expressions. Normalization fixes the magnitude of spherical harmonics, but phases remain to be assigned according to some convention. It has become traditional in atomic structure calculations to adopt the *Condon and Shortley phase convention*[2] such that the function with maximum value for m is

$$Y_{\ell\ell}(\theta,\phi) = \sqrt{\frac{(2\ell+1)!}{4\pi(\ell!)^2}}\left[-e^{i\phi/2}\sin\theta\right]^\ell. \qquad (A.15)$$

---

[2] Other phase conventions will be found, e.g. the inclusion of a factor i^ℓ. These affect the phases of dipole moments.

This function is the normalized solution to the eigenvalue equations

$$\hat{L}_z Y_{\ell\ell} \equiv -\mathrm{i}\,\frac{\partial}{\partial\phi}Y_{\ell\ell}(\theta,\phi) = \ell Y_{\ell\ell}(\theta,\phi), \tag{A.16}$$

$$(\hat{L}_x + \mathrm{i}\,\hat{L}_y)Y_{\ell\ell} \equiv \exp(-\mathrm{i}\phi)\left[\mathrm{i}\cot\theta\frac{\partial}{\partial\phi} - \frac{\partial}{\partial\theta}\right]Y_{\ell\ell}(\theta,\phi) = 0. \tag{A.17}$$

The remaining functions for this value of ℓ are then obtained by repeated action of the lowering operator $\hat{L}_x - \mathrm{i}\,\hat{L}_y$.

Parity. The replacement of angles θ, ϕ by the values $-\theta, -\phi$ produces the function

$$Y_{\ell m}(-\theta, -\phi) = Y_{\ell,-m}(\theta,\phi). \tag{A.18}$$

Reflection through the origin, as produced by the *parity operator* $\hat{\mathcal{P}}$, replaces the angles (θ, ϕ) by $(\pi - \theta, \pi + \phi)$ and hence replaces $\cos\theta$ by $-\cos\theta$. Under this transformation the spherical harmonic $Y_{\ell m}$ alters sign by the factor $(-1)^\ell$. Thus the spherical harmonic $Y_{\ell m}$ is an eigenstate of parity as well as of $\hat{\mathbf{L}}^2$ and $\hat{L}_z$:

$$\hat{\mathcal{P}}Y_{\ell m}(\theta,\phi) = Y_{\ell m}(\pi - \theta, \pi + \phi) = (-1)^\ell Y_{\ell m}(\theta,\phi). \tag{A.19}$$

The parity of $Y_{\ell m}$ is *even* if ℓ is an even integer, and the parity is *odd* if ℓ is an odd integer.

A.1.2 Angular momentum and rotations

The simple differential expression for $\hat{L}_z$ in spherical coordinates makes possible an interpretation of the operator $\hat{L}_z$ as the generator of infinitesimal rotations about the z axis. Consider a function of plane polar coordinates $F(r, \phi) \equiv F(\phi)$. We can think of this function as a set of values affixed to a rigid mesh, the system framework. The laboratory coordinates form a second rigid framework. A rotation may be viewed in two ways: *actively*, as a rotation of the rigid system mesh by an angle φ about a fixed point, holding the laboratory coordinates fixed, or *passively*, as a rotation of the coordinates by the inverse angle $-\varphi$, holding the system mesh fixed.

In the active view, rotation produces from the function $F(\phi)$ a new *rotated* function $F'(\phi)$. In the passive view this same rotation appears as a shift of coordinates, $\phi \to \phi' = \phi + \varphi$. The equivalence of the two views is expressed by the rule

$$F'(\phi) = F(\phi') = F(\phi - \varphi). \tag{A.20}$$

Assuming that the function is differentiable, we can express the angular displacement as a Taylor series:

$$F(\phi - \varphi) = F(\phi) - \varphi\frac{\partial F(\phi)}{\partial\phi} + \cdots. \tag{A.21}$$

We can express the partial derivative as the angular momentum operator to write this Taylor series:

$$F(\phi - \varphi) = F(\phi) - (\mathrm{i}\varphi)\hat{L}_z F(\phi) + \frac{(\mathrm{i}\varphi\hat{L}_z)^2}{2!}F(\phi) + \cdots. \tag{A.22}$$

This series defines the exponential of operator $i\varphi\hat{L}_z$:

$$F'(\phi) = F(\phi - \varphi) = \exp(-i\varphi\hat{L}_z)F(\phi). \tag{A.23}$$

This formula expresses the passive view: the operator $\exp(-i\phi\hat{L}_z)$ rotates the coordinates of the function $F(\phi)$ to the angle $\phi' = \phi - \varphi$. In the active view the operator $\exp(-i\varphi\hat{L}_z)$ acting upon the function F produces the rotated function $F'(\phi)$.

We can extend these results to more general rotations. More generally we can express the effect of rotation by an angle φ about an axis specified by the unit vector $\hat{\mathbf{u}}$ as

$$F'(\mathbf{r}) = \exp(-i\varphi\,\hat{\mathbf{u}}{\cdot}\hat{\mathbf{L}})\,F(\mathbf{r}) = F(\mathbf{r}'). \tag{A.24}$$

The argument $\mathbf{r}'$ is obtained from $\mathbf{r}$ by rotation through angle $-\varphi$ around the axis defined by the unit vector $\hat{\mathbf{u}}$.

A.1.3 Matrix representations: Spin matrices

The commutator properties of angular momentum operators $\hat{J}_1, \hat{J}_2, \hat{J}_3$ can be implemented using square matrices S_1, S_2, S_3, also denoted S_x, S_y, S_z, known in this context as (three) *spin matrices*. By definition these three matrices satisfy cyclic commutator relations

$$[S_1, S_2] = iS_3, \qquad [S_2, S_3] = iS_1, \qquad [S_3, S_1] = iS_2. \tag{A.25}$$

Any one of the matrices, conventionally chosen to be $S_3 \equiv S_z$, together with the matrix

$$S^2 = (S_1)^2 + (S_2)^2 + (S_3)^2 = (S_x)^2 + (S_y)^2 + (S_z)^2, \tag{A.26}$$

can be brought to diagonal form by a unitary transformation. The matrix S^2 is $s(s+1) = N(N-1)/4$ times the N-dimensional unit matrix, while the spin matrix S_z is diagonal, with elements running from $-s$ by integer steps to $+s$,

$$S^2 = s(s+1)\begin{bmatrix} 1 & 0 & \cdots & 0 \\ 0 & 1 & \cdots & 0 \\ \vdots & \vdots & \ddots & \vdots \\ 0 & 0 & \cdots & 1 \end{bmatrix}, \qquad S_z = \begin{bmatrix} -s & 0 & \cdots & 0 \\ 0 & -s+1 & \cdots & 0 \\ \vdots & \vdots & \ddots & \vdots \\ 0 & 0 & \cdots & +s \end{bmatrix}. \tag{A.27}$$

Spin eigenvectors. The eigenvectors χ_μ of the spin matrices S^2 and S_z have the property

$$S^2\chi_\mu = s(s+1)\chi_\mu, \qquad S_3\chi_\mu = S_z\chi_\mu = \mu\chi_\mu, \tag{A.28}$$

for $\mu = -s, -s+1, \ldots, +s$. Thus the eigenvectors χ_μ are column vectors of dimension $N = 2s + 1$. Conversely, square matrices of dimension N are associated with spin $s = (N-1)/2$. By definition χ_μ is an eigenvector of the spin matrix $S_3 = S_z$. The action on this vector of the other two spin matrices is

$$[S_1 \pm iS_2]\chi_\mu \equiv [S_x \pm iS_y]\chi_\mu = \sqrt{(s \mp \mu)(s+1 \pm \mu)}\,\chi_{\mu\pm1}. \tag{A.29}$$

The spin eigenvectors χ_μ are just the basis vectors ψ_n renumbered, with labels running from $\mu = -s$ to $+s$ in integer increments rather than from $n = 1$ to N. The connections are

$$\psi_1 = \chi_{-s} = \begin{bmatrix} 1 \\ 0 \\ \vdots \\ 0 \end{bmatrix}, \quad \psi_2 = \chi_{-s+1} = \begin{bmatrix} 0 \\ 1 \\ \vdots \\ 0 \end{bmatrix}, \quad \dots, \quad \psi_N = \chi_{+s} = \begin{bmatrix} 0 \\ 0 \\ \vdots \\ 1 \end{bmatrix}. \tag{A.30}$$

Two dimensions: Spin one-half. For dimension $N = 2$ the spin matrices are those for spin one-half, $s = 1/2$, traditionally taken to be

$$S_1 = S_x = \frac{1}{2} \begin{bmatrix} 0 & 1 \\ 1 & 0 \end{bmatrix}, \quad S_2 = S_y = \frac{1}{2} \begin{bmatrix} 0 & -i \\ i & 0 \end{bmatrix}, \quad S_3 = S_z = \frac{1}{2} \begin{bmatrix} 1 & 0 \\ 0 & -1 \end{bmatrix}. \tag{A.31}$$

The matrices S_j are half the *Pauli spin matrices* $\hat{\sigma}_j$,

$$S_j = \tfrac{1}{2}\hat{\sigma}_j, \quad j = 1, 2, 3 \text{ or } x, y, z. \tag{A.32}$$

The two eigenvectors are relabeled basis vectors of the two-state system,

$$\chi_{-1/2} = \psi_1 = \begin{bmatrix} 1 \\ 0 \end{bmatrix}, \quad \chi_{+1/2} = \psi_2 = \begin{bmatrix} 0 \\ 1 \end{bmatrix}. \tag{A.33}$$

Functions of spin matrices. The spin one-half matrices have the property that the product of two spin matrices is another spin matrix (or the unit matrix), as in the formulas

$$S_x S_x = S_y S_y = S_z S_z = \frac{1}{4}\mathbf{1}, \quad S_x S_y = \frac{i}{2}S_z. \tag{A.34}$$

The matrix S^2 is 3/4 times the two-dimensional unit matrix. We can use these properties to evaluate series definitions of functions of spin matrices, such as the exponential $\exp(-i\varphi S_j)$. For example, we have

$$\begin{aligned} \exp(-i\vartheta S_y) &= 1 - (i\vartheta)S_y + \frac{(i\vartheta)^2}{2!}(S_y)^2 - \frac{(i\vartheta)^3}{3!}(S_y)^3 + \cdots \\ &= 1\left[1 - \frac{\vartheta^2}{2^2 2!} + \cdots\right] - 2iS_y\left[\frac{\vartheta}{2} - \frac{\vartheta^3}{2^3 3!} + \cdots\right] \\ &= S_0\cos(\vartheta/2) - 2iS_y\sin(\vartheta/2) = \begin{bmatrix} \cos(\vartheta/2) & -\sin(\vartheta/2) \\ \sin(\vartheta/2) & \cos(\vartheta/2) \end{bmatrix}. \end{aligned} \tag{A.35}$$

This particular function has application as a matrix representation of a rotation operator. It also expresses the dynamics of resonant excitation with a constant RWA Hamiltonian, expressible as $W = \Omega S_y$. The effect of time evolution is then obtained by evaluating the matrix $\exp(-i\Omega t S_y)$. Equation (A.35), with $\vartheta = \Omega t$, provides the time-evolution matrix.

Three dimensions: Spin one. For dimension $N = 3$ the spin matrices are those for unit spin, $S = 1$. In a basis in which S_z is diagonal the three matrices are

$$S_x = \frac{1}{\sqrt{2}}\begin{bmatrix} 0 & 1 & 0 \\ 1 & 0 & 1 \\ 0 & 1 & 0 \end{bmatrix}, \quad S_y = \frac{1}{\sqrt{2}}\begin{bmatrix} 0 & -i & 0 \\ i & 0 & -i \\ 0 & i & 0 \end{bmatrix}, \quad S_z = \begin{bmatrix} 1 & 0 & 0 \\ 0 & 0 & 0 \\ 0 & 0 & -1 \end{bmatrix}. \quad (A.36)$$

The eigenstates of S_z, associated with eigenvalues $-1, 0, +1$, are just the three-state basis vectors, relabeled:

$$\chi_{-1} = \psi_1 = \begin{bmatrix} 1 \\ 0 \\ 0 \end{bmatrix}, \quad \chi_0 = \psi_2 = \begin{bmatrix} 0 \\ 1 \\ 0 \end{bmatrix}, \quad \chi_{+1} = \psi_3 = \begin{bmatrix} 0 \\ 0 \\ 1 \end{bmatrix}. \quad (A.37)$$

These states need not have any association with any atomic angular momentum; it is only necessary that there be three of them.

For dimension $N = 3$ the spin matrices by themselves are not sufficient to express the nine elements of any arbitrary matrix; we require another six independent matrices. A simple procedure is to take products of pairs of the three basic matrices, say $T_{ij} = S_i S_j$. However, other possibilities, characterized by symmetries, also prove useful, such as the combinations $S_i S_j \pm S_j S_i$.

Spin-one eigenvectors and Cartesian unit vectors. In three dimensions S^2 is equal to 2 times the three-dimensional unit matrix. Hence all three-dimensional vectors, and more generally all three-dimensional vector fields, are eigenstates of S^2; all are spin-one states. A simple example of three-dimensional vectors is the set of three Cartesian unit vectors $\mathbf{e}_x$, $\mathbf{e}_y$ and $\mathbf{e}_z$. From these we can construct the complex unit vectors

$$\mathbf{e}_0 = \mathbf{e}_z, \qquad \mathbf{e}_{\pm 1} = \mp \frac{1}{\sqrt{2}}\left[\mathbf{e}_x \pm i\mathbf{e}_y\right]. \quad (A.38)$$

These are all eigenvectors of the matrix S_z and of the matrix S^2,

$$S_z \mathbf{e}_q = q\,\mathbf{e}_q, \qquad S^2 \mathbf{e}_q = 2\mathbf{e}_q, \quad q = +1, 0, -1. \quad (A.39)$$

The matrices $S_{\pm 1}$ have the following action upon the spin eigenvectors:

$$\sqrt{2} S_{\pm 1} \mathbf{e}_q = \mp \sqrt{(2 \pm q)(1 \mp q)}\,\mathbf{e}_{q \pm 1}. \quad (A.40)$$

The connection between the two sets of unit vectors is a unitary transformation,

$$
\begin{bmatrix} \mathbf{e}_{+1} \\ \mathbf{e}_0 \\ \mathbf{e}_{-1} \end{bmatrix} = \frac{1}{\sqrt{2}} \begin{bmatrix} -1 & -i & 0 \\ 0 & 0 & \sqrt{2} \\ 1 & -i & 0 \end{bmatrix} \begin{bmatrix} \mathbf{e}_x \\ \mathbf{e}_y \\ \mathbf{e}_z \end{bmatrix} \equiv \mathsf{U} \begin{bmatrix} \mathbf{e}_x \\ \mathbf{e}_y \\ \mathbf{e}_z \end{bmatrix},
$$

$$
\begin{bmatrix} \mathbf{e}_x \\ \mathbf{e}_y \\ \mathbf{e}_z \end{bmatrix} = \frac{1}{\sqrt{2}} \begin{bmatrix} -1 & 0 & 1 \\ i & 0 & i \\ 0 & \sqrt{2} & 0 \end{bmatrix} \begin{bmatrix} \mathbf{e}_{+1} \\ \mathbf{e}_0 \\ \mathbf{e}_{-1} \end{bmatrix} \equiv \mathsf{U}^{-1} \begin{bmatrix} \mathbf{e}_{+1} \\ \mathbf{e}_0 \\ \mathbf{e}_{-1} \end{bmatrix}.
$$

(A.41)

The connection between a matrix M in the spherical basis and that matrix M' in the Cartesian basis is (see [Ros57])

$$ \mathsf{M}' = \mathsf{U}^* \mathsf{M} \mathsf{U}^T, \qquad \mathsf{M} = \mathsf{U}^T \mathsf{M}' \mathsf{U}^*, \tag{A.42} $$

where U^T denotes the transpose of U and U^* denotes the complex conjugate of U. When we apply this transformation to the spin-one matrices of eqn. (A.36) the resulting spin-one matrices are

$$
\mathsf{S}'_x = \begin{bmatrix} 0 & 0 & 0 \\ 0 & 0 & -i \\ 0 & i & 0 \end{bmatrix}, \qquad \mathsf{S}'_y = \begin{bmatrix} 0 & 0 & -i \\ 0 & 0 & 0 \\ i & 0 & 0 \end{bmatrix}, \qquad \mathsf{S}'_z = \begin{bmatrix} 0 & -i & 0 \\ i & 0 & 0 \\ 0 & 0 & 0 \end{bmatrix}. \tag{A.43}
$$

Whereas the matrices of eqn. (A.36) refer to spherical components, the matrices of eqn. (A.43) refer to Cartesian components. The two definitions offer alternative sets of angular momentum matrices. We can recognize that the transformation from Cartesian to spherical vectors diagonalizes the matrix S'_z.

The spin matrices for $S = 1$ are useful for expressing various three-dimensional vector manipulations. For arbitrary three-dimensional matrix $\mathbf{A}$ we evaluate the product $(\mathbf{S} \cdot \mathbf{A})$ as

$$ (\mathbf{S} \cdot \mathbf{A}) = \sum_{j=x,y,z} \mathsf{S}'_j A_j = i \begin{bmatrix} 0 & -A_z & A_y \\ A_z & 0 & -A_x \\ -A_y & A_x & 0 \end{bmatrix}. \tag{A.44} $$

Then vector multiplication of two vectors $\mathbf{B}$ and $\mathbf{A}$ can be written

$$ \mathbf{A} \times \mathbf{B} = -i (\mathbf{S} \cdot \mathbf{A}) \mathbf{B}. \tag{A.45} $$

We can use this property to write the curl of a vector field as

$$ \nabla \times \mathbf{A}(\mathbf{r}) = -i (\mathbf{S} \cdot \nabla) \mathbf{B}(\mathbf{r}). \tag{A.46} $$

We can also use this relationship to rewrite an equation of motion. If the vector $\mathbf{C}(t)$ obeys the linear equation

$$ \frac{d}{dt} \mathbf{C}(t) = -i \mathsf{W}(t) \mathbf{C}(t), \tag{A.47} $$

and the matrix $\mathsf{W}(t)$ is expressible as the scalar product with the spin vector,

$$\mathsf{W}(t) = \mathbf{S} \cdot \mathbf{V}(t) = V_x(t)\mathsf{S}_x + V_y(t)\mathsf{S}_y + V_z(t)\mathsf{S}_z, \tag{A.48}$$

then eqn. (A.47) can be written as a torque equation,

$$\frac{d}{dt}\mathbf{C}(t) = \mathbf{V}(t) \times \mathbf{C}(t). \tag{A.49}$$

Functions of spin-one matrices. The square of each of the matrices S'_j is a diagonal matrix. For example,

$$(\mathrm{i}\,\mathsf{S}'_x)^2 = \begin{bmatrix} -1 & 0 & 0 \\ 0 & -1 & 0 \\ 0 & 0 & 0 \end{bmatrix}. \tag{A.50}$$

Each matrix also has the property

$$(\mathrm{i}\,\mathsf{S}'_j)^3 = -(\mathrm{i}\,\mathsf{S}'_j). \tag{A.51}$$

It follows that all powers of S'_j are expressible as two matrices, say $\mathrm{i}\,\mathsf{S}'_j$ and $(\mathrm{i}\,\mathsf{S}'_j)^2$, together with the unit matrix $\mathbf{1}$. (Recall that for spin one-half only the unit matrix and a single power S_j are needed.) This property simplifies the evaluation of series definitions of functions of the spin matrices. For example, the series definition of the exponential gives the result (compare with eqn. (A.35))

$$\exp(-\mathrm{i}\,\vartheta\,\mathsf{S}'_j) = \mathbf{1} - \frac{\vartheta}{1!}(\mathrm{i}\,\mathsf{S}'_j) + \frac{\vartheta^2}{2!}(\mathrm{i}\,\mathsf{S}'_j)^2 - \frac{\vartheta^3}{3!}(\mathrm{i}\,\mathsf{S}'_j)^3 + \cdots$$

$$= \mathbf{1} + (\mathrm{i}\,\mathsf{S}'_j)^2 - (\mathrm{i}\,\mathsf{S}'_j)\sin\vartheta - (\mathrm{i}\,\mathsf{S}'_j)^2\cos\vartheta. \tag{A.52}$$

A special case is

$$\exp(-\mathrm{i}\,\vartheta\,\mathsf{S}'_z) = \begin{bmatrix} \cos\vartheta & -\sin\vartheta & 0 \\ \sin\vartheta & \cos\vartheta & 0 \\ 0 & 0 & 1 \end{bmatrix}. \tag{A.53}$$

This is the spin-one counterpart of eqn. (A.35).

A.1.4 Rotation matrices

In any dimension the spin matrices S, when exponentiated, act to rotate the unit vectors. More generally, the angular momentum operator $\hat{\mathbf{J}}$ is the generator of rotations, meaning that rotation by angle ϕ about an axis defined by the unit vector $\mathbf{u}$ is produced by the operator

$$\hat{R}(\mathbf{u}, \phi) = \exp(-\mathrm{i}\,\phi\,\mathbf{u}\cdot\hat{\mathbf{J}}), \tag{A.54}$$

where

$$\mathbf{u}\cdot\hat{\mathbf{J}} = u_x\hat{J}_x + u_y\hat{J}_y + u_z\hat{J}_z. \tag{A.55}$$

When treating three-dimensional vectors (examples of spin 1) the angular momentum is a spin matrix and the rotation operator is a 3×3 matrix. We can evaluate the exponential in the following way. We express the Cartesian coordinates of the unit vector as

$$u_x = \sin\theta \cos\phi, \qquad u_y = \sin\theta \sin\phi, \qquad u_z = \cos\theta, \qquad (A.56)$$

and define the three-dimensional matrices

$$
\mathsf{P} = \begin{bmatrix} u_x^2 & u_x u_y & u_x u_z \\ u_x u_y & u_y^2 & u_y u_z \\ u_x u_z & u_y u_z & u_z^2 \end{bmatrix}, \quad
\mathsf{Q} = \begin{bmatrix} 0 & -u_z & u_y \\ u_z & 0 & -u_x \\ -u_y & u_x & 0 \end{bmatrix}, \quad
\mathsf{I} = \begin{bmatrix} 1 & 0 & 0 \\ 0 & 1 & 0 \\ 0 & 0 & 1 \end{bmatrix}.
$$
$$(A.57)$$

Then the rotation matrix is

$$\mathsf{R}(\mathbf{u}, \phi) = \mathsf{P} + (\mathsf{I} - \mathsf{P})\cos\phi + \mathsf{Q}\sin\phi. \qquad (A.58)$$

This construction is useful for evaluating the effect on the Bloch vector of a rotation by the torque vector.

Euler angles. When treating general angular momentum states it is convenient to parametrize a rotation differently, by means of *Euler angles* α, β, γ. These are defined as being produced by a sequence of three rotations, the first about the z axis by angle γ, the second about the new y' axis by angle β, and the third about the original z axis by angle α. The rotation operator is

$$\hat{R}(\alpha, \beta, \gamma) = \exp(-i\alpha \hat{J}_z)\exp(-i\beta \hat{J}_y)\exp(-i\gamma \hat{J}_z). \qquad (A.59)$$

Evaluated between states of well-defined angular momenta this produces the function

$$
\begin{aligned}
\mathcal{D}_{M,M'}^{(J)}(\alpha, \beta, \gamma) &= \langle JM | \hat{R}(\alpha, \beta, \gamma) | JM' \rangle \\
&= \exp(iM\alpha - iM'\gamma)\langle JM | \exp(-i\beta \hat{J}_y) | JM' \rangle \\
&\equiv \exp(-iM\alpha - iM'\gamma)d_{MM'}^{(J)}(\beta),
\end{aligned}
$$
$$(A.60)$$

in which the *reduced rotation matrix* $d_{MM'}^{(J)}(\beta)$ is expressible in terms of sines and cosines of β. Table A.1 lists values of the elements of the reduced rotation matrix for $J = \frac{1}{2}$ and $j = 1$.

A.2 Angular momentum coupling

In systems having several degrees of freedom, perhaps involving several particles, there occur independent sets of operators whose commutation properties mark them as examples of angular momentum.

A common example is the intrinsic spin and orbital angular momentum of an atomic electron, or the vector nature and spatial dependence of a vector field. Such situations, involving

Table A.1. *Values of reduced rotation matrix* $d_{MM'}^{(J)}(\beta)$

$$J = \tfrac{1}{2}$$

	$M' = +\tfrac{1}{2}$	$-\tfrac{1}{2}$
$M = +\tfrac{1}{2}$	$\cos(\beta/2)$	$-\sin(\beta/2)$
$-\tfrac{1}{2}$	$\sin(\beta/2)$	$\cos(\beta/2)$

$$J = 1$$

	$M' = +1$	0	-1
$M = +1$	$\dfrac{1+\cos\beta}{2}$	$-\dfrac{\sin\beta}{\sqrt{2}}$	$\dfrac{1-\cos\beta}{2}$
0	$\dfrac{\sin\beta}{\sqrt{2}}$	$\cos\beta$	$\dfrac{-\sin\beta}{\sqrt{2}}$
-1	$\dfrac{1-\cos\beta}{2}$	$\dfrac{\sin\beta}{\sqrt{2}}$	$\dfrac{1+\cos\beta}{2}$

two sets of angular momentum operators $\hat{\mathbf{J}}(A)$ and $\hat{\mathbf{J}}(B)$, permit complete description by product states

$$|J_1 M_1, J_2 M_2\rangle^{AB} \equiv |J_1 M_1\rangle^A |J_2 M_2\rangle^B, \qquad (A.61)$$

where

$$\hat{\mathbf{J}}(a)^2 |JM\rangle^a = J(J+1)|JM\rangle^a, \qquad \hat{J}_z(a)|JM\rangle^a = M|JM\rangle^a, \quad a = A, B. \qquad (A.62)$$

It often proves useful to introduce, in place of these *uncoupled* basis states, superpositions that are eigenstates of the total angular momentum [Sho90, §19.1]

$$\hat{\mathbf{J}} = \hat{\mathbf{J}}(A) + \mathbf{J}(\hat{B}), \quad \text{meaning} \quad \hat{J}_i = \hat{J}_i(A) + \hat{J}_i(B), \quad i = x, y, z. \qquad (A.63)$$

Such *coupled* basis states are constructed with the aid of Clebsch–Gordan (CG) coefficients [Bie81a; Zar88; Lou06; Auz10],

$$|J_1 J_2 J M\rangle^{AB} = \sum_{M_1 M_2} |J_1 M_1\rangle^A |J_2 M_2\rangle^B \, (J_1 M_1, J_2 M_2 | J M), \qquad (A.64)$$

or three-j symbols,

$$|J_1 J_2 J M\rangle^{AB} = \sum_{M_1 M_2} |J_1 M_1\rangle^A |J_2 M_2\rangle^B \, (-1)^{J_1 - J_2 + M} \sqrt{2J+1} \begin{pmatrix} J_1 & J_2 & J \\ M_1 & M_2 & -M \end{pmatrix}. \qquad (A.65)$$

The inverse relationship expresses an uncoupled state as a superposition of coupled states,

$$|J_1 M_1\rangle^A |J_2 M_2\rangle^B = \sum_{JM} |J_1 J_2 J M\rangle^{AB} (-1)^{J_1 - J_2 + M} \sqrt{2J+1} \begin{pmatrix} J_1 & J_2 & J \\ M_1 & M_2 & -M \end{pmatrix}.$$

(A.66)

The coupling coefficients incorporate the following constraints:

$$|J_1 - J_2| \le J \le J_1 + J_2, \qquad M = M_1 + M_2. \tag{A.67}$$

This coupling procedure has use for any pair of eigenstates of angular momentum operators, e.g. for combinations of spherical unit vectors with spherical harmonics to form spherical vector fields (see App. D.2), for combinations of spin matrices to construct symmetry-adapted basis functions, and for ensembles of two-state systems (see Chap. 16.1).

A.2.1 *Rotations and irreducible tensors*

The rotation matrices provide a powerful means of extending the notion of angular momentum to encompass operators more elaborate than the angular momentum vector $\hat{\mathbf{J}}$, and to classify operators constructed from various vectors. For this purpose we define an *irreducible tensor* [Sho90, Chap. 20] [Fan63; Var88; Auz10] $\hat{T}^{(k)}$ of order k to be a set of $2k+1$ tensor components $\hat{T}_q^{(k)}$, combinations of functions, operators, or matrices, which under coordinate rotation transform amongst themselves according to the rule[3]

$$\hat{T}_{q'}^{(k)}(\mathcal{R}') = \sum_q \hat{T}_q^{(k)}(\mathcal{R}) \mathcal{D}_{qq'}^{(k)}(\alpha, \beta, \gamma). \tag{A.68}$$

Here (α, β, γ) are the Euler angles of the rotation that takes reference frame $\mathcal{R}'$ into frame $\mathcal{R}$. Examples include spherical harmonics (tensors of order ℓ),

$$Y_{\ell m'}(\theta', \phi') = \sum_m Y_{\ell m}(\theta, \phi) \mathcal{D}_{m,m'}^{(\ell)}(\alpha, \beta, \gamma), \tag{A.69}$$

and any arbitrary three-dimensional vectors $\mathbf{A}$ when expressed in terms of the three complex spherical components A_q,

$$A_{q'}' = \sum_q A_q \mathcal{D}_{qq'}^{(1)}(\alpha, \beta, \gamma) = \sum_q \mathcal{D}_{q'q}^{(1)}(\alpha\beta\gamma)^* A_q. \tag{A.70}$$

Angular momentum, linear momentum, and position vectors all fall in this class and so they all form examples of irreducible tensors of order one. Another important example of irreducible tensors is provided by complex spherical unit vectors $\mathbf{e}_q$ in three dimensions. Under rotation they obey a transformation appropriate to unit angular momentum: they have an intrinsic angular momentum (or spin) of unity. To express this property we need a notation such as $\mathbf{e}_q(\mathcal{R})$ that makes explicit the coordinate system produced by rotation $\mathcal{R}$

---

[3] This definition establishes a connection amongst components having a common value k, but it does not prescribe either the normalization or the phase of the complete set of $2k+1$ components.

from the reference frame of $\mathbf{e}_q \equiv \mathbf{e}_q(\hat{\mathbf{r}})$. Then we can write the rotation of the vector as the transformation

$$\mathbf{e}_{q'}(\mathcal{R}) = \sum_q \mathbf{e}_q \, \mathcal{D}^{(1)}_{qq'}(\mathcal{R}). \tag{A.71}$$

We use this property to evaluate the scalar product of two helicity unit vectors:

$$\langle q, \mathcal{R}|q', \mathcal{R}'\rangle = \mathbf{e}_q(\mathcal{R})^* \cdot \mathbf{e}_{q'}(\mathcal{R}') = \mathcal{D}^{(1)}_{qq'}(\alpha, \beta, \gamma). \tag{A.72}$$

The Euler angles α, β, γ are those needed to rotate between the two reference frames. Because these unit vectors are also examples of irreducible spherical tensors they are particularly suited to problems in which angular momentum is important.

Commutators and irreducible tensors. The definition of irreducible tensors in terms of rotation properties emphasizes elements of the rotation matrix. An alternative, but equivalent, definition takes the commutation properties of spherical harmonics as the basis for generalization. With this approach an irreducible tensor operator $\hat{\mathbf{T}}^{(k)}$ is defined to be a set of $2k+1$ component operators $\hat{T}_q^{(k)}$ that obey the following commutation relations with the angular momentum operators $\hat{J}_{+1}, \hat{J}_0, \hat{J}_{-1}$:

$$\sqrt{2}\left[\hat{J}_{\pm 1}, \hat{T}_q^{(k)}\right] = \mp\sqrt{k(k+1)-q(q\pm 1)}\,\hat{T}_{q\pm 1}^{(k)},$$
$$\left[\hat{J}_0, \hat{T}_q^{(k)}\right] = q\,\hat{T}_q^{(k)}. \tag{A.73}$$

These relations can be expressed in terms of the CG coefficient,

$$\left[\hat{J}_p, \hat{T}_q^{(k)}\right] = \sqrt{k(k+1)}(1p, kq|kp+q)\,\hat{T}_{p+q}^{(k)}. \tag{A.74}$$

It is easy to verify that, with $k=1$, the three components of the angular momentum operator obey this rule. The three components of the position operator and the momentum operator obey this rule when the operator $\hat{\mathbf{J}}$ is the orbital angular momentum.

The commutator definition leads directly to relationships amongst matrix elements of tensor operators, sometimes termed Racah algebra, after the work of Racah [Bie65; Bie81b]. The following section presents those results, embodied in the Wigner–Eckart theorem.

A.2.2 The Wigner–Eckart theorem

Just as we can couple basis states amongst themselves and can couple pairs of operators, we can proceed still further and couple operators with basis states. Thus, from the irreducible tensor operator $\hat{T}_q^{(k)}$ and the basis state $|\alpha J M\rangle$ we can form the coupled combination (a *coupled state*)

$$|\beta K Q\rangle = \sum_{qM} \hat{T}_q^{(k)}|\alpha J M\rangle \, (JM, kq|KQ). \tag{A.75}$$

The labels α and β here represent any additional quantum numbers needed for the complete definition of the states. (For example, β must identify the particular tensor operator $\hat{\mathbf{T}}^{(k)}$.)

We can invert this expression and write

$$\hat{T}_q^{(k)}|\alpha J M\rangle = \sum_{KQ} |\beta K Q\rangle\,(J M, kq|K Q). \tag{A.76}$$

From this formula we can obtain the matrix elements of the operator $\hat{T}_q^{(k)}$ in an angular momentum basis:

$$\langle \alpha' J' M' | \hat{T}_q^{(k)} | \alpha J M \rangle = \sum_{KQ} \langle \alpha' J' M' | \beta K q \rangle\,(J M, kq|K q). \tag{A.77}$$

Because angular momentum states are orthogonal the summation over KQ picks out the values $J'M'$. We thereby find that the matrix elements of $\hat{T}_q^k$ are proportional to a CG coefficient, which contains the entire dependence upon magnetic quantum numbers, i.e. the dependence upon orientation geometry. This proportionality is the content of the *Wigner–Eckart theorem*. When we extract the CG coefficient we are left with a factor independent of magnetic quantum numbers. The factorization is the Wigner–Eckart theorem [Edm57; Mes62]

$$\langle \alpha' J' M' | \hat{T}_q^{(k)} | \alpha J M \rangle = (\alpha' J' \| T^{(k)} \| \alpha J)\,(-1)^{J+k-M'} \begin{pmatrix} J & k & J' \\ M & q & -M' \end{pmatrix}, \tag{A.78}$$

where $(\alpha' J' \| T^{(k)} \| \alpha J)$ is termed the *reduced matrix element* of the operator $\hat{T}_q^{(k)}$. This definition, taken with the normalization property of three-j symbols, associates the absolute square of the reduced matrix element with a summation over all of the projection quantum numbers M, q, M',

$$|(\alpha' J' \| T^{(k)} \| \alpha J)|^2 = \sum_{MMq'} |\langle \alpha' J' M' | \hat{T}_q^{(k)} | \alpha J M \rangle|^2. \tag{A.79}$$

Elementary reduced matrix elements. The simplest examples of reduced matrix elements occur when the angular momentum states are single-particle states and the tensor operator is a simple single-particle operator (the following section discusses more complicated cases). In these cases the reduced matrix element can be evaluated by considering any specific case of eqn. (A.78) for which the three-j symbol does not vanish. This is conveniently chosen to be the matrix element of $T_0^{(k)}$ and the "stretched" state $|\alpha J M\rangle = |\alpha J J\rangle$. One obtains the reduced matrix element by multiplying the particular "stretched" matrix element by the appropriate three-j symbol. The Wigner–Eckart theorem provides a connection between any arbitrary matrix element and either the reduced matrix element or the "stretched" element.

Three-j selection rules. For the three-j symbol to be nonzero the six arguments must satisfy the following conditions.

1. The numbers, $2J$, $2k$, and $2J'$ are non-negative integers.
2. The numbers $2M$, $2q$, and $2M'$ are integers (possibly negative) bounded by $|M| \leq J$, $|q| \leq k$, and $|M'| \leq J'$.
3. The numbers $-M$, q, and M' must sum to zero.
4. The numbers J, k, and J' must form sides of a triangle having integer perimeter. Thus the difference of any two of these numbers cannot exceed the third. In particular we have the bound $|J - J'| \leq k$.

The first two rules simply restate the properties of angular momentum quantum numbers. The last two conditions express *selection rules*: they specify which values of JM can be linked by the operator $\hat{T}_q^{(k)}$. Selection rules are often expressed as conditions governing changes of quantum numbers between initial and final states, as in $\Delta J = J' - J$. With this convention selection rule 4 becomes $\Delta J \leq |k|$.

A.3 Hyperfine linkages

A quantum state that has electronic angular momentum J has $2J + 1$ magnetic sublevels, distinguished by quantum number M_J. In the absence of an external field these are degenerate. When the atom nucleus has also angular momentum (spin), customarily denoted I, this adds an additional degree of freedom, with $2I + 1$ nuclear orientations, distinguished by a projection quantum number M_F. This section, based on [Sho90, §21.1], summarizes the theory used to describe such interacting parts of a quantum system.

The interaction energy of the nuclear electric and magnetic moments with the electric and magnetic fields created by the moving electrons removes part of the degeneracy. The magnetic moments associated with the angular momenta $\mathbf{I}$ and $\mathbf{J}$ produce a spin-spin interaction. When $I > 1$ the nucleus has an electric quadrupole moment, and this can interact with the local electric-field gradient. The results of these interactions are observable as a *hyperfine splitting* of the energy levels; transitions between these nondegenerate levels appear as *hyperfine structure* in the spectra [Kop58; Sob72; Eme06; And06; Wyb07; Auz10].

The Hamiltonian responsible for the hyperfine interaction [Sho90, §21.1],

$$V^{HF} = V^{NM1} + V^{NE2}, \tag{A.80}$$

has two contributions, one from magnetic moments and the other from electric moments, often written as

$$
\begin{aligned}
V^{NM1} &= hA\,\mathbf{T}^{(1)}(\mathbf{I}) \cdot \mathbf{T}^{(1)}(\mathbf{J}), \quad \text{magnetic dipole,} \\
V^{NE2} &= hB\,\mathbf{T}^{(2)}(\mathbf{I}) \cdot \mathbf{T}^{(2)}(\mathbf{J}), \quad \text{electric quadrupole,}
\end{aligned}
\tag{A.81}
$$

where $h = 2\pi\hbar$ is the Planck constant. Here A and B are parameters, dimensionally frequencies, derived from the structure of the specific atom and $\mathbf{T}^{(k)}(\mathbf{J})$ is a dimensionless irreducible tensor of order k constructed from angular momentum $\mathbf{J}$. As examples, the

0 components of the first and second rank tensors are

$$T_0^{(1)}(\mathbf{J}) = J_z, \qquad T_0^{(2)}(\mathbf{J}) = \frac{1}{\sqrt{6}}\left[2(J_z)^2 - (J_x)^2 - (J_y)^2\right]. \qquad (A.82)$$

To express a quantum state involving the separate nuclear and atomic degrees of freedom we can use product states $|I, M_I\rangle|J, M_J\rangle$. The hyperfine interactions have linkages between such different states; they are not diagonal. However, each of these interactions is a scalar, and so the total angular momentum of the system, combining electronic and nuclear angular momenta, customarily denoted F, is not altered by hyperfine interactions. The quantized orientations of this vector are labeled by quantum number M_F. The overall quantum state, in the presence of hyperfine interaction, carries the labels J, I, F, M_F.

Electric dipole transitions between hyperfine sublevels obey similar selection rules as do transitions in the absence of nuclear spin (see Sec. 12.2): F can change by at most one unit, and M_F obeys the same selection rules as does M_J. That is, F may change by 0 or ± 1 (but $0 \leftrightarrow 0$ transitions do not occur) as may M_F.

Figure A.1 illustrates some linkage patterns that occur for linear polarization when hyperfine interactions are present [Sho81b; Kuh98]. In the left column are examples for excitation $J = 1 \rightarrow 0$, while the right column shows examples for excitation $J = 1 \rightarrow 1$. In the top row (*a*) there is no nuclear spin; sublevels bear labels M_j. The middle row (*b*) shows the effect of including the additional degeneracy of nuclear spin $I = 1$, but without including the hyperfine interaction. Here each sublevel of given M_J is associated with three possible orientations of the nuclear spin. This does not change during an electric-dipole radiative

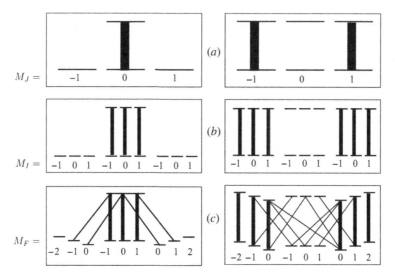

Figure A.1 Linkage pattern for $J = 1 \rightarrow 0$ (left columns) and $J = 1 \rightarrow 1$ (right columns) with linear polarization. (*a*) Without nuclear spin. (*b*) With nuclear spin $I = 1$ but negligible hyperfine interaction. (*c*) With nuclear spin $I = 1$ and hyperfine interaction. [Redrawn from Fig. 1 of B. W. Shore and M. A. Johnson, Phys. Rev. A, **23**, 1608–1610 (1981) with permission.]

interaction. The bottom row shows the quantum labels needed in the presence of hyper-fine interaction. The hyperfine interaction shifts the energies and mixes the quantum states. Strong lines show linkages allowed in the limit of zero hyperfine interaction, weak lines show transitions that are allowed by first-order perturbation theory.

In the uncoupled basis the matrix elements of the dipole moment are an application of the Wigner–Eckart theorem, eqn. (12.3)

$$\langle I M_I, J M_J | d_q | I' M_I', J M_J' \rangle \tag{A.83}$$

$$= \delta(I, I')\, \delta(M_I, M_I')\, (-1)^{J-M} \begin{pmatrix} J & 1 & J' \\ -M_J & q & M_{J'} \end{pmatrix} (J||d||J'),$$

where $\delta(n, m)$ is the Kronecker delta and $(J||d||J')$ is the reduced matrix element of the dipole transition moment, see App. H. The radiative interaction involves electrons, not nuclear spins, and so the quantum numbers I, M_i do not change. With this choice of basis one can treat separately each M_I value, so that the matrices do not involve all $(2I+1)(2J+1)$ sublevels simultaneously. Alternatively, in the coupled basis the required expression is [Sho90, §21.1]

$$\langle I J F M | d_q | I' J' F' M' \rangle = \delta(I, I')\, (-1)^{F-M} \begin{pmatrix} F & 1 & F' \\ -M & q & M' \end{pmatrix} \tag{A.84}$$

$$\times (-1)^{I+J+F+1} \sqrt{(2F+1)(2F'+1)} \begin{Bmatrix} J & 1 & J' \\ F' & I & F \end{Bmatrix} (J||d||J'),$$

where $\{:::\}$ is a six-j symbol [Bie81a; Zar88; Auz10]. The atomic Hamiltonian H^{at} must, in either case, include the hyperfine interaction V^{HF}. The choice of coupling scheme does not affect the final results, although the computations differ: if one chooses the *coupled* scheme, then V^{HF} is already diagonal, whereas the diagonalization of V^{HF} in the *uncoupled* scheme produces the coupled states as eigenstates.

Appendix B

The multipole interaction

Quantum mechanics deals with energies – kinetic, potential, and interaction. To formulate a theory of radiation affecting quantum structures it is necessary to have expressions for the energy of this interaction. The starting point is the Lorentz force acting upon the various constituents of an atom or molecule: The Hamiltonian (or Lagrangian) must reproduce this force. Two general approaches are found in the literature of electrodynamics. In much of the traditional work on quantum electrodynamics (QED) the magnetic field is introduced as the curl of a vector potential $\mathbf{A}$ and the interaction is appropriate to a point charge: Both the Coulomb field that binds the electrons to nuclei and the laser field that excites them are part of the full field [Bia75; Coh89]. This approach, well suited to relativistic calculations of energy levels and to free particles, obtains an interaction proportional to the product of particle momentum $\mathbf{p}$ and the vector potential $\mathbf{A}$. An alternative approach treats the Coulomb field as producing bound structures whose properties are expressed as multipole moments [Fiu63; Pow78; Cra84; Pow85]. This has been the traditional formalism for treating laser-induced excitation and quantum manipulation. This appendix, based on [Sho90, §9.15], summarizes the derived multipole interaction energy.

B.1 The bound-particle interaction

We convert the Lorentz force acting on a single point–charge e at position $\mathbf{r}$, moving with velocity $\mathbf{v}$,

$$\mathbf{F}(\mathbf{r}) = \mathsf{e}\mathbf{E}(\mathbf{r}) + (\mathsf{e}/c)\mathbf{v} \times \mathbf{B}(\mathbf{r}), \tag{B.1}$$

into the energy of a bound particle by evaluating the work done by the force in moving the charge first to the stationary coordinate origin at point $\mathbf{R}$, and then to its moving location at point $\mathbf{r}$ relative to $\mathbf{R}$. The energy has two parts, $V^{elec}(t)$ and $V^{magn}(t)$, corresponding to the two contributions to the Lorentz force of eqn. (B.1). From the electric field we obtain the energy

$$V^{elec}(t) = V^{E0} - \mathsf{e} \int_0^{\mathbf{r}} d\mathbf{r} \cdot \mathbf{E}(\mathbf{R}+\mathbf{r}, t), \tag{B.2}$$

where V^{E0} is the energy of charge e in an electrostatic potential $\phi(\mathbf{R})$,

$$V^{E0} = \mathsf{e}\phi(\mathbf{R}). \tag{B.3}$$

459

The energy of a moving point charge in a magnetic field is similarly expressible as an integral,

$$V^{magn}(t) = -e \int_0^{\mathbf{r}} d\mathbf{r} \cdot \mathbf{v} \times \mathbf{B}(\mathbf{R} + \mathbf{r}, t). \tag{B.4}$$

We apply these expressions to fields that do not vary appreciably over the small distances that are appropriate for expressing $\mathbf{r}$ within an atom by employing a series expansion of each field about the origin $\mathbf{R}$, located at the center of mass. The necessary expansion is a three-dimensional generalization of the simple one-dimensional Taylor-series expansion of a scalar function $f(x)$ about a value x (a Maclaurin series if $x = 0$),

$$f(x+s) = f(x) + \frac{s}{1!} \frac{d}{dx} f(x) + \frac{s^2}{2!} \frac{d^2}{dx^2} f(x) + \cdots . \tag{B.5}$$

Generalized to components of a three-dimensional vector field $\mathbf{f}(\mathbf{R})$ the expansion takes the form

$$f_i(\mathbf{R} + \mathbf{r}) = \sum_{n=0}^{\infty} \frac{1}{n!} \left[\sum_{j=1}^{3} s_j \frac{\partial}{\partial R_j} \right]^n f_i(\mathbf{R}) = \sum_{n=0}^{\infty} \frac{1}{n!} \left(\mathbf{r} \cdot \frac{\partial}{\partial \mathbf{R}} \right)^n f_i(\mathbf{R}). \tag{B.6}$$

Here subscripts i and j refer to the three Cartesian components of the vectors and $\partial/\partial\mathbf{R}$ denotes the vector partial derivative operator involving position $\mathbf{R}$.

Electric and magnetic energies. When we substitute this expansion into the expression for electric energy and carry out the integration over λ we obtain the result

$$V^{elec}(t) = V^{E0} - e \sum_{n=1}^{\infty} \frac{1}{n!} \left(\mathbf{r} \cdot \frac{\partial}{\partial \mathbf{R}} \right)^{n-1} \mathbf{r} \cdot \mathbf{E}(\mathbf{R}, t). \tag{B.7}$$

A similar expansion and integration for the magnetic field leads to the expression

$$V^{magn}(t) = -e \sum_{n=1}^{\infty} \frac{1}{(n+1)!} \left(\mathbf{r} \cdot \frac{\partial}{\partial \mathbf{R}} \right)^{n-1} (\mathbf{r} \times \mathbf{P}) \cdot \mathbf{B}(\mathbf{R}, t). \tag{B.8}$$

The factor $\mathbf{r} \times \dot{\mathbf{r}}$ that appears here can be reexpressed in terms of the linear momentum $\mathbf{p} \simeq m\dot{\mathbf{r}} \equiv m\mathbf{v}$ for a particle of mass m and, in turn, in terms of the particle angular momentum $\hbar\boldsymbol{\ell}$ relative to the coordinate origin $\mathbf{R}$.[1] By introducing the dimensionless angular momentum vector $\boldsymbol{\ell}$ through the definition

$$\hbar\boldsymbol{\ell} = \mathbf{r} \times \mathbf{p} \simeq m\mathbf{r} \times \dot{\mathbf{r}}, \tag{B.9}$$

we can write the magnetic energy of a point particle as

$$V^{magn}(t) = -\frac{e\hbar}{m} \sum_{n=1}^{\infty} \frac{1}{(n+1)!} \left(\mathbf{r} \cdot \frac{\partial}{\partial \mathbf{R}} \right)^{n-1} \boldsymbol{\ell} \cdot \mathbf{B}(\mathbf{R}, t). \tag{B.10}$$

---

[1] The formula $\mathbf{p} = m\mathbf{v}$ neglects an electromagnetic component of momentum that is responsible for various diamagnetic effects, see [Sho90, §9.15] or [Cra84]. It also neglects a variation of mass with velocity that is required by the special theory of relativity, see [Sho90, §7.1]

The first term of this expansion is the energy of a magnetic moment $\mathbf{m}_\ell$ in the magnetic field at the center of mass,

$$V^{magn}(t) = -\mathbf{m}_\ell \cdot \mathbf{B}(\mathbf{R}, t) + \cdots. \tag{B.11}$$

The magnetic moment is directly proportional to the (orbital) angular momentum and inversely proportional to the mass,

$$\mathbf{m}_\ell = \frac{e\hbar}{2m}\boldsymbol{\ell}. \tag{B.12}$$

The combination of parameters $e\hbar/2m$ (or $e\hbar/2mc$ in Gaussian units) containing the dimensions of the magnetic moment is known as the *magneton*. In particular, when the mass m is that of the electron, m_e, this is the *Bohr magneton*, μ_B, while for nucleons the mass is taken as the proton mass, m_p, and the result is the *nuclear magneton*, μ_N.

$$\mu_B = \frac{e\hbar}{2m_e}, \qquad \mu_N = \frac{m_e}{m_p}\mu_B. \tag{B.13}$$

The Bohr magneton can be written $\mu_B = \alpha e a_0/2$, where $\alpha \simeq 1/137$ is the Sommerfeld fine-structure constant. Hence matrix elements of magnetic moments tend to be smaller than those of electric moments by $\alpha/2 \simeq 1/274$, when expressed in atomic units.

These equations express the energy associated with the presence of a single moving point charge of mass m in the presence of imposed electric and magnetic fields. Whereas the strength of the electric interaction is set by the electric charge e, the strength of the magnetic interaction is set by the combination of values $e\hbar/m$. This difference generally makes magnetic interactions smaller than electric interactions by two orders of magnitude [Sho90, §2.9].

Spin. Electrons, protons, neutrons, and many atomic nuclei (those with nonzero spin) possess intrinsic magnetic moments that supplement the magnetic interaction presented above. These magnetic moments may be thought of as permanently attached to the particle. As evidenced by the neutron, they may occur with neutral as well as charged particles. Like the magnetic interaction of orbital motion, the magnitude of the intrinsic moment is inversely proportional to the particle mass. The direction of the moment is independent of the motion of the particle and may be taken as the direction of an intrinsic angular momentum (spin) vector $\mathbf{s}$, the *spin* of the particle, and expressed in units of $\hbar$. (For atomic nuclei the collective nucleon spin is usually written $\mathbf{I}$.) It is customary to express the spin magnetic moment as

$$\mathbf{m}_s = \frac{e\hbar}{2m}g_s\mathbf{s}, \tag{B.14}$$

where $|e|$ is the electron charge in coulombs, m is the particle mass in kg, and the spin $|\mathbf{s}|$ is dimensionless, either an integer or half an odd integer. (The spin is $1/2$ for electrons, protons, and neutrons.) The dimensionless *g-factor* g_s particularizes the formula to a specific type of particle. For electrons $g_s \simeq 2.002$.

In common with any magnetic moment, the intrinsic (spin) magnetic moment has an energy of orientation in a magnetic field. Because the spin magnetic moment is attached to the particle, this energy may be written in a form analogous to eqn. (B.7),

$$V^{spin}(t) = -\frac{e\hbar}{2m} g_s \sum_{n=1}^{\infty} \frac{1}{n!} \left(\mathbf{r} \cdot \frac{\partial}{\partial \mathbf{R}} \right)^{n-1} \mathbf{s} \cdot \mathbf{B}(\mathbf{R}, t). \tag{B.15}$$

As can be seen from the occurrence of a common factor $e\hbar/m$, spin-dependent interactions are comparable in magnitude to interactions originating with orbital motion. Both the spin-dependent energy and the energy of orbital motion are commonly treated as magnetic interactions through the replacement of eqn. (B.10) with

$$V^{magn}(t) = -\frac{e\hbar}{2m} \sum_{n=1}^{\infty} \left[\frac{2}{n+1} g_\ell \boldsymbol{\ell} + g_s \mathbf{s} \right] \frac{1}{n!} \left(\mathbf{r} \cdot \frac{\partial}{\partial \mathbf{R}} \right)^{n-1} \cdot \mathbf{B}(\mathbf{R}, t). \tag{B.16}$$

The orbital g-factor g_ℓ is e/e, the ratio of particle charge to the unit of charge e.

B.2 The multipole moments

We generalize the resulting single-charge formulas to aggregates of point charges e_α at positions $\mathbf{r}(\alpha)$ around a common origin $\mathbf{R}$. The results are expressions for various multipole moments of charge and current distributions.

Electric multipoles. The interaction energy is the sum of interactions of the individual charges. The electric contribution to this energy can be written as a series involving successively higher derivatives of the field,

$$V^{elec}(t) = \mathcal{Q}^{E0} \phi(\mathbf{R}) - \sum_i \mathcal{Q}_i^{E1} E_i(\mathbf{R}, t) - \sum_{ij} \mathcal{Q}_{ij}^{E2} \frac{\partial}{\partial R_j} E_i(\mathbf{R}, t) + \cdots$$

$$= V^{E0} + V^{E1}(t) + V^{E2}(t) + \cdots . \tag{B.17}$$

The coefficients of this expansion are multipole moments of the electric charge distribution, starting with the monopole (a scalar, the total charge), dipole (a vector), and quadrupole (a tensor),

$$\mathcal{Q}^{E0} = \sum_\alpha e_\alpha, \qquad \mathcal{Q}_i^{E1} = \sum_\alpha e_\alpha r_i(\alpha), \qquad \mathcal{Q}_{ij}^{E2} = \frac{1}{2} \sum_\alpha e_\alpha r_i(\alpha) r_j(\alpha). \tag{B.18}$$

Here $r_i(\alpha)$ is the ith Cartesian component of particle α. Each term V^{En} of this series is the product of a factor that depends only upon the spatial distribution of electric charge within the atom – the *electric multipole moment* in brackets – and a factor that expresses either the electrostatic potential, the electric field, or spatial derivatives of the electric field. The series is termed the *electric multipole* expansion of the energy.

Magnetic multipoles. The magnetic interaction also appears as a series, the *magnetic multipole expansion,*

$$V^{magn}(t) = -\sum_i \mathcal{M}_i^{M1} B_i(\mathbf{R}, t) - \sum_{ij} \mathcal{M}_{ij}^{M2} \frac{\partial}{\partial R_j} B_i(\mathbf{R}, t) + \cdots$$

$$= V^{M1}(t) + V^{M2}(t) + \cdots, \tag{B.19}$$

where, with the inclusion of spin angular momentum, the first two coefficients are [Sho90, §9.1]

$$\mathcal{M}_i^{M1} = \sum_\alpha \left(\frac{e_\alpha \hbar}{2m_\alpha} \right) [\ell_i(\alpha) + 2s_i(\alpha)], \tag{B.20}$$

$$\mathcal{M}_{ij}^{M2} = \sum_\alpha \left(\frac{e_\alpha \hbar}{2m_\alpha} \right) \left[\frac{2}{3} \ell_i(\alpha) r_j(\alpha) + 2s_i(\alpha) r_j(\alpha) \right].$$

Again we find a succession of terms that factor into atomic variables (positions and angular momenta) and spatial derivatives of a field.

The sum proceeds over all charges, both the electrons as well as neutrons and protons within the atomic nuclei. Because these nucleons are some 2000 times more massive than the electrons, the dominant contribution to the atomic magnetic moment comes from electrons.

With the neglect of nuclear contributions, and upon taking the g-factors to be those of an electron, $g_\ell = -1$, $g_s = -2$, we write the magnetic moment as

$$\mathbf{m} = -\mu_B [\mathbf{L} + 2\mathbf{S}], \tag{B.21}$$

where $\mathbf{L}$ and $\mathbf{S}$ are the total electronic orbital and spin angular momenta, respectively, and m_e is the electron mass.

Magnetic moments. The connection between magnetic moment $\mathbf{m}$ and an angular momentum $\hbar \mathbf{J}$ is often written

$$\mathbf{m} = \gamma \hbar \mathbf{J}. \tag{B.22}$$

The proportionality constant γ is the *gyromagnetic ratio.* For orbital electron motion this ratio is $\gamma_{orb} = -e/2m_e$, whereas for electron spin the ratio is (apart from a small radiative correction) twice this value, $\gamma_{spin} = 2\gamma_{orb} = -e/m_e$.

Many atoms possess permanent magnetic moments, and these give rise to a magnetic-field alignment energy V^{magn} in a static magnetic field. The resulting *Zeeman shift* is the magnetostatic counterpart of the electrostatic Stark shift caused by V^{elec}. Magnetic-dipole transition moments also exist, and these interact with radiation to produce M1 transitions. Because magnetic moments are smaller than electric moments by roughly the ratio of electron speed to the speed of light, these M1 transitions are generally much weaker than E1 transitions.

B.3 Examples

As an example, consider the various interactions between an atom and a linearly polarized plane wave of constant amplitude. Let the polarization axis (the direction of the E field) be z and the propagation direction be x, so that the electric and magnetic fields are [Sho90, §9.1]

$$\mathbf{E}(\mathbf{r}, t) = \mathbf{e}_z \mathcal{E} \cos(kx - \omega t), \qquad \mathbf{B}(\mathbf{r}, t) = -\mathbf{e}_y \frac{\mathcal{E}}{c} \cos(kx - \omega t). \tag{B.23}$$

This represents a traveling wave for which the mean-square value of the electric field, averaged over time, is $|\mathcal{E}|^2/2$ and for which the intensity (irradiance) is $I = c\epsilon_0 |\mathcal{E}|^2/2$. These fields are to be taken at the center of mass, $z = 0$, when evaluating the multipole interaction. The electric dipole (E1), magnetic dipole (M1), and electric quadrupole (E2) interactions arising from the electrons (as contrasted with nucleons in the nuclei) are

$$V^{E1}(t) = -\mathbf{d} \cdot \mathbf{E}(0, t) \qquad = -\hat{d}_z \mathcal{E} \cos(\omega t),$$

$$V^{M1}(t) = -\mathbf{m} \cdot \mathbf{B}(0, t) \qquad = -\tfrac{1}{2}\alpha e a_0 \left(\hat{L}_y + 2\hat{S}_y \right) \mathcal{E} \cos(\omega t), \tag{B.24}$$

$$V^{E2}(t) = -\tfrac{1}{6}\mathbf{Q}^{(2)} : \nabla \mathbf{E}(0, t) = \tfrac{1}{6} Q_{xz} k \mathcal{E} \sin(\omega t).$$

Here the scaling involves the Sommerfeld fine-structure constant $\alpha \approx 1/137$ and the Bohr radius $a_0 \approx 53$ pm. The atomic moments $\hat{d}_z$, $\hat{m}_y$, and $\hat{Q}_{xz}$ are Cartesian components of collective atomic properties. When we interpret these expressions with quantum theory the atomic variables become quantum-mechanical operators, which lead in turn to matrices of transition moments.

B.4 Induced moments

The electric field acts to distort the charge distribution of electrons within an atom or molecule. The altered structure is expressible as an induced multipole moment. This too can produce transitions. The most important is the *induced dipole moment*, proportional to the electric field. This is responsible for the induced-dipole (or Raman) interaction

$$V^{pol}(t) = -\tfrac{1}{2}\mathbf{E}(0, t) \cdot \mathbf{X}(\omega) \cdot \mathbf{E}(0, t) = -\tfrac{1}{2} X_{zz}(\omega) \mathcal{E}^2 \cos^2(\omega t). \tag{B.25}$$

Here $\mathbf{X}(\omega)$ is the frequency-dependent (dynamic) polarizability tensor.[2] The field-dependent factor of this interaction is proportional to the square of the electric field, i.e. to the intensity of the light. This expression can also be written

$$V^{pol}(t) = -\tfrac{1}{4} X_{zz}(\omega) \mathcal{E}^2 [1 + \cos(2\omega t)], \tag{B.26}$$

from which we recognize that the interaction varies periodically with frequency 2ω, i.e. twice the carrier frequency.

---

[2] Customarily the polarizability tensor here denoted $\mathbf{X}$ is symbolized by α. To avoid overuse of the letter alpha I use X.

B.5 Irreducible tensor form

When expressed in terms of irreducible tensor components these formulas become

$$V^{E1}(t) = -d_0^{(1)}\mathcal{E}\cos(\omega t),$$

$$V^{M1}(t) = \frac{i}{\sqrt{2}}\left[m_{+1}^{(1)} + m_{-1}^{(1)}\right]\mathcal{E}\cos(\omega t), \tag{B.27}$$

$$V^{E2}(t) = \frac{1}{\sqrt{24}}\left[Q_{+1}^{(2)} - Q_{-1}^{(2)}\right]\frac{\omega\mathcal{E}}{c}\sin(\omega t),$$

The quadrupole tensor $Q_q^{(2)}$ that occurs with V^{E2} is defined as

$$Q_q^{(2)}(\mathbf{r}) = r^2 C_q^{(2)}(\hat{\mathbf{r}}) \quad \text{or} \quad Q_0^{(2)}(\mathbf{r}) = \tfrac{1}{2}\left[3z^3 - r^2\right]. \tag{B.28}$$

The induced dipole moment, responsible for Raman transitions, is

$$V^{pol}(t) = -\tfrac{1}{2}\mathbf{E}(0,t)\cdot\mathbf{X}(\omega)\cdot\mathbf{E}(0,t)$$

$$= -\frac{\sqrt{3}}{12}\left[X_0^{(0)} - \sqrt{2}X_0^{(2)}\right]\mathcal{E}^2[1+\cos(2\omega t)]. \tag{B.29}$$

In these examples our choice of z axis, along the direction of the electric field, provides a simple expression with only one component for the electric dipole moment, but the other multipoles require linear superpositions of irreducible components. Note that the induced interaction V^{pol} involves a coherent superposition of tensors of two different orders.

B.6 Rabi frequencies

More generally than eqn. (B.23), consider the plane-wave field

$$\mathbf{E} = \mathbf{e}\mathcal{E}\cos(\mathbf{k}\cdot\mathbf{r} - \omega t), \qquad \mathbf{B} = \mathbf{e}'\frac{\mathcal{E}}{c}\cos(\mathbf{k}\cdot\mathbf{r} - \omega t). \tag{B.30}$$

The unit vector $\mathbf{e}'$ associated with the magnetic field may be written

$$\mathbf{e}' = \hat{\mathbf{k}} \times \mathbf{e} = -i\sqrt{2}\mathbf{X}(\hat{\mathbf{k}}, \mathbf{e})^{(1)}, \tag{B.31}$$

where $\hat{\mathbf{k}} = \mathbf{k}/k$ is a unit vector in the propagation direction and $\mathbf{X}(\mathbf{a}, \mathbf{b})^{(K)}$ is the tensor of order K constructed (see [Sho90, §20.4]) by coupling the two unit vectors $\mathbf{a}$ and $\mathbf{b}$,

$$\mathbf{X}(\mathbf{a}, \mathbf{b})_Q^{(K)} = \sum_{qq'}(1q', 1q|K\,Q)(-1)^{q'}\mathbf{a}_{-q'}\mathbf{b}_q. \tag{B.32}$$

When we construct the RWA Hamiltonian for these fields we are led to define Rabi frequencies such as

$$\hbar\Omega_{ab}^{(E1)} = -\mathcal{E}^* \sum_q (-1)^q \, \mathsf{X}(\hat{\mathbf{k}}, \mathbf{e})_{-q}^{(1)} \, \langle a|d_q|b\rangle,$$

$$\hbar\Omega_{ab}^{(M1)} = \mathrm{i}\sqrt{2}\,\frac{\mathcal{E}^*}{c} \sum_q (-1)^q \, \mathsf{X}(\hat{\mathbf{k}}, \mathbf{e})_{-q}^{(1)} \, \langle a|m_q|b\rangle,$$

$$\hbar\Omega_{ab}^{(E2)} = -\frac{\mathrm{i}\omega}{\sqrt{24}}\,\frac{\mathcal{E}^*}{c} \sum_q (-1)^q \, \mathsf{X}(\hat{\mathbf{k}}, \mathbf{e})_{-q}^{(2)} \, \langle a|Q_q^{(2)}|b\rangle, \qquad (\text{B.33})$$

$$\hbar\Omega_{ab}^{(pol)} = -\frac{1}{2}(\mathcal{E}^*)^2 \sum_q (-1)^q \, \mathsf{X}(\mathbf{e}, \mathbf{e})_{-q}^{(2)} \, \langle a|X_q^{(2)}|b\rangle.$$

The irreducible polarizability tensors $\mathsf{X}(\mathbf{a}, \mathbf{b})_Q^{(K)}$ embody the geometric properties of the particular electromagnetic field – the relationship of the field unit vector $\mathbf{e}$ and the propagation direction. The factors such as $\langle a|d_q|b\rangle$ express the intrinsic properties of the two atomic states a and b. The combination of tensor and matrix elements, together with the field amplitude $\mathcal{E}$, evaluate to give the Rabi frequency for a particular transition and particular field.

B.7 Angular momentum selection rules

From the Wigner–Eckart theorem we deduce various selection rules for the matrix elements of the RWA Hamiltonian expressed by the preceding Rabi frequencies when the quantum states are eigenstates of angular momentum operators. For atoms (as contrasted with molecules) the eigenvalues of the total electronic orbital angular momentum $\mathbf{L}$ and total spin $\mathbf{S}$, along with the eigenvalues of their vector sum $\mathbf{J} = \mathbf{L} + \mathbf{S}$, serve to partially label the quantum states. The quantum numbers of these operators have the following selection rules, discussed in texts on angular momentum [Zar88; Lou06; Bie09; Auz10].

The electric-dipole interaction involves the matrix element of an order-one tensor, for which the Wigner–Eckart theorem reads

$$\langle \alpha' J' M'|d_q|\alpha J M\rangle = (-1)^{J+1-M'} \begin{pmatrix} J & 1 & J' \\ M & q & -M' \end{pmatrix} (\alpha' J'||d||\alpha J). \qquad (\text{B.34})$$

Embodied in the three-j symbol are the requirements that J, J', and unity must form the three sides of a triangle. This means that J can change by no more than 1, as expressed in the selection rules

$$E1, M1: \qquad J = J' \text{ or } J = J' \pm 1, \qquad (\text{B.35a})$$

Starting from $J = 0$ it is not possible create the triangle if $J' = 0$, and so we have the restriction

$$E1, M1: \qquad \text{not } J = 0 \leftrightarrow J' = 0. \qquad (\text{B.35b})$$

This same selection rule on J applies to magnetic dipole transitions.

For the electric quadrupole moment the Wigner–Eckart theorem reads

$$\langle \alpha' J' M' | Q_q^{(2)} | \alpha J M' \rangle = (\alpha' J' \| Q \| \alpha J) \, (-1)^{J+2-M'} \begin{pmatrix} J & 2 & J' \\ M & q & -M' \end{pmatrix}. \tag{B.36}$$

The angular momenta J and J' must here form a triangle with the integer 2. The resulting selection rules on quadrupole radiation are

$$E2: \qquad J = J' \text{ or } J = J' \pm 1, \text{ or } J = J' \pm 2, \tag{B.37a}$$

restricted by the requirements

$$E2: \qquad \text{not } J = 0 \leftrightarrow J' = 0 \text{ or } 1. \tag{B.37b}$$

The electric multipole moments, e.g. the electric dipole and quadrupole moments, are independent of spin. Thus they do not alter the spin quantum numbers

$$E1, E2: \qquad S = S'. \tag{B.38}$$

To evaluate the matrix elements of the electron magnetic moment operator we must express its separate dependence on orbital and spin angular momentum by means of Landé g-factors,

$$\mathbf{m} = -\mu_B \sum_i [g_L \boldsymbol{\ell}(i) + g_S \boldsymbol{s}(i)] = -\mu_B (\mathbf{L} + 2\mathbf{S}). \tag{B.39}$$

Neither $\mathbf{L}$ nor $\mathbf{S}$ induce changes to L or S. We evaluate each of these terms separately using Racah algebra [Bie65; Bie81b],

$$\langle \alpha' L' S' J' | \mathbf{L} | \alpha L S J \rangle = \delta_{SS'} \delta_{LL'} \sqrt{L(L+1)} \sqrt{2L+1} \tag{B.40}$$

$$\times (-1)^{L+S+J'+1} \sqrt{(2J+1)(2J'+1)} \begin{Bmatrix} J & 1 & J' \\ L & S & L \end{Bmatrix},$$

$$\langle \alpha' L' S' J' | \mathbf{S} | \alpha L S J \rangle = \delta_{SS'} \delta_{LL''} \sqrt{S(S+1)} \sqrt{2S+1} \tag{B.41}$$

$$\times (-1)^{L+S+J'+1} \sqrt{(2J+1)(2J'+1)} \begin{Bmatrix} J & 1 & J' \\ S & L & S \end{Bmatrix}.$$

Appendix C
Classical radiation

Throughout this monograph the effect of radiation upon a quantum system has been presented as an interaction between an atomic moment (usually the electric dipole moment) and a pulse of nearly periodic electromagnetic radiation, typically the electric field of that radiation. The strength of the interaction of such a field with an individual atom or molecule is typically parametrized by the field intensity (or radiation irradiance) and a transition moment (or oscillator strength). When the atoms form macroscopic aggregates, through which the laser radiation must pass, the laser-induced alteration of atomic states produces new fields that subtract from or add to the original field [Sho90, Chap. 12]. As a result, pulses propagating through matter become altered. The pulse amplitude will, at first, decrease as energy is absorbed by the atoms, but more dramatic effects can occur that drastically alter the shape of the pulse as it travels through greater thicknesses of matter [Sho90, Chap. 9]. Furthermore, new frequencies may be generated – the phenomena of nonlinear optics [Rei84; Gae06; Boy08].

Prior to the advent of laser radiation there was little interest in short pulses, and the equations describing radiation dealt with the flow of energy through matter that could absorb or divert the radiation, embodied in the theory of radiative transfer [Cha60; Tuc75; Ryb85; Car07]. Such descriptions characterized the radiation–matter interaction by a static complex-valued index of refraction, whose imaginary part produced absorption while the real (dispersive) part altered the propagation velocity [Bre32; Mea60; Bor99]. As it became possible to create pulses whose duration was comparable to, or shorter than, the response time of the material (i.e. the Rabi period for two-state excitation), it was necessary to treat in more detail the coupling between matter and radiation [Fra63; Ics69]. This appendix summarizes the underlying theory, developed first in the nineteenth century by Maxwell [Max54]. Appendix D discusses the traditional approach, devised by Dirac in the twentieth century [Dir27; Dir58], for fitting electromagnetic theory within the structure of quantum theory, i.e. quantizing the radiation field.

C.1 The Lorentz force; Maxwell's equations

The Lorentz force. The electromagnetic field receives operational definition from the forces that it exerts on test charges and currents. The fundamental forces are expressed as the

Lorentz force acting on a point charge e at position **r** and moving with velocity **v**,

$$\mathbf{F}(\mathbf{r}, t) = e\mathbf{E}(\mathbf{r}, t) + (e/c)\mathbf{v} \times \mathbf{B}(\mathbf{r}, t). \qquad (C.1)$$

The vector field $\mathbf{E}(\mathbf{r}, t)$ acting on the charge e is the *electric field*, or simply E field, expressed (in SI units) in volts per meter (or, equivalently, newtons per coulomb) for charges in coulombs. The vector field $\mathbf{B}(\mathbf{r}, t)$, referred to in the present monograph as the *magnetic field* or B field, is known variously as the magnetic induction or magnetic flux density; it is expressed in tesla (or, equivalently, webers per square meter or volt-seconds per square meter). From this force law, or from the energy involved as the force does work on a particle, come the equations of motion for charged and moving objects, including quantum systems of electrons and nuclei.

Maxwell's equations. The fields themselves originate with charges and currents. Away from these sources the fields obey *Maxwell's equations* [Str41; DeG69; Pow82; Jac99]The first two vector Maxwell's equations (in SI units) express dynamical changes (to simplify topography, I omit the arguments $\mathbf{r}, t$ on fields),

$$\nabla \times \mathbf{E} + \frac{\partial}{\partial t}\mathbf{B} = 0, \qquad (C.2a)$$

$$\nabla \times (\mathbf{B} - \mu_0\mathbf{M}) - \mu_0\frac{\partial}{\partial t}(\epsilon_0\mathbf{E} + \mathbf{P}) = \mu_0\mathbf{j}'. \qquad (C.2b)$$

To these dynamical equations we add Maxwell's equations of constraint:

$$\nabla \cdot \mathbf{B} = 0, \qquad \nabla \cdot (\epsilon_0\mathbf{E} + \mathbf{P}) = \rho'. \qquad (C.2c)$$

The field $\mathbf{j}'$ expresses the current density (amperes per square meter) of any currents of free charge; the field ρ' is the density of free charge (in coulombs per square meter). Both of these fields are absent in the treatment of radiation propagating through neutral matter. The *polarization field* **P** expresses (in coulombs per square meter) density of polarization while *magnetization field* **M** expresses (in amperes per meter) the density of magnetization (Amperian currents). The universal constants ϵ_0 (the permittivity of free space) and μ_0 (the permeability of free space) occur as a consequence of using SI units. The square root of the product of μ_0 and ϵ_0 is the inverse of the speed of light in vacuum, c, and the square root of their ratio is the *impedance of free space*, Z_0,

$$\sqrt{\mu_0\epsilon_0} = \frac{1}{c}, \qquad \sqrt{\frac{\mu_0}{\epsilon_0}} = Z_0 \approx 137 \text{ ohm}. \qquad (C.3)$$

These equations, subject to boundary conditions and supplemented with equations that relate the polarization and magnetization to the electric and magnetic fields (constitutive relationships) form the starting point for all calculations of electromagnetic phenomena.

Maxwell's equations are often expressed in terms of two further vector fields, the *electric displacement field* **D**, also called the electric induction, expressed in units of coulombs per square meter (or equivalently, newtons per volt-meter) and the *magnetizing field* **H** (also called the magnetic field intensity),[1] expressed in units of amperes per meter. Using these auxiliary fields, defined as

$$D = \epsilon_0 E + P, \qquad H = \frac{1}{\mu_0} B - M, \tag{C.4}$$

we write Maxwell's equations in the simpler form:

$$\nabla \times E + \frac{\partial}{\partial t} B = 0, \quad \nabla \cdot B = 0,$$

$$\nabla \times H - \frac{\partial}{\partial t} D = j', \quad \nabla \cdot D = \rho'. \tag{C.5}$$

C.2 Wave equations

Our interest is with time-varying fields, as contrasted with electrostatics or magnetostatics, and so the key equations are the two dynamical Maxwell's equations (C.2a) and (C.2b). These combine to produce the single equation

$$\nabla \times \nabla \times E + \frac{1}{c^2} \frac{\partial^2}{\partial t^2} E = -\mu_0 \frac{\partial}{\partial t} \left[\frac{\partial}{\partial t} P + \nabla \times M + j' \right]. \tag{C.6}$$

Given a solution to this equation for E, we can determine the companion field B by integrating eqn. (C.2a). Alternatively, we can deal with an equation for B,

$$\nabla \times \nabla \times B + \frac{1}{c^2} \frac{\partial^2}{\partial t^2} B = \mu_0 \nabla \times \left[\frac{\partial}{\partial t} P + \nabla \times M + j' \right]. \tag{C.7}$$

In the absence of free-charge currents ($j' = 0$), as is the case for neutral matter assumed in the present monograph, the time and spatial variations of the polarization field P and magnetization field M act as the sole sources for the electric and magnetic fields E and B.

These equations for E and B have, as possible solutions, a variety of functions that represent traveling waveforms; they are termed *wave equations*. More specifically they are *inhomogeneous* wave equations as a result of the nonzero right-hand sides. Paragraphs below discuss these wave properties.

Several approximations are commonly made when using the wave equation for quantum optics. The first of these treats the electric and magnetic fields near the atom as transverse plane waves. Any vector field F can be expressed as the sum of a *lamellar* (longitudinal) part F^L whose curl vanishes and a *solenoidal* (transverse) part F^T whose divergence vanishes[2]

$$F = F^L + F^T, \qquad \nabla \times F^L = 0, \qquad \nabla \cdot F^T = 0. \tag{C.8}$$

---

[1] In much of the literature the field B is termed the *magnetic induction vector*, while $H = B/\mu$ is called the magnetic field.

[2] For plane waves the solenoidal (divergenceless) fields are transverse to the propagation direction whereas the lamellar fields are directed longitudinally, along the propagation axis. Because we usually deal with plane waves it is customary to refer to radiation fields as transverse.

By applying this decomposition to the electric field **E** and polarization field **P** and using the relationship

$$\nabla \times \mathbf{F} = \nabla(\nabla \cdot \mathbf{F}) - \nabla^2 \mathbf{F}, \tag{C.9}$$

we obtain, in the absence of free currents and magnetization, the inhomogeneous wave equation

$$\left[\nabla^2 - \frac{1}{c^2}\frac{\partial^2}{\partial t^2}\right]\mathbf{E}^T = \frac{1}{\epsilon_0 c^2}\frac{\partial^2}{\partial t^2}\mathbf{P}^T. \tag{C.10}$$

It is the solenoidal part of the electric field $\mathbf{E}^T$ that constitutes the propagating radiation, and it is this field that is to be determined here. However, the field acting on the atom, and parametrized as the Rabi frequency, is the total field. For typographical simplicity I omit the label T upon this field; the field **E** and its source field **P** are to be understood as transverse.

C.2.1 Radiation in free space; intensity

In the absence of matter (i.e. in free space) the fields **P**, **M**, ρ', and $\mathbf{j}'$ that characterize matter all vanish and the two dynamical Maxwell's equations read

$$\nabla \times \mathbf{E} + \frac{\partial}{\partial t}\mathbf{B} = 0, \qquad \nabla \times \mathbf{B} - \frac{1}{c^2}\frac{\partial}{\partial t}\mathbf{E} = 0. \tag{C.11}$$

Each of these fields is solenoidal, meaning divergenceless,

$$\nabla \cdot \mathbf{E} = 0, \qquad \nabla \cdot \mathbf{B} = 0. \tag{C.12}$$

From these there follow two homogeneous wave equations for fields in free space,

$$\left[\nabla^2 - \frac{1}{c^2}\frac{\partial^2}{\partial t^2}\right]\mathbf{E} = 0, \qquad \left[\nabla^2 - \frac{1}{c^2}\frac{\partial^2}{\partial t^2}\right]\mathbf{B} = 0. \tag{C.13}$$

Only one of these equations is needed; the present text works with the **E**-field wave equation. The companion **B** field obtains from the appropriate dynamical Maxwell's equation (C.11).

The Poynting vector. From the free-space Maxwell's equations there also follows the equation

$$\nabla \cdot (\mathbf{E} \times \mathbf{B}) = \mathbf{B} \cdot (\nabla \times \mathbf{E}) - \mathbf{E} \cdot (\nabla \cdot \mathbf{E}) = -\frac{1}{2}\frac{\partial}{\partial t}\left[\mathbf{B}^2 + \mu_0\epsilon_0\mathbf{E}^2\right]. \tag{C.14}$$

This has the form of a *continuity equation* equating the divergence of a vector field to the time rate of change in a scalar field. To express this more clearly we define the *Poynting vector* field

$$\mathbf{S} = \frac{1}{\mu_0}\mathbf{E} \times \mathbf{B} = \mathbf{E} \times \mathbf{H} \tag{C.15}$$

and the two scalar fields

$$u^E = \frac{\epsilon_0}{2}\mathbf{E}^2, \qquad u^B = \frac{1}{2\mu_0}\mathbf{B}^2. \tag{C.16}$$

In free space these two scalar fields are equal; their total can be expressed in terms of the electric field as

$$u = u^E + u^B = \epsilon_0 \mathbf{E}^2. \tag{C.17}$$

Then eqn. (C.14) reads

$$\nabla \cdot \mathbf{S} = -\frac{\partial}{\partial t} u. \tag{C.18}$$

This continuity equation, known as Poynting's theorem, has the interpretation of a change in density (of the scalar field u) related to the divergence of a flux (the vector field $\mathbf{S}$). From this we identify u as the electromagnetic energy density and $\mathbf{S}$ as the field momentum density and the flux (power per unit area) of electromagnetic energy.

The two vector fields $\mathbf{E}$ and $\mathbf{B}$ are characterized by magnitude and direction. In free space they are perpendicular to each other and to the Poynting vector,

$$\mathbf{E} \cdot \mathbf{B} = 0, \qquad \mathbf{E} \cdot \mathbf{S} = 0, \qquad \mathbf{B} \cdot \mathbf{B} = 0. \tag{C.19}$$

These three vectors therefore define an orthogonal coordinate system that can provide local Cartesian axes x, y, z.

Irradiance, intensity. In free space the electromagnetic energy density at location $\mathbf{r}$ and time t is

$$u(\mathbf{r}, t) = \epsilon_0 |\mathbf{E}(\mathbf{r}, t)|^2. \tag{C.20}$$

When characterizing radiation fields, whose time variation typically oscillates at optical frequencies, our interest lies with the magnitude of energy flow averaged over optical cycles. The time average of the Poynting vector, over many cycles, is the *irradiance* (power per unit area) $I(\mathbf{r}, t)$,

$$I(\mathbf{r}, t) = \{|\mathbf{S}(\mathbf{r}, t)|\}_{av}. \tag{C.21}$$

Here the symbol $\{\cdots\}_{av}$ denotes the cycle average. To accommodate both traveling waves (as in a laser beam) and standing waves (as in a cavity) we use the laser *intensity*, defined as $c\epsilon_0$ times the cycle-averaged electric field,

$$I(\mathbf{r}, t) = c\epsilon_0 \{|\mathbf{E}(\mathbf{r}, t)|^2\}_{av}. \tag{C.22}$$

For traveling waves this is equal to the irradiance.

Effect of matter. In the presence of polarization and magnetization fields the energy densities become

$$u^E = \frac{1}{2}\mathbf{D} \cdot \mathbf{E} = \frac{\epsilon_0}{2}|\mathbf{E}|^2 + \frac{1}{2}\mathbf{P} \cdot \mathbf{E}, \qquad u^B = \frac{1}{2}\mathbf{H} \cdot \mathbf{B} = \frac{1}{2\mu_0}|\mathbf{B}|^2 - \frac{1}{2}\mathbf{M} \cdot \mathbf{B}. \tag{C.23}$$

The new terms express the interaction between the electromagnetic field and matter. The electric contribution of the polarization field is the energy density

$$u^{pol} = \frac{1}{2}\mathbf{P} \cdot \mathbf{E}. \tag{C.24}$$

In addition to these three contributions to energy density, the distributed atoms responsible for the fields $\mathbf{P}$ and $\mathbf{M}$ contribute an energy density u^{at}, evaluated from the density of internal excitation.

C.2.2 Monochromatic fields

Because the free-space wave equations are separable in space and time coordinates it is useful to express a free field as the product of a spatial function and a time variation. The mathematically simplest fields are those that are *monochromatic*, having a single frequency ω. To facilitate the mathematics it is useful to express the real-valued fields $\mathbf{E}$ and $\mathbf{B}$ in terms of complex functions of space and time, through a decomposition into a positive- and a negative-frequency part[3] of the real-valued fields $\mathbf{E}$ and $\mathbf{B}$ as

$$\begin{aligned} \mathbf{E}(\mathbf{r}, t) &= \mathbf{E}^{(+)}(\mathbf{r})\, e^{-\mathrm{i}\omega t} + \mathbf{E}^{(-)}(\mathbf{r})\, e^{+\mathrm{i}\omega t}, \\ \mathbf{B}(\mathbf{r}, t) &= \mathbf{B}^{(+)}(\mathbf{r})\, e^{-\mathrm{i}\omega t} + \mathbf{B}^{(-)}(\mathbf{r})\, e^{+\mathrm{i}\omega t}. \end{aligned} \tag{C.25}$$

In free space the spatial portions of these constructions obey the *vector Helmholtz equation*,

$$[\nabla^2 + (\omega/c)^2]\mathbf{E}^{(\pm)}(\mathbf{r}) = 0, \qquad [\nabla^2 + (\omega/c)^2]\mathbf{B}^{(\pm)}(\mathbf{r}) = 0, \tag{C.26}$$

and the paired fields are related by the equations

$$\nabla \times \mathbf{E}^{(+)}(\mathbf{r}) = \mathrm{i}\omega \mathbf{B}^{(+)}(\mathbf{r}), \qquad \nabla \times \mathbf{B}^{(+)}(\mathbf{r}) = -\mathrm{i}\frac{\omega}{c^2}\mathbf{E}^{(+)}(\mathbf{r}). \tag{C.27}$$

C.2.3 Plane waves

The Laplacian operator ∇^2 of the Helmholtz equation is separable in various coordinate systems, and for each there exist well-studied analytic solutions in which the fields maintain constant value over surfaces defined by constant values of one coordinate. In spherical coordinates r, θ, ϕ the surfaces of constant radius r (spherical shells) appear as *spherical waves* (see App. D.2.2). In Cartesian coordinates x, y, z the elementary solutions are *plane waves* in which the functions have constant values over a plane, such as are defined by a given value of the scalar product of the position vector $\mathbf{r}$ with a vector $\mathbf{k}$,

$$\mathbf{k} \cdot \mathbf{r} = k_x x + k_y y + k_z z. \tag{C.28}$$

The surfaces may be either stationary (standing waves) or moving (traveling waves). The direction of the vector defined by the Cartesian components k_x, k_y, k_z is the optical axis or, for traveling waves, the *propagation axis*.

Standing waves. A simple example of a standing plane wave is the product function

$$\mathbf{E}(\mathbf{r}, t) = f(\omega t)\, \mathbf{F}(\mathbf{k} \cdot \mathbf{r}), \tag{C.29}$$

---

[3] The signs in the exponent are those that are traditional in physics, where a forward-traveling wave is written as $\exp(\mathrm{i}kx - \mathrm{i}\omega t)$.

where f is a scalar function and $\mathbf{F}$ is a vector function of their arguments. The $\mathbf{B}$ field is

$$\mathbf{B}(\mathbf{r}, t) = -\frac{i}{\omega} \nabla \times \mathbf{F}(\mathbf{k} \cdot \mathbf{r}) \int^t dt' \, f(\omega t'). \tag{C.30}$$

These fields are constant over planes of constant $\mathbf{k} \cdot \mathbf{r}$. To satisfy the wave equation (C.13) the two functions must separately satisfy the equations

$$\nabla^2 \mathbf{F} = -k^2 \mathbf{F}, \qquad \frac{d^2}{dt^2} f = -(c\omega)^2 f. \tag{C.31}$$

That is, f must be periodic, with frequency ω, and $\mathbf{F}$ must be a solution to the Helmholtz equation. The two separation constants of these equations must satisfy the relationship

$$\omega = ck. \tag{C.32}$$

Traveling waves. The light emerging from a laser, and forming a beam, is an example of a traveling wave. A plane-wave electric field traveling toward increasing $\mathbf{r}$ has the form

$$\mathbf{E}(\mathbf{r}, t) = \mathbf{F}(\mathbf{k} \cdot \mathbf{r} - \omega t), \tag{C.33}$$

where $\mathbf{F}(\xi)$ is an arbitrary vector function of the scalar argument $\xi = \mathbf{k} \cdot \mathbf{r} - \omega t$. The field is constant over moving planes of constant ξ.

This field is periodic in time, with angular frequency ω and, for fixed time t, is periodic in space along the propagation axis. The spatial period is the *wavelength* λ, related to the magnitude k of the propagation vector through the equation

$$k \equiv |\mathbf{k}| = \sqrt{k_x^2 + k_y^2 + k_z^2} = 2\pi/\omega. \tag{C.34}$$

The distance between nodal planes, where the function vanishes, is $\lambda/2$.

In order that the wave equation (C.13), or the Helmholtz equation (C.26), be satisfied for a given frequency ω the magnitude of the propagation vector $\mathbf{k}$ must have the value

$$k = \omega/c, \qquad \text{so} \qquad k = 2\pi/\lambda. \tag{C.35}$$

Thus the plane wave surfaces move toward increasing $\mathbf{k}$ with velocity c.

To ensure that Maxwell's equations (C.11) hold, both the electric field and the magnetic field must be transverse, i.e. perpendicular to the propagation vector $\mathbf{k}$ that defines the orientation of the planes of constant field,

$$\mathbf{k} \cdot \mathbf{E}(\mathbf{r}, t) = \mathbf{k} \cdot \mathbf{B}(\mathbf{r}, t) = 0. \tag{C.36}$$

The Poynting vector is aligned with the $\mathbf{k}$ vector, which therefore defines the direction of energy flow; $\mathbf{k}$ is termed the *propagation vector* (or wavevector).

Plane waves, either standing or traveling, are solutions to Maxwell's equations, but because the fields extend over infinite transverse planes their total energy is infinite. Therefore they can only represent realizable fields within a limited volume.

Example. An example of a traveling plane-wave electric field is

$$\mathbf{E}(\mathbf{r}, t) = \mathbf{e}\mathcal{E} \cos(\mathbf{k} \cdot \mathbf{r} - \omega t), \tag{C.37}$$

where $\mathbf{e}$ is a real unit vector specifying the direction of the electric vector and $\mathcal{E}$ is a real number that provides the magnitude (and dimensions) of the field. The field of eqn. (C.37) has constant value over planes perpendicular to the vector $\mathbf{k}$; it is a traveling plane wave of frequency ω that travels in the direction of increasing $\mathbf{k}$.

For the field to satisfy Maxwell's equation $\nabla \cdot \mathbf{E} = 0$ the field direction, specified by the unit vector $\mathbf{e}$, must be perpendicular to the propagation direction,

$$\mathbf{k} \cdot \mathbf{e} = 0, \tag{C.38}$$

i.e. the field must be transverse. For example, when $\mathbf{k}$ lies along the z axis (so $k_x = k_y = 0$, $k_z = \omega/c$) then the field of eqn. (C.37) has the form

$$\mathbf{E}(\mathbf{r}, t) = \mathbf{E}(z, t) = \mathbf{e}\mathcal{E} \cos(kz - \omega t), \tag{C.39}$$

and the unit vector $\mathbf{e}$ must lie somewhere in the x, y plane. This electric field is accompanied by the magnetic field

$$\mathbf{B}(\mathbf{r}, t) = \frac{1}{ck} \mathbf{k} \times \mathbf{E}(\mathbf{r}, t) \tag{C.40}$$

that is perpendicular to both $\mathbf{E}$ and $\mathbf{k}$. The magnitudes of the two fields are related as

$$|\mathbf{E}(\mathbf{r}, t)| = c|\mathbf{B}(\mathbf{r}, t)|. \tag{C.41}$$

The intensity (irradiance) of this beam is

$$I(\mathbf{r}, t) = c\epsilon_0 \{|\mathbf{E}(\mathbf{r}, t)|^2\}_{av} = \tfrac{1}{2} c\epsilon_0 \{|\mathcal{E}|^2\}_{av}. \tag{C.42}$$

The electromagnetic energy density of this field,

$$u(\mathbf{r}, t) = \tfrac{1}{2}\epsilon_0 |\mathcal{E}|^2, \tag{C.43}$$

is constant in time and uniform in space. The numerical connection between the electric-field amplitude $\mathcal{E}(t)$ of eqn. (C.37) and the intensity $I(t)$ is

$$|\mathcal{E}(t)|\,[\text{V/cm}] = 27.4\sqrt{I(t)\,[\text{W/cm}^2]}. \tag{C.44}$$

Complex values. We generalize the expression of eqn. (C.37) to allow complex-valued functions by rewriting that field as

$$\mathbf{E}(\mathbf{r}, t) = \text{Re}\left\{\mathbf{e}\mathcal{E} \exp(\mathrm{i}\mathbf{k} \cdot \mathbf{r} - \mathrm{i}\omega t)\right\}. \tag{C.45}$$

Here both the field amplitude $\mathcal{E}$ and the unit vector $\mathbf{e}$ can be complex, thereby facilitating the treatment of circular polarization. Alternatively, we allow the amplitude to be complex, $\breve{\mathcal{E}}$, and use the positive- and negative-frequency parts of that plane wave,

$$\mathbf{E}^{(+)}(\mathbf{r}) = \tfrac{1}{2}\mathbf{e}\breve{\mathcal{E}} \exp(\mathrm{i}\mathbf{k} \cdot \mathbf{r}), \qquad \mathbf{E}^{(-)}(\mathbf{r}) = \tfrac{1}{2}\mathbf{e}^*\breve{\mathcal{E}}^* \exp(-\mathrm{i}\mathbf{k} \cdot \mathbf{r}). \tag{C.46}$$

C.2.4 Beams: The paraxial wave equation

The most common form of laser radiation used to produce atomic or molecular excitation is a directed beam – a field that is restricted in the transverse direction while transporting energy and momentum along a straight line. It is useful to consider the fields in such a beam, for they are what an atomic beam experiences as it moves across a laser beam. The free-space wave equation, expressed in Cartesian coordinates, reads

$$\left[\frac{\partial^2}{\partial x^2} + \frac{\partial^2}{\partial y^2} + \frac{\partial^2}{\partial z^2} - \frac{1}{c^2} \frac{\partial^2}{\partial t^2} \right] \mathbf{E}(x, y, z, t) = 0. \tag{C.47}$$

To describe a monochromatic directed beam, propagating in the z direction, we introduce a plane-wave carrier of frequency ω and write the positive-frequency part of the field as

$$\mathbf{E}^{(+)}(x, y, z, t) = \frac{1}{2} \mathbf{e} \breve{\mathcal{E}}(x, y, z) \, e^{ikz - i\omega t}, \tag{C.48}$$

where $k = \omega/c$. We assume that the complex-valued envelope function $\breve{\mathcal{E}}$ varies much more slowly than does the carrier exponential, as expressed by the inequality

$$\left| \frac{\partial^2 \breve{\mathcal{E}}}{\partial z^2} \right| \ll k \left| \frac{\partial \breve{\mathcal{E}}}{\partial z} \right|. \tag{C.49}$$

The result is the *paraxial wave equation*

$$\left[\frac{\partial^2}{\partial x^2} + \frac{\partial^2}{\partial y^2} + 2ik \frac{\partial}{\partial z} \right] \breve{\mathcal{E}}(x, y, z) = 0. \tag{C.50}$$

Solutions to this equation provide descriptions of a variety of laser beams, e.g. Sec. 3.7. Over small transverse distances, comparable to the size of an atom, they can be regarded as plane waves, independent of the transverse coordinates x, y.

C.3 Frequency components

Coherent excitation requires fields that maintain well-defined phase relationships during the excitation process. The simplest example of such a field is a monochromatic (cw) field, such as specified by eqn. (C.37) with constant envelope $\mathcal{E}$ and phase φ. Somewhat more generally we consider a *stationary* field, for which the intensity is independent of time, $I(t) = I(0)$, while the amplitude and phase undergo random (stochastic) fluctuations. Such steady radiation forms the basis for traditional spectroscopy and incoherent excitation.

To connect these definitions with the distribution of frequencies present in the field we introduce the Fourier transform (FT) of the electric field:[4]

$$\widetilde{\mathbf{E}}(\omega) = \frac{1}{2\pi} \int_{-\infty}^{+\infty} dt \, e^{+i\omega t} \, \mathbf{E}(t). \tag{C.51}$$

---

[4] The signs of the exponential factor $\pm \omega t$, and the placement of the factor 2π, here follow the usage common in quantum mechanics. Many authors use the opposite sign convention and different placement of the 2π.

The inverse transform is

$$\mathbf{E}(t) = \int_{-\infty}^{-\infty} d\omega \, e^{-i\omega t} \, \widetilde{\mathbf{E}}(\omega). \tag{C.52}$$

Because the original field $\mathbf{E}(t)$ is real, the FT has the property $\widetilde{\mathbf{E}}(\omega)^* = \widetilde{\mathbf{E}}(-\omega)$.

It proves useful to separate these transform pairs into complex-valued positive and negative frequency parts, writing

$$\widetilde{\mathbf{E}}(\omega) = \widetilde{\mathbf{E}}^{(+)}(\omega) + \widetilde{\mathbf{E}}^{(-)}(\omega), \tag{C.53}$$

with corresponding separation of the positive- and negative-frequency parts of the field,

$$\mathbf{E}(t) = \mathbf{E}^{(+)}(t) + \mathbf{E}^{(-)}(t). \tag{C.54}$$

That is, the positive frequency part has the construction[5]

$$\mathbf{E}^{(+)}(t) = \int_{0}^{+\infty} d\omega \, e^{-i\omega t} \, \widetilde{\mathbf{E}}(\omega). \tag{C.55}$$

As an example, the positive frequency part (the *analytic signal*) of the field in eqn. (C.37) is

$$\mathbf{E}^{(+)}(t) = \mathbf{e} \, \check{\mathcal{E}}(t) \, e^{-i\omega_c t}. \tag{C.56}$$

Here both $\mathbf{e}$ and $\check{\mathcal{E}}(t)$ may be complex-valued. The positive and negative frequency components are, for classical fields, complex conjugates of each other, $\mathbf{E}^{(-)}(t) = \mathbf{E}^{(+)}(t)^*$. The formula for the intensity, expressed in terms of positive and negative frequency parts, reads

$$I(\mathbf{r}, t) = 2c\epsilon_0 \, \{\mathbf{E}^{(-)}(\mathbf{r}, t) \cdot \mathbf{E}^{(+)}(\mathbf{r}, t)\}_{av}. \tag{C.57}$$

C.3.1 Spectral distribution

When treating incoherent processes it is the intensity, rather than the field amplitude, that holds interest. The analysis of frequency content proceeds through the Fourier transform of the electric field envelope, eqn. (C.51). To treat stationary stochastic fields we introduce the frequency distribution of the radiation energy through the *spectral-density function*

$$S(\omega) \equiv \lim_{\delta \to 0} \int_{\omega-\delta}^{\omega+\delta} d\omega' \, \{\widetilde{\mathcal{E}}(\omega)^* \cdot \widetilde{\mathcal{E}}(\omega')\}_{av}. \tag{C.58}$$

The connection between time and frequency domains occurs through the field *autocorrelation function*, defined as

$$\Gamma(\tau) \equiv \lim_{T \to \infty} \int_{-T}^{+T} dt \, \mathcal{E}(t)^* \cdot \mathcal{E}(t+\tau) = \{\mathcal{E}(t)^* \cdot \mathcal{E}(t+\tau)\}_{av}. \tag{C.59}$$

---

[5] This definition, with the present convention for the signs in $\exp(\pm i\omega t)$, leads to a quantized field description in which positive frequency parts correspond to photon annihilation, while negative frequency parts are associated with photon creation, see App. D.1. For a classical field the relationship $\mathbf{E}(t) = 2\text{Re} \, \mathbf{E}^{(+)}(t)$ holds.

According to the Wiener–Khinchin theorem the spectral density and the autocorrelation function are, for radiation from a stationary stochastic process, related by Fourier transforms

$$S(\omega) = \frac{1}{2\pi} \int_{-\infty}^{+\infty} d\tau \; e^{i\omega\tau} \Gamma(\tau), \qquad \Gamma(\tau) = \int_{-\infty}^{+\infty} d\omega \; e^{-i\omega\tau} S(\omega). \tag{C.60}$$

The distribution of energy with frequency ω, is expressed by the formula

$$I(t) = \frac{\epsilon_0}{c} \{\mathcal{E}(t)^2\}_{av} = \frac{\epsilon_0}{c} \{\mathcal{E}(0)^2\}_{av} = \frac{\epsilon_0}{c} \int_{-\infty}^{+\infty} d\omega \; S(\omega). \tag{C.61}$$

We introduce the frequency distribution of the radiation through the definition

$$\widetilde{I}(\omega) = \frac{c\epsilon_0}{\pi} \{\widetilde{\mathbf{E}}^{(-)}(\omega) \cdot \widetilde{\mathbf{E}}^{(+)}(\omega)\}_{av}, \tag{C.62}$$

thereby parametrizing the amplitude (but not the phase) of the FT of the electric field. The total of this quantity is the integral over all frequencies,

$$\int_{-\infty}^{+\infty} d\omega \widetilde{I}(\omega) = \frac{c\epsilon_0}{\pi} \int_{-\infty}^{+\infty} d\omega \int_{-\infty}^{+\infty} dt \int_{-\infty}^{+\infty} dt'$$

$$\times \exp[-i\omega(t-t')] \{\mathbf{E}^{(-)}(t) \cdot \mathbf{E}^{(+)}(t')\}_{av}. \tag{C.63}$$

The integral over frequency appearing on the right-hand side produces the Dirac delta of eqn. (9.38),

$$\int_{-\infty}^{+\infty} d\omega \; e^{-i\omega(t-t')} = 2\pi \delta(t-t'), \tag{C.64}$$

thereby providing the relationship

$$\int_{-\infty}^{+\infty} d\omega \; \widetilde{I}(\omega) = 2c\epsilon_0 \int_{-\infty}^{+\infty} dt \{\mathbf{E}^{(-)}(t) \cdot \mathbf{E}^{(+)}(t)\}_{av} = \int_{-\infty}^{+\infty} dt \; I(t). \tag{C.65}$$

The quantity $\widetilde{I}(\omega)$ therefore has the interpretation of the radiation energy per unit area per unit frequency interval, i.e. the spectrum.

C.3.2 Coherence

Contemporary discussions of coherence typically draw upon the work of Glauber [Gla63a], who proposed the use of correlation functions to describe both classical and quantized electromagnetic fields [Sho90, §1.9]. The correlation function of order n is defined as the averaged expectation value of $2n$ field variables, evaluated at a set of n pairs of space-time coordinates $x \equiv (\mathbf{r}, t)$, and ordered with positive-frequency parts to the right,

$$G^{(n)}(x_1, \ldots, x_{2n}) = \{E^{(-)}(x_1) \cdots E^{(-)}(x_n) E^{(+)}(x_{n+1}) \cdots E^{(+)}(x_{2n})\}_{av}. \tag{C.66}$$

Here the scalars $E^{(\pm)}(x_n)$ can be taken as any components of the field vectors; complete specification should therefore include, as subscripts on G, a set of indices needed to specify these components, i.e. G is a tensor of order $2n$.

A special case is the first-order correlation function

$$G^{(1)}(x_1, x_2) = \{E^{(-)}(x_1)E^{(+)}(x_2)\}_{av}. \tag{C.67}$$

A field is said to have *first-order coherence* if this function is expressible as the product of two factors,

$$G^{(1)}(x_1, x_2) = \mathcal{E}(x_1)^* \mathcal{E}(x_2). \tag{C.68}$$

The normalized first-order coherence, defined as

$$g^{(1)}(t, t') = G^{(1)}(t, t')/\sqrt{G^{(1)}(t, t)G^{(1)}(t', t')}, \tag{C.69}$$

takes the value unity when the field is first-order coherent; this occurs only within a range of time differences $t - t' \equiv \tau$ not much longer than a characteristic time scale, the coherence time.

To simplify the description we introduce the normalized autocorrelation function

$$g(\tau) = \frac{\{E(\tau)E(0)\}_{av}}{\{|E(0)|^2\}_{av}}. \tag{C.70}$$

A long coherence time leads to a correlation function that remains nonzero for a large range of τ. The function $g(\tau)$ is defined to have peak value unity, $g(0) = 1$, for $\tau = 0$, and thus the time integral of $g(\tau)$ (a temporal area) provides a measure of the duration of coherence, i.e. the (first-order) coherence time.

We further define, from the spectral density function, the normalized *line profile* [Sho90, §1.8], centered around an appropriate central frequency ω_c, as

$$\mathsf{g}(\omega_c - \omega) = S(\omega)/\int_{-\infty}^{+\infty} d\omega\, S(\omega). \tag{C.71}$$

By construction, this function has unit area,

$$\int_{-\infty}^{+\infty} d\omega\, \mathsf{g}(\omega_c - \omega) = 1, \tag{C.72}$$

and thus the peak value $\mathsf{g}(0)$ provides a measure of the inverse of the spectral width, i.e. the duration of coherence, as parametrized by the *correlation time*

$$\tau_c = \pi \mathsf{g}(0). \tag{C.73}$$

A broad spectrum has a short correlation time, a long correlation time is associated with a narrow spectrum. A bandwidth-limited (or transform-limited) pulse is one in which the pulse duration is not longer than the correlation time, so that the product of pulse duration and spectral width is as small as possible.

The correlation time, or its frequency counterpart the spectral line width, does not uniquely characterize a radiation beam. An important aspect of coherent excitation, as described by the TDSE, is that

the spectrum alone does not uniquely define the dynamics.

For full characterization of the radiation it is necessary to extend the notion of correlation to include nth-order correlation functions. For proper treatment of coherent excitation it is necessary to work with the explicit time dependence of the field, not with Fourier transforms.

C.4 The influence of matter

The influence of matter upon radiation enters Maxwell's equations through the electric polarization field **P** and the magnetization field **M**. It is customary in optics (including quantum optics) to idealize the matter as a continuous homogeneous distribution of identical atoms, each having electric dipole moment $\langle \mathbf{d} \rangle$ and magnetic moment $\langle \mathbf{m} \rangle$. That is, all measurable properties of individual atoms (e.g. their ability to absorb and emit light) are spread out over regions that, though microscopic, contain many atoms. With this approximation the polarization and magnetization fields are the expectation values,

$$\mathbf{P} = \mathcal{N}\langle \mathbf{d} \rangle, \qquad \mathbf{M} = \mathcal{N}\langle \mathbf{m} \rangle, \tag{C.74}$$

where $\mathcal{N}$ is the number density of atoms. It is through these macroscopic fields that the effects of atomic excitation alter the radiation.

As with the electric and magnetic fields we separate the polarization and magnetization fields into positive- and negative-frequency parts, in the form

$$\mathbf{P} = \mathbf{P}^{(+)} + \mathbf{P}^{(-)}, \qquad \mathbf{M} = \mathbf{M}^{(+)} + \mathbf{M}^{(-)}. \tag{C.75}$$

Although the full field vectors **P** and **M** have real-valued components, the components of the positive-frequency portions are complex valued.

C.4.1 The linear and nonlinear polarization fields

For weak steady-state monochromatic radiation (as contrasted with transient pulsed radiation) or for steady nonresonant radiation, the atomic dipole moment expectation value $\langle \mathbf{d} \rangle$ is an induced moment directly proportional to the electric field **E**. Hence the macroscopic polarization field **P** is also linearly related to the electric field. The proportionality is explained quantitatively with the Lorentz oscillator model of an atom or molecule, applicable so long as the field is weak and monochromatic, and if there is sufficient time for the atom to relax into equilibrium with the excitation and its other surroundings. The proportionality constant (the polarizability) depends upon the frequency of the radiation and upon atomic properties but is constant in time. For intense fields and for transient phenomena the linear approximation for **P** is not adequate. The remaining portion of the polarization field is termed the nonlinear portion. One therefore writes the polarization field as two parts:

$$\mathbf{P} = {}^{lin}\mathbf{P} + {}^{nl}\mathbf{P}. \tag{C.76}$$

The connection between the vector ${}^{lin}\mathbf{P}$ and the vector **E** is generally a frequency-dependent tensor relationship, proportional to the atomic number density $\mathcal{N}$. To quantify this proportionality one introduces the following quantities:

α	polarizability (susceptibility per atom), denoted elsewhere as X
$\chi = \mathcal{N}\alpha$	linear electric susceptibility
$\epsilon = \epsilon_0[1 + \chi]$	electric permittivity (dielectric constant)
μ	magnetic permeability
$\eta = c\sqrt{\mu\epsilon}$	refractive index

These are defined through the expressions

$$^{lin}\mathbf{P}^{(+)} = \mathcal{N}\alpha\mathbf{E}^{(+)} = \epsilon_0\chi\mathbf{E}^{(+)} = (\epsilon - \epsilon_0)\mathbf{E}^{(+)} \tag{C.77}$$

relating positive-frequency components of the fields $^{lin}\mathbf{P}$ and $\mathbf{E}$. In general both χ and ϵ will be complex-valued frequency-dependent tensors. The complex-valued refractive index $\check{\eta}$ is related to these as

$$\check{\eta}^2 = \frac{\mu\epsilon}{c^2} = \frac{\mu\epsilon}{\mu_0\epsilon_0} = \frac{\mu}{\mu_0}(1 + \chi). \tag{C.78}$$

This presentation suggests that the nonlinear portion of the polarization field is the part that does not vary linearly with the electric field. However, there is no reason why one cannot incorporate a portion of the linear response into χ and leave the remainder in $^{nl}\mathbf{P}$. Alternatively, one could place some nonlinearity into χ. It is only necessary to have a well-defined prescription for separating out a portion of the linear response. The separation may be made, for example, on the basis of fast and slow response time of the atom. Alternatively, one may consider material composed of several species of atoms, only one of which is near resonant with the light. The remaining atoms serve as a host in a solid, or as a buffering background in a vapor, and their optical properties are expressed by means of a refractive index. The properties of the near-resonant atoms then derive from what is here denoted as the polarization field $^{nl}\mathbf{P}$.

C.4.2 The refractive index

Taking the linear response to be parametrized as in eqn. (C.77) we write the linear polarization field as

$$\frac{1}{\epsilon_0 c^2}\frac{\partial^2}{\partial t^2}{}^{lin}\mathbf{P}^{(+)} = \left[\frac{\epsilon}{\epsilon_0} - 1\right]\frac{1}{c^2}\frac{\partial^2}{\partial t^2}\mathbf{E}^{(+)}, \tag{C.79}$$

and obtain, for nonmagnetic material ($\mu = \mu_0$), the inhomogeneous wave equation

$$\left[\nabla^2 - \frac{\check{\eta}^2}{c^2}\frac{\partial^2}{\partial t^2}\right]\mathbf{E}^{(+)} = \frac{1}{\epsilon_0 c^2}\frac{\partial^2}{\partial t^2}{}^{nl}\mathbf{P}^{(+)}. \tag{C.80}$$

In the absence of the nonlinear polarization field (i.e. for the traditional realm of steady response to a weak field) this is the homogeneous wave equation,

$$\left[\nabla^2 - \frac{\check{\eta}^2}{c^2}\frac{\partial^2}{\partial t^2}\right]\mathbf{E}^{(+)} = 0, \tag{C.81}$$

but now with a complex-valued refractive index $\check{\eta}$,

$$\check{\eta} = \sqrt{\epsilon/\epsilon_0} = \sqrt{1+\chi} = n + i\kappa, \tag{C.82}$$

that varies with frequency and whose imaginary part κ produces attenuation of a traveling wave. It is this equation, with suitable spatial variation of the refractive index, that underlies all of conventional optics: the reflection from mirrors, the focusing properties of lenses, the dispersive properties of slits and gratings. All of these phenomena rely on the linearity of the response of matter to radiation. Interfaces between different materials, such as air and glass, produce spatial discontinuities of $\check{\eta}$.

Implicit in the linearity of eqn. (C.77) is the assumption of steady response of the atoms to idealized monochromatic radiation impinging on uniform matter that has no resolvable granularity. Under these circumstances one can model the matter as a collection of individual atoms, each of which responds to the continuous-wave field but which are sampled by the field as though distributed continuously. The formalism of scattering theory [Sho67; Sho69; Ish78] then provides a satisfactory description of the refractive index, and presents the frequency dependence of the attenuation as a succession of distinct resonances – the dark spectral lines seen in absorption spectra.

All such approaches to the modeling of the refractive index are based on steady-state descriptions of atomic response that fail for rapidly changing fields, such as the first few cycles of a pulse. To treat such situations it is necessary to model the detailed time-dependent changes induced in individual atoms by the field – to solve equations of motion for the bound electrons that are responsible for the macroscopic properties of matter. It is these equations, particularly the TDSE, with which the present monograph deals.

C.4.3 Tensor properties; anisotropy

The susceptibility and dielectric constant introduced above as scalars are, for some materials, tensors, as introduced through the relationships amongst Cartesian components χ_{ij} of the *susceptibility tensor* or ϵ_{ij} of the *dielectric tensor*,

$$P_i^{(+)} = \sum_j \epsilon_0 \chi_{ij} E_j^{(+)} = \sum_j \epsilon_0 [\epsilon_{ij} - \delta_{ij}] E_j^{(+)}, \qquad i,j = x,y,z. \tag{C.83}$$

The off-diagonal elements of these tensors are responsible for the anisotropic optical properties of crystals, in which there occur differing propagation velocities along different crystal axes. They may include nonlinear effects.

C.5 Pulse-mode expansions

The transverse electric field within nonmagnetic matter satisfies the inhomogeneous wave equation

$$\left[\nabla^2 - \frac{1}{c^2}\frac{\partial^2}{\partial t^2}\right] \mathbf{E}^{(\pm)}(\mathbf{r},t) = \frac{1}{\epsilon_0 c^2}\frac{\partial^2}{\partial t^2}\mathbf{P}^{(\pm)}(\mathbf{r},t), \tag{C.84}$$

in which the polarization field $\mathbf{P}$ expresses the macroscopic density of dipole moments.

Our objective will be the description of solenoidal traveling-wave solutions to the transverse inhomogeneous wave equation. We are particularly interested in pulses of laser radiation, i.e. radiation that has a well-defined direction of propagation and carrier frequency. We therefore introduce a set of appropriate pulsed-mode fields to express the electric field at position $\mathbf{r}$ as a sum of discrete traveling waves characterized by unit vectors $\mathbf{e}_\lambda$, envelopes $\mathcal{E}_\lambda$, and phases ξ_λ, each identified by a discrete label λ,

$$\mathbf{E}^{(+)}(\mathbf{r},t) = \sum_\lambda \mathbf{E}_\lambda^{(+)}(\mathbf{r},t) = \frac{1}{2}\sum_\lambda \mathbf{e}_\lambda \mathcal{E}_\lambda(\mathbf{r},\tau)\exp(i\xi_\lambda), \qquad \text{(C.85a)}$$

with a similar expansion for the polarization field,

$$\mathbf{P}^{(+)}(\mathbf{r},t) = \sum_\lambda \mathbf{P}_\lambda^{(+)}(\mathbf{r},t) = \frac{1}{2}\sum_\lambda \mathbf{e}_\lambda \mathcal{P}_\lambda(\mathbf{r},\tau)\exp(i\xi_\lambda). \qquad \text{(C.85b)}$$

The dominant dependence upon position $\mathbf{r}$ and time t is presumed to be contained in the carrier-wave phase $\xi_\lambda \equiv \xi_\lambda(\mathbf{r},t)$. The carrier propagation vector (or wavevector) $\mathbf{k}_\lambda$ and the (instantaneous) frequency ω_λ are defined as space- and time-derivatives of the phase ξ_λ:

$$\mathbf{k}_\lambda = \nabla\xi_\lambda, \qquad \omega_\lambda = -\frac{\partial}{\partial t}\xi_\lambda. \qquad \text{(C.86)}$$

The pulse envelopes $\mathcal{E}_\lambda(\mathbf{r},\tau)$ and $\mathcal{P}_\lambda(\mathbf{r},\tau)$, possibly complex, contain the remainder of the time and space dependence, assumed to be slowly varying. We require the amplitude functions to be such that each separate mode λ satisfies the inhomogeneous wave equation (C.84)

$$\left[\nabla^2 - \frac{1}{c^2}\frac{\partial^2}{\partial t^2}\right]\mathbf{E}_\lambda^{(+)} = \frac{1}{\epsilon_0 c^2}\frac{\partial^2}{\partial t^2}\mathbf{P}_\lambda^{(+)}. \qquad \text{(C.87)}$$

C.5.1 The slowly varying envelope approximation (SVEA)

The simplest approach (and the one followed here) is to require that $\mathbf{k}_\lambda$ and ω_λ be independent of time and position, so that all second derivatives of the phase ξ_λ vanish:

$$\frac{\partial^2}{\partial t^2}\xi_\lambda = 0, \qquad \nabla^2\xi_\lambda = 0, \qquad \nabla\frac{\partial}{\partial t}\xi_\lambda = 0. \qquad \text{(C.88)}$$

That is, we write the carrier phase, just as with a Fourier decomposition in space and time, as

$$\xi_\lambda = \mathbf{k}_\lambda \cdot \mathbf{r} - \omega_\lambda t, \qquad \text{(C.89)}$$

where $\mathbf{k}_\lambda$ and ω_λ are independent of position and time but may depend on the mode. This assumption still leaves the precise relationship between the magnitude of the wavevector k_λ and ω_λ undefined; in general this connection can be written in the form

$$ck_\lambda = \eta_\lambda(\omega_\lambda)\omega_\lambda, \qquad \text{(C.90)}$$

thereby defining a (frequency-dependent) index of refraction $\eta(\omega)$, required at the value $\eta_\lambda = \eta(\omega_\lambda)$.

The derivatives needed in the wave equation involve two functions: the exponentiated phase ξ_λ, and the envelope functions $\mathcal{E}_\lambda(\mathbf{r}, t)$ and $\mathcal{P}_\lambda(\mathbf{r}, t)$. To apply eqn. (C.87) we require the results

$$\nabla^2 \mathcal{E}_\lambda(\mathbf{r}, t) \exp(i\xi_\lambda) = \exp(i\xi_\lambda) \left[-(k_\lambda)^2 + 2i\mathbf{k}_\lambda \cdot \nabla + \nabla^2 \right] \mathcal{E}_\lambda(\mathbf{r}, t),$$

$$\frac{\partial^2}{\partial t^2} \mathcal{E}_\lambda(\mathbf{r}, t) \exp(i\xi_\lambda) = \exp(i\xi_\lambda) \left[-(\omega_\lambda)^2 - 2i\omega_\lambda \frac{\partial}{\partial t} + \frac{\partial^2}{\partial t^2} \right] \mathcal{E}_\lambda(\mathbf{r}, t), \qquad \text{(C.91)}$$

$$\frac{\partial^2}{\partial t^2} \mathcal{P}_\lambda(\mathbf{r}, t) \exp(i\xi_\lambda) = \exp(i\xi_\lambda) \left[-(\omega_\lambda)^2 - 2i\omega_\lambda \frac{\partial}{\partial t} + \frac{\partial^2}{\partial t^2} \right] \mathcal{P}_\lambda(\mathbf{r}, t).$$

We require that the dominant spatial and temporal variations should occur in the exponential carriers; specifically we require that the envelopes change slowly over an optical cycle or a wavelength (the *slowly varying envelope approximation* (SVEA)), as expressed by the inequalities (recall App. C.24),

$$\left| \frac{\partial}{\partial z} \mathcal{F} \right| \ll |k\mathcal{F}|, \qquad \left| \frac{\partial}{\partial t} \mathcal{F} \right| \ll |\omega\mathcal{F}|, \qquad \text{(C.92)}$$

where $\mathcal{F}$ is any of the envelope functions $\mathcal{P}_\lambda$ or $\mathcal{E}_\lambda$. In this approximation we neglect the second derivatives (but not the first derivatives), and obtain the equation

$$\left[\mathbf{k}_\lambda \cdot \nabla + \frac{\omega_\lambda}{c^2} \frac{\partial}{\partial t} - \frac{i}{2c^2} [(\omega_\lambda)^2 - (ck_\lambda)^2] \right] \mathcal{E}_\lambda(\mathbf{r}, t) = \frac{\omega_\lambda}{2c^2\epsilon_0} \left[\omega_\lambda + 2i \frac{\partial}{\partial t} \right] i \mathcal{P}_\lambda(\mathbf{r}, t). \qquad \text{(C.93)}$$

The first derivatives of $\mathcal{P}_\lambda$ are here retained for consistency in treating linear response.

The polarization fields $\mathcal{P}_\lambda$ have two effects: they alter electric field envelopes that are already present (by modifying the group velocity and introducing absorption) and they can serve as sources of fields that were not originally present. It is useful to consider separately the effects that occur transiently, during the relatively short time that a pulsed laser field is changing, and the effects that are present for long time intervals, during which the atoms reach a steady-state equilibrium with any fields. They then produce a linear response.

We separate the linear and nonlinear response by writing

$$\mathcal{P}_\lambda(z, t) = \epsilon_0 \chi_\lambda \mathcal{E}_\lambda(z, t) + {}^{nl}\mathcal{P}_\lambda(z, t) \qquad \text{(C.94)}$$

and we consider plane-wave propagation along the z axis. The equation then becomes, with $k_\lambda = \check{\eta}_\lambda \omega_\lambda / c$,

$$\left[\pm \frac{\partial}{\partial z} + \frac{[1 + \chi_\lambda]}{\check{\eta}_\lambda} \frac{1}{c} \frac{\partial}{\partial t} - i\omega_\lambda \frac{[1 + \chi_\lambda - (\check{\eta}_\lambda)^2]}{2c\check{\eta}_\lambda} \right] \mathcal{E}_\lambda(z, t)$$

$$= i \frac{1}{2c\check{\eta}_\lambda \epsilon_0} \left[\omega_\lambda + 2i \frac{\partial}{\partial t} \right] {}^{nl}\mathcal{P}_\lambda(z, t). \qquad \text{(C.95)}$$

where the $\pm$ sign refers to forward traveling waves (plus) or backward waves (minus). We use the relationship

$$(\check{\eta}_\lambda)^2 = 1 + \chi_\lambda \tag{C.96}$$

and obtain the mode-field envelope propagation equation

$$\left[\pm \frac{\partial}{\partial z} + \frac{\check{\eta}_\lambda}{c} \frac{\partial}{\partial t} \right] \mathcal{E}_\lambda(z,t) = +\mathrm{i} \frac{\omega_\lambda}{2c\check{\eta}_\lambda \epsilon_0} {}^{nl}\mathcal{P}_\lambda(z,t). \tag{C.97}$$

This is the key equation for treating propagation of the electric field envelope and, in turn, propagation of the Rabi frequency. The atomic effects upon the field appear both as an altered, and complex-valued, refractive index, and as a polarization envelope that expresses deviations from the linear response.

To proceed we require a prescription for the time-varying induced dipole moment, from which to evaluate the electric polarization field. When the atomic response is weak, the induced time-varying dipole moment is proportional to the incident electric field. The effects then are observable through attenuation and can be incorporated into an index of refraction. Larger responses can occur when the field is resonant with a transition linked to the ground state of the atoms.

As the field becomes stronger it will produce significant coherent excitation. This, in turn, will produce a variety of nonlinear effects in the propagation equation. In particular, if the atoms are prepared in a coherent superposition of ground and excited states, then a number of unusual effects can occur [Scu92; Pas02; Har97b; Fle05]. Chapter 21 mentions some.

Failure of the SVEA. The SVEA is of use for pulses whose duration extends over many optical cycles, and for which numerical integration over separate cycles is therefore impractical. At one time, pulses that were shorter than any relaxation time – and for which coherence was therefore a dominant feature – were known as *ultrashort pulses*. Subsequent techniques have made possible few-cycle pulses, in which there are only two or three cycles of the optical frequency. Obviously the SVEA and RWA are not appropriate for treating excitation by such pulses.

C.5.2 Group velocity

When the response of the atoms to the field can be approximated by a linear response, through the susceptibility χ_λ, then in the absence of a nonlinear polarization field the envelope equation for a forward-traveling wave reads

$$\left[\frac{\partial}{\partial z} + \frac{\check{\eta}_\lambda}{c} \frac{\partial}{\partial t} \right] \mathcal{E}_\lambda(z,t) = 0. \tag{C.98}$$

This equation has solutions of the form

$$\mathcal{E}_\lambda(z,t) = F(z - ct/\check{\eta}_\lambda), \tag{C.99}$$

where $F(\xi)$ is any complex-valued function of its argument ξ. This represents an envelope function that moves steadily, without changing shape, toward larger z, with the *group velocity*

$$v_\lambda = \left(\frac{\partial k_\lambda}{\partial \omega}\right)^{-1} = \frac{c}{\check{\eta}_\lambda} = \frac{c}{\sqrt{1+\chi_\lambda}}, \qquad (\text{C}.100)$$

set by the refractive index or the susceptibility, evaluated at the carrier frequency ω_λ. The imaginary part of the complex-valued refractive index $\check{\eta}_\lambda$ is responsible for attenuating the envelope.

Appendix D
Quantized radiation

To treat particles and fields consistently within quantum theory it is necessary that, in addition to treating positions and momenta of the particles as noncommuting operators, the electric and magnetic fields must also be regarded as being noncommuting operators $\hat{\mathbf{E}}$ and $\hat{\mathbf{B}}$. The following sections, based on [Sho90, Chap. 9], outline the conventional approach to the resulting theory, quantum electrodynamics (QED) [Bia75; Coh89], and the consequent quantization of the radiation field [Fre06].

Free space. The notion of radiation as a dynamical system rests on the idealization of electromagnetic fields in the absence of any matter – far outside the distributions of charges and currents that are the source of the radiation. A satisfactory starting point for subjecting the radiation field to constraints of quantum mechanics is the free-space wave equation together with the solenoidal constraint that ensures a transverse field. For the electric field these equations are

$$\left[\nabla^2 - \frac{1}{c^2}\frac{\partial^2}{\partial t^2}\right]\hat{\mathbf{E}}(\mathbf{r}, t) = 0, \qquad \nabla \cdot \hat{\mathbf{E}}(\mathbf{r}, t) = 0. \tag{D.1}$$

The solutions are separable in space and time; we take the spatial part to be a classical field, leaving the time dependence to become quantum-mechancial operators. Specifically, we consider a field of frequency ω and write it as the product of a dimensionless spatial (vector) function $\mathbf{U}(\mathbf{r})$, a dimensionless time variation $\hat{a}(t)$, and a complex-valued numerical factor $\breve{\mathcal{E}}$ that provides dimensions for the electric field. Upon separating the field into positive- and negative-frequency parts,

$$\hat{\mathbf{E}}(\mathbf{r}, t) = \hat{\mathbf{E}}^{(+)}(\mathbf{r}, t) + \hat{\mathbf{E}}^{(-)}(\mathbf{r}, t), \tag{D.2}$$

we use the construction

$$\hat{\mathbf{E}}^{(+)}(\mathbf{r}, t) = \breve{\mathcal{E}}\,\mathbf{U}(\mathbf{r})\,\hat{a}(t), \qquad \hat{\mathbf{E}}^{(-)}(\mathbf{r}, t) = \breve{\mathcal{E}}^*\,\mathbf{U}(\mathbf{r})^*\,\hat{a}^\dagger(t). \tag{D.3}$$

A comparable construction holds for the magnetic field. We require that the vector field $\mathbf{U}(\mathbf{r})$ obey the Helmholtz equation

$$[\nabla^2 + (\omega/c)^2]\mathbf{U}(\mathbf{r}) = 0. \tag{D.4}$$

It follows that the time-dependent factors obey the harmonic oscillator equations

$$\left[\frac{d^2}{dt^2}+\omega^2\right]\hat{a}(t)=0, \qquad \left[\frac{d^2}{dt^2}+\omega^2\right]\hat{a}^\dagger(t)=0. \tag{D.5}$$

It is the quantities $\hat{a}(t)$ and $\hat{a}^\dagger(t)$ that describe the field dynamics; it is these variables that must incorporate the quantum-mechanical characteristics of the field.

D.1 Field quantization

Fields as oscillators. The application of quantum theory to the radiation field proceeds by making a connection with the traditional one-dimensional harmonic oscillator of mass m, coordinate $\hat{q}$, and conjugate momentum $\hat{p}$. The Hamiltonian for that system is

$$\hat{H}^{osc}=\frac{\hat{p}(t)^2}{2m}+\frac{m}{2}\omega^2\hat{q}(t)^2. \tag{D.6}$$

From this Hamiltonian come the equations of motion (Hamilton's equations)

$$\frac{d}{dt}\hat{q}(t)=\frac{1}{m}\hat{p}(t), \qquad \frac{d}{dt}\hat{p}(t)=-m\omega^2\hat{q}(t), \tag{D.7}$$

recognizable as the harmonic oscillator equations

$$\left[\frac{d^2}{dt^2}+\omega^2\right]\hat{q}(t)=0, \qquad \left[\frac{d^2}{dt^2}+\omega^2\right]\hat{p}(t)=0. \tag{D.8}$$

The required connection between the single-coordinate harmonic oscillator and the single-mode electromagnetic field comes by making the identification

$$\hat{a}(t)=\frac{1}{\sqrt{2m\hbar\omega}}[m\omega\hat{q}(t)+\mathrm{i}\,\hat{p}(t)], \qquad \hat{a}^\dagger(t)=\frac{1}{\sqrt{2m\hbar\omega}}[m\omega\hat{q}(t)-\mathrm{i}\,\hat{p}(t)]. \tag{D.9}$$

The parameter m appearing here is only needed to provide appropriate units; it can be regarded as an arbitrary constant.

Quantization. Once a system is described by Hamilton's equations, through generalized coordinates and momenta, quantum theory is invoked by requiring that each position and its associated momentum should satisfy equal-time commutation relations

$$[\hat{q}(t),\,\hat{p}(t)]=\mathrm{i}\,\hbar. \tag{D.10}$$

This requirement should hold for both a massive particle and for the electromagnetic field mode. For the latter the quantum condition becomes the equal-time commutator

$$[\hat{a}(t),\hat{a}^\dagger(t)]\equiv\hat{a}(t)\hat{a}^\dagger(t)-\hat{a}^\dagger(t)\hat{a}(t)=1. \tag{D.11}$$

We take the time dependence of the free-field operators to be in exponentials,

$$\hat{a}(t)=\hat{a}\exp(-\mathrm{i}\omega t), \qquad \hat{a}^\dagger(t)=\hat{a}^\dagger\exp(\mathrm{i}\omega t), \tag{D.12}$$

leaving the commutation properties, and hence the quantum nature of the field, to be carried by the time-independent operators $\hat{a}$ and $\hat{a}^\dagger$. These must obey the commutation relations

$$[\hat{a}, \hat{a}^\dagger] = 1. \qquad (D.13)$$

It is with this relationship that the field of eqn. (D.3) becomes a quantum field.

D.1.1 Photon-number states

Having ensured that the electric field has characteristics consistent with quantum theory, i.e. it is *quantized*, we next identify suitable basis states. Historically these were taken to be eigenstates of the *number operator*

$$\hat{N} = \hat{a}^\dagger \hat{a}. \qquad (D.14)$$

This operator has, as eigenvalues, the non-negative integers $n = 0, 1, 2, \ldots$ that serve as labels for the eigenstates of the number operator, denoted as $|n\rangle$,

$$\hat{N}|n\rangle = n|n\rangle. \qquad (D.15)$$

Eigenstates of the Hermitian operator $\hat{N}$ that have different eigenvalues are orthogonal; we take them to be normalized,

$$\langle n|n'\rangle = \delta_{nn'}. \qquad (D.16)$$

The action of the operators $\hat{a}$ and $\hat{a}^\dagger$ on such states is

$$\hat{a}|n\rangle = \sqrt{n}\,|n-1\rangle, \qquad \hat{a}^\dagger|n\rangle = \sqrt{n+1}\,|n+1\rangle. \qquad (D.17)$$

That is, the operator $\hat{a}^\dagger$ increases by one the number of field increments n (it is a *creation operator*) while the operator $\hat{a}$ diminishes this number by one (it is an *annihilation operator*). The set of photon states $|n\rangle$ for a single mode form an infinite-dimensional Hilbert space known as a (single-mode) *Fock space*.

Multiple modes. The solutions to the Helmholtz equation become completely defined only with the specification of spatial boundary conditions, typically that the field should vanish over some surfaces in Cartesian, spherical, or cylindrical coordinates. Each set of boundary conditions leads to a set of orthogonal fields, labeled by some set of indices. These are *mode fields* $\mathbf{U}_\lambda(\mathbf{r})$, each of which satisfies the Helmholtz equation,

$$[\nabla^2 + (\omega_\lambda/c)^2]\mathbf{U}_\lambda(\mathbf{r}) = 0. \qquad (D.18)$$

Using these mode fields we express the (quantized) electric field as

$$\hat{\mathbf{E}}^{(+)}(\mathbf{r}, t) = \sum_\lambda \check{\mathcal{E}}_\lambda \mathbf{U}_\lambda(\mathbf{r})\hat{a}_\lambda \exp(-i\omega_\lambda t) = \left[\hat{\mathbf{E}}^{(-)}(\mathbf{r}, t)\right]^\dagger. \qquad (D.19)$$

Each mode of the field, identifiable by label λ in eqn (D.19), has time-dependent operators $\hat{a}_\lambda(t)$ and $\hat{a}_\lambda^\dagger(t)$ that embody the dynamics. Their commutators generalize eqn. (D.11),

$$[\hat{a}_\lambda(t), \hat{a}_{\lambda'}^\dagger(t)] = \delta_{\lambda\lambda'}, \qquad (D.20)$$

and the associated number states $|n_\lambda\rangle$ define single-mode fields.

The next few paragraphs consider just one of these mode fields. Starting with Sec. D.2 the discussion treats details of the mode fields and multiple modes, specifically plane waves and spherical waves.

The vacuum. For the electromagnetic field the eigenvalue n, which denotes the number of excitation quanta of the single-mode harmonic oscillator, has the interpretation of photon number: the state $|n\rangle$ is a field state, of the mode field $\mathbf{U}(\mathbf{r})$, in which there are n photons. The operator $\hat{a}$ removes (annihilates) one photon while the operator $\hat{a}^\dagger$ increases the number of photons by one (it creates a photon). The lowest eigenvalue $n = 0$ defines a (photon) *vacuum state* $|0\rangle$ in which no photons are present. This state has the properties

$$\hat{a}|0\rangle = 0, \qquad \hat{a}^\dagger|0\rangle = |1\rangle. \tag{D.21}$$

The photon creation and annihilation operators provide the dynamical description of the electromagnetic field. Their use in eqns. (D.3) or (D.19) means that the positive-frequency part of the field diminishes the number of photons whereas the negative-frequency part increments the field by one photon in each accessed mode. The photon vacuum state, $|vac\rangle$, defined as the terminus of all photon decrements, therefore is defined by the equation

$$\hat{\mathbf{E}}^{(+)}(\mathbf{r}, t)|vac\rangle = 0. \tag{D.22}$$

The vacuum state $|vac\rangle$ has no photons in any mode, and so it may be expressed as a product of null-photon states $|0_\lambda\rangle$ for a complete set of modes. The state produced from the photon vacuum by the action of the negative-frequency part, $\hat{\mathbf{E}}^{(-)}(\mathbf{r}, t)|vac\rangle$, is therefore a one-photon state. Its properties depend upon the amplitudes $\mathcal{E}_{1\lambda}$ that specify the field and the choice of mode fields $\mathbf{U}_\lambda(\mathbf{r})$ chosen for a basis. Because the fields have quantum properties, the ordering of positive- and negative-frequency parts matters:

$$\hat{\mathbf{E}}^{(+)}(\mathbf{r}, t)\hat{\mathbf{E}}^{(-)}(\mathbf{r}, t) \neq \hat{\mathbf{E}}^{(-)}(\mathbf{r}, t)\hat{\mathbf{E}}^{(+)}(\mathbf{r}, t). \tag{D.23}$$

Expectation values. Because the operators $\hat{a}$ and $\hat{a}^\dagger$ change the number of photons, it follows that when the system is in a photon-number state $|n\rangle$, these operators have null expectation value,

$$\langle n|\hat{a}|n\rangle = 0, \qquad \langle a|\hat{a}^\dagger|n\rangle = 0, \tag{D.24}$$

as do all powers of these operators,

$$\langle n|\hat{a}\hat{a}\cdots|n\rangle = 0, \qquad \langle n|\hat{a}^\dagger\hat{a}^\dagger\cdots|n\rangle = 0. \tag{D.25}$$

The photon-number states are easily used in calculations, and they were used almost exclusively for representing fields until laser fields became of interest. However, they have noticeable nonclassical properties. For example, when the system is in a number state $|n\rangle$ the expectation value of the electric field of eqn. (D.3) vanishes,

$$\langle n|\hat{\mathbf{E}}(\mathbf{r}, t)|n\rangle = 0, \tag{D.26}$$

although the expectation value of $|\hat{\mathbf{E}}(\mathbf{r}, t)|^2$, and hence the intensity, does not vanish,

$$\langle n | \, |\hat{\mathbf{E}}(\mathbf{r}, t)|^2 |n\rangle = n |\mathbf{U}(\mathbf{r}, t)|^2 \, |\check{\mathcal{E}}|^2. \tag{D.27}$$

Furthermore the field phase, a variable conjugate to the photon number, is not specified. To obtain a quantum state in which the field expectation value does not vanish and in which phase can be determined it is necessary to take coherent superpositions of number states as in App. D.1.2.

Correlation functions for quantum fields. The statistical properties of fields are characterizable by various correlation functions, as discussed in Sec. C.3.2. Extension to a quantum mechanical formulation of radiation involves the replacement of classical ensemble (or time) averages $\{\cdots\}_{av}$ by quantum-mechanical expectation values $\langle\cdots\rangle$. The Glauber tensor correlation function of order n [Gla63b], is

$$G^{(n)}_{i\ldots j\ldots k}(x_1\cdots x_{2n}) = \langle \hat{E}^{(-)}_i(x_1)\cdots \hat{E}^{(-)}_j(x_n)\hat{E}^{(+)}_{j+1}(x_{n+1})\cdots \hat{E}^{(+)}_k(x_{2n})\rangle. \tag{D.28}$$

Here x_n denotes a space-time argument $\mathbf{r}_n t_n$ and the subscripts $i\ldots j\ldots k$ denote Cartesian components of the vectors. The field operators appearing inside the expectation value are *normally ordered*: all positive-frequency parts (and hence all annihilation operators) appear to the right of any negative-frequency parts (and their creation operators).

When the system (field or field plus atoms) is in a pure quantum state $\Psi(t)$, then we can evaluate the expectation value by taking matrix elements of field operators. For example, the first-order correlation function is

$$G^{(1)}_{ij}(x_1, x_2) = \langle \Psi(t)|\hat{E}^{(-)}_i(x_1)\hat{E}^{(+)}_j(x_2)|\Psi(t)\rangle. \tag{D.29a}$$

More generally, we require a density matrix $\hat{\rho}$ to describe the system. The needed algorithm then becomes

$$G^{(1)}_{ij}(x_1, x_2) = \mathrm{Tr}\left[\hat{\rho}\hat{E}^{(-)}_i(x_1)\hat{E}^{(+)}_j(x_2)\right]. \tag{D.29b}$$

D.1.2 Coherent states

It follows from the basic construction of eqn (D.19) and the properties of photon annihilation and creation operators that the electric field has null expectation value in a state with definite photon number. To circumvent this disadvantage of photon-number states one can introduce eigenstates of the positive-frequency field $\hat{\mathbf{E}}^{(+)}(\mathbf{r}, t)$, say

$$\hat{\mathbf{E}}^{(+)}(\mathbf{r}, t)|\cdots\rangle = {}^c\mathbf{E}(\mathbf{r}, t)|\cdots\rangle, \tag{D.30}$$

where ${}^c\mathbf{E}(\mathbf{r}, t)$ is a complex-valued classical vector field – a solution to Maxwell's equations in free space. This we can accomplish by introducing single-mode eigenstates $|\alpha_\lambda\rangle$ of the annihilation operator $\hat{a}_\lambda$,

$$\hat{a}_\lambda|\alpha_\lambda\rangle = \alpha_\lambda|\alpha_\lambda\rangle. \tag{D.31}$$

This definition leads to the complex-valued classical field

$$^c\mathbf{E}(\mathbf{r}, t) = \sum_\lambda \check{\mathcal{E}}_\lambda \mathbf{U}_\lambda(\mathbf{r}) \alpha_\lambda \exp(-i\omega_\lambda t) \tag{D.32}$$

as the eigenvalue of the positive-frequency field operator $\hat{\mathbf{E}}^{(+)}(\mathbf{r}, t)$. The annihilation-operator eigenvalues α_λ are essentially Fourier coefficients in a mode expansion of the positive-frequency part of a classical field.

Such states were first used by Schrödinger [Sch26]. Their contemporary application in quantum optics was pioneered by Glauber [Gla63a], and so they are sometimes termed *Glauber states*. More commonly they are termed *coherent states*. This latter name expresses an important property of the nth order correlation function of the field, eqn. (D.28). For coherent states this correlation function reduces to the product of $2n$ separate factors

$$G^{(n)}_{i_1\ldots j\ldots k}(x_1 \cdots x_{2n}) = E_i^*(x_1) \cdots E_j^*(x_n) E_{j+1}(x_{n+1}) \cdots E_k(x_{2n}). \tag{D.33}$$

Such properties make these states *fully coherent*.

Single-mode coherent states. The Glauber coherent states [Gla63a; Kib68; Per86; Buz90; Jex95], are defined, for complex-valued parameter α, as eigenstates of the annihilation operator $\hat{a}$,

$$\hat{a}|\alpha\rangle = \alpha|\alpha\rangle. \tag{D.34}$$

The eigenvalue α that labels the coherent state is a complex number whose magnitude squared represents the mean photon number $\bar{n}$ (related in turn to intensity) and whose complex value provides information on the phase of the wave, ϕ:

$$\alpha = \sqrt{\bar{n}} \exp(i\phi). \tag{D.35}$$

These states can be constructed from the number states as the summation

$$|\alpha\rangle = \sum_{n=0}^{\infty} |n\rangle \frac{\alpha^n}{\sqrt{n!}} \exp(-\tfrac{1}{2}|\alpha|^2). \tag{D.36}$$

That is, we can readily verify that the action of the operator $\hat{a}$ upon this sum is equivalent to multiplication by α, and that the construction gives a state with unit norm:

$$\langle\alpha|\alpha\rangle = 1. \tag{D.37}$$

Although the coherent states are normalized, they are not orthogonal; their overlap is

$$\langle\alpha|\beta\rangle = \exp\left(\alpha^*\beta - \tfrac{1}{2}|\alpha|^2 - \tfrac{1}{2}|\beta|^2\right). \tag{D.38}$$

From this expression we recognize that states with very disparate values of α are, for practical purposes, orthogonal. The lack of exact orthogonality occurs because the coherent states are *overcomplete*: there are more coherent states than there are number states (the number states are denumerable and the continuum of coherent states is nondenumerable), and the number states are already complete.

The Glauber transformation. The construction formula of eqn. (D.36) exhibits a single-mode coherent state as a Poisson distribution of photon numbers about a mean value $\bar{n} = |\alpha|^2$. The coherent state $|\alpha\rangle$ is obtained from the photon vacuum state $|0\rangle$ by multiple application of creation operators using the formula

$$|n\rangle = \frac{(\hat{a}^\dagger)^n}{\sqrt{n!}}|0\rangle. \tag{D.39}$$

We can then recognize the summation of eqn. (D.36) as the expansion produced by an exponential operator $\hat{D}(\alpha)$ acting on the vacuum state,

$$|\alpha\rangle = \exp(\alpha\hat{a}^\dagger)\exp(-\tfrac{1}{2}|\alpha|^2)|0\rangle \equiv \hat{D}(\alpha)|0\rangle. \tag{D.40}$$

This equation expresses the production of the coherent state $|\alpha\rangle$ from the photon-number vacuum state $|0\rangle$ by means of an operator that displaces the eigenvalue from the (vacuum) value 0 to the value α. This displacement operator, often termed the *Glauber transformation*, $\hat{D}(\alpha)$, is commonly written

$$\hat{D}(\alpha) = \exp(\alpha\hat{a}^\dagger - \alpha^*\hat{a}). \tag{D.41}$$

The equivalence of these two forms for $\hat{D}(\alpha)$ rests upon application of the *Baker–Hausdorff identity*,

$$\exp(\hat{A})\exp(\hat{B}) = \exp\left(\hat{A} + \hat{B} + \tfrac{1}{2}[\hat{A}, \hat{B}]\right), \tag{D.42}$$

valid for operators $\hat{A}$ and $\hat{B}$ that commute with their commutator $[\hat{A}, \hat{B}]$.

Coherent-state expectation values. For a coherent state $|\alpha\rangle$ the creation and annihilation operators have the expectation values

$$\langle\alpha|\hat{a}|\alpha\rangle = \alpha, \qquad \langle\alpha|\hat{a}^\dagger|\alpha\rangle = \alpha^*, \tag{D.43}$$

and the number operator has the expectation value

$$\bar{n} \equiv \langle\alpha|\hat{N}|\alpha\rangle = \langle\alpha|\hat{a}^\dagger\hat{a}|\alpha\rangle = |\alpha|^2. \tag{D.44}$$

This result means that the expectation value of the field defined by eqn. (D.3) is

$$\langle\alpha|\hat{\mathbf{E}}(\mathbf{r}, t)|\alpha\rangle = \tfrac{1}{2}\sqrt{\bar{n}}\,\breve{\mathcal{E}}\,\mathbf{U}(\mathbf{r})\,e^{i\phi} + \tfrac{1}{2}\sqrt{\bar{n}}\,\breve{\mathcal{E}}^*\,\mathbf{U}(\mathbf{r})^*\,e^{-i\phi}. \tag{D.45}$$

Thus the expectation value of the field, in a coherent state, is a classical field. We choose the value α to fix the amplitude and phase of this classical field by the requirement that the positive-frequency part of the single-mode classical field be

$$^c\mathbf{E}(t)(+) = \mathcal{E}\alpha\exp(-i\omega t + \phi), \tag{D.46}$$

where $\mathcal{E}$ is the single-photon field strength for the particular mode. For a cubic cavity it is $\mathcal{E} = \sqrt{2\hbar\omega/\epsilon_0 L^3}$; more generally it is obtainable from the expression

$$\mathcal{E}_\lambda = \sqrt{2\hbar\omega_\lambda/\epsilon_0}|\mathbf{U}_\lambda(0)|. \tag{D.47}$$

Quite generally we have the exact result

$$\langle \alpha | \hat{\mathbf{E}}(\mathbf{r}, t) | \alpha \rangle = {}^c\mathbf{E}(\mathbf{r}, t), \qquad (D.48)$$

where ${}^c\mathbf{E}(\mathbf{r}, t)$ is a classical (unquantized) field. Fields expressed by coherent states are the fields that are generated by classical currents [Kib68].

Examples. For example, the field operator for a monochromatic traveling plane wave

$$\hat{\mathbf{E}}(z, t) = \tfrac{1}{2}\mathbf{e}(x)\mathcal{E}\left[\exp(-\mathrm{i}kz)\hat{a}^\dagger(t) + \exp(\mathrm{i}kz)\hat{a}(t)\right] \qquad (D.49)$$

has the expectation value

$$\langle \alpha | \hat{\mathbf{E}}(z, t) | \alpha \rangle = \mathbf{e}(x)\mathcal{E}\sqrt{\bar{n}}\cos(kz - \omega t + \phi). \qquad (D.50)$$

Similarly for the quantized standing wave

$$\hat{\mathbf{E}}(z, t) = \mathbf{e}(x)\mathcal{E}\sin(kz)\left[\hat{a}(t) + \hat{a}^\dagger(t)\right]/\sqrt{2} \qquad (D.51)$$

we obtain the formula

$$\langle \alpha | \hat{\mathbf{E}}(z, t) | \alpha \rangle = \mathbf{e}(x)\mathcal{E}\sqrt{\bar{n}/2}\sin(kz)\cos(\omega t - \phi). \qquad (D.52)$$

These examples illustrate how expressions used in Chap. 3, presented for classical fields, remain applicable for a quantized field.

D.1.3 Phase states

Photon number and phase cannot both be sharply defined simultaneously for a quantum field: a state for which there is a definite number of photons has a completely undetermined phase. The construction of a photon-number operator is an immediate consequence of the basic quantization procedure. To construct a Hermitian operator whose eigenvalues correspond to observable phases other approaches are necessary. The following paragraphs summarize one of these.

Photon-increment states. To treat laser pulses we require states with large mean photon number $\bar{n}$, so that the expectation value of the photon-number operator $\hat{N} = \hat{a}^\dagger\hat{a}$ is

$$\langle \hat{a}^\dagger\hat{a} \rangle = \bar{n} \gg 1. \qquad (D.53)$$

For this purpose it is useful to express the photon number n, the eigenvalue of $\hat{N}$, as an increment m to the mean number $\bar{n}$. That is, we let

$$n = \bar{n} + m, \qquad (D.54)$$

and we write a photon-number statevector as

$$|n\rangle = |\bar{n}, m\rangle. \qquad (D.55)$$

Our interest lies with those states for which the increment m is much smaller than the mean photon number, $m \ll \bar{n}$. We next introduce a phase ϕ as an alternative to the number increment m, and we introduce (incremental) *phase states* $|\bar{n}, \phi\rangle$ as an alternative to the basis states $|\bar{n}, m\rangle$ through the definitions [Car68; Bia76; Ber91]

$$|\bar{n}, \phi\rangle = \sum_{m=-\bar{n}}^{\infty} |\bar{n}, m\rangle \frac{\exp(\mathrm{i}m\phi)}{\sqrt{2\pi}}, \qquad |\bar{n}, m\rangle = \int_0^{2\pi} d\phi |\bar{n}, \phi\rangle \frac{\exp(-\mathrm{i}m\phi)}{\sqrt{2\pi}}. \tag{D.56}$$

The connection between these representations gives us a photon wavefunction, a function of the phase ϕ:

$$\langle \phi | m \rangle = \frac{\exp(-\mathrm{i}m\phi)}{\sqrt{2\pi}}. \tag{D.57}$$

This photon wavefunction has some similarities with the wavefunctions one uses for freely moving point particles (i.e. linear-momentum wavefunctions) although the photon wavefunction is a function of phase, rather than position. Like the momentum state of a particle this function is not localizable.

Operators in the phase-increment representation. Just as with the point particle, where we represent the momentum operator $\hat{p}$ acting upon a wavefunction as a differential operator $\hat{p} \rightarrow -\mathrm{i}\hbar\frac{\partial}{\partial x}$, so too can we represent the photon increment m as a differential operator acting upon functions of phase:

$$\hat{m} \rightarrow \mathrm{i}\frac{\partial}{\partial \phi}. \tag{D.58}$$

In turn we can represent the number operator as

$$\hat{N} = \hat{a}^\dagger \hat{a} \rightarrow \bar{n} + \mathrm{i}\frac{\partial}{\partial \phi} \tag{D.59}$$

and we can represent the annihilation and creation operators, acting on functions of ϕ and $\bar{n}$, as

$$\hat{a} \rightarrow \exp(\mathrm{i}\phi)\sqrt{\bar{n} + \mathrm{i}\frac{\partial}{\partial \phi}}, \qquad \hat{a}^\dagger \rightarrow \sqrt{\bar{n} + \mathrm{i}\frac{\partial}{\partial \phi}}\exp(-\mathrm{i}\phi). \tag{D.60}$$

These formulas maintain the commutator relationship $\left[\hat{a}, \hat{a}^\dagger\right] = 1$. They define a *phase-increment representation* for the quantized but free electromagnetic field – a representation in which the phase angle ϕ replaces the photon increment m about a large mean photon number $\bar{n}$. Note that not all values of m represent physical states: the increment m cannot be smaller than the negative of the mean number $\bar{n}$. For typical laser fields, with large $\bar{n}$, the increments m are small and this restriction poses no difficulty.

The large-$\bar{n}$ approximation. When $\bar{n}$ becomes large we can approximate the square-root operators by a series expansion in powers of $1/(\bar{n})$:

$$\sqrt{\bar{n} + i\frac{\partial}{\partial\phi}} = \sqrt{\bar{n}}\left[1 + \frac{i}{2\bar{n}}\frac{\partial}{\partial\phi} + \cdots\right]. \tag{D.61}$$

Often only the first term, $\sqrt{\bar{n}}$, need be retained. This large-$\bar{n}$ approximation yields the expressions

$$\hat{a} \to \exp(i\phi)\sqrt{\bar{n}}, \qquad \hat{a}^\dagger \to \sqrt{\bar{n}}\exp(-i\phi), \tag{D.62}$$

and, in turn, leads to the matrix elements

$$\langle\bar{n},\phi|\hat{a}|\bar{n},\phi\rangle = \sqrt{\bar{n}}\exp(i\phi). \tag{D.63}$$

This approximation yields operators $\hat{a}$ and $\hat{a}^\dagger$ that commute. Although the approximations of eqns. (D.62) are inadequate for the number operator $\hat{a}^\dagger\hat{a}$, which must include a phase derivative in order to maintain the correct commutation relation $[\hat{a},\hat{a}^\dagger] = 1$, they suffice for treating the field operators $\hat{\mathbf{E}}$ and $\hat{\mathbf{B}}$. The latter operators, being linearly dependent upon the elementary creation and annihilation operators, can be approximated by functions of $\bar{n}$ and ϕ, without including phase derivatives. As an example, the electric field operator

$$\hat{\mathbf{E}} = \tfrac{1}{2}(\mathbf{e}\mathcal{E}_1\hat{a} + \mathbf{e}^*\mathcal{E}_1^*\hat{a}^\dagger) \tag{D.64}$$

becomes, when acting on a wavefunction for a large-$\bar{n}$ state, the function

$$\mathbf{E}(\phi) = \tfrac{1}{2}\sqrt{\bar{n}}\left[\mathbf{e}\mathcal{E}_1\exp(i\phi) + \mathbf{e}^*\mathcal{E}_1^*\exp(-i\phi)\right]. \tag{D.65}$$

The combination $\mathcal{E}_1\exp(i\phi)$ represents a classical complex-valued electric field amplitude.

D.2 Mode fields

The traditional theoretical description of photons presented above rests upon the introduction of a set of normal modes for the electromagnetic field. These modes are complete sets of vector fields $\mathbf{U}_\lambda(\mathbf{r})$ that satisfy the vector Helmholtz equation,

$$\nabla\nabla\cdot\mathbf{U}_\lambda(\mathbf{r}) - \nabla\times\nabla\times\mathbf{U}_\lambda(\mathbf{r}) + (k_\lambda)^2\mathbf{U}_\lambda(\mathbf{r}) = 0, \tag{D.66}$$

together with suitable boundary conditions (typically a requirement for nodal surfaces along suitable coordinates), and are normalized by the requirement

$$\langle\lambda|\lambda'\rangle \equiv \int d\mathbf{r}\,\mathbf{U}_\lambda(\mathbf{r})^*\cdot\mathbf{U}_\lambda(\mathbf{r}) = \delta_{\lambda\lambda'}. \tag{D.67}$$

The label λ that identifies different solutions must specify both the spatial dependence of the field amplitude and the directional characteristics of the vector; it may include both discrete indices and continuous variables. The fields are complete, meaning that we can express the spatial parts of any electric or magnetic field as a superposition of these fields,

$$\mathbf{E}^{(+)}(\mathbf{r}) = \sum_\lambda \check{\mathcal{E}}_\lambda \mathbf{U}_\lambda(\mathbf{r}), \qquad \mathbf{B}^{(+)}(\mathbf{r}) = \sum_\lambda \check{\mathcal{B}}_\lambda \mathbf{U}_\lambda(\mathbf{r}). \tag{D.68}$$

To treat free-space radiation we require vector functions that are solenoidal (divergenceless),

$$\nabla \cdot \mathbf{U}_\lambda(\mathbf{r}) = 0, \tag{D.69}$$

but a complete set of vector fields must include lamellar fields as well.

Mode characterization. The vector Helmholtz equation is separable in several coordinate systems, and each of these provides a discrete set of analytically expressible mode fields, identifiable by integer labels that rank the number of nodes along one of the coordinates. When constructing mode fields for quantization it is useful to construct fields that are associated with a set of commuting operators. The Laplacian ∇^2 is one of these, used for the scalar spatial function as solutions to the scalar Helmholtz equation

$$\left[\nabla^2 + k^2\right] U_k(\mathbf{r}) = 0. \tag{D.70}$$

Because all vector fields are spin-one eigenstates the spin operator $\mathbf{S}^2$ is a second operator. Additional choices take linear or angular momentum operators to classify the spatial functions and spin projections for the vector parts. The following subsections describe some of the choices that have use in treating laser excitation.

D.2.1 Plane waves

Eigenstates of the differential operator

$$\nabla = \mathbf{e}_x \frac{\partial}{\partial x} + \mathbf{e}_y \frac{\partial}{\partial y} + \mathbf{e}_z \frac{\partial}{\partial z} \tag{D.71}$$

are scalar plane waves (standing or traveling). The scalar eigenvalue equation

$$\nabla U_{\mathbf{k}}(\mathbf{r}) = \mathrm{i}\mathbf{k} U_{\mathbf{k}}(\mathbf{r}) \tag{D.72}$$

has plane-wave solutions

$$U_{\mathbf{k}}(\mathbf{r}) = \langle \mathbf{r}|\mathbf{k}\rangle = \frac{1}{\sqrt{(2\pi)^3}} \exp(\mathrm{i}\mathbf{k} \cdot \mathbf{r}), \tag{D.73}$$

identified by the vector eigenvalue $\mathbf{k}$ that specifies the propagation axis; $\hbar\mathbf{k}$ is the momentum associated with a field increment, a plane-wave photon. These fields have the normalization of a Dirac delta,

$$\langle \mathbf{k}|\mathbf{k}'\rangle \equiv \int d\mathbf{r} \, U_{\mathbf{k}}(\mathbf{r}) * U_{\mathbf{k}'}(\mathbf{r}) = \delta(\mathbf{k} - \mathbf{k}') \equiv \delta(k_x - k_x')\delta(k_y - k_y')\delta(k_z - k_z'). \tag{D.74}$$

Vector properties. A simple procedure for constructing a basis of vector fields from scalar fields is to use Cartesian unit vectors $\mathbf{e}_x$, $\mathbf{e}_y$, and $\mathbf{e}_z$ aligned with a fixed laboratory reference frame, to produce the fields

$$\mathbf{U}_{j\mathbf{k}}(\mathbf{r}) = \mathbf{e}_j U_{\mathbf{k}}(\mathbf{r}), \quad j = x, y, z. \tag{D.75}$$

However, this construction does not allow separation into solenoidal and lamellar fields, as is necessary for describing free-space radiation fields. For that purpose we introduce a coordinate system fixed with respect to the propagation vector $\mathbf{k}$. We take this direction to define the first unit vector,

$$\mathbf{e}_Z(\hat{\mathbf{k}}) = \mathbf{k}/k \equiv \hat{\mathbf{k}}. \tag{D.76}$$

We complete a triad of orthonormal vectors by defining

$$\mathbf{e}_Y(\hat{\mathbf{k}}) = \hat{\mathbf{k}} \times \mathbf{e}, \qquad \mathbf{e}_X(\hat{\mathbf{k}}) = \hat{\mathbf{k}} \times \mathbf{e}_Y(\hat{\mathbf{k}}). \tag{D.77}$$

where $\mathbf{e}$ is any unit vector perpendicular to $\mathbf{k}$, say parallel to $\mathbf{E}$.
Using these three unit vectors we can create the vector fields

$$\mathbf{U}_{j\mathbf{k}}(\mathbf{r}) = \mathbf{e}_j(\hat{\mathbf{k}}) U_{\mathbf{k}}(\mathbf{r}), \quad j = X, Y, Z. \tag{D.78}$$

These fields have the orthogonality property

$$\langle j\mathbf{k} | j'\mathbf{k}' \rangle \equiv \int d\mathbf{r}\, \mathbf{U}_{j\mathbf{k}}(\mathbf{r}) \cdot \mathbf{U}_{j'\mathbf{k}'}(\mathbf{r}) = \delta_{jj'}\delta(\mathbf{k} - \mathbf{k}'). \tag{D.79}$$

These are the fields used to describe linearly polarized plane waves. When used as a basis for quantization they lead to photons with well-defined linear momentum $\hbar\mathbf{k}$ and linear polarization (X or Y).

Helicity fields. An alternative choice of unit vectors, used for treating circular polarization, are the helicity vectors. For a given propagation direction $\hat{\mathbf{k}}$ we define these as

$$\mathbf{e}_0(\hat{\mathbf{k}}) = \mathbf{e}_Z(\hat{\mathbf{k}}), \qquad \mathbf{e}_q(\hat{\mathbf{k}}) = -\frac{q}{\sqrt{2}}\left[\mathbf{e}_X(\hat{\mathbf{k}}) + \mathrm{i}\, q\, \mathbf{e}_X(\hat{\mathbf{k}})\right], \quad q = \pm 1. \tag{D.80}$$

These three basis vectors can be combined with any plane-wave scalar field to produce a triad of plane-wave vector basis fields,

$$\mathbf{U}_{q\mathbf{k}}(\mathbf{r}) = \mathbf{e}_q(\hat{\mathbf{k}}) U_{\mathbf{k}}(\mathbf{r}), \qquad q = 0, \pm 1, \tag{D.81}$$

that have the orthogonality property

$$\langle q\mathbf{k} | q'\mathbf{k}' \rangle = \delta_{qq'}\delta(\mathbf{k} - \mathbf{k}'). \tag{D.82}$$

The unit vectors $\mathbf{e}_q(\hat{\mathbf{k}})$ of eqn. (D.80) are also eigenvectors of the projection of spin along the propagation direction, with eigenvalue q,

$$(\mathbf{k} \cdot \mathbf{S})\, \mathbf{e}_q(\hat{\mathbf{k}}) = q\, k\, \mathbf{e}_q(\hat{\mathbf{k}}). \tag{D.83}$$

These vectors are helicity unit vectors. The unit vectors with positive and negative helicity ($q = +1$ and -1) are transverse to the propagation direction, whereas the vector with null helicity, $q = 0$, is longitudinal. Because the basis fields of eqn. (D.81) have plane-wave spatial portions they are eigenfunctions of the operators ∇^2 and ∇. Because their vector properties derive from helicity vectors they are also eigenfunctions of the curl operator,

$$\nabla \times \mathbf{U}_{q\mathbf{k}}(\mathbf{r}) = qk\mathbf{U}_{q\mathbf{k}}(\mathbf{r}). \tag{D.84}$$

The photons associated with these vector fields are plane-wave helicity photons.

D.2.2 Spherical waves

Rather than solve the Helmholtz equation in Cartesian coordinates, thereby obtaining plane-wave solutions, we can use spherical coordinates to obtain solutions that have spherical nodal surfaces. The resulting mode fields are eigenfunctions of the angular momentum operator $\hat{\mathbf{L}}^2$, and the associated photons have orbital angular momentum.

The connection between Cartesian and spherical scalar basis fields, and hence between photons of definite linear momentum and those with definite angular momentum, is provided by the Rayleigh expansion

$$\exp(\mathrm{i}\mathbf{k} \cdot \mathbf{r}) = 4\pi \sum_{\ell m} \mathrm{i}^\ell \, j_\ell(kr) \, Y_{\ell m}(\hat{\mathbf{k}})^* Y_{\ell m}(\hat{\mathbf{r}}), \tag{D.85}$$

where j_ℓ is a spherical Bessel function of order ℓ, $\hat{\mathbf{r}}$ is a unit vector specifying direction, and $Y_{\ell m}$ is a spherical harmonic. Such expansions are usually termed partial-wave expansions, acknowledging the breakup of a plane wave into constituent partial waves with well-defined orbital angular momentum. From this relationship it follows that the function $U_{k\ell m}(\mathbf{r})$ is, apart from normalization, the Fourier transform of the spherical harmonic whose angle argument defines the direction of the propagation vector $\mathbf{k}$:

$$U_{k\ell m}(\mathbf{r}) = \mathrm{i}^{-\ell} \sqrt{\frac{2k^2}{\pi}} \int \frac{d\hat{\mathbf{k}}}{4\pi} \exp(\mathrm{i}\mathbf{k} \cdot r) Y_{\ell m}(\hat{\mathbf{k}}). \tag{D.86}$$

Vector spherical modes. By combining helicity unit vectors with spherical Bessel functions and spherical harmonics we obtain the set of vector basis fields

$$\mathbf{U}_{k\ell mq}(\mathbf{r}) = \sqrt{\frac{2k^2}{\pi}} \, j_\ell(kr) Y_{\ell m}(\hat{\mathbf{r}}) \, \mathbf{e}_q(\hat{\mathbf{k}}). \tag{D.87}$$

The spherical harmonic $Y_{\ell m}$ is an eigenfunction of the orbital angular momentum operators $\mathbf{L}^2$ and $\hat{L}_z$, while the spherical unit vectors $\mathbf{e}_q$ are eigenvectors of the spin operator $\mathbf{S}^2$ and $\mathbf{k} \cdot \mathbf{S}$. Thus these fields are eigenfunctions of the operators ∇^2, $\mathbf{L}^2$, L_z, $\mathbf{S}^2$, and $\mathbf{k} \cdot \mathbf{S}$ with respective eigenvalues $-k^2$, $\ell(\ell+1)$, m, 2, and qk. They have the orthogonality properties

$$\langle k\ell mq | k'\ell'm'q' \rangle = \delta(k - k')\delta_{\ell\ell'}\delta_{m,m'}\delta_{q,q'}. \tag{D.88}$$

Vector harmonics. The association of a spherical harmonic with a spherical unit vector is an example of an uncoupled angular momentum state. Just as with other pairs of angular momentum operators it is sometimes useful to combine the spherical harmonics with eigenstates to construct vector harmonics, coupled angular momentum functions that are eigenstates of total angular momentum $\mathbf{J}^2 = (\mathbf{L} + \mathbf{S})^2$, and to use these, together with appropriate radial functions, to construct the mode fields

$$\mathbf{U}_{k\ell JM}(\mathbf{r}) = \sum_{mq} (\ell m, 1q | J M) \mathbf{U}_{k\ell mq}(\mathbf{r}). \tag{D.89}$$

These coupled angular momentum vector basis fields provide a complete set of vector fields. However, they do not provide a simple distinction between longitudinal and transverse (or lamellar and solenoidal) fields, nor do they yield simple expressions for the curl operator and the connection between electric and magnetic fields. They are therefore not the most useful set for treating free-space radiation fields, which are purely solenoidal. The following section describes a procedure for making this separation, leading to an alternative set of fields, the Hansen multipoles.

Hansen fields. To construct a set of vector basis fields we only need to have solutions to the scalar Helmholtz equation in some coordinate system. From such a solution $U(\mathbf{r})$ one can construct a triad of independent vector fields by applying, successively, the operators ∇, $\mathbf{L}$, and curl (or, equivalently, the operator $\mathbf{S} \cdot \nabla$). The resulting fields are, apart from normalization factors, the multipole fields of Hansen [Han35]. Given a scalar field $U(\mathbf{r})$ we define the triad of orthogonal vector fields

$$\mathbf{U}_Z(\mathbf{r}) = -\frac{i}{k} \nabla U(\mathbf{r}),$$

$$\mathbf{U}_X(\mathbf{r}) = \frac{\hat{\mathbf{L}}}{|\mathbf{L}|} U(\mathbf{r}), \tag{D.90}$$

$$\mathbf{U}_Y(\mathbf{r}) = -\frac{i}{k} \nabla \times \frac{\hat{\mathbf{L}}}{|\mathbf{L}|} U(\mathbf{r}) \equiv -\frac{i}{k} \nabla \times \mathbf{U}_X(\mathbf{r}) = -\frac{1}{k} \mathbf{S} \cdot \nabla \mathbf{U}_X(\mathbf{r}).$$

The field phases and directions are such that if

$$\mathbf{E}^{(+)}(\mathbf{r}, t) = \mathbf{U}_X(\mathbf{r}) \exp(-ickt) \tag{D.91a}$$

then

$$\mathbf{B}^{(+)}(\mathbf{r}, t) = \mathbf{U}_Y(\mathbf{r}) \exp(-ickt). \tag{D.91b}$$

By construction the X and Y fields are solenoidal, the Z field is lamellar,

$$\begin{aligned}
\nabla \cdot \mathbf{U}_Z = ikU, \qquad & \nabla \times \mathbf{U}_Z = 0, \\
\nabla \cdot \mathbf{U}_X = 0, \qquad & \nabla \times \mathbf{U}_X = ik\mathbf{U}_Y, \\
\nabla \cdot \mathbf{U}_Y = 0, \qquad & \nabla \times \mathbf{U}_Y = -ik\mathbf{U}_X.
\end{aligned} \tag{D.92}$$

The labels X, Y, Z only partially describe these fields; further labels are needed to identify the scalar function $U(\mathbf{r})$ used in their construction. For spherical waves the Hansen fields are $\mathbf{U}_{XkJM}, \mathbf{U}_{YkJM}, \mathbf{U}_{ZkJM}$.

The Hansen fields are eigenfunctions of the operators ∇^2, $\mathbf{S}^2$, $\mathbf{J}^2$, and J_z, but not of helicity nor generally of $\mathbf{L}^2$. Each field has well-defined parity, and can be identified as lamellar or solenoidal. These fields are associated with photons that have definite total angular momentum, J. They are the fields that are emitted or absorbed when an atom in free space undergoes a transition between two states of well-defined angular momentum, e.g. spontaneous emission.

Helicity multipole fields. The combinations $\mp \mathbf{U}_X - i\,\mathbf{U}_Y$ correspond to fields with helicity ± 1. Specifically, we have

$$\mathbf{U}_{kJMq}(\mathbf{r}) = \begin{cases} \mathbf{U}_{ZkJM}(\mathbf{r}) & \text{for } q = 0, \\[2mm] -\frac{q}{\sqrt{2}}[\mathbf{U}_{XkJM}(\mathbf{r}) + iq\mathbf{U}_{YkJM}(\mathbf{r})] & \text{for } q = \pm 1. \end{cases} \tag{D.93}$$

Each helicity multipole with $q = \pm 1$ is a mixture of two parity multipoles. Because atomic states in the absence of external perturbations are usually states with well defined parity, parity conservation produces a single parity multipole rather than a single helicity multipole.

D.2.3 Fields near the origin

When evaluating the interaction of an atom or other localized quantum system with radiation we require the electric or magnetic field at the center of mass, taken as the coordinate origin, $\mathbf{r} = 0$. The strength of the interaction is set, in the dipole approximation, by the value of the field at the coordinate origin. It is therefore desirable to employ spherical coordinates for expressions of mode fields. For small radial distances, $kr \ll 1$, we can express the spherical Bessel function by the first term of a series expansion,

$$j_\ell(x) \simeq \frac{x^\ell}{(2\ell + 1)!!} \equiv \frac{x^\ell}{1 \times 3 \times 5 \times \cdots \times (2\ell + 1)}. \tag{D.94}$$

At the origin, $r = 0$, only the spherical Bessel function $j_0(0) = 1$ is nonzero. We therefore find that the only multipole basis fields that do not vanish at the coordinate origin $r = 0$ are those with $J = 1$. These comprise a lamellar dipole field

$$\mathbf{U}_{Zk1M}(0) = -\frac{i}{\sqrt{3}}\mathbf{U}_{k01M}(0) = -\frac{ik}{\pi\sqrt{6}}\mathbf{e}_M \tag{D.95}$$

and one solenoidal field (source of the electric-dipole radiation field),

$$\mathbf{U}_{Yk1M}(0) = -\frac{i}{k}\nabla \times \mathbf{U}_{Xk1M}(0) = \sqrt{\frac{2}{3}}\mathbf{U}_{k01M}(0) = \frac{k}{\pi\sqrt{3}}\mathbf{e}_M. \tag{D.96}$$

As with other modes, the sum of squared field amplitudes at the origin is

$$\sum_{jJM} |\mathbf{U}_{jkJM}(0)|^2 = \frac{3k^2}{2\pi^2}. \tag{D.97}$$

D.2.4 Connection between multipoles and plane waves

The connection between scalar fields in Cartesian and spherical coordinates may be written in the form

$$U_{k\ell m}(\mathbf{r}) = \int d\hat{\mathbf{k}} \, U_{\mathbf{k}}(\mathbf{r}) \, \langle \mathbf{k} | k\ell m \rangle, \qquad U_{\mathbf{k}}(\mathbf{r}) = \sum_{\ell m} U_{k\ell m}(\mathbf{r}) \, \langle k\ell m | \mathbf{k} \rangle, \tag{D.98}$$

where the transformation coefficients are

$$\langle k\ell m | \mathbf{k} \rangle = \langle \mathbf{k} | k\ell m \rangle^* = \frac{\mathrm{i}^\ell}{k} Y_{\ell m}(\hat{\mathbf{k}})^* = \frac{\mathrm{i}^\ell}{k} \sqrt{\frac{2\ell+1}{4\pi}} \, \mathcal{D}_{m0}^{(\ell)}(\hat{\mathbf{k}}). \tag{D.99}$$

The connection between the multipole fields $\mathbf{U}_{jkJM}(\mathbf{r})$ and the plane-wave field $\mathbf{U}_{\mathbf{k}j}(\mathbf{r})$ is obtained from eqn. (D.86) and the definition (D.87),

$$\mathbf{U}_{jkJM}(\mathbf{r}) = \int d\hat{\mathbf{k}} \mathbf{U}_{\mathbf{k}j}(\hat{\mathbf{r}}) \langle \mathbf{k}j | jkJM \rangle,$$

$$\mathbf{U}_{\mathbf{k}j}(\mathbf{r}) = \sum_{JM} \mathbf{U}_{jkJM}(\mathbf{r}) \langle jkJM | \mathbf{k}j \rangle, \tag{D.100}$$

where j denotes multipole type, X, Y, or Z,

$$\mathbf{U}_{\mathbf{k}Z} = \hat{\mathbf{k}} U_{\mathbf{k}}, \qquad \mathbf{U}_{\mathbf{k}Y} = \hat{\mathbf{k}} \times \mathbf{U}_{\mathbf{k}X}, \qquad \mathbf{U}_{\mathbf{k}X} = \frac{\hat{\mathbf{L}}}{|\mathbf{L}|} U_{\mathbf{k}}. \tag{D.101}$$

That is, a single multipole source produces plane waves in all directions, with intensities governed by the transformation coefficient $\langle JkJM | \mathbf{k}j \rangle$. These coefficients are those given in eqn. (D.99),

$$\langle JkJM | \mathbf{k}j \rangle = \frac{\mathrm{i}^J}{k} Y_{JM}(\hat{\mathbf{k}})^* = \frac{\mathrm{i}^J}{k} \sqrt{\frac{2J+1}{4\pi}} \, \mathcal{D}_{M0}^{(J)}(\hat{\mathbf{k}}). \tag{D.102}$$

These coefficients provide the connection between photons that have linear momentum and photons that have orbital angular momentum. Their absolute squares are proportional to the angular distribution of intensity, see App. D.5.3.

D.2.5 Cavity modes; standing waves

The fields emerging as a directed beam from a laser or which expand outward from an atom source travel freely through space and can be regarded as traveling waves – either plane waves or spherical waves. By contrast the fields within a cavity are constrained by walls whose effects must appear as boundary conditions supplementing the vector Helmholtz equation. Idealized as perfectly reflecting surfaces these boundaries impose the requirement that the electric field must vanish at the edges of a finite volume (a cavity). This boundary condition allows solutions to the second-order differential equation only for discrete values of the frequency; the solutions form a denumerable discrete set, characterized, for example, by discrete values (eigenvalues) for wavevector magnitude k. When the cavity is large the mode frequencies are closely spaced; in the limit of an infinitely large volume the allowed values form a continuum.

The boundary condition is most easily satisfied when the volume surfaces are those of one of the several coordinate systems in which the Helmholtz equation is separable. Rectangular boxes and the associated Cartesian coordinates are the simplest example, but spherical or cylindrical surfaces also allow straightforward imposition of boundary conditions. For more complicated shapes it is necessary to use numerical techniques to solve the eigenvalue problem.

Standing waves vs. running waves. The simplest illustration of boundary conditions occurs when the one-dimensional traveling wave, hitherto treated with the exponential $\exp(ikx - i\omega t)$ is required to fit between mirror surfaces at $x = 0$ and $x = L$. By requiring that the field vanish at these mirror surfaces we allow only functions of the form $\sin(kx)$, with k restricted to integer multiples of π/L. This restriction on k accompanies a restriction on frequency: ω can only take integer multiples of π/cL.

For given k the two scalar base-fields for this geometry are

$$U_s(x) = \sqrt{\frac{2}{L}}\sin(kx), \qquad U_c(x) = \sqrt{\frac{2}{L}}\cos(kx). \tag{D.103}$$

Only the s mode is require for describing the electric field; the c mode is needed for the magnetic field. The photon annihilation operators associated with these modes are $\hat{a}_s$ and $\hat{a}_c$ and the associated two-mode Fock space needed for electric fields is spanned by the photon-number states $|n_s, n_c\rangle$ in which $n_c = 0$. The positive-frequency part of the field is

$$\mathbf{E}^{(+)}(x, t) = \sqrt{2}\,\mathcal{E}_0 \mathbf{e}\left[\hat{a}_s \sin(kx)\right] \exp(-i\omega t). \tag{D.104}$$

An alternative set of fields for this same k are those of forward and backward traveling waves,

$$U_+(x) = \frac{1}{\sqrt{L}}\exp(ikx), \qquad U_-(x) = \frac{1}{\sqrt{L}}\exp(-ikx). \tag{D.105}$$

The photon annihilation operators associated with these modes are $\hat{a}_+$ and $\hat{a}_-$, and their Fock space has states $|n_+, n_-\rangle$ with no restriction on either n value. For these modes the positive-frequency part of the field is

$$\mathbf{E}^{(+)}(x, t) = \mathcal{E}_0 \mathbf{e}\left[\hat{a}_+ \exp(ikx) + \hat{a}_- \exp(-ikx)\right]e^{-i\omega t}. \tag{D.106}$$

The connection between the two basis fields,

$$U_\pm(x) = \frac{1}{\sqrt{2}}[U_c(x) \pm i U_s(x)], \qquad U_s(x) = \frac{1}{\sqrt{2}}[U_+(x) - i U_-(x)], \tag{D.107}$$

leads to a connection between the two sets of annihilation operators,

$$\hat{a}_\pm = \frac{1}{\sqrt{2}}\left[\hat{a}_c \pm i\hat{a}_s\right], \qquad \hat{a}_s = \frac{1}{\sqrt{2}}\left[\hat{a}_+ - i\hat{a}_-\right]. \tag{D.108}$$

That is, a standing wave is expressible as a coherent superposition of two traveling waves.

Although the connections presented with these equations are mathematically correct, and the two bases appear equally valid for describing fields, the physics is different when mirrors are present, and experiments can distinguish between the two fields [Sho91b]. For example, an atom that is free to move will acquire linear momentum $\hbar\mathbf{k}$ when it absorbs a photon from a traveling wave of wavevector $\mathbf{k}$. A succession of such absorptions, each followed by emission into random directions, will produce a measurable change to the moving atomic center of mass. If the radiation is constrained by mirrors the absorption is of a coherent superposition of traveling waves: because of physical constraints at the mirrors it is not possible to change the number of "forward" photons without a companion change in the number of "backward" photons. There is thus a measurable difference between a standing wave and two oppositely running waves.

D.2.6 Summary

The great variety of solutions to the scalar Helmholtz equation in various coordinate systems permits a corresponding variety of analytically expressed mode functions with which to express an electromagnetic field. Each set of modes leads to a class of photons. For classical fields the coefficients of these mode functions, in a mode expansion of the field, must be determined to fit boundary conditions. In simple cases, when the boundary surfaces coincide with coordinate surfaces, the field will be a single mode, but in general it will appear as a superposition of modes.

Fields generated by quantum-mechanical interaction with an atom have similar properties. For these fields matrix elements of the interaction take the place of boundary conditions in prescribing the field. In general the interaction produces a coherent superposition of modes – the prominent example is the representation of the field by plane waves. By suitably choosing the modes it becomes possible to identify a single mode, typically a dipole field, as the field produced by the atom. This is the utility of the parity (or Hansen) multipole fields: they provide a compact analytic expression for the field produced by a multipole interaction. When quantized they describe the photons produced in spontaneous emission from an atom in free space.

The totality of transverse plane waves at the origin produces the sum of eqn. (D.97),

$$\sum_\lambda |\mathbf{U}_\lambda(0)|^2 = \frac{3k^2}{2\pi^2}. \tag{D.109}$$

Table D.1 provides some examples of useful ways of characterizing the mode fields.

Table D.1. *Examples of mode fields: solutions to the vector Helmholtz equation*

λ	Eigenfunctions of	$\mathbf{U}_\lambda(\mathbf{r})$	$\mathbf{U}_\lambda(0)$
Plane waves			
Linearly polarized plane waves ($j = X, Y, Z$)			
$\mathbf{k}j$	$\mathbf{S}^2, \nabla^2, \nabla$	$\dfrac{1}{\sqrt{(2\pi)^3}}\, \mathbf{e}_j(\hat{\mathbf{k}}) \exp(i\mathbf{k}\cdot r)$	$\dfrac{1}{\sqrt{(2\pi)^3}}\mathbf{e}_j(\hat{\mathbf{k}})$
Helicity plane waves ($q = +1, 0, -1$)			
$\mathbf{k}q$	$\mathbf{S}^2, \nabla^2, \nabla, \mathbf{S}\cdot\nabla$	$\dfrac{1}{\sqrt{(2\pi)^3}}\, \mathbf{e}_q(\hat{\mathbf{k}}) \exp(i\mathbf{k}\cdot r)$	$\dfrac{1}{\sqrt{(2\pi)^3}}\mathbf{e}_q(\hat{\mathbf{k}})$
Spherical waves			
Parity (Hansen) multipoles ($j = X, Y, Z$)			($J = 1$)
$jkJM$	$\mathbf{S}^2, \nabla^2, \mathbf{J}^2, J_z$	$\dfrac{i^{-J}}{k}\int d\hat{\mathbf{k}}\, \mathbf{U}_{\mathbf{k}j}(\mathbf{r}) Y_{JM}(\hat{\mathbf{k}})$	$Y:\ \dfrac{k}{\pi\sqrt{3}}\mathbf{e}_M$
			$Z:\ -\dfrac{ik}{\pi\sqrt{6}}\mathbf{e}_M$
Helicity multipoles ($q = +1, 0, -1$)			($J = 1$)
$kJMq$	$\mathbf{S}^2, \nabla^2, \mathbf{J}^2, J_z, \mathbf{S}\cdot\nabla$	$\dfrac{i^{-J}}{k}\int d\hat{\mathbf{k}}\, \mathbf{U}_{\mathbf{k}q}(\mathbf{r})\, \mathcal{D}^{(J)}_{Mq}(\hat{\mathbf{k}})^*$	$-\dfrac{ik}{\pi\sqrt{6}}\mathbf{e}_M$

D.3 Photon states

In treating laser-induced atomic excitation we deal with a system in which, after some initiation time t_0, an intense radiation field is present. Typically we idealize this radiation as a single mode of the electromagnetic field, e.g. a traveling plane wave. By supposing an intense field we presume that initially a few field modes are occupied by a large number of photons, while the remaining modes are initially unoccupied.

With passing time atoms absorb radiation from the highly populated modes, and spontaneously emit radiation into other directions. As a consequence, additional radiation modes will become occupied. So long as these incremental fields are much weaker than the original field, we make no appreciable error in our description of the radiation by ignoring the weak fields and assuming that the field statevector changes very little, if at all. This assumption, discussed in detail below, leads to a treatment in which the radiation field appears as a given classical force, imposed upon the atom by sources that are unaffected by atomic behavior. However, when we wish to describe the emitted radiation we must consider these weakly populated modes. A complete, consistent quantum theory of matter and radiation, including spontaneous emission, must include a complete set of field modes in expressions for $\hat{\mathbf{E}}$ and for $\hat{H}^R$. Let us examine a simple realization of the quantum states needed for this description.

Fock states . To describe the radiation field we can use a complete set of orthonormal eigenstates of the radiation Hamiltonian H^R, the photon-number states. For this purpose it is often useful to adopt the artifice of enclosing the atom–field system within a cube, of side L, so that the number of field modes is denumerable. This artifice allows us to label the radiation modes by discrete labels $\lambda, \lambda', \lambda'', \ldots$ that specify frequency, propagation direction, and polarization of the modes. Having selected some such suitable set of field modes, we obtain a set of field states by apportioning non-negative integers amongst the modes in all possible ways. The set of such integers, $n, n', n'', \ldots$ (the *occupation numbers*), together with the mode labels $\lambda, \lambda', \lambda'', \ldots$ to which they apply, provide the labels for such a field state, termed a *Fock state*. The totality of such Fock states form the basis of a (multimode) Fock space. Let us denote a typical Fock state of the radiation field by the notation

$$|\ldots n_\lambda, n'_{\lambda'}, n''_{\lambda''}, \ldots\rangle^R.$$

Although the number of distinct modes is (denumerably) infinite, the states of interest are usually those in which only a few modes are occupied. We simplify the notation considerably by listing only those modes for which the occupation number is nonzero.

Multimode Fock space. As a simple example of Fock states, consider just two field modes (out of the infinite set that provide a complete description of spatial structure of the field). The basis states $|n_1, n_2\rangle$, for all non-negative integers n_1 and n_2, provide a complete set of two-mode states. Specific examples include the states

$$|0_1, 0_2\rangle, |1_1, 0_2\rangle, |2_1, 0_2\rangle, \ldots |0_1, 1_2\rangle, |0_1, 2_2\rangle, \ldots |1_1, 1_2\rangle, |1_1, 2_2\rangle, \ldots.$$

Each of these field states has unit norm, and different states are orthogonal. The complete set of all Fock states form an example of an infinite-dimensional Hilbert space, a multimode Fock space.

The radiation density matrix. The Fock states are examples of single "pure" quantum states, in which the mode energies are precisely known. Any linear combination of such states is itself a pure quantum state. The notion of a single pure quantum state is, of course, a simplified idealization of more general descriptors of a system. Chapter 6 discussed the distinction between pure quantum states and more general "mixed" states. There we noted that conventional procedures usually prepare a sample of atoms in mixtures of states, rather than pure states. As we saw there, the density matrix, or statistical matrix, provides a compact formalism for expressing the most general initial conditions, ranging from a single quantum state (or coherent superposition of states) to a distribution of states specified by probabilities. Given any complete set of states $|a\rangle$ and a set of probabilities $\mathrm{p}(a)$, one constructs the statistical operator

$$\hat{\rho} = \sum_{a} |a\rangle \mathrm{p}(a) \langle a|. \tag{D.110}$$

The density matrix ρ has as its elements ρ_{nm} the matrix elements $\langle \Psi_n | \hat{\rho} | \psi_m \rangle$ of the operator $\hat{\rho}$. (Note that the basis states Ψ_n used to define the elements of the matrix ρ need not be the same states used to define the operator $\hat{\rho}$.) The Fock states of the radiation field provide a simple means of extending the notion of density matrices from atoms to radiation fields. We construct the operator

$$\hat{\rho} = \sum_{\lambda, \lambda', \dots} |\dots n_\lambda, n'_{\lambda'}, \dots\rangle \mathrm{p}(\dots n_\lambda, n'_{\lambda'}, \dots) \langle \dots n_\lambda, n'_{\lambda'}, \dots|. \tag{D.111}$$

The summation extends over all modes of the radiation field. Then matrix elements of $\hat{\rho}$ are elements of the radiation-field density matrix.

This equation is but one of many useful expressions for the density matrix of the radiation field. It is based upon the photon-number representation, in which we deal with quantum states having sharply defined single-mode energy. Subsequent sections discuss other single-mode states. Each of these representations offers a different expression for the radiation-field density matrix.

D.4 The free-field radiation Hamiltonian

Having expressed the electric field operator in terms of basis fields and photon creation and annihilation operators, we can next evaluate the energy operator for the free field – the free-field Hamiltonian. The most direct procedure starts from the classical expression for the field energy as the integrated electromagnetic field energy density $u(\mathbf{r}, t)$. For radiation fields in free space the electric and magnetic fields make equal contributions to the energy and we have, for the energy of the classical free field, the expression

$$H^R(t) = \int d\mathbf{r}\, u(\mathbf{r}, t) = \frac{\epsilon_0}{2} \int d\mathbf{r} \left[\mathbf{E}(\mathbf{r}, t)^2 + c^2 \mathbf{B}(\mathbf{r}, t)^2 \right] = \epsilon_0 \int d\mathbf{r}\, \mathbf{E}(\mathbf{r}, t)^2. \tag{D.112}$$

To convert this expression into one appropriate to quantum fields we must deal with noncommuting field operators, notably the positive- and negative-frequency parts of the field.

Symmetric ordering. A common approach is to replace the real-valued classical fields by Hermitian operators, sums of positive- and negative-frequency parts. This procedure produces the operator

$$\hat{H}^{R'}(t) = \epsilon_0 \int d\mathbf{r}\, \hat{\mathbf{E}}(\mathbf{r}, t) \cdot \hat{\mathbf{E}}(\mathbf{r}, t)$$

$$= \epsilon_0 \int d\mathbf{r} \left[\hat{\mathbf{E}}^{(-)}(\mathbf{r}, t) \cdot \mathbf{E}^{(+)}(\mathbf{r}, t) + \hat{\mathbf{E}}^{(+)}(\mathbf{r}, t) \cdot \mathbf{E}^{(-)}(\mathbf{r}, t) \right]. \tag{D.113}$$

From the mode expansion

$$\mathbf{E}^{(+)}(\mathbf{r}, t) = \oint_\lambda \mathcal{E}_\lambda \mathbf{U}_\lambda(\mathbf{r}) a_\lambda(t) = \left[\mathbf{E}^{(+)}(\mathbf{r}, t)\right]^\dagger \tag{D.114}$$

and the mode normalization

$$\langle \lambda | \lambda' \rangle \equiv \int d\mathbf{r} \, \mathbf{U}_\lambda(\mathbf{r})^* \cdot \mathbf{U}_{\lambda'}(\mathbf{r}) = \delta_{\lambda\lambda'} \tag{D.115}$$

we evaluate the spatial integral as the time-independent operator

$$\hat{H}^{R'} = \epsilon_0 \oint_\lambda \oint_{\lambda'} \left[\hat{a}_\lambda^\dagger(t) \hat{a}_{\lambda'}(t) + \hat{a}_{\lambda'}(t) \hat{a}_\lambda^\dagger(t)\right] \delta_{\lambda,\lambda'} |\mathcal{E}_\lambda|^2$$

$$= \epsilon_0 \oint_\lambda |\mathcal{E}_\lambda|^2 \left[2\hat{a}_\lambda^\dagger \hat{a}_\lambda + 1\right]. \tag{D.116}$$

We can obtain the desired expression for field energy as a sum of photon energies by defining the frequency to be $\omega_\lambda = ck_\lambda$ and choosing the normalizing constant $\mathcal{E}_\lambda$ such that

$$|\mathcal{E}_\lambda|^2 = \hbar c k_\lambda / 2\epsilon_0 \equiv \hbar \omega_\lambda / 2\epsilon_0. \tag{D.117}$$

With this choice of normalization and definition of ω_λ the free-field Hamiltonian $\hat{H}^{R'}$ of eqn. (D.116) is the sum of two parts,

$$\hat{H}^{R'} = \hat{H}^R + \hat{H}^{vac}, \tag{D.118}$$

each of which is a weighted sum over all possible modes of the radiation field, i.e. a complete set of solenoidal solutions to the vector Helmholtz equation:

$$\hat{H}^R = \oint_\lambda \hbar \omega_\lambda \hat{N}_\lambda, \qquad \hat{H}^{vac} = \frac{1}{2} \oint_\lambda \hbar \omega_\lambda. \tag{D.119}$$

The energy of the radiation field obtained as the expectation value of the Hamiltonian $\hat{H}^{R'}$ is

$$\langle \hat{H}^{R'} \rangle = \oint_\lambda \hbar \omega_\lambda \bar{n}_\lambda + \frac{1}{2} \oint_\lambda \hbar \omega_\lambda \equiv \langle \hat{H}^R \rangle + \langle \hat{H}^{vac} \rangle. \tag{D.120}$$

The first portion $\langle \hat{H}^R \rangle$ has an obvious interpretation as the sum of photon energies, with the energy of mode λ being the energy $\hbar \omega_\lambda \equiv \hbar c k_\lambda$ of a single photon and $\bar{n}_\lambda$ being the mean number of photons in that mode. The second sum $\langle \hat{H}^{vac} \rangle$ takes no account of photon distributions. It represents a vacuum energy that occurs even in the absence of photons or occupied modes. Because this vacuum energy is always present, and is constant for a given mode decomposition (i.e. a fixed geometry for the Helmholtz equation), it is customary to express energies relative to this zero point. This choice of energy scale corresponds to subtracting a vacuum energy from all energy eigenvalues, or to considering the field Hamiltonian to be the operator $\hat{H}^R$. This choice of Hamiltonian, when expressed in terms of electric field operators, becomes

$$\hat{H}^R = \oint_\lambda \hbar \omega_\lambda \hat{a}_\lambda^\dagger \hat{a}_\lambda = 2\epsilon_0 \int d\mathbf{r} \, \hat{\mathbf{E}}^{(-)}(\mathbf{r}, t) \cdot \hat{\mathbf{E}}^{(+)}(\mathbf{r}, t). \tag{D.121}$$

It is customary to adopt the Hamiltonian $\hat{H}^R$ rather than $\hat{H}^{R'}$ as the radiation field Hamiltonian, an approach I have followed.

Normal ordering. The vacuum energy made its appearance because we obtained the quantum Hamiltonian by direct replacement of real-valued fields by Hermitian operators. The vacuum energy occurs in the quantum theory of radiation whenever we deal with the symmetric expressions $\hat{a}^\dagger\hat{a} + \hat{a}\hat{a}^\dagger$. However, this choice for introducing a quantum Hamiltonian is not essential. We may with equal justification deal with the operator $\hat{H}^R$. This choice is equivalent to requiring that prior to the introduction of quantum conditions all expressions should be normally ordered, with positive-frequency parts $\mathbf{E}^{(+)}$ and $a(t)$ appearing to the right of negative frequency parts $\mathbf{E}^{(-)}$ and $a^\dagger(t)$. That is, instead of $\hat{H}^{R'}$ from eqn. (D.113) we take the radiation Hamiltonian to be $\hat{H}^R$ of eqn. (D.121). This procedure eliminates the vacuum energy. As explained by Milonni and Shi [Mil94], it does not eliminate any observable physical effects; it merely alters our interpretation of those effects.

D.5 Interpretation of photons

Quantum theory endows particles, such as electrons and atomic nuclei, with wavelike properties.[1] It also discretizes increments of fields as quanta. For the electromagnetic field these incremental quanta are photons; they represent the smallest increments by which a field can change. This discreteness, and reference to photons, occurs in several ways. The following sections discuss the notion of a photon as it applies to coherent excitation.

D.5.1 Photons as mode-field increments

The theory of quantized radiation presented above introduces discrete increments of a chosen set of field modes. The field increments are referred to as photons; the number states $|n\rangle$ with which we describe the field are regarded as photon-number states. The choice of modes is arbitrary, typically taken for convenience in fitting boundary conditions to nodal planes of coordinates. Each coordinate system, each choice of field modes, leads to a specific type of photon – standing-wave, traveling-wave, plane-wave, multipole, etc. Although a specific experimental apparatus may be best described by a particular mode choice, it is always possible to express one mode as a coherent superposition of alternative modes. One choice of photon, as a field increment, can therefore be regarded as a linear superposition of alternative types of photons. Theory makes no distinction between types of photons when summing over a complete set, but experimental conditions may offer a preferred choice. The following paragraphs discuss definitions of photons from an experimental approach.

---

[1] More precisely, the Schrödinger equation, which describes particle probability distribution, has wavelike solutions.

D.5.2 Photons as absorbed quanta

Perhaps the simplest operational definition of a photon is based upon the photoelectric effect. In its simplest idealization, one observes individual electrons ejected from a photosensitive surface illuminated by radiation. Each appearance of an electron, of energy E, can be attributed to the absorption of a quantum of electromagnetic energy $\hbar\omega = E - E_0$, where E_0 is the binding energy of the electron to the surface (the *work function*). This energy quantum is the photon.

This definition readily extends to observations of photoelectrons ejected from single free atoms, or even to photoexcitation of a single atom or molecule. In the latter situation one must be in a regime for which rate equations apply, so that the process is unidirectional (i.e. without Rabi oscillations). It is also possible to regard the appearance of latent images in photographic emulsions as indicators of photon absorption.

These definitions introduce idealized photon detectors that register the destruction of a photon. One infers that, prior to this destruction, a photon was present. The act of detection localizes the electromagnetic energy and so localizes the photon in space and time, but the destructive nature of the measurement precludes any further observation of this particular photon.

By passing the radiation through suitable frequency filters prior to detection, it is possible to obtain photons with well-defined energy $\hbar\omega$. Typical photon detectors respond to radiation traveling within a small solid angle. The photons defined by a detector of infinitesimal size are photons with well-defined linear momentum $\hbar\mathbf{k}$. By inserting polarizing elements into the optical train one may restrict detection to photons with well-defined helicity.

D.5.3 Photons as emitted quanta

A second type of operational definition can be based upon the act of spontaneous emission. Prepare an atom in an excited state at time $t = 0$. With passing time the atom will spontaneously emit one quantum of electromagnetic energy, $\hbar\omega = E_2 - E_1$, as it decays from state 2 to state 1. This energy, and the concomitant electromagnetic field, may be regarded as a single photon. The single-photon quantum state corresponding to such emission is produced by the action of the negative-frequency part of the electric field operator acting upon the photon vacuum:

$$\hat{\mathbf{E}}^{(-)}(\mathbf{r}, t)|vac\rangle.$$

The electric field produced in this way has the angular distribution of a multipole field (see [Sho90, §9.2]), but has a dependence upon radial distance r that vanishes for times earlier than $t = r/c$, the time taken for a light signal to reach distance r. The subsequent exponential decay of excitation probability, with rate coefficient A (the Einstein A) and lifetime $\tau = 1/A$, leads to an exponentially declining field at any position. Thus the positive-frequency part of the field has the form

$$\mathbf{E}^{(+)}(\mathbf{r}, t) = \text{constant} \times \frac{\mathbf{e}}{r}\exp\left[-\mathrm{i}\left(\omega - \frac{\mathrm{i}A}{2}\right)\left(t - \frac{r}{c}\right)\right]\Theta(ct - r), \qquad (\text{D.122})$$

where **e** is a unit vector and $\Theta(x)$ is the Heaviside step function,

$$\Theta(x) = \begin{cases} 0 & x < 0 \\ \frac{1}{2} & x = 0 \\ 1 & x > 0 \end{cases} . \tag{D.123}$$

Angular distributions. Although photons spontaneously emitted from an isolated atom have well-defined energy (within the lifetime-imposed uncertainty), they do not have well-defined linear momentum. Instead, they acquire well-defined angular momentum from the atom (see [Sho90, Sec. 19.5]). Conventional radiation detectors record energy flowing in a well-defined direction, and so they observe linear-momentum photons.[2] The change of photon basis, from angular momentum to linear momentum, is expressed by an overlap integral such as $\langle \mathbf{k}.q' | qkJM \rangle$. The square of this amplitude provides the probability of observing a photon in direction $\hat{\mathbf{k}}$ given that it was emitted with angular momentum J, i.e. it gives the relative intensity distribution over solid angle from an emission source.

Conversely, a laser beam provides linear-momentum photons, while a free atom absorbs angular-momentum photons. Again the overlap integral is needed to evaluate the interaction strength for various geometric arrangements of the two coordinate systems.

D.5.4 Wavepacket photons

Although it may be tempting to visualize a photon, like a massive particle, as a localized concentration of energy and momentum, such pictures can be misleading. The free-space photon must be regarded as a massless particle, traveling always at the velocity of light, and hence there is no reference frame in which it can be viewed at rest. However, when radiation enters matter it is no longer a free field. The propagating electromagnetic disturbance combines both an **E** field and a **P** field into a dark-state polariton [Fle02] that propagates at a group velocity different from c, so-called slow light. This velocity can be made small or even zero. The field increment, the remnant of the free-field photon, then comes to rest [Phi01; Fle02; Liu01; Luk03].

Various forms of traveling waves are appropriate candidates for single-photon fields. From laser sources the radiation typically is confined to a beam in the transverse direction, although it is unlocalized in the longitudinal direction. Authors have discussed approaches to defining a photon wavefunction that has similarities with particle wavefunctions [Bia96; Coo82].

We can construct wavepacket pulses of radiation, but those fields are constructed by superposing many modes of the field and so they are not the single modes whose occupation number here defines a photon. Nevertheless, we can construct, by superposing single-photon modes of slightly different frequency and wavevector, fields that are localized into

---

[2] A detector capable of distinguishing single-photon emissions of angular momentum would require interferometic superpositions of linear-momentum fields from several directions.

wavepackets [Tit65]. Because such wavepacket states are constructed entirely from single-photon states they represent localized single-photon states.

Similarly we can define multiple-photon wavepackets as superpositions of multiple-photon modes. By employing such wavepacket pulses as basis states we can provide a photon interpretation of short pulses of laser radiation. If the laser radiation has carrier frequency ω_L and a time-varying intensity envelope $I(t)$, then a pulse through area spatial A can be regarded as comprising, on average, a mean number of photons $\bar{n}$ where

$$\hbar\omega_L\bar{n} = \int dA \int dt \, I(t). \tag{D.124}$$

The pulse field comprises a distribution of photon-number states, whose mean photon number is $\bar{n}$. Thus $\bar{n}$ need not be an integer, and may in fact be less than one photon per pulse.

Appendix E

Adiabatic states

Adiabatic states, by definition, are instantaneous orthonormal eigenstates of a Hamiltonian. In the present monograph that Hamiltonian is the RWA Hamiltonian $\hbar \mathsf{W}(t)$ and the adiabatic states are eigenvectors $\boldsymbol{\Phi}_\nu(t)$ of this matrix,

$$\mathsf{W}(t)\, \boldsymbol{\Phi}_\nu(t) = \varepsilon_\nu(t)\, \boldsymbol{\Phi}_\nu(t). \tag{E.1}$$

These slowly varying states, normalized at each time t by the requirement

$$\langle \boldsymbol{\Phi}_\nu(t) | \boldsymbol{\Phi}_{\nu'}(t) \rangle = \delta_{\nu\nu'}, \tag{E.2}$$

generalize the constant eigenstates (dressed states) of the constant RWA Hamiltonian matrix discussed in Sec. 8.3.4, allowing pulse amplitudes and detunings that vary slowly with time.

Equation (E.1) fixes the direction of a vector in Hilbert space and eqn. (E.2) fixes the magnitude, but these say nothing about the absolute phase. This is initially arbitrary, at time $t = 0$. Chapter 20 discusses the measurability of this phase and its definition at later times.

E.1 Terminology

Adiabatic and diabatic states. The terminology of "adiabatic" and "diabatic" dates from early work on the theory of reactive scattering [Mot65; Nik74]: *diabatic* states refer to colliding atoms that move toward each other along potential energy surfaces that do not allow excitation, whereas the *adiabatic* states incorporate the full interaction energy. The dependence of the interatomic potential energy on separation distance translates to a time dependence for moving particles, and hence the adiabatic states are time dependent.

Within the present context the diabatic states correspond to Hilbert-space unit vectors $\boldsymbol{\psi}_n$ or $\boldsymbol{\psi}'_n(t)$ while the adiabatic states correspond to the instantaneous eigenvectors $\boldsymbol{\Phi}_\nu(t)$ of the slowly varying RWA Hamiltonian matrix $\mathsf{W}(t)$.

Dressed and bare states. An alternative terminology originated for the treatment of atoms interacting with quantized radio-frequency fields [Coh92] and was subsequently used for various calculations with cw laser radiation. The atom–field interaction is then constant. The eigenstates of the resulting full Hamiltonian (describing free atoms and free photons

Adiabatic states

interacting) were termed "dressed states", contrasting with the "bare states" of the nonin-
teracting atoms and fields. In that situation the Hamiltonian was independent of time, and
the relationship between dressed and bare states was constant.

The verbal similarity between the words *adiabatic* and *diabatic* is a potential source
of unfortunate confusion, and therefore many speakers and authors prefer the terms *bare
state* for the diabatic basis states $\psi_n(t)$ that are eigenstates of an unperturbed Hamiltonian.
They use the term *dressed state* to describe an eigenstate of the full Hamiltonian; when the
Hamiltonian changes with time these are the adiabatic states.

Adiabatic following (AF) and rapid adiabatic passage (RAP). When the temporal changes
of the RWA Hamiltonian are sufficiently slow (adiabatic) then the statevector $\Psi(t)$ will
remain fixed, apart from a phase change, in a rotating Hilbert space $\mathcal{H}'$ whose basis vectors
are adiabatic eigenvectors $\Phi_\nu(t)$. In particular, if the statevector is initially aligned with
one of the adiabatic eigenvectors then it will remain aligned with this vector even as that
eigenvector moves in stationary Hilbert space $\mathcal{H}$, as recorded by changes of the components
of $\Psi(t)$ along the reference vectors $\psi'_n(t)$: the statevector will undergo *adiabatic following*
(AF).

It may happen that, while the statevector undergoes AF, the composition of the followed
adiabatic vector $\Phi(t)$ changes, from alignment with a basis vector $\psi'_1(t)$ to alignment with
a different vector $\psi'_2(t)$. There will occur a passage of population from state 1 to state 2,
i.e. *adiabatic passage* will occur (a special case of adiabatic following). In order that the
overall process take place coherently, so that a statevector description is applicable, it must
occur more rapidly than any decoherence time; in this sense it is *rapid adiabatic passage*
(RAP).

Complete (coherent) population return (CPR). When the detuning is constant and nonzero
and changes of the Rabi frequency occur slowly, then the system will return to its initial state
after a coherent excitation pulse. The more slowly the change occurs, the more complete is
the population return.

Curve crossings. Three classes of energies occur in treatments of adiabatic processes. Each
of these has a time dependence that, when plotted, becomes a *curve* of energy versus time.
The following classes of curves occur.

Bare: The energies of the undisturbed Hamiltonian, in the absence of any interaction
between the atom and any fields, are the constants (or *unperturbed energies E_n*). For the
two-state system the curves are the horizontal lines at

$$E_1, \qquad E_2.$$

These curves can never cross, although they can coincide (when the energies are
degenerate).

Diabatic: The diagonal elements of the RWA Hamiltonian $\hbar W_{nn}(t)$ are *diabatic energies*. For the two-state system considered here these are the curves

$$0, \qquad \hbar\Delta(t).$$

The diabatic energy curves are associated with diabatic states (bare states). Two diabatic curves will coincide at moments when $W_{nn}(t) = W_{mm}(t)$ and *can cross*. For the two-state system the diabatic curves are independent of the Rabi frequency $\Omega(t)$ and will cross whenever the detuning $\Delta(t)$ changes sign.

Adiabatic: The results of diagonalizing the RWA Hamiltonian matrix $\hbar W(t)$ are *adiabatic energies* $\hbar\varepsilon_k(t)$. For the two-state system considered here these are the curves

$$\hbar\varepsilon_-(t) = -\hbar\sqrt{\Omega(t)^2 + \Delta(t)^2}, \qquad \hbar\varepsilon_+(t) = +\hbar\sqrt{\Omega(t)^2 + \Delta(t)^2}.$$

The present monograph labels adiabatic eigenvalues in accord with their numerical value. By definition $\varepsilon_+(t)$ is the larger of the two-state eigenvalues. With this definition two adiabatic eigenvalues *never cross*. They will be equal (degenerate) only when the elements of the RWA Hamiltonian matrix vanish; the matrix is then singular and has only a single eigenvector. The moment when two diabatic curves cross but the adiabatic curves do not touch is often termed an "*avoided crossing*".

Alternative labels. In labeling adiabatic states I have identified them with their eigenvalue ordering. This makes possible the unambiguous identification of adiabatic energy curves. By definition, adiabatic curves may touch but never cross.

Alternative schemes are possible. For example, the adiabatic state can bear the label of the dominant component, when there is one, or can bear the label of the state to which it connects in some asymptotic limit. With such a convention it is not possible, without analytic expressions for the eigenvectors or numerical analysis of their components, to assign unambiguous labels to portions of eigenvalue curves. Nor can one say that with adiabatic evolution the statevector will remain aligned with an adiabatic state bearing the same label at all times, as is the case with the labeling scheme adopted here.

E.2 Adiabatic evolution

Adiabatic evolution of the statevector occurs when it remains fixed in a reference frame tied to adiabatic states. Considerable literature deals with adiabatic evolution [Hio83; Ore84a; Sho91a; Sho92; Smi92; Pil93; Gue02; Yat02b; Kra07]. Some articles are specific for two-state systems [Dyk60; Gri73; Loy74; Rob82] while others treat the three-state lambda system [Kuk89; Elk95; Vit96]. To understand the conditions that allow this description it is useful to consider first the simpler of these.

Two states. For the two-state system the adiabatic states obey the equation

$$W(t)\,\Phi_\pm(t) = \varepsilon_\pm(t)\,\Phi_\pm(t), \tag{E.3}$$

where the two-state RWA Hamiltonian matrix is

$$W(t) = \frac{1}{2} \begin{bmatrix} 0 & \Omega(t)\,e^{-i\varphi} \\ \Omega(t)\,e^{i\varphi} & 2\Delta(t) \end{bmatrix}.$$

(E.4)

The eigenvalues of this matrix are

$$\varepsilon_\pm(t) = \frac{1}{2}[\Delta(t) \pm \Upsilon(t)], \qquad \Upsilon(t) \equiv \sqrt{\Delta(t)^2 + |\Omega(t)|^2}.$$

(E.5)

The two eigenvectors, expressed in the rotating frame, can be expressed, when $\Upsilon(t) \neq 0$,

$$\Phi_- = \frac{1}{\sqrt{2\Upsilon(t)}} \begin{bmatrix} \sqrt{\Upsilon(t) + \Delta(t)} \\ -e^{i\varphi}\sqrt{\Upsilon(t) - \Delta(t)} \end{bmatrix}, \qquad \varepsilon_+ = \frac{1}{2}[\Delta(t) + \Upsilon(t)]$$

$$\Phi_+ = \frac{1}{\sqrt{2\Upsilon(t)}} \begin{bmatrix} \sqrt{\Upsilon(t) - \Delta(t)} \\ e^{i\varphi}\sqrt{\Upsilon(t) + \Delta(t)} \end{bmatrix}, \qquad \varepsilon_- = \frac{1}{2}[\Delta(t) - \Upsilon(t)]$$

(E.6)

As discussed in Secs. 8.3.4 and 8.4.3, the normalized eigenvectors and their eigenvalues can also be written as

$$\Phi_+(t) = \sin(\theta/2)\,\psi_1 + \cos(\theta/2)\,e^{i\varphi}\,\psi_2'(t), \quad \varepsilon_+(t) = \Upsilon\cos^2(\theta/2),$$

$$\Phi_-(t) = \cos(\theta/2)\,\psi_1 - \sin(\theta/2)\,e^{i\varphi}\,\psi_2'(t), \quad \varepsilon_-(t) = -\Upsilon\sin^2(\theta/2).$$

(E.7)

where

$$\cos\theta(t) = \Delta(t)/\Upsilon(t), \qquad \sin\theta(t) = \Omega(t)/\Upsilon(t).$$

(E.8)

The angle φ establishes the orientation of the diabatic states in Hilbert space and so the construction of the adiabatic states, expressed with the framework of diabatic states, is completely determined by the mixing angle $\theta(t)$. In turn, this angle is controllable by prescribing the pulsed changes of Rabi frequency $\Omega(t)$ and detuning $\Delta(t)$. Thus the adiabatic states can be considered known. What remains to be done is to associate the unknown statevector $\Psi(t)$ with the adiabatic states, and to understand the conditions under which that connection can remain constant while the adiabatic states change.

Adiabatic basis. For a two-state system we can write the statevector in the adiabatic basis as

$$C(t) = A_+(t)\,\Phi_+(t) + A_-(t)\,\Phi_-(t) \equiv \sum_k A_k(t)\Phi_k(t).$$

(E.9)

Evolution will be adiabatic if the coefficients $A_k(t)$ are constant. The following paragraphs discuss the conditions needed for that result.

The time derivative of the construction (E.9) is

$$\frac{d}{dt}C(t) = \sum_k \Phi_k(t)\frac{d}{dt}A_k(t) + \sum_k A_k(t)\frac{d}{dt}\Phi_k(t).$$

(E.10)

From the TDSE we obtain the requirement that this derivative should be

$$\frac{d}{dt}\mathbf{C}(t) - i\mathbf{W}(t) \sum_k A_k(t) \mathbf{\Phi}_k(t) = -i \sum_k \varepsilon_k(t) A_k(t) \mathbf{\Phi}_k(t). \tag{E.11}$$

The time derivatives of the adiabatic states (the nonadiabatic couplings) are expressible as

$$\frac{d}{dt}\mathbf{\Phi}_\pm(t) = \pm \frac{1}{2}\frac{d}{dt}\theta(t)\,\mathbf{\Phi}_\mp(t), \tag{E.12}$$

so these undergo interchange at a rate dependent on the time derivative of the mixing angle. From eqn. (E.10) we obtain the following equation, basically the RWA Schrödinger equation in an adiabatic basis,

$$\frac{d}{dt}\mathbf{A}(t) = -i\mathbf{W}^A(t)\,\mathbf{A}(t), \tag{E.13}$$

where $\mathbf{A}(t)$ is a column vector of the adiabatic coefficients and $\mathbf{W}^A(t)$ is the RWA Hamiltonian in the adiabatic basis,

$$\mathbf{A}(t) = \begin{bmatrix} A_-(t) \\ A_+(t) \end{bmatrix}, \quad \mathbf{W}^A(t) = \begin{bmatrix} \varepsilon_-(t) & \frac{1}{2}i\frac{d}{dt}\theta(t) \\ -\frac{1}{2}i\frac{d}{dt}\theta(t) & \varepsilon_+(t) \end{bmatrix}. \tag{E.14}$$

If the nonadiabatic coupling $\frac{d}{dt}\theta(t)$ is negligible, meaning the RWA Hamiltonian changes very slowly (adiabatically), then the projections of the statevector onto the adiabatic eigenvectors change only with phases,

$$A_\pm(t) = \exp\left[-i\int_0^t dt\,\varepsilon_\pm(t')\right] A_\pm(0). \tag{E.15}$$

E.2.1 Adiabatic conditions

The statevector will remain in a fixed superposition of adiabatic states, i.e. the coefficients $a_k(t)$ will remain constant, if the off-diagonal terms of $\mathbf{W}^A(t)$ are smaller than the difference of the diagonal terms,

$$\frac{1}{2}\left|\frac{d}{dt}\theta(t)\right| \ll |\varepsilon_+(t) - \varepsilon_-(t)| = |\Upsilon(t)|. \tag{E.16}$$

Under these conditions the time evolution will be adiabatic and the statevector will remain aligned with an initial fixed superposition of adiabatic states – or a single such state. This is the condition of adiabatic following.

To relate this adiabaticity condition directly to the two functions that define the pulse we write the derivative of the angle $\theta(t)$ as

$$\frac{d}{dt}\theta = \frac{d}{dt}\arctan[\Omega(t)/\Delta(t)]. \tag{E.17}$$

Using elementary calculus and the chain rule we write

$$\frac{d}{dt}\theta = \frac{1}{1 + [\Omega(t)/\Delta(t)]^2} \times \frac{d}{dt}[\Omega(t)/\Delta(t)] = \left[\Delta\frac{d}{dt}\Omega - \Omega\frac{d}{dt}\Delta\right]/\Upsilon^2. \tag{E.18}$$

Thus the *adiabatic condition* for the statevector to remain fixed in a reference frame of adiabatic states is

$$\left| \Delta \frac{d}{dt} \Omega - \Omega \frac{d}{dt} \Delta \right| \ll \left[\Omega(t)^2 + \Delta(t)^2 \right]^{3/2}. \tag{E.19}$$

When this condition is fulfilled, by suitable construction of the pulses, then the evolution will be adiabatic. In particular, a detuning sweep will induce a transition between bare diabatic states.

Chirped detuning. Near $\Delta = 0$ the adiabatic condition, eqn. (E.16), becomes

$$\left| \frac{d}{dt} \Delta \right| \ll (\Omega_0)^2. \tag{E.20}$$

That is, the rate of change in detuning must be small compared with (Ω_0).[2] This condition emerges from the exact analytic results of the Landau–Zener model of diabatic curve crossing (or adiabatic avoided crossings): to be adiabatic, and hence produce a transition, the evolution must be such that the LZ parameter $\eta = (2\Omega_0)/|\dot{\Delta}|$ should be very much larger than 1.

However, the detuning cannot change too slowly. During the course of the pulse, as Δ passes through zero, there must be a change greater than the separation of eigenvalues, or

$$\left| \tau \frac{d}{dt} \Delta \right| > |\Omega_0|. \tag{E.21}$$

We therefore find that, in addition to requiring a large Rabi angle, adiabatic passage requires that the detuning change fit within the bounds

$$(\Omega_0)^2 \gg \left| \frac{d}{dt} \Delta \right| > \Omega_0/\tau. \tag{E.22}$$

Constant detuning. When the detuning is constant the adiabatic condition (E.19) becomes

$$\left| \Delta \frac{d}{dt} \Omega(t) \right| \ll |\Delta^2 + |\Omega(t)|^2|^{3/2}. \tag{E.23}$$

This inequality imposes a constraint on the rate of change in the dimensionless variable $x = \Omega(t)/\Delta$ with respect to the dimensionless variable $y = \Delta t$. For detuning much larger than the Rabi frequency the condition is $|dx/dy| \ll 1$.

E.2.2 Superadiabatic states

The transformation to the adiabatic basis diminishes the off-diagonal elements of the matrix responsible for temporal changes of Hilbert-space coordinates, i.e. the RWA Hamiltonian is replaced by the adiabatic Hamiltonian in the representation of the TDSE. One can repeat this process by introducing a *superadiabatic basis* that further diminishes to off-diagonal

elements when changes occur slowly to the Hamiltonian [Ber90; Lim91; Fle99; Dre99].
The transformed coordinates again obey a set of coupled linear equations,

$$\frac{d}{dt}\mathbf{A}'(t) = -i\mathbf{W}'(t)\mathbf{A}'(t). \tag{E.24}$$

When the pulse properties change slowly, the matrix $\mathbf{W}'(t)$ has small off-diagonal elements.

This procedure, of introducing a succession of transformations that bring the effective Hamiltonian matrix into successively more diagonally dominant form can lead to expressions that help design pulses that produce more complete population transfer. However, it eventually will produce larger off-diagonal elements.

E.3 The Dykhne–Davis–Pechukas (DDP) formula

Adiabatic time evolution, and the attendant adiabatic following by the statevector of an adiabatic state offers several advantages over alternative scenarios for quantum-state manipulation. Typically techniques based on adiabatic following are complete only in some limit, such as very large temporal pulse area. Analyses of the nonadiabatic coupling or, more generally, the analytic properties of adiabatic eigenvalues, offer opportunities to improve adiabatic passage by designing pulse shapes that are particularly effective in minimizing the nonadiabaticity for given pulse area.

We start by introducing explicit phases to present the description of adiabatic evolution in a Dirac picture,

$$\boldsymbol{\Psi}(t) = A_+(t)\,e^{-i\zeta_+(t)}\boldsymbol{\Phi}_+(t) + A_-(t)\,e^{-i\zeta_+(t)}\boldsymbol{\Phi}_-(t) \equiv \sum_k A_k(t)\,e^{-i\zeta_k(t)}\boldsymbol{\Phi}_k(t)(t), \tag{E.25}$$

where the phases are obtained from the adiabatic energies,

$$\frac{d}{dt}\zeta_k(t) = \varepsilon_k(t). \tag{E.26}$$

The coefficients obey the equations

$$\frac{d}{dt}A_\pm(t) = \pm i\,K(t)\,e^{\pm i\,D(t)}\,A_\mp(t), \tag{E.27}$$

where the integrated eigenvalue difference is

$$D(t) = \int_0^t dt'[\varepsilon_+(t') - \varepsilon_-(t')] = \int_0^t dt'\sqrt{\Omega(t')^2 + \Delta(t')^2} \tag{E.28}$$

and the nonadiabatic coupling is

$$K(t) = \langle \boldsymbol{\Phi}_-(t)|\frac{d}{dt}\boldsymbol{\Phi}_+(t)\rangle. \tag{E.29}$$

Define the angle $\theta(t)$ by the relationship

$$\tan\theta(t) = \frac{W_{12}(t)}{W_{22}(t) - W_{11}(t)} = \frac{\Omega(t)}{\Delta(t)}. \tag{E.30}$$

Then the nonadiabatic coupling is expressible as the derivative of $\theta(t)$,

$$K(t) = \frac{1}{2}\frac{d}{dt}\theta(t). \tag{E.31}$$

Next we consider the effect of a pulse upon a system that is initially in one adiabatic state, say Φ_-, as defined by the coefficients

$$A_-(-\infty) = 1, \qquad A_+(-\infty) = 0. \tag{E.32}$$

Then the probability that a pulse, upon completion, will produce a (nonadiabatic) transition to the other adiabatic state is

$$\mathcal{P}^{non} = |A_+(\infty)|^2. \tag{E.33}$$

To evaluate this transition probability we allow t to be complex and assume that the RWA Hamiltonian, embodied in the Rabi frequency $\Omega(t)$ and detuning $\Delta(t)$, is an analytic function of this complex variable. For real values of t the eigenvalues and the integral $D(t)$ are real, but for complex t these are complex.

The usual requirement for maintaining adiabatic evolution while the diabatic energies cross (thereby producing a transition between bare states) is that the eigenvalues remain well separated while the diabatic curves cross. If the eigenvalues were to become equal then any crossing of diabatic curves would fail to produce a transition: the system point would not follow the adiabatic energy curve, and the evolution would be nonadiabatic. Thus the possibility of degenerate adiabatic energies (referred to in much of the literature as a "curve crossing") signals a failure of adiabatic evolution.

Assume that on the real time axis there are no energy degeneracies (no "curve crossings") and that there are degeneracies at $t \to \pm\infty$. If the pulses are rather complicated there may be many degeneracies in the complex t plane. Our interest is in the degeneracy (the "crossing point") that lies closest to the real-t axis, because this one dominates the nonadiabatic transition.

Define *level lines* to be contour lines of the imaginary part of the integral $D(t)$, i.e. lines where

$$\text{Im } D(t) = \mathcal{C} = \text{ constant}. \tag{E.34}$$

The real-t axis is the contour line with $\mathcal{C} = 0$. As we examine contours that lie further from the real-t axis we will eventually encounter a degeneracy of the eigenvalues. Let t_0 be the complex time of the degeneracy nearest the real axis – this is the degeneracy for which the value of Im $D(t)$ is smallest. In the semi-classical limit (as $\hbar \to 0$) or adiabatic limit (slowly changing time-dependent Hamiltonian that is an analytic function of time) the following approximation holds [Dyk60; Dav76b]:

$$A_-(\infty) \approx \exp[i\,D(t_0)]. \tag{E.35}$$

The probability of a nonadiabatic transition (and hence a failure of adiabatic following) is then the exponential

$$\mathcal{P}^{non} \approx \exp[-2 \, \mathrm{Im} \, D(t_0)]. \tag{E.36}$$

Remarkably, this *Dykhne–Davis–Pechukas* (DDP) formula is independent of the nonadiabatic coupling. One can use this result to design functions that minimize $\mathcal{P}^{non}$ and which therefore provide statevector changes that are the most adiabatic, see Sec. 8.4.8.

Appendix F

Dark states; the Morris–Shore transformation

A simple redefinition of the quantization axis can change the linkage pattern of a linearly polarized field from one in which only pairs of state are linked (e.g. quantization axis along the linear polarization direction) to one in which several states are linked, with consequences such as those discussed in Sec. 12.4. The change of linkage pattern accompanies a redefinition of the basis states. It is natural to wonder whether a suitable change of basis can reduce other linkage patterns into ones in which only pairs of states are linked. If so, then one might make use of the very simple analytic solutions for two-state systems in treatments of more elaborate systems. Such simplification is possible under some conditions [Sho08]. The procedure involves a change of basis states known as a *Morris–Shore (MS) transformation* [Mor83; Vit00a; Vit03; Kis04; Iva06; Ran06].

F.1 The Morris–Shore transformation

An interaction pattern comprising multiple linkages can, when the pattern satisfies appropriate conditions, be replaced by a set of independent two-state interactions. The necessary conditions are [Mor83]

I. Two sets of states:
   Set A (initial, ground), with N_A elements.
   Set B (excited), with N_B elements.
II. There are no couplings within the A set or the B set, only couplings between A and B. (The graph corresponding to this linkage pattern is therefore bipartite.)
III. The two sets share common diagonal elements. In the RWA these are the two detunings Δ_A and Δ_B.

These conditions produce a RWA Hamiltonian having the structure

$$\mathsf{W} = \begin{bmatrix} \Delta_A \mathbf{1}_A & \mathsf{V} \\ \mathsf{V}^\dagger & \Delta_B \mathbf{1}_B \end{bmatrix} \begin{matrix} A \\ B \end{matrix}, \tag{F.1}$$

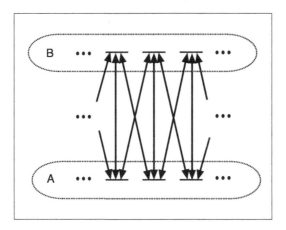

Figure F.1 Example of degenerate transitions amenable to Morris–Shore transformation, showing the A and B sets of states.

where $\mathbf{1}_A$ and $\mathbf{1}_B$ are square unit matrices, of dimension N_A and N_B respectively, and V is a rectangular matrix, of dimension $N_A \times N_B$. The following discussion assumes that W remains constant; the generalization to adiabatic change is straightforward.

Figure F.1, a generalization of Fig. 15.20(a), shows an example of transitions between degenerate magnetic sublevels that fit this pattern, for elliptical polarization and arbitrary orientation of the quantization axis. The initial state, part of set A, is one of the degenerate ground sublevels; the excited sublevels are set B.

Given this Hamiltonian, we introduce a Hilbert-space coordinate transformation

$$\widetilde{C}_j(t) = \sum_j U_{jn} C_n(t) \tag{F.2}$$

such that the resulting TDSE,

$$\frac{d}{dt}\widetilde{\mathbf{C}}(t) = -\mathrm{i}\widetilde{\mathsf{W}}\widetilde{\mathbf{C}}(t), \tag{F.3}$$

involves a Hamiltonian matrix $\widetilde{\mathsf{W}}$ that has the structure of uncoupled 2×2 blocks together with a unit matrix $\mathbf{1}_S$ of dimension $N_S = |N_A - N_B|$,

$$\widetilde{\mathsf{W}} = \begin{bmatrix} \mathsf{w}^{(1)} & \mathbf{0} & \mathbf{0} & \cdots & \mathbf{0} \\ \mathbf{0} & \mathsf{w}^{(2)} & \mathbf{0} & \cdots & \mathbf{0} \\ \mathbf{0} & \mathbf{0} & \mathsf{w}^{(3)} & \ddots & \mathbf{0} \\ \vdots & \vdots & \vdots & \ddots & \mathbf{0} \\ \mathbf{0} & \mathbf{0} & \mathbf{0} & \mathbf{0} & \Delta\,\mathbf{1}_S \end{bmatrix}. \tag{F.4}$$

The Δ is the detuning associated with the larger set of states: It is Δ_A if $N_A > N_B$ and is Δ_B if $N_B > N_a$. It is absent if $N_A = N_B$. The individual 2×2 blocks each have the form

$$\mathbf{w}^{(j)} = \begin{bmatrix} \Delta_A & \frac{1}{2}\bar{\Omega}^{(j)} \\ \frac{1}{2}\bar{\Omega}^{(j)} & \Delta_B \end{bmatrix}, \tag{F.5}$$

in which every block has the same pair of detunings, Δ_A and Δ_B, but the Rabi frequencies $\bar{\Omega}^{(j)}$ differ from block to block.

The transformation to this block-diagonal form $\widetilde{\mathsf{W}} = \mathsf{U}\mathsf{W}\mathsf{U}^\dagger$ is by means of a matrix

$$\mathsf{U} = \mathsf{G} \begin{bmatrix} \mathsf{A} & 0 \\ 0 & \mathsf{B} \end{bmatrix} \mathsf{G}^{-1} \qquad \text{where} \qquad \begin{array}{l} \mathsf{A} \text{ acts on } A \text{ states (ground),} \\ \mathsf{B} \text{ acts on } B \text{ states (excited),} \\ \mathsf{G} \text{ is a permutation matrix.} \end{array} \tag{F.6}$$

The two transformation matrices A and B are required to diagonalize the square matrices $\mathsf{V}\mathsf{V}^\dagger$ and $\mathsf{V}^\dagger\mathsf{V}$, respectively:

$$\mathsf{A}\mathsf{V}\mathsf{V}^\dagger\mathsf{A}^\dagger = \text{diagonal}, \qquad \mathsf{B}^\dagger\mathsf{V}^\dagger\mathsf{V}\mathsf{B} = \text{diagonal}.$$

The elements of these diagonal matrices are the squares of the Rabi frequencies $\Omega^{(j)}$. The signs of the Rabi frequencies are obtained only by evaluating $\mathsf{U}\mathsf{W}\mathsf{U}^\dagger$.

F.2 Bright and dark states

The result of this MS transformation is a set of independent two-state systems plus N_S uncoupled states. By assumption, the (stable) ground state is in set A, while set B includes at least one excited state. Because there exists a radiative link into the excited state, it will be possible for it to decay by spontaneous emission, thereby producing a fluorescence signal. Set B thus comprises *bright states*: with these there occur periodic population transfers into an excited state, signaled by fluorescence. Depending on the linkage pattern, set A may include some excited states (as occurs in Fig. F.4(*a*) below), and set B may include some stable sublevels.

The uncoupled states are unaffected by the radiation. If these are in set A they have no connection to any excited state and therefore cannot produce a fluorescence signal; they are termed *dark states* [Alz79; Ari96; Mil98; Wyn99; Kis01b]. If the unconnected states are in set B they do not include the initial state, and so these states remain unpopulated. They are *spectator states*.

Figure F.2 illustrates these two cases. The number of coupled pairs of states is $N_<$, the lesser of the two dimensions N_A and N_B. Of each pair, one state (a *bright* state) comprises a superposition of states from set A. The other state comprises a superposition of states from set B (excited states).

Display convention. Although the ordering of energy levels is not significant for the construction of the RWA Hamiltonian, it is customary to incorporate this information into

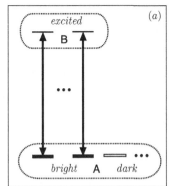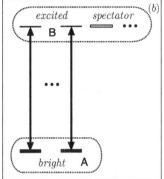

Figure F.2 Results of the MS transformation. (*a*) More lower states, $N_B < N_A$; have $N_A - N_B$ dark states. (*b*) More upper states, $N_B < N_A$; have $N_B - N_A$ spectator states. [Redrawn from Fig. 81 of B. W. Shore, Acta Physica Slovaka, **58**, 243–486 (2008) with permission.]

displays of linkage patterns. With the introduction of bright and dark states the energy association becomes even less significant. We can place the horizontal lines of the linkage patterns (the nodes of the graph of the linkages) as we please. In Fig. F.2 the positions of the dark and spectator states are placed, arbitrarily, on a line with the active states.

Population trapping. When $N_A > N_B$ there is at least one dark state. Any population that is initially in such a state will remain trapped there, unaffected by the coherent excitation. Such states have been termed population-trapping states [Gra78b; Rad82b; Dal82; Hio88; Ari96].

Population can remained trapped, immune to excitation, only so long as the Hamiltonian remains constant. Any change of phase, for example, may alter the dark-state construction and allow excitation. The example of optical pumping in Sec. 12.4 that left a portion of the population unexcited relied on maintenance of the quantization axis. Unpolarized light does not preserve the field phase needed to maintain a specific orientation on the Poincaré sphere, and hence it will eventually produce complete population loss from the stable magnetic sublevels.

Decoherence-free subspace. In traditional applications of STIRAP the initial and final states of the 1–2–3 chain are either stable or metastable (long-lived though excited), and only the intermediate state 2 undergoes spontaneous emission. States 1 and 3 therefore are not affected by the decoherence that intrudes both by population loss and by recycling of population as optical pumping: Their dynamics is free from any incoherent effects, or decoherence, apart from unavoidable randomness in the laser fields. Such states, termed here dark states, form what is termed a *decoherence-free subspace* of the full Hilbert space [Lid98; Bei00; Lid01a; Lid01b]. By restricting the statevector to such a subspace one prolongs the phase relationships between statevector components, as required for quantum information processing; see Sec. 5.7.

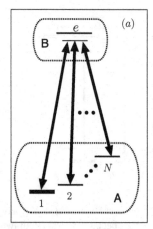 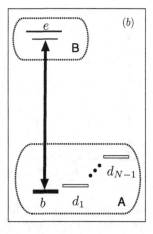

Figure F.3 (*a*) A general fan-type linkage pattern. (*b*) Equivalent linkages. There is one bright state, one excited state, and $N - 1$ dark states, shown here (arbitrarily) at the positions of the original states. [Redrawn from Fig. 82 of B. W. Shore, Acta Physica Slovaka, **58**, 243–486 (2008) with permission.]

F.3 Fan linkages

The simplest example of a MS transformation occurs with the tripod linkage of the three-state system. There is one uncoupled state after the MS transformation; it may be either a dark state (if it is in the set that includes the initial state, so that the linkage has the lambda pattern) or a spectator state (as happens when the initial state is the center of the chain, so the linkage forms a vee pattern). A generalization of this occurs when the A set comprises several states, all linked to a single excited state, in a "fan" pattern. The four-state tripod is an example of this. Figure F.3 illustrates a generalization in which there are N states in set A: there is a single bright state and $N - 1$ dark states. All dynamics of the excited state proceeds through the single bright state.

F.4 Chain linkages

The MS transformation allows simplification of resonant excitation of any degenerate two-level angular momentum system, by elliptically polarized light for a laser beam oriented at an arbitrary angle to the quantization axis, as shown in Fig. F.1. Such a system has a linkage pattern similar to that shown in Fig. 15.20(*a*). The MS transformation simplifies this linkage into $2J_< + 1$ independent two-state links, as would occur if the excitation had been linearly polarized along the quantization axis; see Fig. 15.20(*b*).

The MS transformation is not restricted to linkages derived from degenerate angular momentum sublevels. Other situations may also satisfy the requirements for application of a MS transformation. Figure F.4 shows two examples of resonant chain linkages that fit this pattern. In both of these examples the MS transformation produces a pair of independent

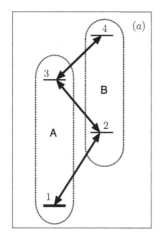 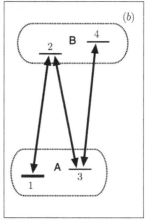

Figure F.4 Examples of resonant chains amenable to Morris–Shore transformation, showing the A and B sets of states. (*a*) A vertical chain, in which each link proceeds to higher energy. (*b*) A horizontal chain, in which even-numbered states have higher energies than odd-numbered states.

two-state systems; there are no dark or spectator states. As mentioned above, the geometry of the pattern is irrelevant, only the graph matters.

F.5 Generalizations

Three-state blocks. A generalization of the MS transformation makes possible, under certain conditions, a reduction in which the final result includes not only independent single- and two-state blocks of the transformed Hamiltonian, but also three-state blocks: Both two-state and three-state excitation chains arise from the set of transformed A states [Ran06].

Propagation equations. The restructuring of the Hamiltonian discussed here has wider application, whenever one encounters a set of ordinary differential equations in the form

$$\frac{d}{dx}\mathbf{C}(x) = -i\mathbf{W}\mathbf{C}(x), \tag{F.7}$$

for variables $C_n(x)$ and constant matrix $\mathbf{W}$. When $\mathbf{W}$ has the appropriate structure, the MS transformation will reduce the number of variables for which solutions are required by introducing "dark" combinations that remain fixed. When x denotes time, the dark superpositions represent constants of the motion. Such is the case in treatments of dark-state polaritons [Zim08].

Appendix G

Near-periodic excitation; Floquet theory

For all but the very shortest pulses, the laser field that produces excitation endures for many optical cycles. Therefore it is reasonable to regard this, in first approximation, as producing a Hamiltonian that is periodic in time [Sho90, §4.2]. This appendix discusses the theory by which we treat such periodic Hamiltonians.

Let the period be $\tau = 2\pi/\omega$. That is, the Hamiltonian has the property $H(t) = H(t + \tau)$. Such a periodic function of time is expressible as a Fourier series[1]

$$H(t) = H^{(0)} + H^{(1)} \exp(i\omega t) + H^{(-1)} \exp(-i\omega t) + \cdots, \qquad (G.1)$$

where the matrices $H^{(m)}$ are constant. For the periodic variation originating with the field $E(t) = e\mathcal{E} \cos(\omega t)$ only two terms contribute, $m = \pm 1$, but more general situations may occur. This Hamiltonian is to be used with the Schrödinger equation

$$\frac{d}{dt} \Psi(t) = -\frac{i}{\hbar} H(t) \Psi(t). \qquad (G.2)$$

Sets of coupled ordinary differential equations with exactly periodic coefficients, of which this Schrödinger equation is a special example, have properties described by what is sometimes termed *Floquet theory* [Shi65; Sal74; Bar77; Hio83; Chu85; Dre99], commemorating *Floquet's theorem*, which asserts that quasiperiodic solutions exist.

G.1 Floquet's theorem

Consider a set of N ordinary differential equations involving a vector of unknowns $Y(t)$, that satisfy the differential equation

$$\frac{d}{dt} Y(t) = -i M(t) Y(t). \qquad (G.3)$$

---

[1] I have here chosen a single fundamental frequency ω as the basis for the periodicity. In some situations the periodicity results from combinations of multiple commensurable periods. It is possible to introduce multiple periodicities and multiple frequencies $\omega_1, \omega_2, \ldots$ but the notation becomes quite cumbersome.

When the $N \times N$ matrix $\mathsf{M}(t)$ is periodic, with period $\tau \equiv 2\pi/\omega$, then Floquet's theorem asserts that a vector of solutions $\mathbf{Y}(t)$ can be found with the form

$$\mathbf{Y}(t) = \exp(-\mathrm{i}\, Zt)\,\mathbf{y}(t), \tag{G.4}$$

where the vector $\mathbf{y}(t)$ is periodic, with period τ. This latter periodicity means that $\mathbf{y}(t)$ is expressible as a Fourier series:

$$\mathbf{y}(t) = \sum_{m=-\infty}^{\infty} \mathbf{v}(m)\exp(\mathrm{i}\,m\omega t). \tag{G.5}$$

Here $\mathbf{v}(m)$ and $\mathbf{y}(t)$, like the original unknown $\mathbf{Y}(t)$, are N-component vectors. Note that although $\mathbf{y}(t)$ is periodic, the construct $\mathbf{Y}(t)$ is instead *quasiperiodic*: the absolute square of $\mathbf{Y}(t)$ is periodic although $\mathbf{Y}(t)$ is not.

The exponent Z is termed variously the *Floquet exponent* or the *quasienergy*. There are as many of these, and as many possible solutions of the form (G.4), as there are dimensions of the original Hamiltonian, i.e. the number of basis states N. Let us denote by Z_ν one of the N Floquet exponents, associated with a solution

$$\mathbf{Y}(\nu; t) = \exp(-\mathrm{i}\, Z_\nu t)\,\mathbf{y}(\nu; t), \tag{G.6}$$

where $\mathbf{y}(\nu; t)$ is periodic and therefore expressible as

$$\mathbf{y}(\nu; t) = \sum_{m=-\infty}^{\infty} \mathbf{v}(\nu, m)\exp(\mathrm{i}\,m\omega t). \tag{G.7}$$

The most general solution to eqn. (G.3) is a linear superposition of these vectors,

$$\mathbf{Y}(t) = \sum_{\nu} c_\nu \mathbf{Y}(\nu; t), \tag{G.8}$$

with the constants c_ν chosen such that $\mathbf{Y}(t)$ satisfies the required initial conditions, say at $t = 0$: they are solutions to the set of equations

$$\sum_{\nu} c_\nu \mathbf{Y}(\nu; 0) = \mathbf{Y}(0), \tag{G.9}$$

where $\mathbf{Y}(0)$ is a known vector.

Floquet's theorem assures the existence of a solution to eqn. (G.3) in the form (G.4), but it does not provide that solution. Nonetheless, the deduced properties of the solutions can often be very useful.

Application to Schrödinger equation. From Floquet's theorem we conclude that, for an N-dimensional periodic Hamiltonian, we can write the statevector $\mathbf{\Psi}(t)$ in the form

$$\mathbf{\Psi}(t) = \sum_{\nu=1}^{N} \exp(-\mathrm{i}\, Z_\nu t) \sum_{n=1}^{N} \sum_{m=-\infty}^{\infty} c_n(\nu, m)\exp(\mathrm{i}\,m\omega t)\,\boldsymbol{\psi}_n. \tag{G.10}$$

The N basis vectors ψ_n here account for the individual vector components, the exponentials $\exp(-i Z_\nu t)$ bring the N Floquet exponents Z_ν, while the remaining terms provide the required periodicities. The constants $c_n(\nu, m)$ are to be determined so that the statevector satisfies the TDSE and appropriate initial conditions.

G.2 Example: Two states

A simple illustration of the preceding formalism occurs for a two-state atom, acted on by a linearly polarized sinusoidal interaction of frequency ω [Sho90, §4.2]. The only nonzero elements of the Hamiltonian matrix are

$$H_{11}^{(0)} = E_1 = 0, \qquad H_{22}^{(0)} = E_2 = \hbar\omega_{12}, \qquad H_{21}^{(1)} = H_{21}^{(-1)} = \tfrac{1}{2}\hbar\Omega. \quad (G.11)$$

For the periodic Hamiltonian the Rabi frequency Ω (and Bohr frequency ω_{12}) are independent of time; more generally they are slowly varying.

There are two basis states in the atomic Hilbert space, ψ_1 and ψ_2, and so the statevector construction reads

$$\Psi(t) = c_1(t)\,\psi_1 + c_2(t)\,\psi_2, \qquad (G.12)$$

with coefficients expressible as

$$c_1(t) = \exp(-i Z_1 t) \sum_{even\ n=-\infty}^{\infty} a_n \exp(in\omega t) + \exp(-i Z_2 t) \sum_{odd\ m=-\infty}^{\infty} a_m \exp(im\omega t),$$

$$(G.13)$$

$$c_2(t) = \exp(-i Z_1 t) \sum_{odd\ m=-\infty}^{\infty} b_m \exp(im\omega t) + \exp(-i Z_2 t) \sum_{even\ n=-\infty}^{\infty} b_n \exp(in\omega t).$$

When the Hamiltonian is exactly periodic the real-valued Floquet exponents Z_1 and Z_2 and the complex-valued amplitudes a_m and b_m are independent of time. When the Hamiltonian is slowly varying, these parameters also vary with time.

The Floquet exponents; quasienergies. From the requirement that the construction above satisfy the TDSE we obtain the infinite set of algebraic equations

$$0 = [m\omega - Z]\,a_m + \tfrac{1}{2}\Omega\,[b_{m-1} + b_{m+1}],$$
$$0 = [\omega_0 + (m+1)\omega - Z]\,b_{m+1} + \tfrac{1}{2}\Omega\,[a_m + a_{m+2}].$$

$$(G.14)$$

These homogeneous equations have solutions only for selected values of Z, namely those that produce a null value for the determinant of the coefficients of the unknowns a_m and b_m. We may regard these equations as an eigenvalue equation for the Floquet Hamiltonian, i.e. the infinite matrix implied by eqns. (G.14). Thus the Floquet exponents Z_ν are the eigenvalues of the Floquet Hamiltonian.

These equations make no assumption about the excitation frequency ω. It need not be close to resonance and it need not be much larger than the Rabi frequency. The equations as written are, of course, restricted to a two-state atom and to a single-frequency field, but even so they have no simple closed-form analytic solutions.

G.3 Floquet theory and the RWA

The construction of eqn. (G.12) can be rewritten to exhibit the two Floquet exponentials as

$$\Psi(t) = \exp(-i Z_1 t)\,\Phi_1(t) + \exp(-i Z_2 t)\,\Phi_2(t), \tag{G.15}$$

in which the vector properties are embodied in the two combinations of basis vectors

$$\Phi_1(t) = \sum_{\substack{\text{even } n=-\infty}}^{\infty} a_n \exp(i n\omega t)\psi_1 + \sum_{\substack{\text{odd } m=-\infty}}^{\infty} b_m \exp(i m\omega t)\psi_2,$$
$$\Phi_2(t) = \sum_{\substack{\text{odd } m=-\infty}}^{\infty} a_m \exp(i m\omega t)\psi_1 + \sum_{\substack{\text{even } n=-\infty}}^{\infty} b_n \exp(i n\omega t)\psi_2. \tag{G.16}$$

This construction is similar to that used with the RWA. That expression, when written in terms of dressed states $\Phi_\pm(t)$, reads

$$\Psi(t) = c_+\Phi_+(t) + c_-\Phi_-(t). \tag{G.17}$$

Each dressed state has a simple exponential time dependence, through a factor $\exp(-i\varepsilon_\pm t)$ involving the dressed-state eigenvalues $\varepsilon_\pm$. In the RWA the formulas read

$$c_1(t) = \exp(-i\varepsilon_+ t)\,a_0 + \exp(-i\varepsilon_- t)\,a_1,$$
$$c_2(t) = \exp(-i Z_+ t)\,b_1 + \exp(-i\varepsilon_- t)\,b_0. \tag{G.18}$$

In the RWA each basis state ψ_n is accompanied by two frequency components, at the eigenvalues ε_+ and ε_- of the 2×2 RWA Hamiltonian matrix. In the more general expression of eqn. (G.10) these two eigenvalues become two Floquet exponents Z_1 and Z_2; the values $\varepsilon_\pm$ are approximations to these. Furthermore, each of the characteristic frequencies Z_1 and Z_2 of the more general expression has an infinite set of associated frequencies $Z_n + m\omega$ differing by integer multiples of the frequency ω: The set of Floquet exponents form a pair of infinite sequences.

Equation (G.10) presents the general form of a solution to the problem of a two-state atom subject to a periodic off-diagonal interaction. There remains the problem of determining the Floquet exponents Z_1 and Z_2 and the various constant coefficients a_m and b_m, for a given set of parameters that define the Hamiltonian. For the two-state atom these parameters are the Bohr frequency, the field carrier frequency, and the Rabi frequency.

G.4 Floquet theory and the Jaynes–Cummings model

The Floquet treatment of a periodic field, using Fourier exponentials to form a Hilbert space, has much in common with a treatment involving photons, approached with the aid of a Fock space [Shi65]. The prototype system is the Jaynes–Cummings model (JCM) discussed in Sec. 22.2.3.

The JCM is similar to, but not identical with, the Floquet description of monochromatic excitation. The difference between the two approaches is in the succession of Rabi frequencies. In the classical limit of large mean photon number there is no significant difference, but for small photon number the difference becomes significant. In particular, when the mean photon number is very small, as can occur for an atom in a small cavity, and the photon distribution is that of a coherent state, the damped oscillations exhibit a revival, evidence of the discreteness of the photon numbers [Nar81].

To emphasize the similarity of the two approaches we express the photon number as a deviation from the (large) mean number $\bar{n}$ and write[2]

$$\Psi(t) = \sum_{m=-\infty}^{\infty} C_{a,\bar{n},m}(t), \psi_a(t) \phi_{\bar{n}+m}. \tag{G.19}$$

The counterpart of the Floquet state having harmonic $m = 0$ is the photon state having mean photon number $\bar{n}$; other harmonics are deviations from this mean. The Rabi frequency becomes $\Omega_m = \sqrt{\bar{n}+|m|}\ \Omega_1$. In the limit of large photon number there is little difference between successive blocks of the JCM Hamiltonian, other than the diagonal elements, and both approaches deal with the same equations.

G.5 Near-periodic excitation; adiabatic Floquet theory

Generally the Hamiltonian of interest is not exactly periodic; the matrices $H^{(m)}$ vary with time, though they do so more slowly than the basic frequency ω. Equation (G.1) then becomes

$$H(t) = \sum_{m=-\infty}^{\infty} H^{(m)}(t) \exp(im\omega t). \tag{G.20}$$

This is not an exact Fourier series. However, because we have a series decomposition of $H(t)$ it is natural to look for a similar series for the statevector $\Psi(t)$. Specifically, let us propose the expansion

$$\Psi(t) = \sum_{a} \sum_{m=-\infty}^{\infty} C_{a,m}(t) \psi_a \exp(im\omega t), \tag{G.21}$$

where the $C_{a,m}(t)$ are expansion coefficients to be determined. The label a specifies the atomic basis state; the sum on m specifies the harmonic.

---

[2] The infinite summation limits merely indicate large positive and negative values of m. In practice, relatively few values of m are needed.

This construction of the statevector is tantamount to enlarging the Hilbert space by introducing basis vectors[3]

$$\phi_m(t) \equiv \exp(im\omega t) \tag{G.22}$$

and writing the statevector as

$$\Psi(t) = \sum_a \sum_{m=-\infty}^{\infty} C_{a,m}(t) \psi_a \phi_m(t). \tag{G.23}$$

The enlarged *Floquet space* has an infinite number of dimensions (because it treats an infinite set of harmonics, each requiring a dimension) but, like the original Hilbert space descriptive of the atom, usually only a few of them are needed. When the Hamiltonian is strictly periodic, so that $H^{(m)}(t)$ is constant, the coefficients $C_{a,m}(t)$ are also constant, and eqn. (G.24) is a Fourier series. More generally the terms $H^{(m)}(t)$ vary slowly with time, as do therefore the amplitudes $C_{a,m}(t)$.

Probabilities. From the statevector construction of eqn. (G.21) it follows that the probability amplitude for state a is a series,

$$C_a(t) \equiv \langle \psi_a | \Psi(t) \rangle = \sum_m \phi_m(t) \, C_{a,m}(t) = \sum_m \exp(im\omega t) \, C_{a,m}(t). \tag{G.24}$$

The probability of observing the atom in state a at time t is therefore the square of an infinite sum of harmonically modulated time-varying amplitudes:

$$P_a(t) = |C_a(t)|^2 = \left| \sum_m \exp(im\omega t) \, C_{a,m}(t) \right|^2. \tag{G.25}$$

Thus interference can occur between different harmonics.

Satisfying the TDSE. To obtain equations for the amplitudes $C_{a,m}(t)$ we substitute the expansion (G.21) into the Schrödinger equation (7.2) and equate to zero the coefficients of each factor $\psi_a \exp(im\omega t)$. This procedure produces an infinite set of coupled equations,

$$\frac{d}{dt} C_{a,m}(t) = -im\omega C_{a,m}(t) - \frac{i}{\hbar} \sum_b \sum_n H_{ab}^{(n)}(t) \, C_{b,m-n}(t). \tag{G.26}$$

These equations are an exact transcription of the original Schrödinger equation, valid for any number of atomic states and an arbitrary number of harmonics in the interaction. There is, as yet, no RWA approximation, only the assumption that any time variation of $H_{ab}^{(n)}(t)$ be slow compared with the period $2\pi/\omega$.

---

[3] The functions $\phi_m(t)$ form an infinite dimensional Hilbert space corresponding to a second degree of freedom.

As with the traditional approach to the TDSE we can express the set of coupled ordinary differential equations in matrix form, as

$$\frac{d}{dt} C_{a,m}(t) = -i \sum_{b,n} K_{ab}^{(n)}(t) C_{b,m-n}(t). \tag{G.27}$$

The infinite matrix of these coefficients, $K_{ab}^{(n)}(t)$, is sometimes termed the *Floquet Hamiltonian*. It differs from the usual Hamiltonian in having the terms $m\omega$ on the diagonal. When the original Hamiltonian is strictly periodic the Floquet Hamiltonian is constant; more generally it varies slowly with time.

Although the equations are denumerably infinite in number, they are amenable to solution by generalization of any of the techniques that are usable for the two-state case. The following paragraphs comment on some properties of the equations and their solutions.

G.6 Example: Two states

To understand the implications of the adiabatic Floquet theory, consider a two-state atom, for which we write the statevector as of eqn. (G.12) as

$$\Psi(t) = \psi_1 \sum_{m=-\infty}^{\infty} \exp(im\omega t) C_{1,m}(t)$$
$$+ \psi_2 \sum_{m=-\infty}^{\infty} \exp(im\omega t) C_{2,m}(t). \tag{G.28}$$

To clarify this construction let us use the notation $A_m(t)$ for harmonic components of state 1, and $B_m(t)$ for components of state 2,

$$C_{1,m}(t) = A_m(t), \qquad C_{2,m}(t) = B_m(t). \tag{G.29}$$

The equations of motion derived from eqn. (G.26) for a two-state atom, acted on by a linearly polarized sinusoidal interaction of frequency ω [Sho90, §4.2] are the infinite sets

$$\frac{d}{dt} A_m(t) = -i\,[m\omega]\,A_m(t) - i\frac{\Omega}{2}\,[B_{m-1}(t) + B_{m+1}(t)],$$
$$\frac{d}{dt} B_{m+1}(t) = -i\,[(m+1)\omega - \omega_0]\,B_{m+1}(t) - i\frac{\Omega}{2}\,[A_m(t) + A_{m+2}(t)]. \tag{G.30}$$

The matrix of coefficients appearing here form the elements of the Floquet Hamiltonian. It has as diagonal elements the frequencies $m\omega$ and the detunings $m\omega - \omega_0$ and as off-diagonal elements half the Rabi frequency Ω.

These equations separate into two independent sets of coupled equations. One set applies to the sequence … A_{-2}, B_{-1}, A_0, B_1, A_2, …, while the other set applies to the sequence ... $B_{-2}, A_{-1}, B_0, A_1, B_2, \ldots$. For each set the Floquet Hamiltonian has the matrix representation

$$
\begin{bmatrix}
\ddots & \vdots & \vdots & \vdots & \vdots & \vdots & \\
\cdots & (m+2)\omega & \tfrac{1}{2}\Omega & 0 & 0 & 0 & \cdots \\
\cdots & \tfrac{1}{2}\Omega & (m+1)\omega+\omega_0 & \tfrac{1}{2}\Omega & 0 & 0 & \cdots \\
\cdots & 0 & \tfrac{1}{2}\Omega & m\omega & \tfrac{1}{2}\Omega & 0 & \cdots \\
\cdots & 0 & 0 & \tfrac{1}{2}\Omega & (m-1)\omega+\omega_0 & \tfrac{1}{2}\Omega & \cdots \\
\cdots & 0 & 0 & 0 & \tfrac{1}{2}\Omega & (m-2)\omega & \cdots \\
& \vdots & \vdots & \vdots & \vdots & \vdots & \ddots
\end{bmatrix}.
$$

In particular, a small portion of the set of coupled equations, for even-integer A_n, has the coefficient matrix (labels at the right identify the states)

$$
\begin{bmatrix}
\boxed{\begin{matrix} 2\omega & \tfrac{1}{2}\Omega \\ \tfrac{1}{2}\Omega & 2\omega+\Delta \end{matrix}} & \begin{matrix} 0 & 0 \\ \tfrac{1}{2}\Omega & 0 \end{matrix} & \begin{matrix} 0 & 0 \\ 0 & 0 \end{matrix} \\
\begin{matrix} 0 & \tfrac{1}{2}\Omega \\ 0 & 0 \end{matrix} & \boxed{\begin{matrix} 0 & \tfrac{1}{2}\Omega \\ \tfrac{1}{2}\Omega & \Delta \end{matrix}} & \begin{matrix} 0 & 0 \\ \tfrac{1}{2}\Omega & 0 \end{matrix} \\
\begin{matrix} 0 & 0 \\ 0 & 0 \end{matrix} & \begin{matrix} 0 & \tfrac{1}{2}\Omega \\ 0 & 0 \end{matrix} & \boxed{\begin{matrix} -2\omega & \tfrac{1}{2}\Omega \\ \tfrac{1}{2}\Omega & -2\omega+\Delta \end{matrix}}
\end{bmatrix}
\begin{matrix} A_{+2} \\ B_{+1} \\ A_0 \\ B_{-1} \\ A_{-2} \\ B_{-3} \end{matrix}.
$$

Each 2×2 block is an example of the RWA equations, differing only by the addition of $2m\omega$ to both diagonal elements – a redefinition of the energy zero point (and an overall phase). In turn, the couplings between the different blocks represent counter-rotating terms in the rotating wave picture.[4] When the interaction $\tfrac{1}{2}\Omega$ is much smaller than 2ω the individual blocks will be nearly independent, and the RWA will apply. The Floquet Hamiltonian can then be approximated by any one of the blocks, for example

$$
\frac{d}{dt}\begin{bmatrix} A_m \\ B_{m-1} \end{bmatrix} = -i \begin{bmatrix} m\omega & \tfrac{1}{2}\Omega \\ \tfrac{1}{2}\Omega & m\omega+\Delta \end{bmatrix}\begin{bmatrix} A_m \\ B_{m-1} \end{bmatrix}. \tag{G.31}
$$

These are just the usual RWA equations. Corrections to the RWA obtain by including successively more harmonics, as expressed by enlarging the Floquet Hamiltonian matrix.

This approach, through a truncated Floquet Hamiltonian, can be used whenever the Hamiltonian elements (i.e. the Rabi frequency and detuning) are slowly varying. At the expense of solving a larger number of coupled equations, one can obtain solutions to the

---

[4] These couplings are absent when the excitation is by circularly polarized light; the matrix then consists of independent blocks.

TDSE without restriction upon either the strength of the interaction or the magnitude of the detuning.[5]

G.7 Adiabatic Floquet energy surfaces

Just as the slowly varying RWA Hamiltonian has slowly varying adiabatic states that can be used to describe the statevector, so too does a slowly varying Floquet Hamiltonian lead to slowly varying generalizations of the basic Floquet states. These, and the Floquet exponents, become slowly varying functions of time. Several papers describe the use of such states [Dre99; Yat02b; Gue03; Dre99].

During adiabatic evolution the statevector remains aligned with a fixed superposition of these adiabatic states, perhaps with a single one. Just as we identify two-state behavior from plots of the two simple adiabatic energies, so too can we use plots of the more general Floquet eigenvalues for this purpose. Instead of dealing with two curves, one for each of the two adiabatic energies, we have an infinite replication of pairs of curves, each set being offset by frequency ω. To understand the dynamics we consider a system point as it follows one of these curves. Encounters with curves for other eigenvalues bring opportunity for either adiabatic following (the system point remains associated with the same adiabatic eigenvalue) if the evolution is sufficiently slow, or for a transition as the system point encounters curves for other eigenvalues and moves rapidly through a crossing of diabatic energies.

When the Hamiltonian takes its time dependence from changes of just two functions, say Rabi frequency $\Omega(t)$ and detuning $\Delta(t)$, then the adiabatic eigenvalues form surfaces in a two-dimensional space. However, because the eigenvalues form an infinite set, offset from one another by the periodicity frequency, it is a space of many connected leaves. We can gain understanding of the system dynamics by following the motion of the system point on these surfaces, a generalization of Fig. 8.9 of Sec. 8.4.5 [Yat02b; Gue03]. In this space the system point follows a path over a surface (rather than along an energy curve) that may take it onto a different leaf.

---

[5] However, the assumption of only two essential states fails as the field becomes more intense.

Appendix H

Transitions; spectroscopic parameters

Between every pair of quantum states for which a nonzero electric or magnetic multipole moment exists there can take place a radiative transition. When the quantum states are both discrete, as they are for pairs of bound states, the radiation is discrete – a spectral line. The frequency of that spectral line is set by the difference of the two energies, and can occur in any region of the electromagnetic spectrum. If there is available a source of coherent radiation at that frequency, then coherent quantum-state manipulation is possible.

For many years spectroscopists routinely assembled collections of wavelengths and line strengths (or transition probabilities) for various elements and molecules. The National Bureau of Standards (NBS), now the National Institute for Standards and Technology (NIST), collected, organized, and published much of this data. Their website, www.nist.gov/pml/data/handbook/index.cfm, provides ready access to this information for all the elements and many molecules. Much of this data has appeared in the *Journal of Physical and Chemical Reference Data*, published by the American Institute of Physics (AIP). Diagrams showing the relative positions of energies and the connecting transitions, often called *Grotrian diagrams* [Bas75; Moo68; Lan99], are helpful for presenting the excitation linkages.

H.1 Spectroscopic parameters

From traditional spectroscopic studies come several parameters with which to describe the resonant interaction between light and an atom – one that exists, ideally, in free space and which therefore has degenerate energy levels.

Transition strength. The simplest connection between spectral properties and atomic structure is through the dimensionless *transition strength* $S(1, 2)$, the square of the absolute value of the electric dipole transition moment $\mathbf{d}_{12} \equiv \langle 1|\mathbf{d}|2\rangle$ expressed in atomic units,

$$S(1, 2) = S(2, 1) = |\mathbf{d}_{12}|^2 / (ea_0)^2. \tag{H.1}$$

As indicated, the atomic unit of dipole moment is the product of the electron charge $|e|$ and the Bohr radius,

$$|ea_0| \simeq 2.542 \text{ debye}. \tag{H.2}$$

For transitions between states of angular momentum the transition strength is proportional to the reduced matrix element of the dipole moment,

$$S(J, J') = \frac{|(J||d||J')|^2}{(ea_0)^2}.$$ (H.3)

Einstein A. The Einstein A coefficient that quantifies the rate of spontaneous emission is expressible using the transition strength as

$$A_{21} = \frac{1}{\tau_{AU}} \frac{4}{3} \left(\frac{2\pi a_0}{\lambda_{12}} \right)^3 \frac{S(1, 2)}{\varpi_2},$$ (H.4)

where $\tau_{AU} = a_0/\alpha c \approx 2.42 \times 10^{-17}$ s is the atomic unit of time and $\lambda_{12} = 2\pi c/\omega_{12}$ is the wavelength corresponding to the Bohr frequency ω_{12}. The statistical weight ϖ_n of level n is the angular momentum degeneracy $\varpi_n = (2J_n + 1)$.

Oscillator strength. The dimensionless absorption oscillator strength for a transition of wavelength λ_{12} between degenerate levels having statistical weights ϖ_n is

$$f_{12} = -\frac{\varpi_2}{\varpi_1} f_{21} = \frac{2}{3} \frac{E_2 - E_1}{E_{AU}} \frac{S_{12}}{\varpi_1} = \frac{4\pi}{3\alpha} \frac{a_0}{\lambda} \frac{S(1, 2)}{\varpi_1}.$$ (H.5)

Rabi frequency. The Rabi frequency used throughout this monograph for parametrizing the coherent excitation of a *nondegenerate* system contrasts with the preceding spectroscopic parameters used to characterize incoherent change of *degenerate* systems. Nevertheless we can obtain from A_{21} and f_{12} a root-mean-square Rabi frequency,

$$|\bar{\Omega}_{21}(t)|^2 = \frac{8\pi\alpha a_0^2}{\hbar} I(t) \frac{S(1, 2)}{3} = \frac{\lambda_{12}^2}{4\pi^2\hbar c} I(t) \varpi_2 A_{21} = \frac{2\alpha^2 a_0 \lambda_{21}}{\hbar} I(t) |\varpi_1 f_{12}|.$$ (H.6)

Here α is the Sommerfeld fine-structure constant. We use this parameter to write the Rabi frequency between magnetic sublevels as

$$\Omega_{21}(t) = \bar{\Omega}_{21}(t)\sqrt{3} \sum_q \frac{\mathcal{E}_q(t)e^{i\phi_q}}{|\mathcal{E}(t)|} (-1)^{J_> - M_2 + q} \begin{pmatrix} J_2 & 1 & J_1 \\ -M_2 & q & M_1 \end{pmatrix}.$$ (H.7)

H.2 Relative transition strengths

When degeneracy is present it is necessary to supplement the state labels 1 and 2 by additional quantum numbers, say m, and to express the dipole transition moment as

$$\langle 2|\mathbf{d} \cdot \mathbf{e}_q|1 \rangle \rightarrow \langle 2, m_2|\mathbf{d} \cdot \mathbf{e}_q|1, m_1 \rangle.$$ (H.8)

One then extracts the dependence upon these quantum numbers as factors. Typically the factorization takes the form

$$|\langle 2, m_2|\mathbf{d} \cdot \mathbf{e}_q|1, m_1 \rangle|^2 = S_q(2, m_1; 1, m_1)|\langle 2||d||1 \rangle|^2,$$ (H.9)

where the *reduced matrix element* is defined, apart from a phase, by the relationship

$$|\langle 2||d||1\rangle|^2 = \sum_{qm_1m_2} |\langle 2, m_2|\mathbf{d} \cdot \mathbf{e}_q|1, m_1\rangle|^2, \tag{H.10}$$

so that the factor $S_q(2, m_2; 1, m_1)$ has the normalization

$$\sum_{qm_1m_2} S_q(2, m_1; 1, m_1) = 1. \tag{H.11}$$

Often it is possible to introduce further factorization to separate the dependence on various sets of quantum numbers. When treating atomic spectra the factors are known as line strengths; their evaluation is discussed in numerous texts [Sob72; Cow81; Sob92; Mar06b].

The geometric factor. The simplest example of such factorization, following from the Wigner–Eckart theorem, extracts the dependence on magnetic quantum numbers M from a transition between angular momentum states (e.g. rotational states of a molecule),

$$|\langle 2, J_2M_2|\mathbf{d} \cdot \mathbf{e}_q|1, J_1M_1\rangle|^2 = S_q^{geom}(J_2M_2; J_1M_1) |\langle 2J_2||d||1J_1\rangle|^2. \tag{H.12}$$

The geometric factor S^{geom} contains all the information about atomic or molecular orientation in a laboratory reference frame, through the component q of the polarization vector $\mathbf{e}_q$ and the magnetic quantum numbers M_1 and M_2. For either electric dipole or magnetic dipole radiation the geometric factor is the square of a three-j symbol,

$$S_q^{geom}(J_2M_2; J_1M_1) = \begin{pmatrix} J_2 & 1 & J_1 \\ -M_2 & q & M_1 \end{pmatrix}^2, \qquad \text{for } E1, M1. \tag{H.13}$$

For electric quadrupole radiation it is

$$S_q^{geom}(J_2M_2; J_1M_1) = \begin{pmatrix} J_2 & 2 & J_1 \\ -M_2 & q & M_1 \end{pmatrix}^2, \qquad \text{for } E2. \tag{H.14}$$

Further factorization of the transition strength depends on details of the wavefunctions.

Atomic line strengths. Atomic transitions between electronic states e_2 and e_1 that have well-defined total spin S and total orbital angular momentum L have factorization

$$
\begin{aligned}
|\langle e_2, L_2S_2J_2M_2|&\mathbf{d} \cdot \mathbf{e}_q|e_1, L_1S_1J_1M_1\rangle|^2 \\
&= S_q^{geom}(J_2M_2; J_1M_1) && \text{geometric factor} \\
&\quad \times S^{mult}(L_2S_2J_2; L_1S_1J_1) && \text{multiplet strength} \\
&\quad \times |\langle e_2L_2||d||e_1L_1\rangle|^2 && \text{electronic reduced moment.} \tag{H.15}
\end{aligned}
$$

The multiplet factor is expressible in terms of a six-j symbol,

$$S^{mult}(L_2 S_2 J_2; L_1 S_1 J_1) = (2J_1+1)(2J_2+1) \left\{ \begin{matrix} J_1 & 1 & J_2 \\ L_2 & S_2 & L_1 \end{matrix} \right\}^2 \delta(S_1, S_2). \quad \text{(H.16)}$$

Hyperfine line strengths. When the atomic spectra include hyperfine structure, and the atomic states bear labels J, I, F, M, then the transition strengths factor as

$$|\langle e_2, J_2 I J F_2 M_2 | \mathbf{d} \cdot \mathbf{e}_q | e_1, J_1 I F_1 M_1 \rangle|^2$$

$$= S_q^{geom}(J F_2 M_2; F_1 M_1) \qquad \text{geometric factor}$$

$$\times S^{line}(J_2 F_2; J_1 F_1) \qquad \text{line strength}$$

$$\times |\langle e_2 J_2 ||d|| e_1 J_1 \rangle|^2 \qquad \text{electronic reduced moment,} \quad \text{(H.17)}$$

with line strength

$$S^{line}(J_2 F_2; J_1 F_1) = (2F_1+1)(2F_2+1) \left\{ \begin{matrix} F_1 & 1 & F_2 \\ J_2 & I & J_1 \end{matrix} \right\}^2. \quad \text{(H.18)}$$

Molecular line strengths. The factorization of atomic line strengths relies on the quantum theory of angular momentum. This theory also provides factorization of molecular line strengths, as discussed in texts on molecular spectra [Dra06; Hue06]; an example is [Kuh98]. For the simple case of an idealized vibrating symmetric top, the energy states bear as labels the vibrational quantum number v and the projection K of angular momentum along the symmetry axis. The dipole transition moment then factors as

$$|\langle e_2, v_2 J_2 K_2 M_2 | \mathbf{d} \cdot \mathbf{e}_q | e_1, v_1 J_1 K_1 M_1 \rangle|^2$$

$$= S_q^{geom}(J_2 M_2; J_1 M_1) \qquad \text{geometric factor}$$

$$\times S^{HL}(J_2 K_2; J_1 K_1) \qquad \text{Hönl–London factor}$$

$$\times S^{FC}(e_2 v_2; e_1 v_1) \qquad \text{Franck–Condon factor}$$

$$\times |\langle e_2 J_2 ||d|| e_1 J_1 \rangle|^2 \qquad \text{electronic transition strength.} \quad \text{(H.19)}$$

The Hönl–London factor is expressible as the square of a three-j symbol [Hue06],

$$S^{HL}(J_2 K_2; J_1 K_1) = (2J_2+1)(2J_1+1) \left(\begin{matrix} J_2 & 2 & J_1 \\ -K_2 & q & K_1 \end{matrix} \right)^2, \qquad \text{with } q = K_2 - K_1.$$

$$\text{(H.20)}$$

The Franck–Condon factor S^{FC} is the square of the overlap of vibrational wavefunctions, and depends upon the particular molecule [Hue06].

Nuclear transitions. The protons and neutrons (nucleons) within atomic nuclei are confined particles, and like the electrons of an atom or molecule they therefore have discrete excitation energies. These can often be characterized as excitations of single nucleons. Other nuclear states exhibit collective behavior appropriate to vibrations and rotations of the entire nucleus. However, because the nuclear forces hold the particles within dimensions much smaller than atomic radii, the excitation energies are correspondingly larger. Transitions typically occur as gamma rays, and are not (yet) amenable to controlled manipulation of the sort described in the present monograph.

References

[Ack77] Ackerhalt J.R. and Shore B.W. "Rate equations versus Bloch equations in multiphoton ionization". Phys. Rev. A, **16**, 277–282 (1977).

[Aga90] Agarwal G.S. and Simon R. "Berry phase, interference of light beams, and the Hannay angle". Phys. Rev. A, **42**, 6924–6927 (1990).

[Aha87] Aharonov Y. and Anandan J. "Phase change during a cyclic quantum evolution". Phys. Rev. Lett, **58**, 1593–1596 (1987).

[Ale92] Alekseef A.V. and Sushilov N.V. "Analytic solutions of Bloch and Maxwell-Bloch equations in the case of arbitrary field amplitude and phase modulation". Phys. Rev. A, **46**, 351–355 (1992).

[All87] Allen L. and Eberly J.H. *Optical Resonance and Two Level Atoms* (Dover, New York, 1987).

[All03] Allen L., Barnett S.M. and Padgett M.J. *Optical Angular Momentum* (Institute of Physics, Bristol, 2003).

[Alz79] Alzetta G., Moi L. and Orriols G. "Nonabsoprtion hyperfine resonances in a sodium vapour irradiated by a multimode dye laser". Nuov. Cim. B, **52**, 209–218 (1979).

[Ana87] Anandan J. and Stodolsky L. "Some geometrical considerations of Berry's phase". Phys. Rev. D, **35**, 2597–2600 (1987).

[Ana88] Anandan J. "Non-adiabatic non-abelian geometric phase". Phys. Lett. A, **133**, 171–175 (1988).

[Ana97] Anandan J.S., Christian J. and Wanelik K. "Resource letter GPP-1: Geometric phases in physics". Am. J. Phys., **65**, 178–185 (1997).

[And06] Andreev A. *Atomic Spectroscopy: Introduction to the Theory of Hyperfine Structure* (Springer, New York, 2006).

[Ang01] Angelakis D.G., Paspalakis E. and Knight P.L. "Coherent phenomena in photonic crystals". Phys. Rev. A, **64**, 013801 (2001).

[Ari96] Arimondo E. "Coherent population trapping in laser spectroscopy". Prog. Opt., **35**, 259–356 (1996).

[Arm75] Armstrong L.J., Beers B.L. and Feneuille S. "Resonant multiphoton ionization via the Fano autoionization formalism". Phys. Rev. A, **12**, 1903–1910 (1975).

[Aro06] Aroca R. *Surface-Enhanced Vibrational Spectroscopy* (Wiley, New York, 2006).

[Ash80] Ashkin A. "Applications of laser radiation pressure". Science, **210**, 1081–1087 (1980).

[Ash07] Ashkin A. *Optical Trapping and Manipulation of Neutral Particles Using Lasers* (World Scientific, Hackensack, N.J., 2007).

[Ass98] Assion A., Baumert T., Bergt M. *et al.* "Control of chemical reactions by feedback-optimized phase-shaped femtosecond laser pulses". Science, **282**, 919–922 (1998).

[Aut55] Autler S.H. and Townes C.H. "Stark effect in rapidly varying fields". Phys. Rev., **100**, 703–722 (1955).

[Auz95] Auzinsh M. and Ferber R. (eds.) *Optical Polarization of Molecules* (Cambridge University Press, Cambridge, 1995).

[Auz10] Auzinsh M., Budker D. and Rochester S.M. *Optically Polarized Atoms: Understanding Light–Atom Interactions* (Oxford University Press, Oxford, 2010).

[Bam81] Bambini A. and Berman P.R. "Analytic solutions to the two-state problem for a class of coupling potentials". Phys. Rev. A, **23**, 2496–2501 (1981).

[Ban91b] Band Y.B. and Julienne P.S. "Density matrix calculation of population transfer between vibrational levels of Na_2 by stimulated Raman scattering with temporally shifted laser beams". J. Chem. Phys., **94**, 5291–5298 (1991).

[Bar03] Barabasi A.L. *Linked: How Everything Is Connected to Everything Else and What It Means for Business, Science, and Everyday Life* (Plume, New York, 2003).

[Bar77] Barone S.R., Narcowich M.A. and Narcowich F.J. "Floquet theory and applications". Phys. Rev., **15**, 1109–1125 (1977).

[Bar06] Bartschat K. "Density matrices". In G. Drake (ed.) *Springer Handbook of Atomic, Molecular, and Optical Physics*, 123–134 (Springer, New York, 2006).

[Bar09] Barnett S.M. *Quantum Information* (Oxford University Press, Oxford, 2009).

[Bas75] Bashkin S. and Stoner J.O.J. *Atomic Energy Levels and Grotrian Diagrams* (Elsevier, New York, 1975).

[Bay06a] Baylis W.E. and Drake G.W.F. "Units and constants". In G. Drake (ed.) *Springer Handbook of Atomic, Molecular, and Optical Physics*, 1–8 (Springer, New York, 2006).

[Bay06b] Baylis W.E. "Atomic multipoles". In G.W.F. Drake (ed.) *Springer Handbook of Atomic, Molecular, and Optical Physics*, 221–226 (Springer, New York, 2006).

[Bei00] Beige A., Braun D. and Knight P.L. "Driving atoms into decoherence-free states". New J. Phys., **2**, 22.1–22.15 (2000).

[Ber74] Berman P.R. "Two-level approximation in atomic systems". Am. J. Phys., **42**, 992–997 (1974).

[Ber80] Bergmann K., Hefter U. and Witt J. "State-to-state differential cross sections for rotationally inelastic scattering of Na_2 by He". J. Chem. Phys., **72**, 4777–4790 (1980).

[Ber81] Bernhardt A.F. and Shore B.W. "Coherent atomic deflection by resonant standing waves". Phys. Rev. A, **23**, 1290 (1981).

[Ber84a] Berman P.R. "Theory of photon echoes in dressed states". Opt. Comm., **52**, 225–230 (1984).

[Ber84b] Berry M.V. "Quantal phase factors accompanying adiabatic changes". Proc. Roy. Soc. (London) A, **392**, 45 (1984).

[Ber87] Berry M.V. "The adiabatic phase and Pancharatnam's phase for polarized light". J. Mod. Opt., **34**, 1401–1407 (1987).

[Ber89] Berry M.V. "The quantum phase, five years after". In A. Shapere and F. Wilczek (eds.) *Geometric Phases in Physics*, 7 (World Scientific, Singapore, 1989).

[Ber90] Berry M.V. "Histories of adiabatic quantum transitions". Proc. Roy. Soc. (London) A, **429**, 61–72 (1990).

[Ber91] Bergou J. and Englert B. "Operators of the phase: Fundamentals". Ann. Phys., **209**, 479–505 (1991).

[Ber98] Bergmann K., Theuer H. and Shore B.W. "Coherent population transfer among quantum states of atoms and molecules". Rev. Mod. Phys., **70**, 1003–1023 (1998).

[Bha97] Bhandari R. "Polarization of light and topological phases". Phys. Rep., **281**, 2–64 (1997).

[Bia75] Bialynicki-Birula I. and Bialynicka-Birula Z. *Quantum Electrodynamics* (Pergamon, New York, 1975).

[Bia76] Bialynicki-Birula I. and Bialynicka-Birula Z. "Quantum electrodynamics of intense photon beams. New approximation method". Phys. Rev. A, **14**, 1101–1108 (1976).

[Bia77] Bialynicka-Birula Z., Bialynicki-Birula I., Eberly J.H. and Shore B.W. "Coherent dynamics of N-level atoms and molecules. II. Analytic solutions". Phys. Rev. A, **16**, 2048–2054 (1977).

[Bia96] Bialynicki-Birula I. "Photon wave functions". Prog. Opt., **36**, 245–294 (1996).

[Bic85] Bickel W. and Baily W. "Stokes vectors, Mueller matrices, and polarized scattered light". Am. J. Phys., **53**, 468–478 (1985).

[Bie81a] Biedenharn L.C. and Louck J.D. *Angular Momentum in Quantum Physics* (Addison-Wesley, Reading, Mass., 1981).

[Bie81b] Biedenharn L.C. and Louck J.D. *The Racah–Wigner Algebra in Quantum Theory* (Addison-Wesley, Reading, Mass., 1981).

[Bie09] Biedenharn L.C. and Louck J.D. *Angular Momentum in Quantum Physics: Theory and Application* (Cambridge University Press, Cambridge, 2009).

[Bie65] Biedenharn L.C. and Van Dam H. *Quantum Theory of Angular Momentum* (Academic, New York, 1965).

[Bix68] Bixon M. and Jortner J. "Intramolecular radiationless transitions". J. Chem. Phys., **48**, 715–726 (1968).

[Bla88] Blatt R. and Zoller P. "Quantum jumps in atomic systems". Eur. J. Phys., **9**, 250 (1988).

[Bla08] Blatt R. and Wineland D. "Review article: Entangled states of trapped atomic ions". Nature, **453**, 1008–1015 (2008).

[Blo45] Bloch F. and Rabi I.I. "Atoms in variable magnetic fields". Rev. Mod. Phys., **17**, 237 (1945).

[Blu96] Blum K. *Density Matrix Theory and Applications* (Plenum, New York, 1996), 2nd edition.

[Bok06] Bokhan P.A., Buchanov V.V., Fateev N.V. *et al. Laser Isotope Separation in Atomic Vapor* (Wiley-VCH, Weinheim, 2006).

[Boo08] Boozer A.D. "Stimulated Raman adiabatic passage in a multilevel atom". Phys. Rev. A, **77**, 023411 (2008).

[Bor99] Born M. and Wolf E. *Principles of Optics* (Pergamon, New York, 1999), 7th edition.

[Boy08] Boyd R.W. *Nonlinear Optics* (Academic, New York, 2008), 3rd edition.

[Bra92] Braginsky V.B. and Khalili F.Y. *Quantum Measurement* (Cambridge University Press, Cambridge, 1992).

[Bra99] Brandt H.E. "Positive operator valued measure in quantum information processing". Am. J. Phys., **67**, 434–439 (1999).

[Bre32] Breit G. "Quantum theory of dispersion". Rev. Mod. Phys., **4**, 504–575 (1932).

[Bre36] Breit G. and Wigner E.P. "Capture of slow neutrons". Phys. Rev., **49**, 519 (1936).

[Bre02] Breuer H.P. and Petruccione F. *The Theory of Open Quantum Systems* (Oxford University Press, Oxford, 2002).

[Bri10] Brif C., Chakrabarti R. and Rabitz H. "Control of quantum phenomena: past, present and future". New J. Phys., **12**, 075008 (2010).

[Bri04] Brixner T., Pfeifer T., Gerber G., Wollenhaupt M. and Baumert T. "Optimal control of atomic, molecular and electron dynamics with tailored femtosecond laser pulses". In P. Hannaford (ed.) *Femtosecond Laser Spectroscopy*, 225–266 (Springer, Berlin, 2004).

[Bro02] Brown E. and Rabitz H. "Some mathematical and algorithmic challenges in the control of quantum dynamics phenomena". J. Math. Chem., **1**, 17–63 (2002).

[Bru03] Brumer P.W. and Shapiro M. *Principles of the Quantum Control of Molecular Processes* (Wiley, New York, 2003).

[Bur82] Burkey R.S. and Cantrell C.D. "Solution of the Schrödinger equation for systems driven by an exponential or semiexponential pulse". Opt. Comm., **43**, 64–68 (1982).

[Buz90] Buzek V., Jex I. and Quang T. "K-photon coherent states". J. Mod. Opt., **37**, 159–163 (1990).

[Buz97] Buzek V., Drobny G., Adam G., Derka R. and Knight P.L. "Reconstruction of quantum states of spin systems via the Jaynes principle of maximum entropy". J. Mod. Opt., **44**, 2607–2627 (1997).

[Buz98] Buzek V., Derka R., Adam G. and Knight P.L. "Reconstruction of quantum states of spin systems: from quantum Bayesian inference to quantum tomography". Ann. Phys., **266**, 454–496 (1998).

[Car68] Carrier G.F. and Pearson C.E. *Ordinary Differential Equations* (Blaisdell, Waltham, Mass., 1968).

[Car86] Carroll C.E. and Hioe F.T. "A new class of analytic solutions of the two-state problem". J. Phys. A, **19**, 3579–3597 (1986).

[Car87] Carmichael H.J. "Spectrum of squeezing and photocurrent shot noise: a normally ordered treatment". J. Opt. Soc. Am. B, **4**, 1588–1603 (1987).

[Car88a] Carroll C.E. and Hioe F.T. "Driven three-state model and its analytic solutions". J. Math. Phys., **29**, 487–509 (1988).

[Car88b] Carroll C.E. and Hioe F.T. "Three-state system driven by resonant optical pulses of different shapes". J. Opt. Soc. Am. B, **5**, 1335–1340 (1988).

[Car90] Carroll C.E. and Hioe F.T. "Analytic solutions for three-state systems with overlapping pulses". Phys. Rev. A, **42**, 1522–1531 (1990).

[Car93] Carroll C.E. and Hioe F.T. "Selective excitation via the continuum and suppression of ionization". Phys. Rev. A, **47**, 571–580 (1993).

[Car07] Carroll B.W. and Ostlie D.A. *Introduction to Modern Astrophysics* (Pearson Addison-Wesley, San Francisco, 2007), 2nd edition.

[Car09] Carlin B.P. and Louis T.A. *Bayesian Methods for Data Analysis*. Texts in Statistical Science (CRC Press, Boca Raton, Fla., 2009).

[Cha60] Chandrasekhar S. *Radiative Transfer* (Dover, New York, 1960).

[Cha07] Chakrabarti R. and Rabitz H. "Quantum control landscapes". Int. Rev. Phys. Chem., **26**, 671–735 (2007).

[Chi86] Chiao R.Y. and Wu Y.S. "Manifestations of Berry's topological phase for the photon". Phys. Rev. Lett., **57**, 933–936 (1986).

[Chu85] Chu S.I. "Recent developments in semiclassical Floquet theories for intense field multiphoton processes". Adv. At. Mol. Phys., **21**, 197–253 (1985).

[Chu86] Chu S., Bjorkholm J.E., Ashkin A. and Cable A. "Experimental observation of optically trapped atoms". Phys. Rev. Lett., **57**, 314–317 (1986).

[Coe07] Coello Coello C.A., Lamont G.B. and Van Veldhuisen D.A. *Evolutionary Algorithms for Solving Multi-Objective Problems* (Springer, New York, 2007).

[Coh89] Cohen L. "Time-frequency distributions: A review". Proc. IEEE, **77**, 941–981 (1989).

[Coh92] Cohen-Tannoudji C., Dupont-Roc J. and Grynberg G. *Atom–Photon Interactions: Basic Processes and Applications* (Wiley, New York, 1992).

[Coh98] Cohen-Tannoudji C.N. "Manipulating atoms with photons". Rev. Mod. Phys., **70**, 707–719 (1998).

[Coo79] Cook R.J. and Shore B.W. "Coherent dynamics of N-level atoms and molecules III. An analytically soluble periodic case". Phys. Rev. A, **20**, 539–544 (1979).

[Coo80] Cooper J., Ballagh R.J. and Burnett K. "Zeeman degeneracy effects in collisional intense-field resonance fluorescence". Phys. Rev. A, **22**, 535–544 (1980).

[Coo82] Cooper J., Ballagh R.J., Burnett K. and Hummer D.G. "On redistribution and the equations for radiative transfer". Ap. J., **260**, 299–316 (1982).

[Coo90] Cook R.J. "Quantum jumps". Prog. Opt., **28**, 363–416 (1990).

[Cou95] Cousins R.D. "Why isn't every physicist a Bayesian". Am. J. Phys., **63**, 398–410 (1995).

[Cow81] Cowan R.D. *The Theory of Atomic Structure and Spectra* (University of California Press, Berkeley, 1981).

[Cra84] Craig D.P. and Thirunamachandran T. *Molecular Quantum Electrodynamics: An Introduction to Radiation–Molecule Interactions* (Academic, New York, 1984).

[Cri70] Crisp M.D. "Propagation of small-area pulses of coherent light through a resonant medium". Phys. Rev. A, **1**, 1604–1611 (1970).

[Cri73] Crisp M.D. "Adiabatic-following approximation". Phys. Rev. A, **8**, 2128–2135 (1973).

[Cro09] Cronin A.D., Schmiedmayer J. and Pritchard D.E. "Optics and interferometry with atoms and molecules". Rev. Mod. Phys., **81** (2009).

[Dai94] Dai H.L. and Field R.W. (eds.) *Molecular Dynamics and Spectroscopy by Stimulated Emission Pumping* (World Scientific, Singapore, 1994).

[Dal82] Dalton B.J. and Knight P.L. "The effects of laser field fluctuations on coherent population trapping". J. Phys. B, **15**, 3997–4015 (1982).

[Dav76a] Davies E.B. *Quantum Theory of Open Systems* (Academic, New York, 1976).

[Dav76b] Davis J.P. and Pechukas "Nonadiabatic transitions induced by a time-dependent Hamiltonian in the semiclassical/adiabatic limit: The two-state case". J. Chem. Phys., **64**, 3129–3137 (1976).

[Dav76c] Davydov A.S. *Quantum Mechanics* (Pergamon, Oxford, 1976).

[Dav91] Davis L. *Handbook of Genetic Algorithms* (Van Nostrand Reinhold, New York, 1991).

[Deb05] Debnath L. and Mikusinski P. *Introduction to Hilbert Spaces: With Applications* (Academic, New York, 2005).

[DeG69] de Groot S.R. *The Maxwell Equations* (North-Holland, Amsterdam, 1969).

[Deh90] Dehmelt H. "Experiments with an isolated subatomic particle at rest". Rev. Mod. Phys., **62**, 525–530 (1990).

[Del89] Delone N.B. and Fedorov M.V. "Above threshold ionization". Prog. Quant. Electron., **13**, 267–298 (1989).

[Del00] Delone N.B. and Krainov V.P. *Multiphoton Processes in Atoms* (Springer, Berlin, 2000), 2nd edition.

[Dem64] Demkov Y.N. "Charge transfer at small resonance defects". Sov. Phys. JETP, **18**, 138 (1964).

[Dem69] Demkov Y.N. and Kunike M. "Hypergeometric model for two-state approximation in collision theory" In Russian. Vestn. Leningr. Univ., Ser. 4: Fiz. Khim., **16**, 39 (1969).

[Dem03] Demtröder W. *Laser Spectroscopy: Basic Concepts and Instrumentation* (Springer, Berlin, 2003), 3rd edition.

[Die05] Diestel R. *Graph Theory* (Springer, Heidelberg, 2005), 3rd edition.

[Die06] Diels J.C. and Rudolph W. *Ultrashort Laser Pulse Phenomena: Fundamentals, Techniques, and Applications on a Femtosecond Time Scale* (Academic, San Diego, 2006), 2nd edition.

[Dir27] Dirac P.A.M. "The quantum theory of the emission and absorption of radiation". Proc. Roy. Soc. (London) A, **114**, 243 (1927).

[Dir58] Dirac P.A.M. *The Principles of Quantum Mechanics* (Clarendon, Oxford, 1958).

[Dra06] Drake G.W.F. (ed.) *Springer Handbook of Atomic, Molecular, and Optical Physics* (Springer, New York, 2006).

[Dre99] Drese K. and Holthaus M. "Floquet theory for short laser pulses". Eur. Phys. J. D, **5**, 119–134 (1999).

[Dua10] Duan L.M. and Monroe C. "Colloquium: Quantum networks with trapped ions". Rev. Mod. Phys., **82**, 1209 (2010).

[Dyk60] Dykhne A.M. "Quantum transitions in the adiabatic approximation". Sov. Phys. JETP, **11**, 411–415 (1960).

[Ebe77] Eberly J.H. and Wodkiewicz K. "The time-dependent physical spectrum of light". J. Opt. Soc. Am., **67**, 1252–1260 (1977).

[Ebe82] Eberly J.H., Yeh J.J. and Bowden C.M. "Interrupted coarse-grained theory of quasi-continuum photoexcitation". Chem. Phys. Lett., **86**, 76–80 (1982).

[Ebe91] Eberly J.H., Javanainen J. and Rzazewski K. "Above-threshold ionization". Phys. Rep., **204**, 331–383 (1991).

[Ebe93] Eberly J.H. and Kulander K.C. "Atomic stabilization by super-intense lasers". Science, **262**, 1229–1233 (1993).

[Ebe06] Eberly J.H. and Stroud C.R.J. "Coherent transients". In G. Drake (ed.) *Springer Handbook of Atomic, Molecular, and Optical Physics*, 1065–1076 (Springer, New York, 2006).

[Edm57] Edmonds A.R. *Angular Momentum in Quantum Mechanics* (Princeton University Press, Princeton, 1957).

[Ehl90] Ehlotzky F. "Atomic phenomena in bichromatic laser fields". Phys. Rep., **345**, 175–264 (1990).

[Ein76] Einwohner T.H., Wong J. and Garrison J.C. "Analytical solutions for laser excitation of multilevel systems in the rotating-wave approximation". Phys. Rev. A, **14**, 1452–1456 (1976).

[Ein79] Einwohner T.H., Wong J. and Garrison J.C. "Effects of alternative transition sequences on coherent photoexcitation". Phys. Rev. A, **20**, 940–947 (1979).

[Eke00] Ekert A., Ericsson M., Hayden P. *et al.* "Geometric quantum computation". J. Mod. Opt., **47**, 2501–2513 (2000).

[Elg80] Elgin J.N. "Semiclassical formalism for the treatment of three-level systems". Phys. Lett. A, **80**, 140–142 (1980).

[Elk95] Elk M. "Adiabatic transition histories of population transfer in the lambda system". Phys. Rev. A, **52**, 4017–4022 (1995).

[Eme06] Emery G.T. "Hyperfine structure". In G. Drake (ed.) *Springer Handbook of Atomic, Molecular, and Optical Physics*, 253–260 (Springer, New York, 2006).

[Fai69] Fain V.M. and Khanin Y.I. *Quantum Electronics. 1: Basic Theory* (MIT Press, Cambridge, Mass., 1969).

[Fai87] Faisal F.H.M. *Theory of Multiphoton Processes* (Plenum, New York, 1987).

[Fan57] Fano U. "Description of states in quantum mechanics by density matrix and operator techniques". Rev. Mod. Phys., **29**, 74–93 (1957).

[Fan61] Fano U. "Effects of configuration interaction on intensities and phase shifts". Phys. Rev., **124**, 1866–1878 (1961).

[Fan63] Fano U. and Racah G. *Irreducible Tensorial Sets* (Academic, New York, 1963).

[Fan65] Fano U. and Cooper J.W. "Line profiles in the far-uv absorption spectra of the rare gases". Phys. Rev. **137**, A1364–A1379 (1965).

[Fen75] Feneuille S. and Schweighofer M.G. "Conditions for the observation of the Autler-Townes effect in a two step resonance experiment". J. de Phys., **36**, 781–786 (1975).

[Fer50] Fermi E. *Nuclear Physics* (University of Chicago Press, Chicago, 1950).

[Few93] Fewell M.P. "The absorption of arbitrarily polarized light by atoms and molecules". J. Phys. B, **26**, 1957–1974 (1993).

[Fey57] Feynman R.P., Vernon F.L.J. and Hellwarth R.W. "Geometrical representation of the Schrödinger equation for solving maser problems". J. Appl. Phys., **28**, 49–52 (1957).

[Fil85] Filipovicz P., Meystre P., Rempe G. and Walther H. "Rydberg atoms. A testing ground for quantum electrodynamics". Opt. Acta, **32**, 1105–1123 (1985).

[Fiu63] Fiutak J. "The multipole expansion in quantum theory". Can. J. Phys., **41**, 12–20 (1963).

[Fle99] Fleischhauer M., Unanyan R., Bergmann K. and Shore B.W. "Coherent population transfer beyond the adiabatic limit: Generalized matched pulses and higher-order trapping state". Phys. Rev. A, **59**, 3751–3760 (1999).

[Fle02] Fleischhauer M. and Lukin M.D. "Quantum memory for photons: Dark-state polaritons". Phys. Rev. A, **65**, 002314 (2002).

[Fle05] Fleischhauer M., Imamoglu A. and Marangos J.P. "Electromagnetically induced transparency: Optics in coherent media". Rev. Mod. Phys., **77**, 633–673 (2005).

[For07] Fortagh J. and Zimmermann C. "Magnetic microtraps for ultracold atoms". Rev. Mod. Phys., **79**, 235–289 (2007).

[Fra63] Frantz L.M. and Nodvik J.S. "Theory of pulse propagation in a laser amplifier". J. Appl. Phys., **34**, 2346–2349 (1963).

[Fre06] Freyberger M., Voge K., Schleich W.P. and O'Connell R.F. "Quantized field effects". In G. Drake (ed.) *Springer Handbook of Atomic, Molecular, and Optical Physics*, 1141–1166 (Springer, New York, 2006).

[Gae06] Gaeto A.L. and Boyd R.W. "Nonlinear optics". In G. Drake (ed.) *Springer Handbook of Atomic, Molecular, and Optical Physics*, 1051–1064 (Springer, New York, 2006).

[Gar02] Garraway B.M. and Suominen K.A. "Wave packet dynamics in molecules". Contemp. Phys., **43**, 97–114 (2002).

[Gar10] Garg A. "Berry phases near degeneracies: Beyond the simplest case". Am. J. Phys., **78**, 661 (2010).

[Gau90] Gaubatz U., Rudecki P., Schiemann S. and Bergmann K. "Population transfer between molecular vibrational levels. A new concept and experimental results". J. Chem. Phys., **92**, 5363–5376 (1990).

[Ger97] Gerry C.C. and Knight P.L. "Quantum superpositions and Schrödinger cat states in quantum optics". Am. J. Phys., **65**, 964–974 (1997).

[Gia80] Giacobino E. and Cagnac B. "Doppler-free multiphoton spectroscopy". Prog. Opt., **17**, 85–162 (1980).

[Gla63a] Glauber R.J. "Coherent states of the quantum oscillator". Phys. Rev. Lett., **10**, 84–86 (1963).

[Gla63b] Glauber R.J. "The quantum theory of optical coherence". Phys. Rev., **130**, 2529–2539 (1963).

[Gol06] Goldman S.P. and Cassar M.M. "Atoms in strong fields". In G. Drake (ed.) *Springer Handbook of Atomic, Molecular, and Optical Physics*, 227–234 (Springer, New York, 2006).

[Gol94] Goldner L.S., Gerz C., Spreeuw R.J. *et al.* "Momentum transfer in laser-cooled cesium by adiabatic passage in a light field". Phys. Rev. Lett., **72**, 997–1000 (1994).

[Gol50] Goldstein H. *Classical Mechanics* (Addison-Wesley, Reading, Mass., 1980), 2nd edition.

[Gon80] Gontier Y., Poirier M. and Trahin M. "Multiphoton absorptions above the ionisation threshold". J. Phys. B, **13**, 1381 (1980).

[Gor97] Gordon R.J. and Rice S.A. "Active control of the dynamics of atoms and molecules". Ann. Rev. Phys. Chem., **48**, 601–641 (1997).

[Got66] Gottfried K. *Quantum Mechanics* (Benjamin, New York, 1966).

[Goy83] Goy P., Raimond J.M., Gross M. and Haroch S. "Observation of cavity-enhanced single-atom spontaneous emission". Phys. Rev. Lett., **50**, 1903–1906 (1983).

[Gra78a] Gray H.R. and Stroud C.R. "Autler-Townes effect in double optical resonance". Opt. Comm., **25**, 359–362 (1978).

[Gra78b] Gray H.R., Whitley R.M. and Stroud C. R. J. "Coherent trapping of atomic populations". Opt. Lett., **3**, 218–220 (1978).

[Gra07] Gradshteyn I.S. and Ryzhik I.M. *Tables of Integrals, Series and Products* (Academic, New York, 2007), 7th edition.

[Gri73] Grischkowsky D. "Adiabatic following and slow optical pulse propagation in rubidium vapor". Phys. Rev. A, **7**, 2096–2101 (1973).

[Gro98] Grover L.K. "Quantum computing: The advantages of superposition". Science, **280**, 228 (1998).

[Gry77] Grynberg G. and Cagnac B. "Doppler-free multiphotonic spectroscopy". Rep. Prog. Phys., **40**, 791–841 (1977).

[Gue98] Guérin S., Yatsenko L.P., Halfmann T., Shore B.W. and Bergmann K. "Stimulated hyper-Raman adiabatic passage II. Static compensation of dynamic Stark shifts". Phys. Rev. A, **58**, 4691–4704 (1998).

[Gue02] Guérin S., Thomas S. and Jauslin H.R. "Optimization of population transfer by adiabatic passage". Phys. Rev. A, **65**, 023409 (2002).

[Gue03] Guérin S. and Jauslin H.R. "Control of quantum dynamics by laser pulses: Adiabatic Floquet theory". Adv. Chem. Phys., **125**, 147–267 (2003).

[Hai10] Hairer E. and Wanner G. *Solving Ordinary Differential Equations II: Stiff and Differential-Algebraic Problems* (Springer, Berlin, 2010).

[Ham68] Hammerling P. "Generalized Bloch equations for any multipole". J. Phys. B, **1**, 759–767 (1968).

[Han35] Hansen W.W. "A new type of expansion in radiation problems". Phys. Rev., **47**, 139–143 (1935).

[Han85] Hannay J.H. "Angle variable holonomy in adiabatic excursion of an integrable Hamiltonian". J. Phys. A, **18**, 221 (1985).

[Hap72] Happer W. "Optical pumping". Rev. Mod. Phys., **44**, 169–238 (1972).

[Hap10] Happer W., Jau Y.Y. and Walker T.G. *Optically Pumped Atoms* (Wiley-VCH, Weinheim, 2010).

[Har97] Harris S.E. "Electromagnetically induced transparency". Phys. Today, **50**, 36–42 (1997).

[Har03] Hariharan P. *Optical Interferometry* (Academic, New York, 2003), 2nd edition.

[Har06] Haroche S. and Raimond J.M. *Exploring the Quantum. Atoms, Cavities and Photons* (Oxford University Press, New York, 2006).

[Has75] Hassan S.S. and Bullough R.K. "Theory of the dynamical Stark effect". J. Phys. B, **8**, L147–L152 (1975).

[Hay03] Hay N., Lein M., Velotta R., *et al.* "Investigations of electron wave-packet dynamics and high-order harmonic generation in laser-aligned molecules". J. Mod. Opt., **50**, 561–577 (2003).

[Hec87] Hecht E. *Optics* (Addison-Wesley, Reading, Mass., 1987).

[Hen06] Henkel C. and Wilkens M. "De Broglie Optics". In G. Drake (ed.), *Springer Handbook of Atomic, Molecular, and Optical Physics*, 1125–1140 (Springer, New York, 2006).

[Her50a] Herzberg G. *Molecular Spectra and Molecular Structure I. Spectra of Diatomic Molecules* (Van Nostrand, New York, 1950).

[Her50b] Herzberg G. *Molecular Spectra and Molecular Structure II. Infrared and Raman Spectra of Polyatomic Molecules* (Van Nostrand, New York, 1950).

[Her09] Herman G.T. *Fundamentals of Computerized Tomography. Image Reconstruction from Projections* (Springer, Dordrecht, 2009).

[Hin90] Hinds E.A. "Cavity quantum electrodynamics". In D. Bates and B. Bederson (eds.) *Advances in Atomic, Molecular, and Optical Physics*, volume 28, 237 (Academic, New York, 1991).

[Hio81] Hioe F.T. and Eberly J.H. "N-level coherence vector and higher conservation laws in quantum optics and quantum mechanics". Phys. Rev. Lett., **47**, 838–841 (1981).

[Hio83] Hioe F.T. "Theory of generalized adiabatic following in multilevel systems". Phys. Lett. A, **99**, 150–155 (1983).

[Hio84a] Hioe F.T. "Linear and nonlinear constants of motion for two-photon processes in three-level systems". Phys. Rev. A, **29**, 3434–3436 (1984).

[Hio84b] Hioe F.T. "Solutions of Bloch equations involving amplitude and frequency modulations". Phys. Rev., **30**, 2100–2107 (1984).

[Hio85a] Hioe F.T. and Carroll C.E. "Two-state problems involving arbitrary amplitude and frequency modulations". Phys. Rev. A, **32**, 1541–1549 (1985).

[Hio85b] Hioe F.T. "Gell-Mann dynamic symmetry for N-level quantum systems". Phys. Rev. A, **32**, 2824–2836 (1985).

[Hio87] Hioe F.T. "N-level quantum systems with SU(2) dynamic symmetry". J. Opt. Soc. Am. B, **4**, 1327–1332 (1987).

[Hio88] Hioe F.T. and Carroll C.E. "Coherent population trapping in N-level quantum systems". Phys. Rev. A, **37**, 3000–3005 (1988).

[Hol89] Holstein B.R. "The adiabatic propagator". Am. J. Phys., **57**, 714–720 (1989).

[Hol94] Holthaus M. and Just B. "Generalized pi pulses". Phys. Rev. A, **49**, 1950–1960 (1994).

[Hou58a] Householder A.S. "The approximate solution of matrix problems". J. ACM, **5**, 205–243 (1958).

[Hou58b] Householder A.S. "Unitary triangularization of a nonsymmetric matrix". J. ACM, **5**, 339–342 (1958).

[Hue06] Huestis D.L. "Radiative transition probabilities". In G. Drake (ed.) *Springer Handbook of Atomic, Molecular, and Optical Physics*, 514–534 (Springer, New York, 2006).

[Hus91] Husman M., Schwieters C., Littman M. and Rabitz H. "Molecular-dynamics simulator for optimal control of molecular motion". Am. J. Phys., **59**, 1012–1017 (1991).

[Ics69] Icsevgi A. and Lamb W.E.J. "Propagation of light pulses in a laser amplifier". Phys. Rev., **185**, 517–545 (1969).

[Ish78] Ishimaru A. *Wave Propagation and Scattering in Random Media* (Academic, New York, 1978).

[Iva06] Ivanov P.A., Kyoseva E.S. and Vitanov N.V. "Engineering of arbitrary U(N) transformations by quantum Householder reflections". Phys. Rev. A, **74**, 22323 (2006).

[Iva07] Ivanov P.A., Torosov B.T. and Vitanov N.V. "Navigation between quantum states by quantum mirrors". Phys. Rev. A, **75**, 012323 (2007).

[Iva08] Ivanov P.A. and Vitanov N.V. "Synthesis of arbitrary unitary transformations of collective states of trapped ions by quantum Householder reflections". Phys. Rev. A, **77**, 012335 (2008).

[Jac99] Jackson J.D. *Classical Electrodynamics* (Wiley, New York, 1999), 3rd edition.

[Jav06] Javanainen J. "Cooling and trapping". In G. Drake (ed.) *Springer Handbook of Atomic, Molecular, and Optical Physics*, 1091–1106 (Springer, New York, 2006).

[Jay63] Jaynes E.T. and Cummings F.W. "Comparison of quantum and semiclassical radiation theories with application to the beam maser". Proc. IEEE, **51**, 89 (1963).

[Jay03] Jaynes E.T. *Probability Theory: The Logic of Science* (Cambridge University Press, New York, 2003).

[Jes96] Jessen P.S. and Deutsch I.H. "Optical lattices". *Advances in Atomic Molecular Optical Physics*, volume 37, 1–44 (Academic, New York 1996).

[Jex95] Jex I., Torma P. and Stenholm S. "Multimode coherent states". J. Mod. Opt., **42**, 1377–1386 (1995).

[Joa95] Joannopoulos J.D., Meade R.D. and Winn. J.N. *Photonic Crystals: Molding the Flow of Light* (Princeton University Press, Princeton, N.J., 1995).

[Jon41] Jones R.C. "A new calculus for the treatment of optical systems I. Description and discussion of the calculus". J. Opt. Soc. Am., **31**, 488–493 (1941).

[Jon47] Jones R.C. "A new calculus for the treatment of optical systems V. A more general formulation and description of another calculus". J. Opt. Soc. Am., **37**, 107–110 (1947).

[Jon56] Jones R.C. "A new calculus for the treatment of optical systems VIII. Electromagnetic theory". J. Opt. Soc. Am., **46**, 126–131 (1956).

[Jud92] Judson R.S. and Rabitz H. "Teaching lasers to control molecules". Phys. Rev. Lett., **68**, 1500–1503 (1992).

[Kar04] Karpati A., Kis Z. and Adam P. "Engineering mixed states in a degenerate four-state system". Phys. Rev. Lett., **93**, 193003 (2004).

[Kaz90] Kazantsev A.P., Durdotovich G.I. and Yakovlev V.P. *Mechanical Action of Light on Atoms* (World Scientific, Singapore, 1990).

[Kib68] Kibble T.W.B. "Some applications of coherent states". In M. Levy (ed.) *Cargese Lectures in Physics*, volume 2 (Gordon and Breach, New York, 1968).

[Kis01a] Kis Z. and Stenholm S. "Nonadiabatic dynamics in the dark subspace of a multilevel stimulated Raman adiabatic passage process". Phys. Rev. A, **64**, 063406 (2001).

[Kis01b] Kis Z., Vogel W., Davidovich L. and Zagury N. "Dark SU(2) states of the motion of a trapped ion". Phys. Rev. A, **63**, 1–7 (2001).

[Kis04] Kis Z., Karpati A., Shore B.W. and Vitanov N.V. "Stimulated Raman adiabatic passage among degenerate-level manifolds". Phys. Rev. A, **70**, 053405 (2004).

[Kis05] Kis Z., Vitanov N.V., Karpati A., Barthel C. and Bergmann K. "Creation of arbitrary coherent superposition states by stimulated Raman adiabatic passage". Phys. Rev. A, **72**, 033403 (2005).

[Kni80] Knight P.L. and Milonni P.W. "The Rabi frequency in optical spectra". Phys. Rep., **66**, 21–107 (1980).

[Kni88] Knize R.J., Wu Z. and Happer W. "Optical pumping and spin exchange in gas cells". In *Advances in Atomic and Molecular Physics*, volume 24, 223–267 (Academic, New York 1988).

[Kni90] Knight P.L., Lauder M.A. and Dalton B.J. "Laser-induced continuum structure". Phys. Rep., **190**, 1–61 (1990).

[Kni93] Knight P. and Shore B. "Schrödinger-cat states of the electromagnetic field and multilevel atoms". Phys. Rev. A, **48**, 642–655 (1993).

[Kop58] Kopferman H. *Nuclear Moments* (Academic, New York, 1958).

[Kor03] Korsunsky E.A., Halfmann T., Marangos J.P. and Bergmann K. "Analytical study of four-wave mixing with large atomic coherence". Eur. Phys. J. D, **23**, 167–180 (2003).

[Kra83] Kraus K. *States, Effects, and Operations: Fundamental Notions of Quantum Theory* (Springer, Berlin, 1983).

[Kra07] Král P., Thanopulos I. and Shapiro M. "Colloquium: Coherently controlled adiabatic passage". Rev. Mod. Phys., **79**, 53–77 (2007).

[Kub80] Kubo H. and Nagata R. "Stokes parameters representation of the light propagation equations in inhomogeneous anisotropic, optically active media". Opt. Comm., **34**, 306–308 (1980).

[Kuh98] Kuhn A., Steuerwald S. and Bergmann K. "Coherent population transfer in NO with pulsed lasers: The consequences of hyperfine structure, Doppler broadening and electromagnetically induced absorption". Eur. Phys. J. B, **1**, 57–70 (1998).

[Kuh01] Kuhr S., Alt W., Schrader D., *et al.* "Deterministic delivery of a single atom". Science, **293**, 278–280 (2001).

[Kuh10] Kuhn A. and Ljunggren D. "Cavity-based single-photon sources". Contemp. Phys., **51**, 1–25 (2010).

[Kuk89] Kuklinski J.R., Gaubatz U., Hioe F.T. and Bergmann K. "Adibatic population transfer in a three-level system driven by delayed laser pulses". Phys. Rev. A, **40**, 6741–6744 (1989).

[Kul06] Kulander K.C. and Lewenstein M. "Multiphoton and strong-field processes". In G. Drake (ed.) *Springer Handbook of Atomic, Molecular, and Optical Physics*, 1077–1090 (Springer, New York, 2006).

[Kyo06] Kyoseva E.S. and Vitanov N.V. "Coherent pulsed excitation of degenerate multistate systems: Exact analytic solutions". Phys. Rev. A, **73**, 023420 (2006).

[Kyo07] Kyoseva E.S., Vitanov N.V. and Shore, B.W. "Physical realization of coupled Hilbert-space mirrors for quantum-state engineering". J. Mod. Opt., **54**, 13–15 (2007).

[Lam06] Lambropoulos P. and Petrosyan D. *Quantum Optics and Quantum In-formation: An Introduction* (Springer, New York, 2006).

[Lan32a] Landau L.D. "Zur Theorie der Energieübertragung bei Stössen". Phys. Z. Sowjet., **1**, 88–98 (1932).

[Lan32b] Landau L.D. "Zur Theorie der Energieübertragung II". Phys. Z. Sowjet., **2**, 46–51 (1932).

[Lan77] Langhoff P.W., Epstein S.T. and Karplus M. "Aspects of time-dependent perturbation theory". Rev. Mod. Phys., **44**, 602–644 (1977).

[Lan84] Landau L.D., Pitaevskii L.P. and Lifshitz E. *Electrodynamics of Continuous Media* (Elsevier, New York, 1984), 2nd edition.

[Lan99] Lang K.R. *Astrophysical Formulae. I. Radiation, Gas Processes and High Energy Astrophysics* (Springer, Berlin, 1999), 3rd edition.

[Leb72] Lebedev N.N. *Special Functions and Their Application* (Dover, New York, 1972).

[Lei03] Leibfried D., Blatt R., Monroe C. and Wineland D. "Quantum dynamics of single trapped ions". Rev. Mod. Phys., **75**, 281–325 (2003).

[Let78] Letokhov V. and Makarov A. "Excitation of multilevel molecular systems by laser IR field". Appl. Phys. A, **16**, 47–57 (1978).

[Lev69] Levine R.D. *Quantum Mechanics of Molecular Rate Processes* (Clarendon, Oxford, 1969).

[Lid98] Lidar D.A., Chuang I.L. and Whaley K.B. "Decoherence-free subspaces for quantum computation". Phys. Rev. Lett., **81**, 2594–2597 (1998).

[Lid01a] Lidar D. A., Bacon D., Kempe J. and Whaley K.B. "Decoherence-free subspaces for multiple-qubit errors. I. Characterization". Phys. Rev. A, **63**, 022306 (2001).

[Lid01b] Lidar D.A., Bacon D., Kempe J. and Whaley K.B. "Decoherence-free subspaces for multiple-qubit errors. II. Universal, fault-tolerant quantum computation". Phys. Rev. A, **63**, 022307 (2001).

[Lim91] Lim R. and Berry M.V. "Superadiabatic tracking of quantum evolution". J. Phys. A, **24**, 3255–3264 (1991).

[Lin76] Lindblad G. "On the generators of quantum dynamical semigroups". Commun. Math. Phys., **48**, 119–130 (1976).

[Lin04] Lindinger A., Lupulescu C., Plewicki M. *et al.* "Isotope selective ionization by optimal control using shaped femtosecond laser pulses". Phys. Rev. Lett., **93**, 033001 (2004).

[Liu01] Liu C., Dutton Z., Behroozi C.H. and Hau L.V. "Observation of coherent optical information storage in an atomic medium using halted light pulses". Nature, **409**, 490–493 (2001).

[Lou73] Loudon R. *The Quantum Theory of Light* (Clarendon, Oxford, 1973).

[Lou06] Louck J.D. "Angular momentum theory". In G. Drake (ed.) *Springer Handbook of Atomic, Molecular, and Optical Physics*, 9–74 (Springer, New York, 2006).

[Loy74] Loy M.M.T. "Observation of population inversion by optical adiabatic rapid passage". Phys. Rev. Lett., **32**, 814–817 (1974).

[Lu03] Lu Z.T. and Wendt K.D.A. "Laser-based methods for ultrasensitive trace-isotope analyses". Rev. Sci. Instrum, **74**, 1169–1179 (2003).

[Luk69] Luke Y.L. *The Special Functions and Their Approximations* (Academic, New York, 1969).

[Luk03] Lukin M.D. "Colloquium: Trapping and manipulating photon states in atomic ensembles". Rev. Mod. Phys., **75**, 457–472 (2003).

[Mac03] MacKay D.J.C. *Information Theory, Inference, and Learning Algorithms* (Cambridge University Press, Cambridge, 2003).

[Mac10] Macleod H.A. *Thin Film Optical Filters* (CRC Press, Boca Raton, Fla. 2010), 4th edition.

[Mai91] Mainfray G. and Manus C. "Multiphoton ionization of atoms". Rep. Prog. Phys., **54**, 1333–1372 (1991).

[Maj32] Majorana E. "Atomi orientati in campo magnetico variabile". Nuov. Cim., **9**, 43–50 (1932).

[Mak78] Makarov A.A., Platonenko V.T. and Tyakht V.V. "Interaction of a 'level-band' quantum system with a quasiresonant monochromatic field". Sov. Phys. JETP, **48**, 1044–1052 (1978).

[Mak10] Mäkelä H. and Messina A. "N-qubit states as points on the Bloch sphere". Phys. Sci., **T140**, 014054 (2010).

[Mal97] Malinovsky V.S. and Tannor D.J. "Simple and robust extension of the stimulated Raman adiabatic passage technique to n-level systems". Phys. Rev. A, **56**, 4929–4937 (1997).

[Mar91] Marte P., Zoller P. and Hall J.L. "Coherent atomic mirrors and beam splitters by adiabatic passage in multilevel systems". Phys. Rev. A, **44**, R4118–R4121 (1991).

[Mar95] Martin J., Shore B.W. and Bergmann K. "Coherent population transfer in multilevel systems with magnetic sublevels. II. Algebraic analysis". Phys. Rev. A, **52**, 583–593 (1995).

[Mar96] Martin J., Shore B.W. and Bergmann K. "Coherent population transfer in multilevel systems with magnetic sublevels. III. Experimental results". Phys. Rev. A, **53**, 1556–1569 (1996).

[Mar06a] Maroulis G. *Atoms, Molecules, and Clusters in Electric Fields: Theoretical Approaches to the Calculation of Electric Polarizability* (Imperial College Press, London, 2006).

[Mar06b] Martin W.C. and Wielse W.L. "Atomic spectroscopy". In G. Drake (ed.) *Springer Handbook of Atomic, Molecular, and Optical Physics*, 175–197 (Springer, New York, 2006).

[Mat06] Matsumoto Y. and Watanabe K. "Coherent vibrations of adsorbates induced by femtosecond laser excitation". Chem. Rev., **106**, 4234–4260 (2006).

[Max54] Maxwell J.C. *Treatise on Electricity and Magnetism* (Dover, New York, 1954).

[McC69] McCall S.L. and Hahn E.L. "Self-induced transparency". Phys. Rev., **183**, 457–485 (1969).

[Mea60] Mead C.A. "Theory of the complex refractive index". Phys. Rev., **120**, 854–866 (1960).

[Mec58] Meckler A. "Majorana formula". Phys. Rev., **111**, 1447–1449 (1958).

[Mes62] Messiah A. *Quantum Mechanics* (Wiley, New York, 1962).

[Mes85] Meschede D., Walther H. and Muller G. "One-atom maser". Phys. Rev. Lett., **54**, 551–554 (1985).

[Mes06] Meschede D. and Rauschenbeutel A. "Manipulating single atoms". In *Advances in Atomic Molecular and Optical Physics*, 53, 75 (2006).

[Met99] Metcalf H.J. and Straten P. van der. *Laser Cooling and Trapping* (Springer, New York, 1999).

[Mey89] Meystre P., Schumacher E. and Stenholm S. "Atomic beam deflection in a quantum field". Opt. Comm., **73**, 443–447 (1989).

[Mey01] Meystre P. *Atom Optics* (Springer, New York, 2001).

[Mic81] Michelson A.A. "The relative motion of earth and ether". Am. J. Sci., **22**, 20 (1881).

[Mih78] Mihalas D. *Stellar Atmospheres* (Freeman, San Francisco, 1978).

[Mil83] Milonni P.W., Ackerhalt J.R., Galbraith H.W. and Shih M.L. "Exponential decay, recurrences and quantum-mechanical spreading in a quasicontinuum model". Phys. Rev. A, **28**, 32–39 (1983).

[Mil88] Milonni P.W. and Eberly J.H. *Lasers* (Wiley, New York, 1988).

[Mil91] Milonni P.W. and Singh S. "Some recent developments in the fundamental theory of light". In D. Bates and B. Bederson (eds.) *Advances in Atomic Molecular and Optical Physics*, volume 28, 75 (Academic, New York, 1991).

[Mil92] Milonni P.W. and Thode L.E. "Theory of mesospheric sodium fluorescence excited by pulse trains". Appl. Opt., **31**, 785–800 (1992).

[Mil94] Milonni P.W. *The Quantum Vacuum* (Academic, Boston, 1994).

[Mil96] Milner V., Chernobrod B.M. and Prior Y. "Coherent-population trapping in a highly degenerate open system". Europhys. Lett., **34**, 557–562 (1996).

[Mil98] Milner V. and Prior Y. "Multilevel dark states: Coherent population trapping with elliptically polarized incoherent light". Phys. Rev. Lett., **80**, 940 (1998).

[Mil99] Milner V., Chernobrod B.M. and Prior Y. "Arbitrary orientation of atoms and molecules via coherent population trapping by elliptically polarized light". Phys. Rev. A, **60**, 1293 (1999).

[Mly81] Mlynek J. and Lange W. "High-resolution coherence spectroscopy using pulse trains". Phys. Rev. A, **24**, 1099–1102 (1981).

[Mof08] Moffitt J.R., Chemla Y.R., Smith S.B. and Bustamante C. "Recent advances in optical tweezers". Ann. Rev. Biochem., **77**, 205–228 (2008).

[Mol69] Mollow B.R. and Miller M.M. "The damped driven two-level atom". Ann. Phys., **52**, 464–478 (1969).

[Moo68] Moore C.E. and Merrill P.W. *Partial Grotrian Diagrams of Astrophysical Interest* (U.S. Government Printing Off, Washington, D.C., 1968).

[Mor53] Morse P.M. and Feshbach H. *Methods of Theoretical Physics* (McGraw-Hill, New York, 1953).

[Mor83] Morris J.R. and Shore B.W. "Reduction of degenerate two-level excitation to independent two-state systems". Phys. Rev. A, **27**, 906–912 (1983).

[Mot65] Mott N.F. and Massey H.S.W. *The Theory of Atomic Collisions* (Clarendon Press, Oxford, 1965).

[Moy99] Moya-Cessa H., Wallentowitz S. and Vogel W. "Quantum-state engineering of a trapped ion by coherent-state superpositions". Phys. Rev. A, **59**, 2920 (1999).

[Mue48] Mueller H. "The foundations of optics". J. Opt. Soc. Am., **38**, 661–669 (1948).

[Muk99] Mukamel S. *Principles of Nonlinear Optical Spectroscopy* (Oxford University Press, Oxford, 1999).

[Mur06] Murr K., Nussmann S., Puppe T. *et al.* "Three-dimensional cavity cooling and trapping in an optical lattice". Phys. Rev. A, **73**, 063415 (2006).

[Nak94] Nakajima T., Elk M., Jian Z. and Lambropoulos P. "Population transfer through the continuum". Phys. Rev. A, **50**, R913–R916 (1994).

[Nar81] Narozhny N.B., Sanchez-Mondragon J.J. and Eberly J.H. "Coherence versus incoherence: collapse and revival in a single quantum model". Phys. Rev. A, **23**, 236–246 (1981).

[New68] Newton R.G. and Young B. "Measurability of the spin density matrix". Ann. Phys. (NY), **49**, 393–402 (1968).

[Nie00] Nielsen M.A. and Chuang I.L. *Quantum Computation and Quantum Information* (Cambridge University Press, New York, 2000).

[Nik74] Nikitin E.E. *Theory of Elementary Atomic and Molecular Processes in Gases* (Clarendon, Oxford, 1974).

[Nus05] Nussmann S., Hijlkema M., Weber B. *et al.* "Submicron positioning of single atoms in a microcavity". Phys. Rev. Lett., **95**, 173602 (2005).

[Obe06] Oberst M., Klein J. and Halfmann T. "Enhanced four-wave mixing in mercury isotopes, prepared by Stark-chirped rapid adiabatic passage". Opt. Comm., **264**, 463–470 (2006).

[Olv10] Olver F.W.J., Lozier D.W., Boisvert R.F. and Clark C.W. (eds.) *NIST Handbook of Mathematical Functions* (Cambridge University Press, Cambridge, 2010).

[Omo77] Omont A. "Irreducible components of the density matrix. Application to optical pumping". Prog. Quant. Electron., **5**, 69–138 (1977).

[Ore84a] Oreg J., Hioe F.T. and Eberly J.H. "Adiabatic following in multilevel systems". Phys. Rev. A, **29**, 690–697 (1984).

[Ore84b] Oreg J. and Goshen S. "Spherical modes in N-Level systems". Phys. Rev. A, **29**, 3205–3207 (1984).

[Ore92] Oreg J., Bergmann K., Shore B.W. and Rosenwaks S. "Population transfer by delayed pulses in a four-state system". Phys. Rev. A, **45**, 4888–4896 (1992).

[Pal06b] Paldus J. "Perturbation theory". In G. Drake (ed.) *Springer Handbook of Atomic, Molecular, and Optical Physics*, 101–114 (Springer, New York, 2006).

[Pan56] Pancharatnam S. "Generalized theory of interference, and its applications". Proc. Ind. Acad. Sci., **A 44**, 247–262 (1956).

[Pan98] Pan J.W. and Zeilinger A. "Greenberger-Horne-Zeilinger-state analyzer". Phys. Rev. A, **57**, 2208 (1998).

[Pas00] Paspalakis E. and Knight P.L. "Spontaneous emission properties of a quasi-continuum". Opt. Comm., **179**, 257–265 (2000).

[Pas02] Paspalakis E. and Kis Z. "Pulse propagation in a coherently prepared multilevel medium". Phys. Rev. A, **66**, 025802 (2002).

[Pea01] Pearson B.J., White J.L., Weinacht T.C. and Bucksbaum P.H. "Coherent control using adaptive learning algorithms". Phys. Rev. A, **63**, 063412 (2001).

[Pea06] Peach G. "Collisional broadening of spectral lines". In G. Drake (ed.) *Springer Handbook of Atomic, Molecular, and Optical Physics*, 875–890 (Springer, New York, 2006).

[Pec10] Pechen A., Brif C., Wu R., Chakrabarti R. and Rabitz H. "General unifying features of controlled quantum phenomena". Phys. Rev. A, **82**, 030101 (2010).

[Per86] Perelomov A. *Generalized Coherent States* (Springer, Heidelberg, 1986).

[Per95] Peres A. *Quantum Theory: Concepts and Methods* (Kluwer Academic, Dordrecht, 1995).

[Per06] Peralta-Conde A., Brandt L. and Halfmann T. "Trace isotope detection enhanced by coherent elimination of power broadening". Phys. Rev. Lett., **97**, 243004 (2006).

[Phi01] Phillips D.F., Fleischhauer A., Mair A. and Walsworth R.L. "Storage of light in atomic vapor". Phys. Rev. Lett., **86**, 783–786 (2001).

[Pil93] Pillet P., Valentin C., Yuan R.L. and Yu J. "Adiabatic population transfer in a multilevel system". Phys. Rev. A, **48**, 845–848 (1993).

[Ple98] Plenio M. and Knight P. "The quantum-jump approach to dissipative dynamics in quantum optics". Rev. Mod. Phys., **70**, 101–144 (1998).

[Pow78] Power E.A. and Thirunamachandran T. "On the nature of the Hamiltonian for the interaction of radiation with atoms and molecules: (e/mc)p.A, d.E and all that". Am. J. Phys., **46**, 370–378 (1978).

[Pow85] Power E.A. and Thirunamachandran T. "Further remarks on the Hamiltonian of quantum optics". J. Opt. Soc. Am. B, **2**, 1100–1105 (1985).

[Pow82] Power E.A. and Thirunamachandran T. "Maxwell's equations and the multipolar Hamiltonian". Phys. Rev. A, **26**, 1800–1801 (1982).

[Pra03] Präkelt A., Wollenhaupt M., Assion A. *et al.* "Compact, robust and flexible setup for femtosecond pulse shaping". Rev. Sci. Instrum, **74**, 4950–4953 (2003).

[Pre98] Preskill J. "Lecture notes for Physics 229: Quantum Information and Computation". available on-line at www.theory.caltech.edu/people/preskill/ph229/ (1998).

[Pro97] Protopapas M., Keitel C.H. and Knight P.L. "Atomic physics with super-high intensity lasers". Rep. Prog. Phys., **60**, 389–486 (1997).

[Pur46] Purcell E.M. "Spontaneous emission probabilities at radio frequencies". Phys. Rev., **69**, 681 (1946).

[Rab37] Rabi I.I. "Space quantization in a gyrating magnetic field". Phys. Rev., **51**, 652–654 (1937).

[Rab54] Rabi I.I., Ramsey N.F. and Schwinger J. "Use of rotating coordinates in magnetic resonance problems". Rev. Mod. Phys., **26**, 167–171 (1954).

[Rab00a] Rabitz H. and Zhu W. "Optimal control of molecular motion: Design, implementation, and inversion". Acc. Chem. Res., **33**, 572 –578 (2000).

[Rab00b] Rabitz H., Vivie-Riedle R.d., Motzkus M. and Kompa K. "Whither the future of controlling quantum phenomena?" Science, **288**, 824–828 (2000).

[Rab04] Rabitz H. "Coherent control". In B.D. Guenther (ed.) *Encyclopedia of Modern Optics*, 123–134 (Elsevier, New York, 2004).

[Rad82a] Radmore P.M. "Decaying pulse excitation of a two-level atom". Phys. Lett. A, **87**, 285–287 (1982).

[Rad82b] Radmore P.M. and Knight P.L. "Population trapping and dispersion in a three-level system". J. Phys. B, **15**, 561–573 (1982).

[Rai01] Raimond J.M., Brune M. and Haroche S. "Manipulating quantum entanglement with atoms and photons in a cavity". Rev. Mod. Phys., **73**, 565–582 (2001).

[Ram49] Ramsey N.F. "A new molecular beam resonance method". Phys. Rev., **76**, 996 (1949).

[Ram95] Ramakrishna V., Salapaka M.V., Daleh M., Rabitz H. and Pierce A. "Controllability of molecular systems". Phys. Rev. A, **51**, 960–966 (1995).

[Ran05] Rangelov A.A., Vitanov N.V., Yatsenko L.P. *et al.* "Stark-shift-chirped rapid-adiabatic-passage technique among three states". Phys. Rev. A, **72**, 053403 (2005).

[Ran06] Rangelov A.A., Vitanov N.V. and Shore B.W. "Extension of the Morris–Shore transformation to multilevel ladders". Phys. Rev. A, **74**, 053402 (2006).

[Ran08] Rangelov A.A., Vitanov N.V. and Shore B.W. "Population trapping in three-state quantum loops revealed by Householder reflections". Phys. Rev. A, **77**, 033404 (2008).

[Ran09] Rangelov A.A., Vitanov N.V. and Shore B.W. "Rapid adiabatic passage without level crossing". Opt. Comm., **283**, 1346–1350 (2009).

[Ray90] Raymer M.G. and Walmsley I.A. "The quantum coherence properties of stimulated Raman scattering". Prog. Opt., **28**, 183–272 (1990).

[Rei84] Reintjes J.F. *Nonlinear Optical Parametric Processes in Liquids and Gases* (Academic, Orlando, Fla., 1984).

[Ric29] Rice O.K. "Perturbations in molecules and the theory of predissociation and diffuse spectra". Phys. Rev., **33**, 748–759 (1929).

[Ric00a] Rice S.A. and Zhao M. *Optical Control of Molecular Dynamics* (Wiley, New York, 2000).

[Ric00b] Rickes T., Yatsenko L.P., Steuerwald S. *et al.* "Efficient adiabatic population transfer by two-photon excitation assisted by a laser-induced Stark shift". J. Chem. Phys., **113**, 534–546 (2000).

[Ric03] Rickes T., Marangos J.P. and Halfmann T. "Enhancement of third-harmonic generation by Stark-chirped rapid adiabatic passage". Opt. Comm., **227**, 133–142 (2003).

[Rob81] Robinson E.J. "Two-level transition probabilities for asymmetric coupling pulses". Phys. Rev. A, **24**, 2239–2241 (1981).

[Rob82] Robiscoe R.T. "Quasiadiabatic solution to the two-level problem". Phys. Rev. A, **25**, 1178–1180 (1982).

[Rob95] Robinson E.J. "Concerning the 'asymmetrized Rosen-Zener model'". J. Phys. B, **28**, L169 (1995).

[Ros32] Rosen N. and Zener C. "Double Stern–Gerlach experiment and related collision phenomena". Phys. Rev., **40**, 502–507 (1932).

[Ros57] Rose M.E. *Elementary Theory of Angular Momentum* (Wiley, New York, 1957).

[Row75] Rowe D.J. and Ngo-Trong C. "Tensor equations of motion for the excitations of rotationally invariant or charge-independent systems". Rev. Mod. Phys., **47**, 471–485 (1975).

[Ryb85] Rybicki G.B. and Lightman A.P. *Radiative Processes in Astrophysics* (Wiley, New York, 1985).

[Saa06] Saalfrank P. "Quantum dynamical approach to ultrafast molecular desorption from surfaces". Chem. Rev., **106**, 4116–4159 (2006).

[Sal74] Salzman W.R. "Quantum mechanics of systems periodic in time". Phys. Rev. A, **10**, 461–465 (1974).

[Sam88] Samuel J. and Bhandari R. "General setting for Berry's phase". Phys. Rev. Lett., **60**, 2339–2342 (1988).

[San04] Sangouard N., Guérin S., Yatsenko L.P. and Halfmann T. "Preparation of coherent superposition in a three-state system by adiabatic passage". Phys. Rev. A, **70**, 013415 (2004).

[San06] Sangouard N., Yatsenko L.P., Shore B.W. and Halfmann T. "Preparation of nondegenerate coherent superpositions in a three-state ladder system assisted by Stark shifts". Phys. Rev. A, **73**, 043415 (2006).

[Sav96] Savva V., Zelenkov V. and Mazurenko A. "Dynamics of molecular systems with nonequidistant energy levels in a field of laser IR radiation". J. Appl. Spectrosc., **63**, 348–356 (1996).

[Sch26] Schrödinger E. "Der stetige Ubergang von der Mikro- zur Makromechanik". Naturwissenschaften, **28**, 664–666 (1926).

[Sch77] Schwinger J. "The Majorana formula". Trans. N.Y. Acad. Sci. II, **38**, 170–184 (1977).

[Sch01] Schrader D., Kuhr S., Alt W. *et al.* "An optical conveyor belt for single neutral atoms". Appl. Phys. B, **8**, 819–824 (2001).

[Sch03] Schwinger J. and Englert B.G. *Quantum Mechanics: Symbolism of Atomic Measurements* (Springer, Berlin, 2003).

[Scu92] Scully M.O. "From lasers and masers to phaseonium and phasers". Phys. Rep., **219**, 191–201 (1992).

[Sha79] Shampine L.F. and Gear C.W. "A user's view of solving stiff ordinary differential equations". SIAM Review, **21**, 1–17 (1979).

[Sha89] Shapere A. and Wilczek F. (eds.) *Geometric Phases in Physics* (World Scientific, Singapore, 1989).

[Sha07] Shapiro E.A., Milner V., Menzel-Jones C. and Shapiro M. "Piecewise adiabatic passage with a series of femtosecond pulses". Phys. Rev. Lett., **99**, 033002 (2007).

[Shi65] Shirley J.H. "Solution of the Schrödinger equation with a Hamiltonian periodic in time". Phys. Rev., **138**, B979–B987 (1965).

[Sho67] Shore B.W. "Scattering theory of absorption-line profiles and refractivity". Rev. Mod. Phys., **39**, 439–462 (1967).

[Sho68] Shore B.W. "Parametrization of absorption-line profiles". Phys. Rev., **171**, 43–54 (1968).

[Sho69] Shore B.W. "Scattering theory of absorption and autoionization-line profiles". In S. Geltman, K.T. Mahanthappa and W.E. Brittin (eds.) *Lectures in Theoretical Physics*, volume 11 (Gordon and Breach, New York, 1969).

[Sho77] Shore B.W. and Ackerhalt J. "Dynamics of multilevel laser excitation: Three-level atoms". Phys. Rev. A, **15**, 1640–1646 (1977).

[Sho79] Shore B.W. and Cook R.J. "Coherent dynamics of N-level atoms and molecules. IV. Two- and three-level behavior". Phys. Rev. A, **20**, 1958–1964 (1979).

[Sho81a] Shore B.W. "Two-level behavior of coherent excitation of multilevel systems". Phys. Rev. A, **24**, 1413–1418 (1981).

[Sho81b] Shore B.W. and Johnson M.A. "Effects of hyperfine structure on coherent excitation". Phys. Rev. A, **23**, 1608–1610 (1981).

[Sho83] Shore B.W. "Coherence in the quasi-continuum model". Chem. Phys. Lett., **99**, 240–243 (1983).

[Sho84] Shore B.W. "Gating of population flow in resonant multiphoton excitation". Phys. Rev. A, **29**, 1578–1582 (1984).

[Sho87] Shore B.W. and Knight P.L. "Enhancement of high optical harmonics by excess-photon ionisation". J. Phys. B, **20**, 413–423 (1987).

[Sho90] Shore B.W. *The Theory of Coherent Atomic Excitation* (Wiley, New York, 1990).

[Sho91a] Shore B.W., Bergmann K., Oreg J. and Rosenwaks S. "Multilevel adiabatic population transfer". Phys. Rev. A, **44**, 7442–7447 (1991).

[Sho91b] Shore B.W., Meystre P. and Stenholm S. "Is a quantum standing wave composed of two traveling waves?" J. Opt. Soc. Am. B, **8**, 903–910 (1991).

[Sho92] Shore B.W., Bergmann K. and Oreg J. "Coherent population transfer: stimulated Raman adiabatic passage and the Landau–Zener picture". Z. Phys. D, **23**, 33–39 (1992).

[Sho93] Shore B.W. and Knight P.L. "Topical Review. The Jaynes-Cummings model". J. Mod. Opt., **40**, 1195–1238 (1993).

[Sho95a] Shore B.W., Martin J., Fewell M.P. and Bergmann K. "Coherent population transfer in multilevel systems with magnetic sublevels. I. Numerical studies". Phys. Rev. A, **52**, 566–582 (1995).

[Sho95b] Shore B.W. "Examples of counter-intuitive physics". Contemp. Phys., **36**, 15–28 (1995).

[Sho03] Shore B.W. "Coherence and transient nonlinearity in laser probing". Spectrachim. Acta B, **58**, 969–998 (2003).

[Sho08] Shore B.W. "Coherent manipulations of atoms using laser light". Acta Phys. Slovaka, **58**, 243–486 (2008).

[Sho10] Shore B., Rangelov A. and Vitanov N. "Stimulated Raman adiabatic passage with temporal pulselets". Opt. Comm., **283**, 730–736 (2010).

[Sie86] Siegman A.E. *Lasers* (University Science Books, Mill Valley, Calif., 1986).

[Sil08] Silverman M.P. *Quantum Superposition: Counterintuitive Consequences of Coherence, Entanglement, and Interference* (Springer, Berlin, 2008).

[Sjo00] Sjöqvist E., Pati A.K., Ekert A. *et al.* "Geometric phases for mixed states in interferometry". Phys. Rev. Lett., **85**, 2845–2849 (2000).

[Smi92] Smith A.V. "Numerical studies of adiabatic population inversion in multilevel systems". J. Opt. Soc. Am. B, **9**, 1543–1551 (1992).

[Sob72] Sobel'man I.I. *Introduction to the Theory of Atomic Spectra* (Pergamon, New York, 1972).

[Sob92] Sobelman I.I. *Atomic Spectra and Radiative Transitions* (Springer, New York, 1992), 2nd edition.

[Sob95] Sobel'man I.I., Vainshtein L.A. and Yukov E.A. *Excitation of Atoms and Broadening of Spectral Lines* (Springer, New York, 1995), 2nd edition.

[Ste72] Stey G.C. and Gibbard R.W. "Decay of quantum states in some exactly soluble models". Physica, **60**, 1–26 (1972).

[Ste83] Stebbings R.F. and Dunning F.B. *Rydberg States of Atoms and Molecules* (Cambridge University Press, Cambridge, 1983).

[Ste86a] Steinberg S. "Lie series, Lie transformations, and their applications". In J. Sanchez-Mondragon and K.B. Wolf (eds.) *Lie Methods in Optics*, 52–103 (Springer, Berlin, 1986).

[Ste86b] Stenholm S. "The semiclassical theory of laser cooling". Rev. Mod. Phys., **58**, 699–740 (1986).

[Str41] Stratton J.A. *Electromagnetic Theory* (McGraw-Hill, New York, 1941).

[Stu32] Stückelberg C. "Theory of inelastic collisions between atoms". Helv. Phys. Acta, **5**, 369–423 (1932).

[Sty01] Styer D.F., Balkin M.S., Becker K.M. *et al.* "Nine formulations of quantum mechanics". Am. J. Phys., **70**, 288–297 (2001).

[Sus11] Sussman B.J. "Five ways to the nonresonant dynamic Stark effect". Am. J. Phys., **79**, 477–484 (2011).

[Sze04] Szekeres P. *A Course in Modern Mathematical Physics: Groups, Hilbert Space and Differential Geometry* (Cambridge University Press, New York, 2004).

[Tan05] Tannor D.J. *Introduction to Quantum Mechanics: A Time Dependent Perspective* (University Science Books, Sausalito, Calif., 2005).

[Tem93] Temkin R.J. "Excitation of an atom by a train of short pulses". J. Opt. Soc. Am. B, **10**, 830–839 (1993).

[Tha04] Thanopulos I., Král P. and Shapiro M. "Complete control of population transfer between clusters of degenerate states". Phys. Rev. Lett., **92**, 113003 (2004).

[The98] Theuer H. and Bergmann K. "Atomic beam deflection by coherent momentum transfer and the dependence on weak magnetic fields". Eur. Phys. J. D, **2**, 279–289 (1998).

[Tit65] Titulaer U.M. and Glauber R.J. "Correlation functions for coherent fields". Phys. Rev., **140**, B676–B682 (1965).

[Tow10] Townsend D., Sussman B.J. and Stolow A. "A Stark future for quantum control". J. Phys. Chem. A, **115**, 357–373 (2010).

[Tuc75] Tucker W.H. *Radiation Processes in Astrophysics* (MIT Press, Cambridge, Mass., 1975).

[Tur01] Turinici G. and Rabitz H. "Quantum wavefunction controllability". Chem. Phys., **267**, 1–9 (2001).

[Una97] Unanyan R.G., Yatsenko L.P., Bergmann K. and Shore B.W. "Laser-induced adiabatic atomic reorientation with control of diabatic losses". Opt. Comm., **139**, 48–54 (1997).

[Una98] Unanyan R.G., Fleischhauer M., Shore B.W. and Bergmann K. "Robust creation and phase-sensitive probing of superposition states via stimulated Raman adiabatic passage (STIRAP) with degenerate dark states". Opt. Comm., **155**, 144–154 (1998).

[Una99] Unanyan R.G., Shore B.W. and Bergmann K. "Laser-driven population transfer in four-level atoms: Consequences of non-Abelian geometrical adiabatic phase factors". Phys. Rev. A, **59**, 2910–2919 (1999).

[Una00] Unanyan R.G., Vitanov N.V., Shore B.W. and Bergmann K. "Coherent properties of a tripod system coupled via a continuum". Phys. Rev. A, **61**, U402–U410 (2000).

[Una04] Unanyan R.G., Pietrzyk M.E., Shore B.W. and Bergmann K. "Adiabatic creation of coherent superposition states in atomic beams". Phys. Rev. A, **70**, 053404 (2004).

[Van65] Van Vleck J.H. *The Theory of Electric and Magnetic Susceptibilities* (Oxford, London, 1965).

[Var88] Varshalovich D.A., Moskalev A.N. and Khersonskii V.K. *Quantum Theory of Angular Momentum: Irreducible Tensors, Spherical Harmonics, Vector Coupling Coefficients, 3j Symbols* (World Scientific, Singapore, 1988).

[Vas04] Vasilev G.S. and Vitanov N.V. "Coherent excitation of a two-state system by a Gaussian field". Phys. Rev. A, **70**, 053407 (2004).

[Vas06] Vasilev G.S. and Vitanov N.V. "Complete population transfer by a zero-area pulse". Phys. Rev. A, **73**, 023416 (2006).

[Vas09] Vasilev G.S., Kuhn A. and Vitanov N.V. "Optimum pulse shapes for stimulated Raman adiabatic passage". Phys. Rev. A, **80**, 013417 (2009).

[Vas10] Vasilev G.S., Ljunggren D. and Kuhn A. "Single photons made-to-measure". New J. Phys., **2**, 63024 (2010).

[Ved06] Vedral V. *Introduction to Quantum Information Science* (Oxford University Press, New York, 2006).

[Vew03] Vewinger F., Heinz M., Fernandez R.G., Vitanov N.V. and Bergmann K. "Creation and measurement of a coherent superposition of quantum states". Phys. Rev. Lett., **9121**, 3001 (2003).

[Vew07a] Vewinger F., Appel J., Figueroa E. and Lvovsky A.I. "Adiabatic frequency conversion of optical information in atomic vapor". Opt. Lett., **32**, 2771–2773 (2007).

[Vew07b] Vewinger F., Heinz M., Shore B.W. and Bergmann K. "Amplitude and phase control of a coherent superposition of degenerate states. I. Theory". Phys. Rev. A, **75**, 043406 (2007).

[Vew09] Vewinger F., Shore B.W. and Bergmann K. "Superpositions of degenerate atomic quantum states: Preparation and detection in atomic beams". *Advances in Atomic, Molecular and Optical Physics*, volume 58, 113–172 (Academic, New York, 2009).

[Vit92] Vitanov N. and Panev G. "Generalization of the Demkov formula in near-resonant charge transfer". J. Phys. B, **25**, 239–248 (1992).

[Vit94] Vitanov N.V. "Asymmetrized Rosen–Zener model". J. Phys. B, **27**, 1351–1360 (1994).

[Vit95a] Vitanov N. and Knight P. "Coherent excitation of a two-state system by a train of short pulses". Phys. Rev. A, **52**, 2245–2261 (1995).

[Vit95b] Vitanov N.V. "Complete population return in a two-state system driven by a smooth asymmetric pulse". J. Phys. B., **28**, L19–L22 (1995).

[Vit95c] Vitanov N.V. and Knight P.L. "Coherent excitation by asymmetric pulses". J. Phys. B., **28**, 1905–1920 (1995).

[Vit95d] Vitanov N.V. and Knight P.L. "Control of atomic transitions by the symmetry of excitation pulses". Opt. Comm., **121**, 31–35 (1995).

[Vit96] Vitanov N.V. and Stenholm S. "Non-adiabatic effects in population transfer in three-level systems". Opt. Comm., **127**, 215–222 (1996).

[Vit98a] Vitanov N.V. "Analytic model of a three-state system driven by two laser pulses on two-photon resonance". J. Phys. B, **31**, 709–725 (1998).

[Vit98b] Vitanov N.V., Shore B.W. and Bergmann K. "Adiabatic population transfer in multistate systems via dressed intermediate states". Euro. Phys. J. D, **4**, 15–29 (1998).

[Vit99a] Vitanov N.V., Suominen K.A. and Shore B.W. "Creation of coherent atomic superpositions by fractional stimulated Raman adiabatic passage". J. Phys. B, **32**, 4535–4546 (1999).

[Vit99b] Vitanov N.V. and Stenholm S. "Adiabatic population transfer via multiple intermediate states". Phys. Rev. A, **60**, 3820–3832 (1999).

[Vit00a] Vitanov N.V. "Measuring a coherent superposition of multiple states". J. Phys. B, **33**, 2333–2346 (2000).

[Vit00b] Vitanov N.V., Shore B.W., Unanyan R.G. and Bergmann K. "Measuring a coherent superposition". Opt. Comm., **179**, 73–83 (2000).

[Vit01a] Vitanov N.V., Fleischhauer M., Shore B.W. and Bergmann K. "Coherent manipulation of atoms and molecules by sequential laser pulses". In B. Bederson and H. Walther (eds.) *Advances in Atomic Molecular and Optical Physics*, volume 46, 55–190 (Academic, New York, 2001).

[Vit01b] Vitanov N.V., Halfmann T., Shore B.W. and Bergmann K. "Laser-induced population transfer by adiabatic passage techniques". Ann. Rev. Phys. Chem., **52**, 763–809 (2001).

[Vit01c] Vitanov N.V., Shore B.W., Yatsenko L. *et al.* "Power broadening revisited: theory and experiment". Opt. Comm., **199**, 117–126 (2001).

[Vit03] Vitanov N.V., Kis Z. and Shore B.W. "Coherent excitation of a degenerate two-level system by an elliptically polarized laser pulse". Phys. Rev. A, **68**, 063414 (2003).

[Vit05] Vitanov N.V. and Shore B.W. "Quantum transitions driven by missing frequencies". Phys. Rev. A, **72**, 052507 (2005).

[Vit06] Vitanov N.V. and Shore B.W. "Stimulated Raman adiabatic passage in a two-state system". Phys. Rev. A, **73**, 053402 (2006).

[von55] Von Neumann J. *Mathematical Foundations of Quantum Mechanics* (Princeton University Press, Princeton, N.J., 1955).

[Wal94] Walther H. "Single atoms in cavities and traps". In J.C. Dainty (ed.) *Current Trends in Optics* (Academic, London, 1994).

[Wer07] Werschnik J. and Gross E.K.U. "PhD Tutorial: Quantum optimal control theory". J. Phys. B, **40**, R175–R211 (2007).

[Whi76] Whitley R.M. and Stroud C.R.J. "Double optical resonance". Phys. Rev. A, **14**, 1498–1512 (1976).

[Wil84] Wilczek F. and Zee A. "Appearance of gauge structure in simple dynamical systems". Phys. Rev. Lett., **52**, 2111–2114 (1984).

[Win08] Winterfeldt C., Spielmann C. and Gerber G. "Colloquium: Optimal control of high-harmonic generation". Rev. Mod. Phys., **80**, 117 (2008).

[Wol05] Wollenhaupt M., Engel V. and Baumert T. "Femtosecond laser photoelectron spectroscopy on atoms and small molecules: Prototype studies in quantum control". Ann. Rev. Phys. Chem., **56**, 25–56 (2005).

[Wol07] Wollenhaupt M., Assion A. and Baumert T. "Femtosecond laser pulses: Linear properties, manipulation, generation and measurement". In *Springer Handbook of Lasers and Optics*, 937–983 (Springer, New York, 2007).

[Wol10] Wollenhaupt M., Bayer T., Vitanov N.V. and Baumert T. "Three-state selective population of dressed states via generalized spectral phase-step modulation". Phys. Rev. A, **81**, 053422 (2010).

[Won76] Wong J., Garrison J.C. and Einwohner T.H. "Multiple-time scale perturbation theory applied to laser excitation of atoms and molecules". Phys. Rev. A, **13**, 674–687 (1976).

[Wyb07] Wybourne B.G. and Smentek L. *Optical Spectroscopy of Lanthanides: Magnetic and Hyperfine Interactions* (CRC Press, Boca Raton, Fla., 2007).

[Wyn99] Wynands R. and Nagel A. "Precision spectroscopy with coherent dark states". Appl. Phys. B, **68**, 1–25 (1999).

[Xie98] Xie X.S. and Trautman J.K. "Optical studies of single molecules at room temperature". Ann. Rev. Phys. Chem., **49**, 441–480 (1998).

[Yat98] Yatsenko L.P., Guerin S., Halfmann T. *et al.* "Stimulated hyper-Raman adiabatic passage I. The basic problem and examples". Phys. Rev. A, **58**, 4683–4690 (1998).

[Yat97] Yatsenko L.P., Unanyan R.G., Bergmann K., Halfmann T. and Shore B.W. "Population transfer through the continuum using laser-controlled Stark shifts". Opt. Comm., **135**, 406–412 (1997).

[Yat99a] Yatsenko L.P., Vardi A., Halfmann T., Shore B.W. and Bergmann K. "Source of metastable H(2s) atoms using Stark chirped rapid adiabatic passage". Phys. Rev. A, **60**, R4237–R4240 (1999).

[Yat99b] Yatsenko L.P., Halfmann T., Shore B.W. and Bergmann K. "Photoionization suppression by continuum coherence: Experiment and theory". Phys. Rev. A, **59**, 2926–2947 (1999).

[Yat02a] Yatsenko L.P., Vitanov N.V., Shore B.W., Rickes T. and Bergmann K. "Creation of coherent superpositions using Stark-chirped rapid adiabatic passage". Opt. Comm., **204**, 413–423 (2002).

[Yat02b] Yatsenko L.P., Guerin S. and Jauslin H.R. "Topology of adiabatic passage". Phys. Rev. A, **65**, 043407 (2002).

[Yat03] Yatsenko L.P., Shore B.W., Vitanov N.V. and Bergmann K. "Retroreflection-induced bichromatic adiabatic passage". Phys. Rev. A, **68**, 043405 (2003).

[Yoo85] Yoo H.I. and Eberly J.H. "Dynamical theory of an atom with two or three levels interacting with quantized cavity fields". Phys. Rep., **118**, 239–337 (1985).

[Yur77] Yuratich M.A. and Hanna D.C. "Coherent anti-Stokes Raman spectroscopy (CARS) selection rules, depolarization ratios and rotational structure". Mol. Phys., **33**, 671–682 (1977).

[Zak85] Zakrzewski J. "Analytic solutions of the two-state problem for a class of chirped pulses". Phys. Rev. A, **32**, 3748–3751 (1985).

[Zar88] Zare R.N. *Angular Momentum: Understanding Spatial Aspects in Chemistry and Physics* (Wiley, New York, 1988).

[Zen32] Zener C. "Non-adiabatic crossing of energy levels". Proc. Roy. Soc. (London), **A137**, 696–699 (1932).

[Zim08] Zimmer F.E., Otterbach J., Unanyan R., Shore B.W. and Fleischhauer M. "Dark-state polaritons for multi-component and stationary light fields". Phys. Rev. A, **77**, 063823 (2008).

[Zol87] Zoller P., Marte M. and Walls D.F. "Quantum jumps in atomic systems". Phys. Rev. A, **35**, 198–207 (1987).

[Zub96] Zubairy M.S. "Quantum state measurement via Autler–Townes spectroscopy". Phys. Lett. A, **222**, 91–96 (1996).

[Zum87] Zumino B. "Geometry and Physics". Technical report, Lawrence Berkeley Lab (1987).

[Zwa90] Zwanziger J.W., Koenig M. and Pines A. "Berry's phase". Ann. Rev. Phys. Chem., **41**, 601–646 (1990).

Index

absorption coefficient, 80
  instantaneous, 394, 404
adiabatic
  basis, 233–235
  condition, 518
  elimination, 153, 209, 394
    in cavity , 430
  evolution, 118–134, 515
    nondegenerate, 382
  following, 119, 123, 168, 235, 238, 514
  passage, 514
    in cavity, 431
  state, 121, 125, 384, 513–521
algorithm
  evolutionary, 441
  learning, 440,
angle
  Brewster, 21
  Euler, 172, 451
  Hannay, 373
  mixing, 232, 516
  Pancharatnam, 373
angular momentum, 181, 442–456
  and rotations, 445
  commutators, 442
  coupling, 451–453
  degeneracy, 171
  intrinsic, 53
  operators
    Cartesian, 443
    spherical, 444
    orbital, 442, 460
  quantum numbers J, M, 171
anholonomy, *see* holonomy
atom
  as structured particle, 9
  average-, 185, 302, 315
  chemical, 9
  in solid, 48

isolated, 50
  moving, 54
  single, 44–48, 50
  size and shape of, 60
atomic units, 6
Autler–Townes splitting, 213
autocorrelation function, 477
average, 300–304
  -atom, 185, 302, 315
  environmental, 302
  over orientation, 303
  over time, 270–272

bandwidth, 28–30
beam
  Gaussian, 40, 55
  laser, 40–41, 476
  particle, 45
Beer's length, 80, 396, 405
Bessel function, and Jacobi-Anger
    expansion, 54, 105
bichromatic field, 141
bipartite, 287, 338
Bloch
  equations, with relaxation, 152
  sphere, 70–73
  variables, 146–148
  vector, 71, 147
    and density matrix, 312
    and geometric phase, 383
bright-dark basis, 251, 407–408

cavity, 20, 45, 420
  bad (very lossy), 428
  classical, 420–421
  Q factor, 420
  quantum, 421
  three-state atom in, 429–434
  two-state atom in, 423–429
Cayley–Klein parameters, 99, 355

Printed in the United States
By Bookmasters